Summary of Low-Speed Airfoil Data

Volume 3

About the Authors

CHRISTOPHER A. LYON received his B.S. in Aerospace Engineering from the University of Notre Dame in the spring of 1995. Upon his arrival at UIUC the following fall, he joined the LSATs team as project coordinator and began pursuing his M.S. degree. Fueled by a desire to design high performance, yet exotic R/C aircraft, he has researched topics ranging from boundary-layer trips to tailless aircraft. Additionally, he has been involved in R/C aircraft for 14 years, and recently was a design team member and pilot of UIUC's winning entry in the 1996–97 AIAA Student Design/Build/Fly Competition.

ANDY P. BROEREN received his B.S. in Mechanical Engineering from the Milwaukee School of Engineering in 1993. Several months after beginning his graduate study in mechanical engineering at UIUC, he joined Michael Selig and Jim Guglielmo in their efforts to establish the UIUC LSATs. Andy has completed his M.S. degree and is presently working toward a Ph.D. with dissertation research in unsteady airfoil aerodynamics. He has participated for several years in the Society of Automotive Engineers (SAE) Aero-Design competition and most recently was team leader and design team member of UIUC's winning entry in the 1996–97 AIAA Student Design/Build/Fly Competition.

PHILIPPE GIGUÈRE has been involved with the UIUC LSATs since his arrival on campus in September 1994. Since then, he has been working towards a Ph.D. degree with research in horizontal-axis wind turbine blade design and optimization. Other research areas of interest include: low-speed airfoil aerodynamics, wind-tunnel boundary corrections, and optimization methods. Originally from Québec City, Canada, Philippe Giguère performed his previous studies at McGill University (Bachelor degree in Mechanical Engineering–1992) and Université Laval (Master of Science degree and Teaching Diploma–1994). He participated in the SAE Aero-Design competition from 1991–94 and flies full-scale gliders.

ASHOK GOPALARATHNAM received his Bachelor of Technology (1989) and Master of Science (1993) in Aerospace Engineering from the Indian Institute of Technology, Madras. During 1991–94, he worked as a Scientist/Engineer at the National Aerospace Laboratories in India as a member of the Hansa-2 design team (a two-seat all-composite light airplane which first flew in 1993). Since 1995, he has been working towards his Ph.D. at UIUC and has been involved with the UIUC LSATs. His research interests include low-speed aerodynamics, aircraft design, and design methodologies. He flies both full-scale sailplanes and general aviation airplanes and was a design team member of UIUC's winning entry in the 1996–97 AIAA Student Design/Build/Fly Competition.

DR. MICHAEL S. SELIG, an accomplished applied aerodynamicist and airfoil designer, is an Assistant Professor of Aeronautical and Astronautical Engineering at the University of Illinois at Urbana-Champaign. He received his B.S. from the University of Illinois in 1984, his M.S.E. from Princeton University in 1988, and his Ph.D. from the Pennsylvania State University in 1992. His current research areas include low-speed aerodynamics, airfoil design, wind energy, and flight simulation. He teaches courses in applied aerodynamics and aircraft design. He has been active in R/C soaring for over 20 years.

Christopher A. Lyon
Andy P. Broeren
Philippe Giguère
Ashok Gopalarathnam
Michael S. Selig

Department of Aeronautical and Astronautical Engineering
University of Illinois at Urbana-Champaign

Summary of Low-Speed Airfoil Data

Volume 3

SoarTech Publications
Virginia Beach, Virginia

SoarTech Publications
1504 N. Horseshoe Circle
Virginia Beach, Virginia 23451, USA

The cover photograph shows a close-up of surface oil flow visualization results for the E374 (B) with a 0.030-in. thick zigzag trip located at $x/c = 30\%$ and $Re = 200{,}000$. As air moves over the model, fluorescent oil applied to the surface of the model flows in the direction of the local flow velocity. In the process, oil streaks are formed which allow the direction and nature of the local flow (laminar, turbulent, separated) to be identified. For similar flow visualization results, the reader is referred to Chapters 3 and 6.

First Printing, December 1997

Lyon, Christopher Alan (1973–
Summary of Low-Speed Airfoil Data – Volume 3 / by Christopher A. Lyon, Andy P. Broeren, Philippe Giguère, Ashok Gopalarathnam, and Michael S. Selig.

Includes bibliographical references.
1. Aerofoils (Airfoils). 2. Aerodynamics. 3. Airplanes—Models.
I. Model Aviation. II. Title
ISBN 0-9646747-3-4

Contents

Preface

Summary of Low-Speed Airfoil Data – Volume 3 is the third installment in a series of books documenting the ongoing work of the University of Illinois at Urbana-Champaign Low-Speed Airfoil Tests (UIUC LSATs). Being an out-growth of the work documented in *Airfoils at Low Speeds* (*SoarTech 8*)[1] performed by Michael Selig, John Donovan, and David Fraser while at Princeton, the project's purpose remains the same — to develop and test airfoils for low-speed aircraft, in particular, R/C model airplanes. Since the time that *SoarTech 8* was published, however, much of the results have found wider application on unmanned aerial vehicles (UAVs), low-speed propellers, and wind turbines.

Despite a growing interest from people in areas outside R/C aircraft, this effort has continued largely through the generous support of model-aviation enthusiasts. Without their contributions, encouragement, and enthusiasm this work would not have been possible or nearly as rewarding. A complete listing of the program contributors is given in the Acknowledgments section.

The current UIUC LSATs team consists of Michael Selig (Asst. Prof.) as project advisor, Chris Lyon (M.S.) as project coordinator, Andy Broeren (Ph.D.), Philippe Giguère (Ph.D.), and Ashok Gopalarathnam (Ph.D.). This group, together with the help of Christopher J. Fisichella (Ph.D.), required two and a half months in mid-March 1996 (Spring '96 installment) and one and a half months in early-April 1997 (Spring '97 installment) to acquire the data presented in this book.

The Book and Its Organization

Readers familiar with the format of previous volumes should find *Volume 3* easy to navigate. As before, Chapter 1 illustrates the airfoils tested and briefly addresses the flap and boundary-layer trip configurations used. Chapter 2 documents changes made to the testing facility since *Volume 2*, and Chapter 3 provides a comparison between UIUC LSATs data and data obtained by NASA Langley in the Low-Turbulence Pressure Tunnel (LTPT). Chapter 4 discusses the airfoils tested, and Chapter 5 contains the corresponding performance plots, including the new pitching-moment data. Included in Chapter 6, and also new to the *Summary of Low-Speed Airfoil Data* series, is an extensive discussion of boundary-layer trip performance. Following Chapter 6 is a glossary of terms the LSATs team feels are important. Finally, Appendix A includes inviscid velocity distributions for many of the airfoils presented in *Volume 2*, Appendices B and C list tabulated airfoil coordinates and drag polars respectively, Appendix D discusses how the data presented in this book can be obtained on diskette and any restrictions regarding its use, and Appendix E contains the UIUC LSATs Manifesto outlining the scope and purpose of this work.

Acknowledgments

Without the support and enthusiasm provided by many people, the UIUC LSATs would be a shadow of its present self. *To each of these people we are indebted.* In particular, we are especially grateful to the organizations, clubs, businesses, individuals, and t-shirt patrons who provided monetary support for equipment and graduate student salaries. Listed below are those who provided this support on which we so dearly depend. (Shown in parentheses is the fraction of total support received from each category.)

- *Organizations (7%):* British Association of Radio Control Soarers (BARCS) and Academy of Model Aeronautics (with special appreciation to Jerry Rouillard).

- *Businesses (42%):* B^2 Streamlines (Bill & Bunny Kulhman), Design and Manufacturing Ltd. (Jack Spitz and Ernest Trent, Jr.), Dynamic Modeling (Don Edberg), Griffith University, Landing Products (Fred Burgdorf), Muncie Pawn Brokers (Bill Greene), R.C. Hanger (Johnny Berlin), Sailplane Modeler (Wil Byers), R/C Soaring Digest (Jerry & Judy Slates), SoarTech Publications (H.A. Stokely), Top Flight (Don Anderson), and Traplet Publications Limited (Dave Jones).

- *Model Clubs (4%):* Baltimore Area Soaring Society, Clent Soaring Association (England), Portland Area Sailplane Society, Round Valley Radio Control Club, Tidewater Model Soaring Society, and Tri-County Aero Club.

- *Individuals (41%):* Anonymous, Mike D. Adkins, David K. Anderson, Arnold Angelici, Jim Armstrong*, Thomas Atwood, Bruce Baker*, Gary S. Baldwin*, Charles Baltzer, Keith Beggin*, Robert Bender*, Bernard Biales, C. Blake*, Eric L. Blanke, Paul Bly*, Charles L. Botzko*, Arthur Boysen*, Andy Broeren, Joyce C. Broeren*, Chris Burns*, Myron Cagen*, Bruce Carmichael, Bill Cavanaugh, Robert A. Champine*, Erich Chase*, Ben Clerx*, John L. Cranmer, Jr.*, Michael J. Cresanta, Michael D. Denton*, Alfred J. DeRenzis*, Michael W. Derr*, Thom Earle*, Waldron E. Ehrlich*, Stephen J. Fauble*, William S. Friedlander, Dave Garwood*, Laurent Gasser, B.I. Gaston*, Rolf Girsberger, William M. Green*, Charles Griswold, John Haren*, James B. Halbert*, Chuck Hallum*, Bob Harold, Don Harris*, Raymond E. Hatton, Brian L. Henry*, George B. Herider*, Bruce Herider*, Takashi Hoshizaki, Lester Hulett, Alfred C. Inman, John S. Jensen*, Giraud Julien, Gen Katayama*, Daniel Kong*, John F. Krohn*, Michael Lachowski*, Ben Lawless*, Lydia Lazurenko, Nhan T. Le*, Laurent Lebrun*, Dr. Robert M. Livin, Charles R. Lohre, Phillip Lontz, Eric H. Loos*, Steve Lucke, Fred Mallett, Lubos Mitas*, Andy Mitas*, Edward Mitchell*, O.G. Morris, Gilbert C. Morris, Steve Neu*, Nick Neve*, John D. Newell, MD, Kevin Noland, Ted Off, Paul H. Ortman, Scott Pack*, Jean Pailet, Phil Pearson,

Paul J. Penna*, John Ponsford*, Jim Porter*, Ralph I. Prey, Gene Quandt, Peter W. Rickard*, Jerry Robertson*, H.J. Rogers, S. Rossi, David Schenken, Herm Schmidt*, Walt Schmoll*, Allan K. Scidmore*, Jason Seal*, Steven Seim*, Michael S. Selig, Martha I. Selig, Luigi Sellitto, John S. Serafini, Bruno Sigrist*, Richard and Elizabeth Silver*, Norma J. Simon*, Arthur Slagel, A.J. Smith, Robert Stanford*, Larry Storie*, Lynn Stubblefield*, Joe Stute*, Al Sugar, T. Sean Tavares*, Jose M. Tellez*, James K. Truitt, Craig L. Uridil*, Chip Vignolini*, Wayne Walke*, Jess Walls*, Jay I. Welch*, Claude Wertheimer*, Don Westergren*, Dr. K. Wilhelm, Don Woelfel, Jr.*, and Gerald R. Zeigenfuse*.
(* denotes T-Shirt Patron as well)

- *T-Shirt Patrons (6%):* Anonymous, Bob Aberle, Thomas Akers, Bruce Alber, Don Allen, Dave Anderson, Robert L. Barrows, David Beardsley, R.P. "Randy" Beloff, Dennis Bickel, Brent Bills, Robert D. Bodwell, Jeanna Bonello, James Bonk, Eugene S. Boyko, Alan Brocklenhurst, Doug Buchanan, George D. Burns, Ed Byrns, Steve Cameron, Melvin Canfield, Peter S. Carr, Harvey J. Carter, Gary Claiborne, Robert R. Cooke III, Bob Cooper, Thomas J. Cunningham, David E. Darling, Dan Devries, David Dueson, Merrill Farmer, Noel D. Garrett, Philippe Giguère, Sally Gillman-Hanz, Rob Glover, Willy Grundler, Joe Hayes, George Hilliard, Brent Hoover, Mark Howard, Jack Iafret, Gordon Jennings, Douglas Joyce, Robert D. Kidd, Jr., Bruce Kirsten, Eric C. Leclercq, Vincent J. Lemmo, Jr., Harry E. Lyon, Jr., Robert Maldonado, Arthur Markiewicz, Bob Matheson, Tony Matyi, Paul Miller, Shannon Perkins, Carol Pesch, Mike Reed, Sensei John M. Roe, Robert J. Romash, Jonathan Salvatini, Eric Sanders, Toshiro Saruwatari, Charles Schmitz, John Schmoll, Alan Schwerin, Glenn H. Sembroski, Paul Sherman, Steve Siebenaler, Stefan Siemens, Roy L. Simpson, Jr., Scott Smith, Frank Smith, Karl M. Sorensen, Sandee Stedwell, Glenn Strickland, Mike Stump, Jim Thomas, Keith Thomson, David Thomson, Kevin Turner, David Veatch, Rod Watkins, B.J. Weisman, Adam Weston, Richard C. Williamson, Oliver Wilson, Wayne D. Wimbish, Joe Wurts, Wayne N. Yamamoto, and Peter W. Young.

Also greatly appreciated is the time and effort spent by the following individuals constructing the airfoil models tested.

- *Wind-Tunnel Model Builders:* Mark Allen (BW-3, LRN1007, SA7035, refurbished SD7062, SG6040, SG6041, SG6042, SG6043), Michael Bame (E374), Delmar Brengman (Ultra-Sport 1000), Frank Carson (SA7036, SA7038), Robert Cavazos (RG14, S6063), Bob Champine (Avistar, RG15), Ralph Cooney (A18), Bill Devany (Falcon 56 Mk II), Tim Foster (SA7036, SA7038), Mike Fox (Goe 417a, S8052), Brian Halkett (ESA), Peter Illick (SD7080), Mike Lachowski (E231, S7012), Tim Lampe (PT-40, S8036, S8037), Mark Levoe (CG Ultimate), John Raley (USNPS-4), Jerry Robertson (Clark-Y, E387, RG15 flap, S7075, SD7037), Doug Stanley (DH4009, E472), Herk Stokely (SD7037), Tinel Technologies/Yvan Tinel (E387), Jim Thomas (S7075, Trainer 60), D'Anne Thompson (SA7036, SD7037),

Stan Watson (SD7032), Bill Williams (SD7037), and Karl Widiner (SD7062).

Not to be forgotten are those who helped manufacture crucial parts of the UIUC LSATs test apparatus. Among these people, we owe a great deal to Lynn King and Hermann Andresen for their valuable time and effort spent designing and manufacturing the new lift and moment balances detailed in Chapter 2. Additionally, support was also provided by Jack Spitz and Ernest Trent Jr. (Design and Manufacturing, Ltd.). Without their beautiful flap endplate brackets, the flap data presented in this volume would not have been possible. In preparation for *Volume 4*, we would like to thank Yvan Tinel for supplying model construction templates. Finally, Bob Gartrell at Interface Inc. is appreciated for his help in acquiring load cells.

Of those at the University of Illinois, we wish to thank Prof. Michael Philpott for the use of his coordinate measuring machine and Stephen Craggs for its initial setup and support. Carol Winkler and Lizz Rogers are appreciated for their efforts in maintaining our mailing lists and processing donations. Andy Broeren wishes to acknowledge his research advisor, Prof. Michael Bragg, for his patience and support throughout the course of this project. Chris Lyon would like to thank Han Kim for his help in programming the new tunnel drive and Cameron Ninham for his help in deciphering C. Ashok Gopalarathnam would like to acknowledge his wife Godha for her patience and encouragement during the wind-tunnel test phase. We also want to thank Chris Carroll (new UIUC LSATs coordinator) for his inquisitive questioning that spurred us to unravel some of the previously puzzling characteristics of the moment balance used to obtain the pitching-moment data.

Finally, there are those who we feel deserve special mention. It is with a deep sense of gratitude that we thank Herk Stokely (SoarTech Publications) for his diligent and ongoing efforts related to distributing *Airfoils at Low Speeds* and the *Summary of Low-Speed Airfoil Data* series of books. We would also like to acknowledge John Donovan, the late David Fraser, and the many others associated with the Princeton tests. Without their earlier work, these tests would have been exceedingly more difficult. Additionally, a special thanks goes to Jef Raskin for promoting the project and Tom Atwood for his wonderful editorials in *Model Airplane News* supporting the effort. We also want to thank Sal DeFrancesco for continuing to promote the test program by including the UIUC LSATs t-shirt order form in his sales catalog. Also, the articles in QFI by Dave Jones and Martin Simons are greatly appreciated. Also, we are particularly indebted to Jim Guglielmo for his hard work as the original UIUC LSATs project coordinator. Had it not been for his long hours setting up equipment and writing data acquisition code this program would not be as prolific as it is today. The efforts of Cody Robertson in designing the 1997 UIUC LSATs t-shirt and Karen Evans for her assistance in designing the UIUC LSATs logo are also sincerely appreciated. Finally, Lisa Selig's efforts and encouragements are greatly appreciated.

It should be noted that much of the work reported here has benefited from other research activities, in particular, wind-turbine airfoil research sponsored by the University of Illinois at Urbana-Champaign and the DOE National Renewable Energy Laboratory, and high-altitude/low Reynolds number airfoil research sponsored by AeroVironment Inc. Finally, we must apologize to those whom we have inadvertently omitted.

List of Figures

List of Tables

List of Symbols/Acronyms

c	airfoil chord
C_l	airfoil lift coefficient
$C_{l,max}$	maximum lift coefficient
C_d	airfoil drag coefficient
C_m	quarter-chord pitching moment as labeled in Chapter 5 plots
$C_{m,c/4}$	airfoil pitching moment about the quarter-chord point
L/D	aircraft lift-to-drag ratio
Re	Reynolds number based on airfoil chord
s	distance along airfoil surface
V	local velocity on airfoil surface
V_∞	freestream velocity
x	distance along airfoil chord
α	angle of attack
α_{0l}	angle of attack for zero lift
GPMM	Great Planes Model Manufacturing
HAWTs	horizontal axis wind turbines
LSATs	Low-Speed Airfoil Tests
LTPT	Low-Turbulence Pressure Tunnel
R/C	radio controlled
UAV	unmanned aerial vehicle
UIUC	University of Illinois at Urbana-Champaign

Chapter 1

The Airfoils Tested

This volume of *Summary of Low-Speed Airfoil Data* documents the wind-tunnel test results of 37 airfoils (shown in Fig. 1.1). As in previous volumes, the airfoils tested are separated into several groups.

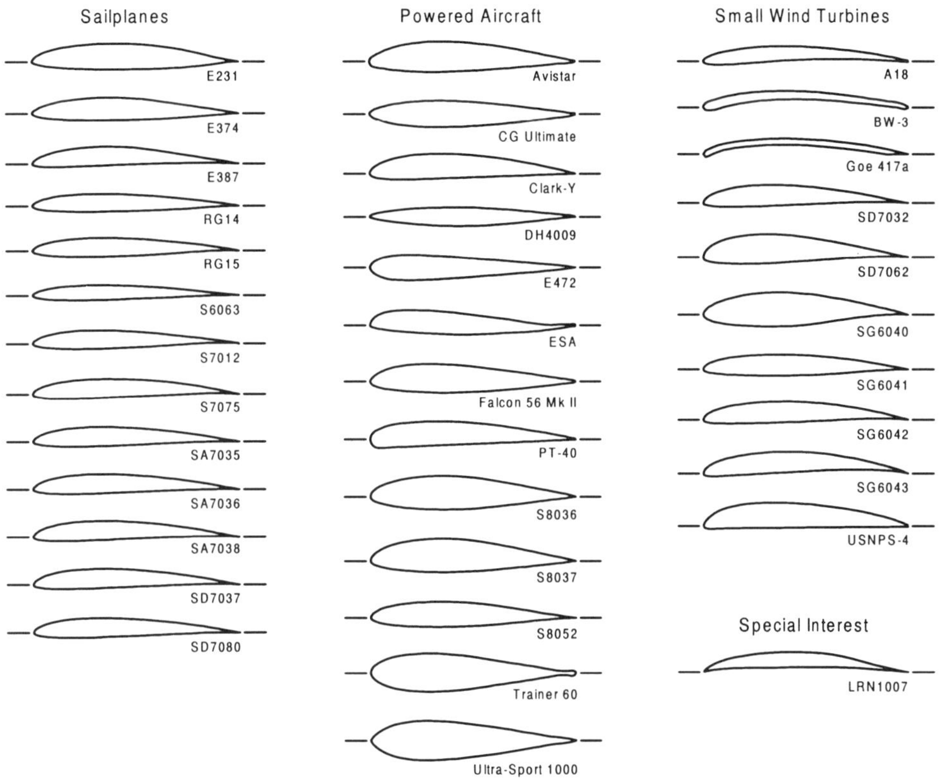

Fig. 1.1 Airfoils tested during Test Series 3 and 4 of the UIUC Low-Speed Airfoil Tests (Spring '96 and Spring '97).

Listed first are airfoils for use in R/C soaring competition. While extensive wind-tunnel data already exists for many of these airfoils, several were re-tested with various trailing-edge flap settings as suggested by R/C soaring enthusiasts. In particular, large flap settings of 15° to 25° were tested to give insight into possible performance improvements during launch, and smaller flap settings of 2.5° are representative of typical control deflections during thermal soaring. Supplementing this data are leading-edge flap settings of $-3°$ and $2.5°$ tested on the popular SD7037. Also, as variations of the popular SD7037, three new airfoils, the SA7035, SA7036, and SA7038, were designed and wind-tunnel tested.

The second group of airfoils listed are those for use on powered R/C aircraft. With a majority of the attention from airfoil designers focused on the design of new R/C soaring airfoils, a large disparity in performance has resulted between sailplane and powered airfoils. With this volume, the UIUC LSATs team has attempted to lay the foundation for reducing this disparity. To be more specific, several airfoils used on popular powered R/C aircraft were tested to provide a benchmark against which new designs can be compared. Serving as an indicator of the possible performance improvements, three new airfoils, the S8036, S8037, and S8052, were designed and tested.

Airfoils for use on small wind turbines comprise the third group of airfoils tested. Among these are four new airfoils (SG6040 – SG6043) specifically designed for variable-speed wind turbines. The reader will note that several of the airfoils listed here were designed specifically for use on aircraft (A18, SD7032, and SD7062). These airfoils, together with three others grouped under the "Sailplane" and "Powered Aircraft" headings (Clark-Y, S7012, and SD7037), were tested with leading-edge roughness to establish their potential for use on small wind turbines.

The last group is comprised of any airfoils that we found difficult to categorize. It should be stressed that while an attempt was made to classify every airfoil, the application of each airfoil is by no means limited to their respective groups. This will be evident from Chapter 4 where several airfoils are discussed in more than one context. The LSATs team naturally encourages experimentation with any airfoil and would appreciate feedback from such undertakings.

Finally, as mentioned in the Preface, this volume contains a rather extensive study of the effect boundary-layer trips have on airfoil performance. Experience, however, has proven such a study to be impractical for every airfoil. Therefore, two candidate airfoils were selected which are representative of a majority of the airfoils used on R/C model aircraft. These were the SD7037 and the E374. Each was tested with various trip configurations (single, multiple), geometries (2-D, 3-D), locations, and heights.

Chapter 2

Measurement Techniques

In order to produce consistent data between *Volume 1*, *2,* and *3,* all experiments were again performed in the UIUC Department of Aeronautical and Astronautical Engineering Subsonic Aerodynamics Laboratory using much of the same equipment and methods detailed in Refs. 2 and 3. Since the acquisition of data for *Volume 2*, changes have been implemented in an effort to improve both the accuracy and speed of the data acquisition process. The following sections detail these modifications.

2.1 Lift Balance

With the aging of any electro-mechanical system, there comes an increased percentage of time spent on maintenance. For the lift balance previously used by the UIUC LSATs, this time had grown beyond practical limits. The decision was therefore made to design a new lift balance that was not only simpler in design, thereby reducing maintenance issues, but also more versatile. What follows is a description of the UIUC LSATs' approach to designing such a device.

Figure 2.1 depicts the new lift measuring apparatus. It consists primarily of a fulcrum supported beam restrained in rotation by a force transducer (strain-gauge load cell). The lift load is applied at the right end of the beam and then transferred to the load cell via a lever arrangement. For correct operation of the lift balance,

Fig. 2.1 New lift balance.

the load cell is required to be in tension. To ensure this, counterweights are applied at the left end of the beam (opposite side of the fulcrum where the lift force acts) with a spring loaded right side to account for the weight of the airfoil model.

Depending on the expected lift range and Reynolds number of the test (60,000 – 500,000), one of three possible load cells (Interface Inc. SM-10, SM-25, and SM-50) were inserted into one of nine possible load cell attachment holes allowing for a variation in the operational range of the lift balance. This feature provides the ability to measure the small forces present at the lower Reynolds numbers while retaining the capability to handle the larger lift forces occurring at the higher Reynolds numbers. In contrast, the absolute resolution of the old lift balance was held constant throughout the measurement range. (Note: Measurements made with the load cells are very accurate and are repeatable to within 0.01% of the rated output.)

While the performance of the new lift balance is excellent over a broad range of operating conditions, there remains room for further improvement, particularly at the lower Reynolds numbers or when unsteady aerodynamic effects are dominant. Because of this, modifications are currently being discussed.

2.2 Moment Balance

In previous volumes, the pitching moment data presented was obtained through computational methods (XFOIL,[4] ISES,[5] or Eppler[6]) and given for only one operating condition. While it was felt that these results were adequate for many applications, their accuracy was suspect over a broad range of angles of attack. For this reason, an apparatus was designed to obtain experimental pitching-moment data over the full range of useful angles of attack.

The resulting device was designed to directly measure pitching moments about the quarter chord of the airfoil ($C_{m,c/4}$). As shown in Fig. 2.2, the moment load is transferred from the airfoil to free member 'A' and then across a load cell (Interface Inc. MB-10) serving as a link with fixed member 'B'. Because the load cell links the free and fixed members, it is subjected to any moment loads the airfoil experiences, thereby allowing the pitching moment of the airfoil to be measured. Also, in a manner similar to that employed on the lift balance, there are two sets of load cell attachment points allowing for an expanded measurement range.

One of the problems encountered during the operation of the moment balance resulted from the addition of two more members between the airfoil and the support structure. Owing to these two new members, increased "play" was introduced into the measurement apparatus. While this was not a factor during normal operations and did not affect angle of attack measurements, vibrations did become excessive during unsteady stall at the higher Reynolds numbers thereby requiring early termination of data acquisition. As is the case for the lift balance, efforts are also underway to improve the current moment balance.

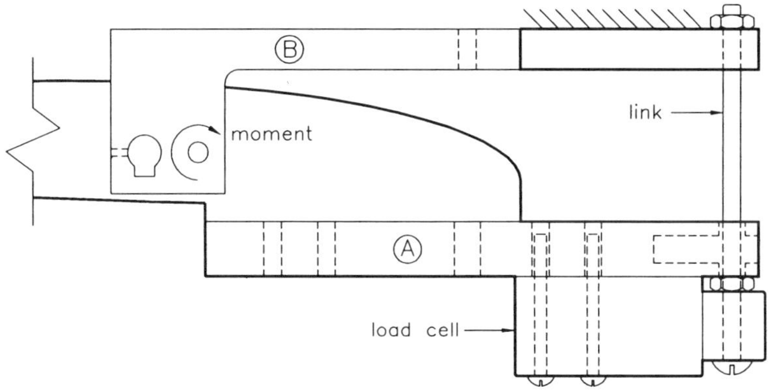

Fig. 2.2 New moment balance.

2.3 Tunnel Speed Controller

The final equipment modification, performed after the Spring '96 tests, was the replacement of the old digital tunnel-fan speed controller with a newer digital version manufactured by ABB Industry Oy. This new controller, combined with a shaft encoder which accurately determines fan RPM, reduced both the time needed to attain specified velocities and the amount of acoustic noise generated by the fan motor. While no turbulence data was acquired with the new controller installed, it is speculated that turbulence levels were reduced at the lower Reynolds numbers. Trends showing larger bubble drag for Spring '97 data as compared with data acquired on the same airfoil during Spring '96 support this conclusion. (See Chapter 3.)

2.4 Data Acquisition and Reduction

Changes regarding the lift balance, moment balance, and tunnel-fan speed controller were not the only modifications made during the collection of data for *Volume 3*. Prior to the beginning of testing, a great deal of time and energy was spent improving the accuracy of the data reduction process. In the course of doing so, the circulation correction discussed in Section 2.3.2 of *Volume 2* was modified and expanded.

Initially, the mathematical expression to account for the effect of airfoil circulation on the freestream velocity when measured close to the model was derived in a manner similar to that done during the Princeton Tests.[1] Following a detailed study of this effect, it was shown that an expansion of the image system used to model the wind-tunnel walls beyond the first image pair was necessary.[7] The new formulation, therefore, now considers an infinite number of images. As an indication of the magnitude of change produced by the new method, Fig. 2.3 shows reduced data for the E387 (C) both with and without the circulation correction. The reader should note that this plot does not show the difference between the old and new circulation corrections. Instead, it indicates the importance of accounting for the circulation effect, especially for $C_l > 1.0$, by comparing corrected and uncorrected data. The difference between the old and new corrections is much less severe.

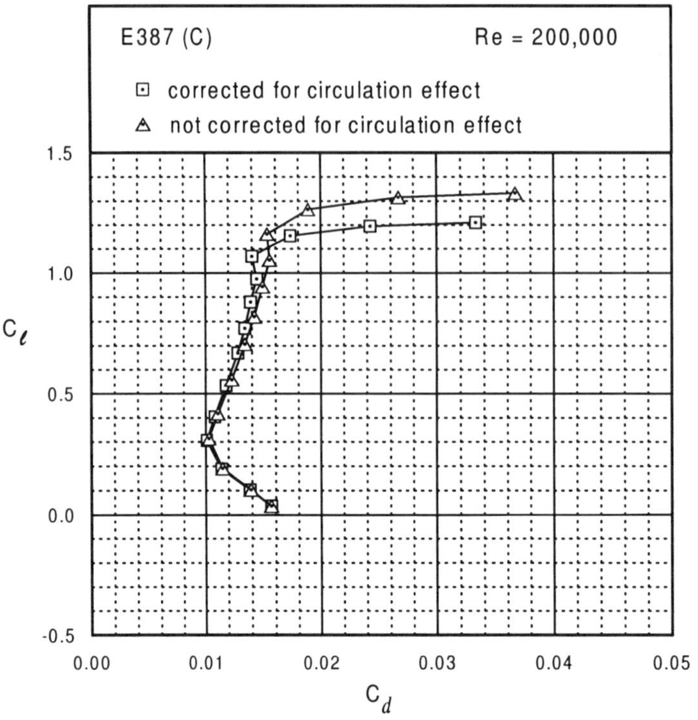

Fig. 2.3 Drag polars for the E387 (C) with and without the necessary circulation correction.

The new formulation for the circulation effect yields a slight improvement in the data reduction process. Consequently, the results obtained from the first two test series, which were reduced using the old formulation, have been re-reduced with the new correction. This new data is available from the UIUC LSATs web site at `http://www.uiuc.edu/ph/www/m-selig`. One should note that while the data of *Volume 1* and *2* has been modified, it is still self-consistent. The old data, therefore, continues to provide a means of comparison between data from the same test series. Finally, the interested reader is invited to consult Ref. 7 for a complete and systematic study of the circulation effect.

While the previously mentioned change impacted the accuracy of the data reduction process, changes were also made which improved data acquisition. Because of limitations reached in the previous programming language (Microsoft QuickBASIC), the decision was made to rewrite the UIUC LSATs codes in Borland C++. In doing so, the data acquisition process was streamlined, leading to improved productivity. It should be noted, however, that this decision was made after data for the Spring '96 installment had already been acquired. Therefore, this change only impacted the data acquired during Spring '97. Finally, it is important to note that changes made to the code only impacted the speed of the data acquisition process. These modifications had no affect on the final results presented in this book.

Chapter 3

Data Validation

Data validation is an important aspect of instilling confidence in any experiment. While fairly comprehensive discussions have been presented in *Volumes 1* and *2* regarding the validity of UIUC LSATs data, the previously mentioned improvements in equipment and data acquisition/reduction procedures demand a repeat of the validation process.

Perhaps the simplest way to validate wind-tunnel data is through comparison with a known standard. Determining exactly what that standard is, however, proves to be difficult. Although no specific facility produces perfect data, there are those considered better than others based on tunnel turbulence level, test-section geometry, and model quality. The Low-Turbulence Pressure Tunnel (LTPT) at NASA Langley Research Center, with its low turbulence, tall test section, and models that mount directly to the test-section walls, produces high quality data.[8] For this reason, data taken in the LTPT will be used as the standard.

3.1 Drag Data

Figure 3.1 shows a comparison between UIUC LSATs and LTPT drag polars for the E387 airfoil. Of the two E387 models tested at UIUC (models C and D), model C was tested at the beginning of both the Spring '96 and Spring '97 tests, while model D was tested only during Spring '97. In an effort to identify wind-tunnel blockage effects, model D was constructed with a 6-in. chord. With the exception of the larger bubble drag at Reynolds numbers below 200,000, the LSATs drag data compares fairly well with that taken at NASA Langley. As later described, the mounting hardware for model D was unfortunately damaged before the data set could be completed.

While comparisons with other facilities provide a great deal of information about wind-tunnel data accuracy, it is often instructive to compare data sets taken at the same facility. Upon comparison of Spring '96 model C data with the Spring '97 data sets, the reader may notice a slight discrepancy in drag through the region of the polar dominated by bubble drag ($0.25 < C_l < 1.0$). This discrepancy, while expected at the lower Reynolds numbers due to large spanwise variations in drag coefficients,[9] continues into the higher Reynolds numbers ($Re > 100{,}000$). It is speculated that the new tunnel-fan speed controller reduced the tunnel turbulence/noise levels. The lower turbulence levels produce larger regions of laminar separation, thereby creating the higher drag coefficients as seen in the Spring '97 data.

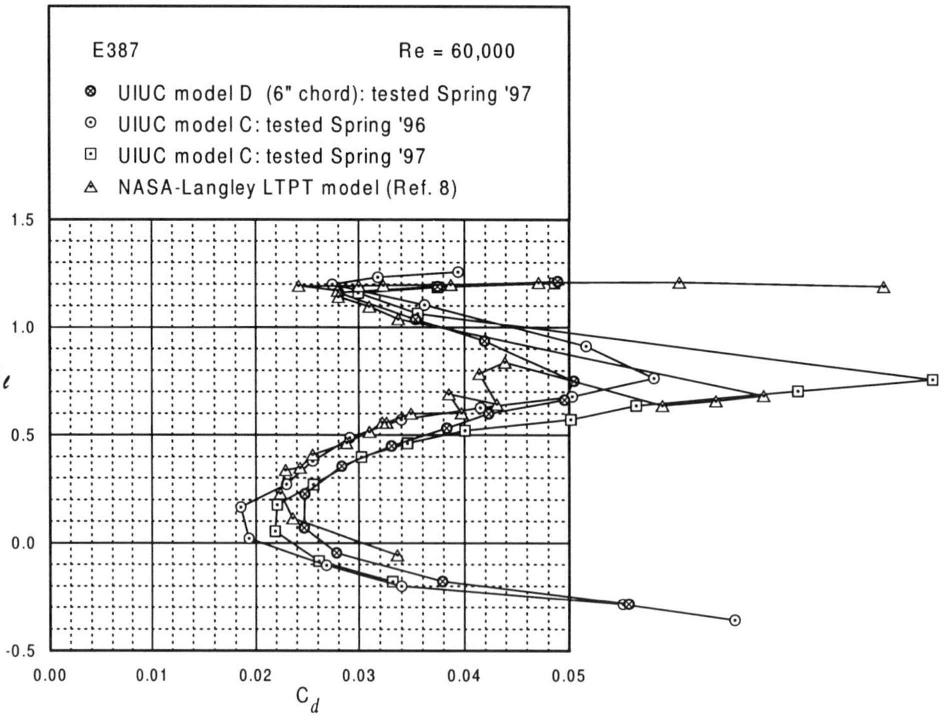

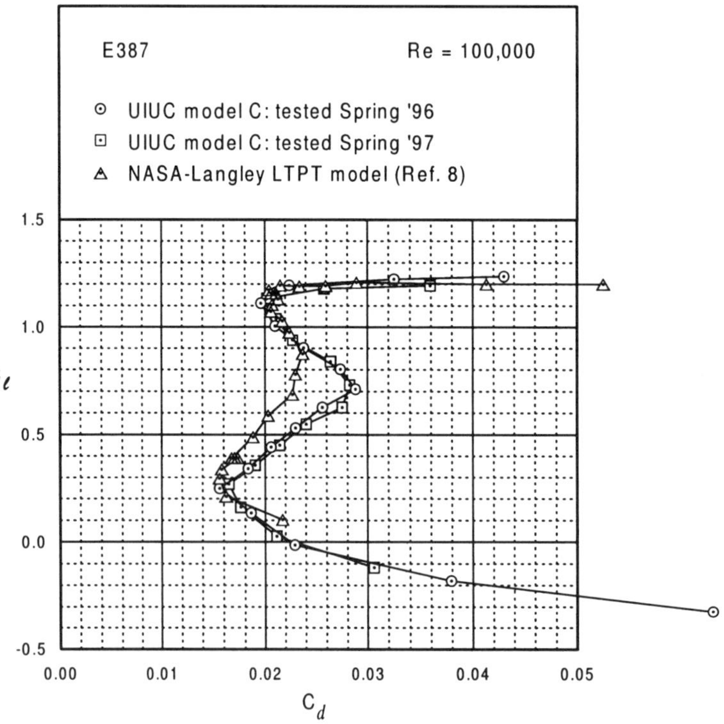

Fig. 3.1 Comparison between UIUC and LTPT E387 drag data ($Re = 60{,}000$, 100,000, 200,000, 300,000, and 460,000).

(figure continues)

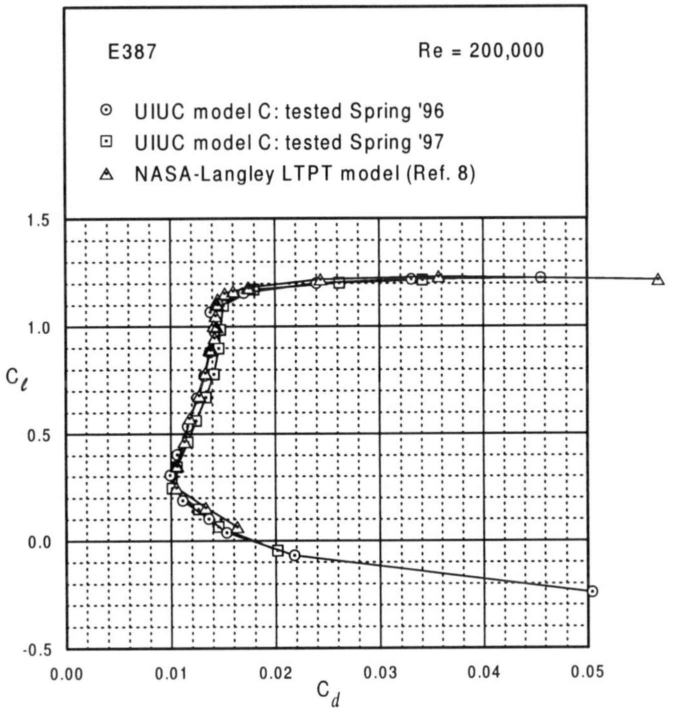

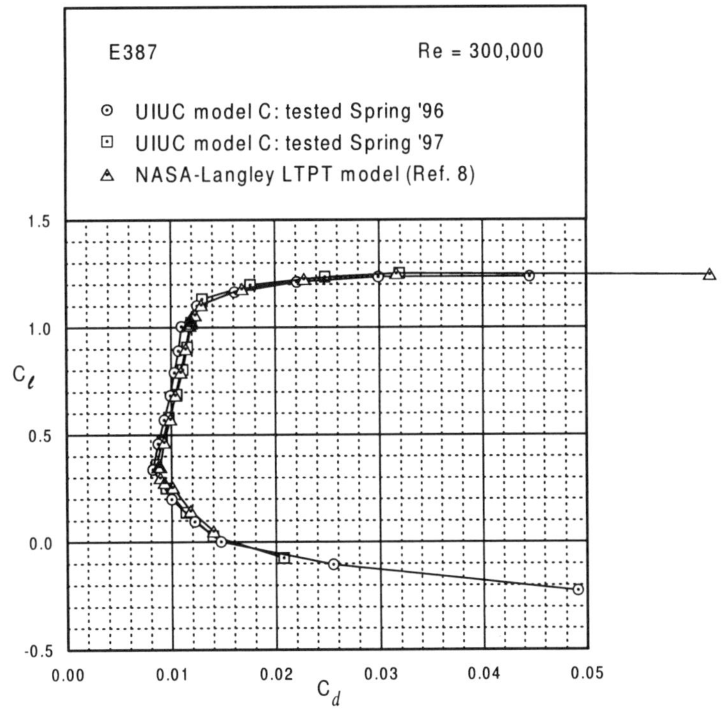

Fig. 3.1 Continued.

(figure continues)

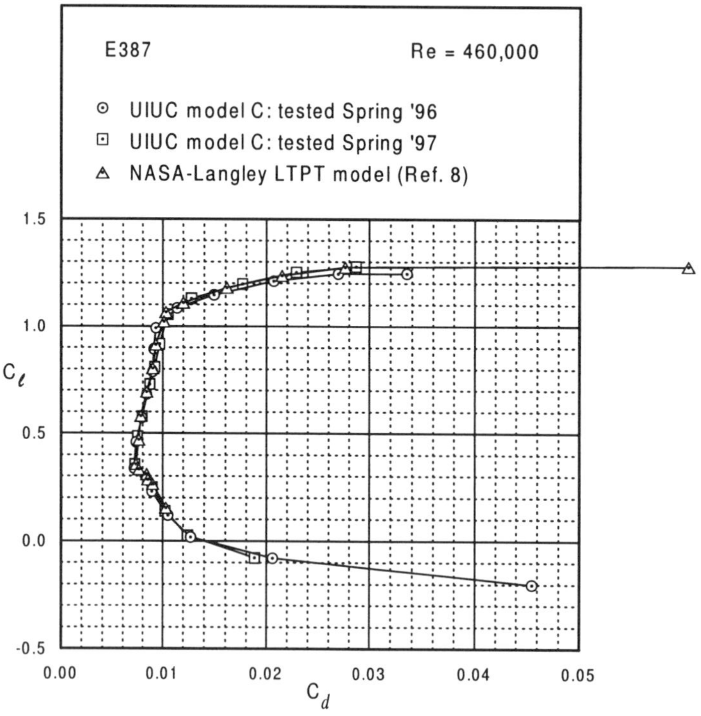

Fig. 3.1 Continued.

3.2 Lift Data

Lift curve comparisons between the UIUC and LTPT data (see Fig. 3.2) show acceptable agreement except at high angles of attack ($\alpha > 12°$), after which point three-dimensional end effects contaminate the data and produce a more gradual stall for the UIUC tests. Model D, with its higher aspect ratio and corresponding smaller end effects, supports the conclusion. To be more specific, because of the reduced end effects, a sharper stall is produced which is more comparable to the LTPT data. This is most evident at Re = 60,000 where a 2° difference exists in stall angle between models D and C.

The effect of increased "slop" in the system as a result of the moment balance can be seen influencing data taken into stall for the higher Reynolds numbers. The forces produced by the unsteady effects occurring at these conditions began to push the limits of the apparatus. As a result of the vibrations in the apparatus, there is some scatter in the lift data in the stall region. In fact, the vibrations generated by model D at Re = 200,000 were so severe that the support hardware for the model was damaged, thereby preventing any further testing of that model. It should be noted that these unsteady effects are not present during drag acquisition because of the limited angle of attack range.

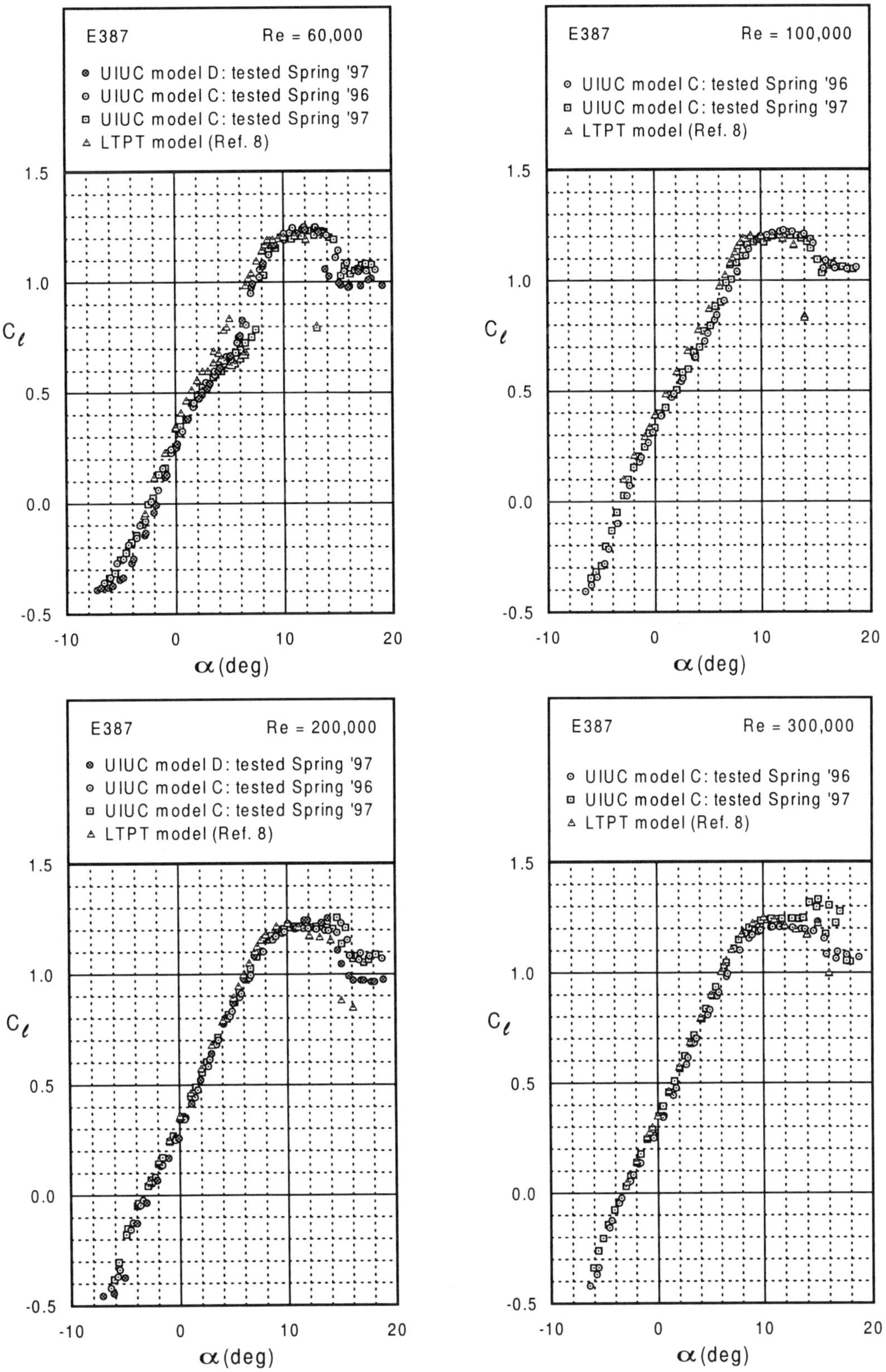

Fig. 3.2 Comparison between UIUC and LTPT E387 lift data (Re = 60,000, 100,000, 200,000, 300,000, and 460,000).

(figure continues)

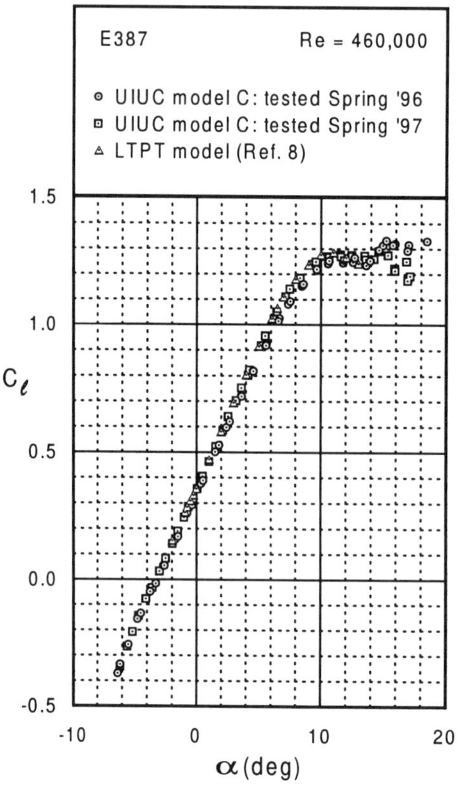

Fig. 3.2 Continued.

3.3 Moment Data

Figure 3.3 compares UIUC LSATs data for both the E387 (C) and (D) models with data reported by NASA Langley. With the exception of some scatter in the data, particularly at the lower Reynolds numbers, agreement is acceptable over a broad range of angles of attack.

3.4 Flow Features

To this point, only performance data (drag polars, lift curves, and moment data) has been validated. For a more complete determination of UIUC LSATs data accuracy, laminar separation and turbulent reattachment locations on model C were obtained using surface oil flow visualization and then compared with those reported by NASA Langley.[8]

The surface oil flow visualization approach made use of a fluorescent pigment (Kent-Moore 28431-1) suspended in a light, household-grade mineral oil that was airbrushed onto the surface of the model. The model was then subjected to 30–45

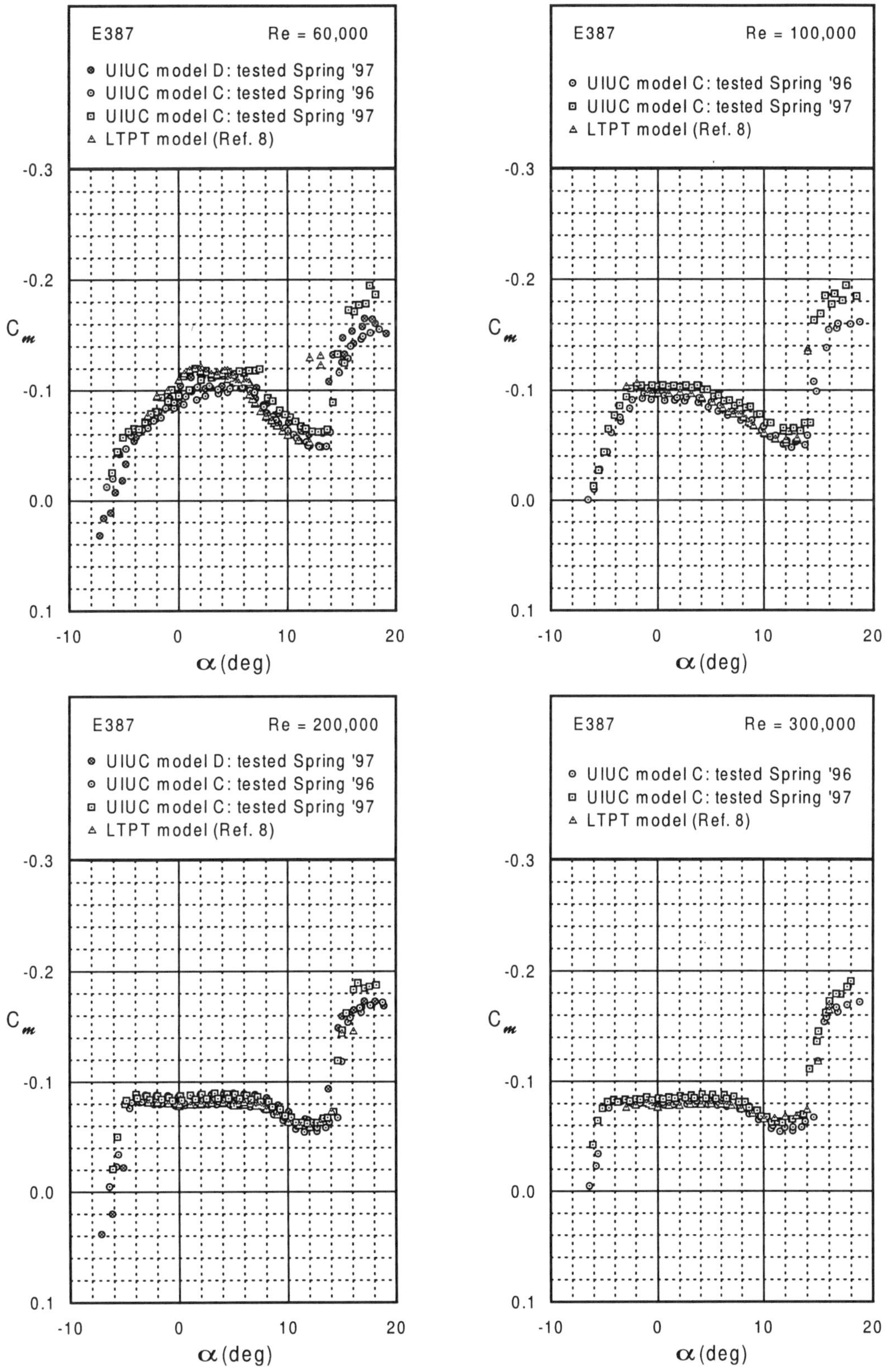

Fig. 3.3 Comparison between UIUC and LTPT E387 moment data (Re = 60,000, 100,000, 200,000, 300,000, and 460,000).

(figure continues)

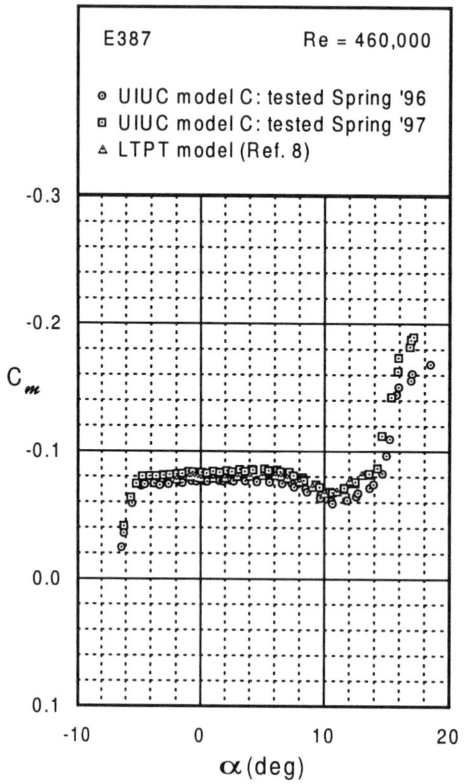

Fig. 3.3 Continued.

minutes of continuous wind-tunnel run time at a fixed angle of attack. During this period, the oil flowed in the direction of the local flow velocity at a rate dependent on the nature of the flow. As a result, discernible regions of laminar, turbulent, and separated flow could be identified.

Figure 3.4 shows a photograph of the fluorescent oil flow for $\alpha = 5°$ and $Re = 200{,}000$ and identifies several important flow regimes. For the purpose of this picture, the span of the model is oriented vertically, flow runs from left to right, and the segmented strip labels chordwise locations in terms of x/c. Laminar flow is seen to exist from the leading edge to approximately $x/c = 37\%$, after which point the flow separates and leaves the oil undisturbed until approximately $x/c = 57\%$. Between $x/c = 57\%$ and the reattachment location at $x/c = 62\%$, a region of reversed flow exists which scrubs any oil present upstream, subsequently forming the oil-accumulation line seen at $x/c = 57\%$. Finally, from the reattachment location rearward, turbulent flow exists. [Note: The oil line just upstream of the trailing edge ($x/c = 96\%$) is produced by a seam in the mylar covering on the surface of the model and does not serve as an indicator of any local flow features.]

The upper-surface flow features, as just described, were obtained over a range of angles of attack for $Re = 200{,}000$ and 300,000. These results are shown in Fig. 3.5

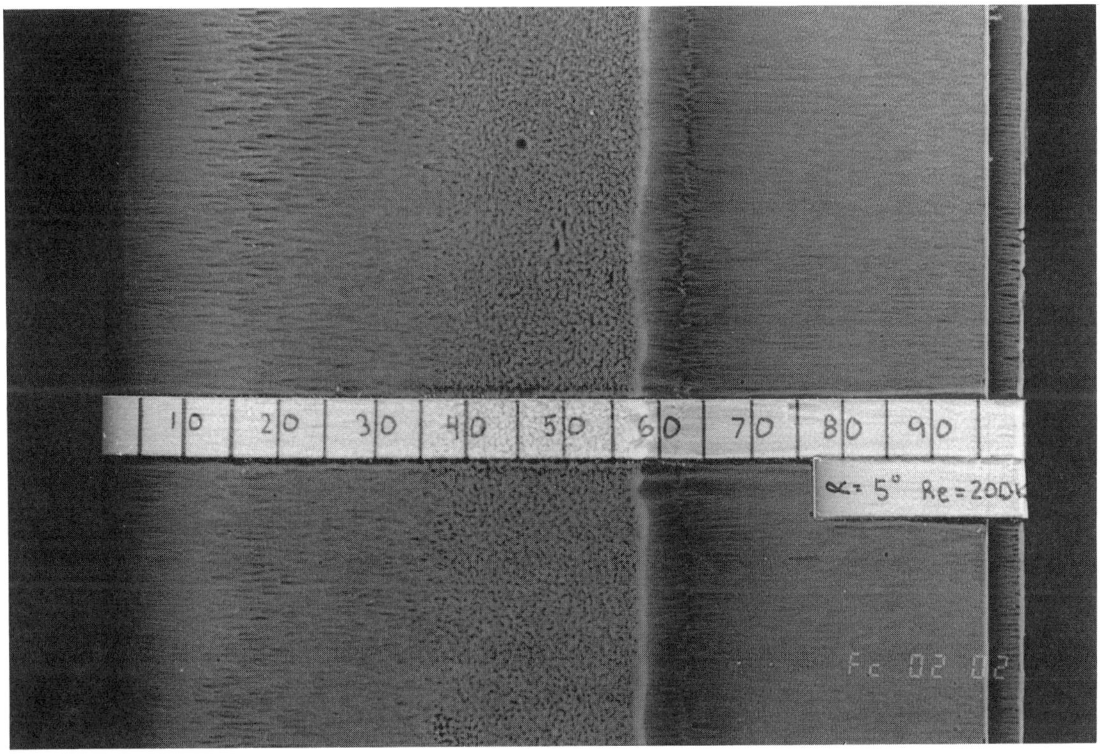

Fig. 3.4 Upper-surface oil flow visualization of major flow features on the E387 (C) ($\alpha = 5°$, Re = 200,000).

and are compared with LTPT data. The agreement in laminar separation and oil-accumulation lines between the data sets is mostly within 1–2% of x/c which is very near the uncertainty of the method. Low skin-friction values occurring below Re = 200,000 prevented acquisition of similar data for Re = 60,000 and 100,000. It should be noted that the feature identified as reattachment by NASA Langley is labeled in Fig. 3.5 as the oil-accumulation line. Owing to an 8-in. chord model, the NASA Langley study required higher dynamic pressures for a given Reynolds number, thereby producing higher skin friction as compared with the current study. As a result, the faint features produced by reattachment can quickly vanish, leaving the prominent oil-accumulation line to be mistaken for reattachment.

3.5 Data Validation Summary

Chapter 3 has made several important points regarding the accuracy of UIUC LSATs data that should instill confidence in the data presented throughout the remainder of this book. First, it was shown that drag data obtained by the UIUC LSATs for the E387 airfoil exhibit excellent agreement with LTPT data for $Re \geq$ 200,000 and only slightly higher bubble drag for $Re \leq$ 100,000. Second, UIUC lift data was

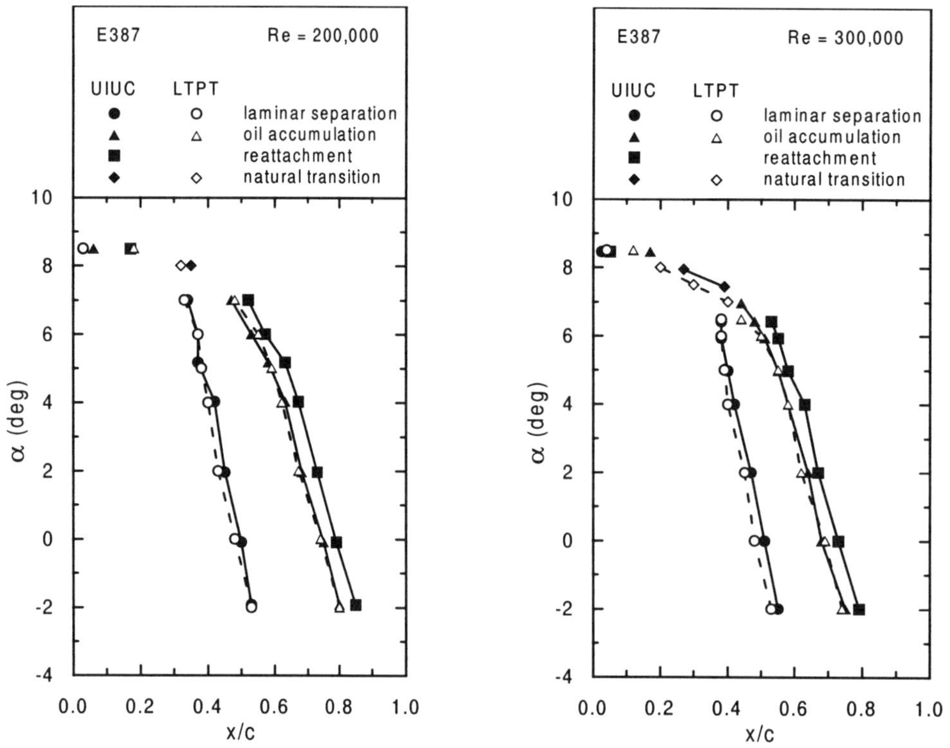

Fig. 3.5 Comparison of major E387 upper-surface flow features between UIUC and LTPT (*Re* = 200,000 and 300,000).

shown to have good agreement with LTPT data for all Reynolds numbers up to stall, after which point three-dimensional end effects and unsteady aerodynamics produced slight discrepancies. Third, UIUC pitching moment data was shown to agree well with LTPT data over a broad range of angles of attack. Lastly, in support of the three previous conclusions, flow visualization results showed excellent agreement between major surface-flow features for UIUC and LTPT data for $Re \geq 200{,}000$.

Chapter 4

Summary of Airfoil Data

Chapter 4, and its companion Chapter 5, are the heart of *Volume 3*. It is in these chapters that data acquired for each airfoil is discussed and plotted. As mentioned in Chapter 1, each airfoil has been grouped into one of four categories: Sailplane, Powered Aircraft, Small Wind Turbines, and Special Interest. The bulk of discussion regarding each airfoil is generally limited to comparisons within these groups. This allows a more straightforward presentation and comparison of the data. On occasion, however, it is useful to compare airfoils across groups. By doing so, the reader gains a better appreciation for the versatility of particular airfoils.

Table 4.1 lists each airfoil along with its thickness, camber, representative quarter chord pitching moment ($C_{m,c/4}$), model construction method (surface finish), model accuracy, and model builder. More detailed descriptions of the airfoil configurations (e.g. flap settings and trip locations) are provided in tables at the beginning of each section and in the profile plots provided in Chapter 5. As an aid to the reader, the following comments apply:

- For airfoils that were previously tested, the discussion is sometimes brief, in which case a more thorough discussion of each airfoil can be found in Refs. 1, 2, or 3.
- The suffixes '(A)', '(B)', etc. refer to multiple versions of a particular airfoil. Owing to construction inaccuracies, it is preferred to test multiple models of the same airfoil for two reasons. First, because each model may vary from the true airfoil geometry in a slightly different way, a measure of the sensitivity of each airfoil to changes in geometry is attained, and second, by having several models constructed, the probability of receiving an extremely accurate model is increased.
- The discussion of each airfoil is based on the *actual* shape of the model and not the designed *true* airfoil geometry. When the average difference between the actual and true airfoil coordinates is large (greater than 0.010 in. for a 12-in. chord), the wind-tunnel data may not be an accurate representation of the true airfoil performance. A better indication of the effects contour inaccuracies may have on performance can be realized from the airfoil accuracy plots in Chapter 5. Differences at the trailing edge behave like camber changes and therefore affect the useful lift range. Inaccuracies along the upper surface can have a large influence on bubble drag, while lower-surface errors are typically not as critical. Finally, if the airfoil is uniformly thicker, the performance will be more indicative of the true airfoil than if the surface error is "wavy."

Table 4.1: Airfoils Sorted by Category

Airfoils for:	Airfoil	Thickness (%)	Camber (%)	$C_{m,c/4}$
Sailplanes	E231	12.34	2.46	−0.040
	E374 (B)†	10.91	2.24	−0.050
	E387 (C)	9.06	3.90	−0.080
	E387 (D)	9.06	3.90	−0.090
	RG14	8.48	1.58	−0.040
	RG15 (C)†	8.92	1.58	−0.060
	S6063	7.05	1.43	−0.020
	S7012 (B)†	8.75	2.02	−0.045
	S7075 (A)†	9.00	3.05	−0.085
	S7075 (B)†	9.00	3.05	−0.077
	SA7035‡	9.20	2.60	−0.060
	SA7036 (A)‡	9.20	2.80	−0.078
	SA7036 (B)‡	9.20	2.80	−0.085
	SA7038‡	9.20	3.30	−0.085
	SD7037 (D)†	9.20	3.02	−0.070
	SD7037 (E)	9.20	3.02	−0.070
	SD7080†	9.15	2.48	−0.050
Powered Aircraft	Avistar	14.52	2.28	−0.050
	CG Ultimate	12.85	0.00	0.000
	Clark-Y (B)	11.72	3.55	−0.080
	DH4009	8.95	0.00	0.000
	E472	12.13	0.00	0.000
	ESA	10.72	1.98	0.040
	Falcon 56 Mk II	13.68	1.63	−0.020
	PT-40 (A)	11.59	2.93	−0.060
	PT-40 (B)	11.59	2.93	−0.060
	S8036‡	16.00	1.89	−0.030
	S8037‡	16.00	2.65	−0.037
	S8052	11.89	1.60	−0.020
	Trainer 60	18.31	0.00	0.000
	Ultra-Sport 1000	18.57	0.00	0.000
Small Wind Turbines	A18†	7.26	3.84	−0.080
	BW-3	5.02	5.65	−0.125
	Goe 417a	3.20	5.88	−0.110
	SD7032 (D)†	9.95	3.66	−0.100
	SD7037 (B)†	9.20	3.02	−0.078
	SD7062 (B)†	13.98	3.97	−0.100
	SG6040‡	16.00	2.50	−0.090
	SG6041‡	10.00	2.00	−0.070
	SG6042‡	10.00	3.80	−0.110
	SG6043‡	10.00	5.50	−0.160
	USNPS-4	11.94	5.00	−0.100
Special Interest	LRN1007 (B)	7.27	5.90	−0.120

† Wind-tunnel models previously tested (see Refs. 1, 2, and/or 3).
‡ New airfoils designed and built for *Volume 3*.

Table 4.1: Continued

Airfoil	Surface Finish	Avg. Diff. (in.)	Builder
E231	smooth	0.0109	M. Lachowski
E374 (B)	smooth	0.0083	M. Bame
E387 (C)	smooth	0.0046	J. Robertson
E387 (D)	smooth	0.0066	Y. Tinel
RG14	smooth	0.0088	R. Cavazos
RG15 (C)	mylar over obechi	0.0101	B. Champine / J. Robertson
S6063	smooth	0.0200	R. Cavazos
S7012 (B)	smooth	0.0066	M. Lachowski
S7075 (A)	smooth	0.0053	J. Robertson
S7075 (B)	smooth	0.0128¶	J. Thomas
SA7035	smooth	0.0044	M. Allen
SA7036 (A)	smooth	0.0082	D. Thompson
SA7036 (B)	smooth	0.0101	T. Foster / F. Carson
SA7038	smooth	0.0036	T. Foster / F. Carson
SD7037 (D)	rough fiberglass	0.0065	H. Stokely / B. Williams
SD7037 (E)	smooth	0.0113	J. Robertson
SD7080	smooth	0.0218	P. Illick
Avistar	varnish over obechi	0.0166*	B. Champine
CG Ultimate	smooth	0.0107*	M. Levoe
Clark-Y (B)	smooth	0.0056	J. Robertson
DH4009	smooth	0.0072	D. Stanley
E472	smooth	0.0056	D. Stanley
ESA	mylar over balsa	0.0122*	B. Halkett
Falcon 56 Mk II	smooth	0.0088*	B. Devaney
PT-40 (A)	mylar over balsa	0.0100*	T. Lampe
PT-40 (B)	open bay	0.0037*¶	T. Lampe
S8036	mylar over balsa	0.0080	T. Lampe
S8037	mylar over balsa	0.0099	T. Lampe
S8052	smooth	0.0063	M. Fox
Trainer 60	smooth	0.0111*	J. Thomas
Ultra-Sport 1000	smooth	0.0301*	D. Brengman
A18	smooth	0.0105	R. Cooney
BW-3	smooth	0.0166*	M. Allen
Goe 417a	smooth	0.0103	M. Fox
SD7032 (D)	mylar over balsa	0.0057	S. Watson
SD7037 (B)	smooth	0.0081	D. Thompson
SD7062 (B)	smooth	0.0068	K. Widiner / M. Allen
SG6040	smooth	0.0089	M. Allen
SG6041	smooth	0.0070	M. Allen
SG6042	smooth	0.0033	M. Allen
SG6043	smooth	0.0065	M. Allen
USNPS-4	smooth	0.0137	J. Raley
LRN1007 (B)	smooth	0.0201¶	M. Allen

* True coordinates were digitized from model plans or an actual wind-turbine blade.
¶ Does not reflect true model accuracy. Refer to model contour plots.

- In some instances, the average difference listed in Table 4.1 may not be indicative of the actual model accuracy for several reasons. Among these are an inaccurate 0° flap setting for the S7075 (B), an open-bay model construction technique resulting in covering sag between ribs for the PT-40 (B), and a sharp leading edge for the LRN1007 (B).
- For several powered-aircraft airfoils, tabulated coordinates were not available. In these cases, coordinates were digitized from model construction drawings and then mathematically smoothed using AFSMO.[10] As a result of the smoothing process, slight abnormalities appear near the trailing edge of several airfoils. In a majority of such instances, these abnormalities were removed during the model construction process.
- When characterizing surface finish qualities in Table 4.1, the term "smooth" refers to a fiberglass surface applied via the vacuum bag method. This is distinguished from a smooth mylar covering that may show imperfections in the underlying layer.
- The models with trailing-edge flaps were hinged on the lower surface with hinge tape and sealed on the upper surface with hinge-gap sealing tape. The SD7037 (E), which had both leading and trailing-edge flaps, used a leading-edge flap that was hinged on the upper surface using a flex-skin technique and sealed on the lower surface using hinge-gap sealing tape. With the exception of the SD7037 (E), which had 15% and 25% leading and trailing-edge flap chords respectively, all trailing-edge flap chords were 21%.
- The inviscid velocity distributions shown for the true airfoils in Chapter 5 (as well as in Appendix A) were calculated using the Eppler code.[6] Flap effects were included when appropriate. For those interested in the effects of viscosity on these distributions, they are encouraged to read *Volume 2*, which contains predicted viscous velocity distributions for several airfoils as well as a brief discussion on their interpretation. With experience, much can be gleaned from both the viscous and inviscid results to help interpret the airfoil drag polars and lift curves.
- The figures in Chapter 5 list nominal Reynolds numbers. Actual Reynolds numbers are available from either the drag polar data listed in Appendix C, the UIUC LSATs web site, or the data distribution diskette. In most cases the difference is typically no larger than $\Delta Re = 100$ to 200.
- Unless otherwise stated, all drag coefficients were obtained by averaging spanwise drag coefficients from four wake surveys spaced 3 in. apart approximately 1.25 chord lengths downstream of the model trailing edge. While these spanwise coefficients are not documented in this volume, they are available from both the UIUC LSATs web site and the data distribution diskette.
- For the lift curves, increasing and decreasing angles of attack are denoted by solid triangles and open circles, respectively.

- For the moment curves, increasing and decreasing angles of attack are denoted by inverted solid triangles and open rectangles, respectively. As has become convention when plotting airfoil pitching-moment data, positive values of $C_{m,c/4}$ are at the bottom of the plot (i.e., the C_m axis is inverted).
- The pitching-moment data presented in Table 4.1 are representative values for the typical operating point of the airfoil. As the reader will notice, many airfoils exhibit large changes in $C_{m,c/4}$ as both angle of attack and Reynolds number change. Because of these large changes, the values listed in Table 4.1 should only be used for comparative purposes. More accurate pitching-moment data than that provided in Table 4.1 is often necessary for detailed stability and control calculations. In such case, the variations in $C_{m,c/4}$ should be accounted for by using the full set of pitching-moment data. This data, as plotted in Chapter 5, can be obtained in tabulated form from either the UIUC LSATs web site or the data distribution diskette.
- Finally, Table 4.2 includes drawings of the various boundary-layer trips used on several models. The locations of these trips along the airfoil chord are drawn together with the airfoil geometries in Chapter 5. Table 4.2, however, does not include those trip geometries tested for Chapter 6. This data is presented separately in Table 6.1.

Table 4.2: Trip Geometries Tested

Trip Name	Airfoils	Trip Geometry†	Trip Heights
zigzag type C	A18 BW-3 Clark-Y S7012 SD7032 SD7037 SD7062 SG6040 SG6041 SG6042 SG6043	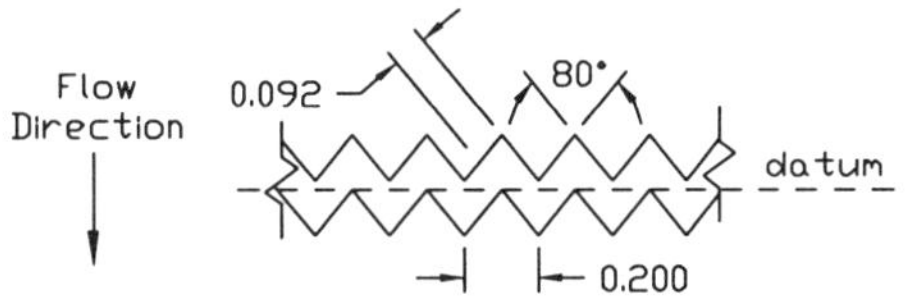	0.005 in. (0.04%c) 0.010 in. (0.08%c) 0.014 in. (0.12%c) 0.018 in. (0.15%c) 0.023 in. (0.19%c)
zigzag type E	S7075	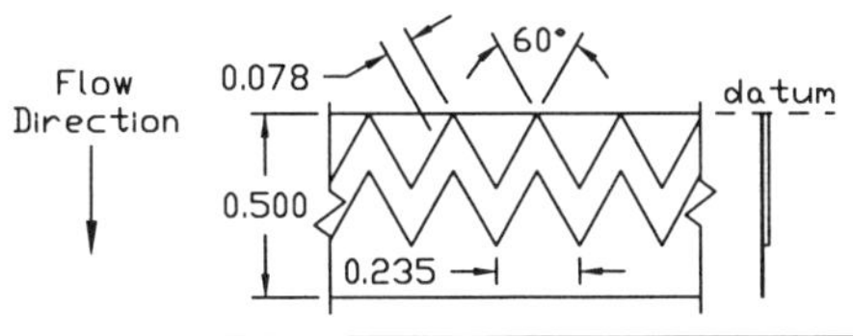	0.021 in. (0.18%c)

† All drawings actual size. All dimensions in inches.

4.1 Airfoils for Sailplanes

Table 4.3: Summary of Airfoil Data for Sailplanes

Airfoil	Configuration	Figures Begin
E231	clean	Fig. 5.29, p. 94
E374 (B)	clean	Fig. 5.33, p. 100
E387 (C)	clean	Fig. 5.37, p. 104
E387 (D)	clean	Fig. 5.41, p. 110
RG14	clean	Fig. 5.73, p. 146
RG15 (C)	clean: 2.5° flap	Fig. 5.77, p. 150
	clean: 15° flap	Fig. 5.81, p. 154
	clean: 20° flap	Fig. 5.83, p. 156
	clean: 25° flap	Fig. 5.85, p. 158
S6063	clean	Fig. 5.87, p. 160
S7012 (B)	clean: 2.5° flap	Fig. 5.95, p. 168
	clean: 15° flap	Fig. 5.98, p. 172
	clean: 20° flap	Fig. 5.100, p. 174
	clean: 25° flap	Fig. 5.102, p. 176
S7075 (A)	clean	Fig. 5.104, p. 178
	trip type E	Fig. 5.108, p. 182
S7075 (B)	clean	Fig. 5.110, p. 186
SA7035	clean	Fig. 5.126, p. 208
SA7036 (A)	clean	Fig. 5.130, p. 214
SA7036 (B)	clean	Fig. 5.134, p. 218
SA7038	clean	Fig. 5.138, p. 222
SD7037 (D)	clean	Fig. 5.150, p. 234
	with upper-surface covering	Fig. 5.154, p. 238
SD7037 (E)	clean: 0° LE flap & 0° TE flap	Fig. 5.156, p. 240
	clean: −3° LE flap & −2.5° TE flap	Fig. 5.160, p. 244
	clean: 2.5° LE flap & 2.5° TE flap	Fig. 5.163, p. 248
	clean: 0° LE flap & 5° TE flap	Fig. 5.166, p. 252
SD7080	clean	Fig. 5.175, p. 264

SD7037 (D) & (E) In R/C soaring competition, most competitors use trailing-edge flaps for launching, thermal soaring, cruising, and landing. By using trailing-edge flaps, the pilot has a more versatile aircraft which performs well over a wider range of operating conditions. While the use of trailing-edge flaps alone can substantially improve performance, the following question attributable to Joe Wurts and Jerry Robertson was asked: Would the combination of leading and trailing-edge flaps outperform the traditional use of trailing-edge flaps alone? To explore this question, the

SD7037 (E) was built to have a 15% leading-edge flap and a 25% trailing-edge flap. While the trailing-edge flap geometry was conventional, the leading-edge flap had a flex-skin hinge on the top surface. To be more specific, the fiberglass skin of the model served as the flap hinge on the upper surface while a typical hinge-gap seal was used on the lower surface.

To date, the preliminary tests are encouraging. Figure 5.164 shows the results for 2.5° of leading-edge flap (nose down) in combination with 2.5° of trailing-edge flap. The most interesting feature of the polars is the relatively broad low-drag range as compared with the conventional trailing-edge flap case shown in Fig. 5.167. For the conventional flap, there is a point in each polar below which the drag markedly increases (e.g. $C_l = 0.95$ for $Re = 100{,}000$ in Fig. 5.167). The higher drag below this point is caused by an upper-surface laminar separation bubble which does not reattach prior to the steep pressure recovery region aft of the flap hinge line. With the larger flap deflections presented in *Volume 2* on the SD7037 (B), the effect is more pronounced (higher drag). To reduce the effect of this adverse pressure gradient, and thereby reduce any increase in drag, it becomes advantageous to limit the trailing-edge flap deflection. Unfortunately, imposing such a limitation also limits the amount of additional lift generated by the flap. By using a leading-edge flap, however, the lost lift can be regained with no additional deflection of the trailing-edge flap. As a result, the drag effect is less severe for the LE/TE flap case.

It should be noted that the polars shown in Figs. 5.164 and 5.167 do not correspond to configurations with the same zero-lift angle of attack. As a result, the polars should not be directly compared. The LE/TE edge flap case produces less lift ($\Delta C_l = -0.1$). If more flap (both leading-edge and trailing-edge flap) had been used in order to yield the same viscous zero-lift angle of attack, the two cases would be comparable. Nevertheless, the effect of more LE/TE flap would be to shift the polars toward higher C_l's with little change in drag. Thus, the SD7037 with a LE/TE flap system is an improvement over the traditional plain flap. Those modelers ambitious enough to design a dual LE/TE flap system will likely be rewarded with higher performance.

For the SD7037 (D), tests were performed with (Fig. 5.152) and without (Fig. 5.154) mylar covering over the otherwise rough fiberglass finish of the model. For the case with covering, it was only affixed to the upper surface of the model. At $Re = 200{,}000$, the covering has little effect, except for the region of $0.0 < C_l < 0.2$ where an increase in drag is observed. (During tests for *Volume 2*, similar results were achieved at low C_l's for the RG15 (C) when tested with both a rough wood-grain finish and a smooth mylar covering.[3]) For the range of $0.2 < C_l < 0.5$, however, there is a slight decrease in drag with covering. These results for the SD7037 (D), and the past results for the RG15 (C), suggest that a rough surface can lead to some drag reduction, most likely through reduced bubble drag. The effect is largest at low C_l's where long lengths of laminar flow are most likely to be influenced by roughness. It can be concluded that distributed roughness, such as that produced by rough fiberglass weave or wood-grain finish, on airfoils at low Reynolds numbers is likely to improve performance.

SA7035, SA7036 (A) & (B), & SA7038 At the 1996 AMA/LSF Nationals, a survey was conducted by Michael Selig and Cameron Ninham with the intention of determining what airfoils were used by those competing in the Unlimited Thermal Soaring Competition. Of the approximately 110 flyers present, responses were obtained from 101 of them. Based on these responses, the most popular airfoil was the SD7037 being used on 40 of the sailplanes, followed by the S3021 and the RG15 each with 8, and then the SD7080 with 6. The remaining responses included over 24 different airfoils. More details of the survey can be found in the Camberline Newsletter No. 4 on the LSATs web site.

The popularity of the SD7037 at the Unlimited Thermal Soaring Competition and elsewhere has inspired an effort to develop a family of airfoils based on the SD7037 which differ in their thicknesses (9.2%, 7.7%, and 6.2%) and useful lift ranges. Such a family of airfoils will enable a sailplane designer to migrate from one airfoil to another when designing sailplanes with different wing loadings, as well as provide aerodynamically-similar airfoils for blending along the wing span from root to tip.

As a first step in this process, three new airfoils have been designed with varying lift ranges — the 9.2% thick Selig/Ashok Gopalarathnam SA7035, SA7036, and the SA7038. These three airfoils, together with the SD7037, are shown co-plotted in

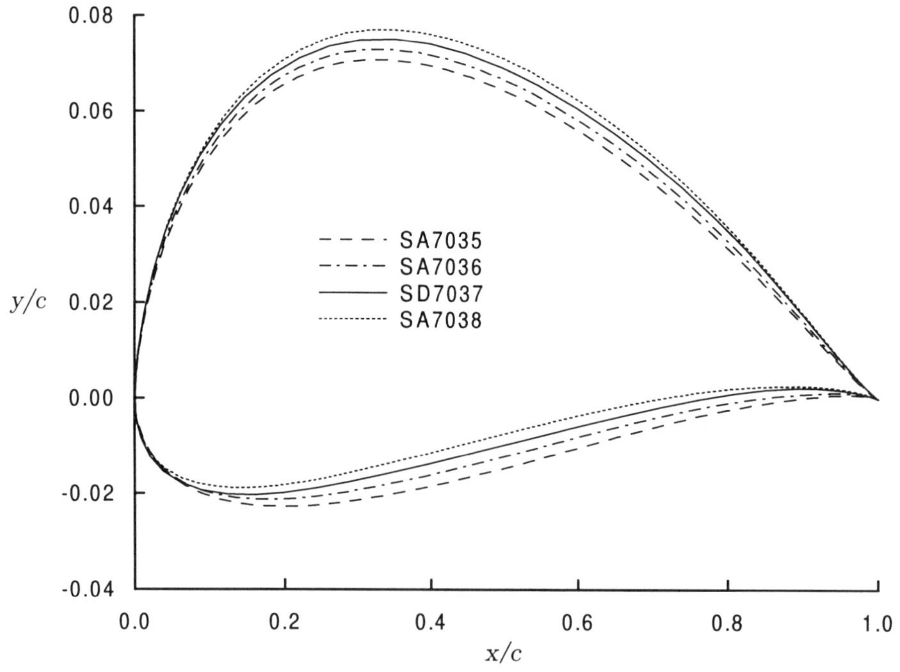

Fig. 4.1 Geometric comparison between the SD7037 and SA series airfoils (y-axis expanded 5× for clarity).

Fig. 4.1. While it may appear that simple camber changes were made to the SD7037 to arrive at the new airfoils, this process was not used. Instead, each of the new airfoils were designed using the inverse design method PROFOIL[11] to achieve the desired lift-range variation. By using PROFOIL to specify the C_l that produces a lower-surface boundary layer just on the verge of laminar separation, the lower corner of the useful lift range for each airfoil was fixed. The shape of the polar from this point onward was then obtained by tailoring the bubble ramp on the upper surface of each airfoil. Thus, not only do the four airfoils share common aerodynamic features, but each airfoil is also optimized for the lift-range over which it was designed.

The model accuracy plots, inviscid velocity distributions, and performance plots for the three new airfoils are shown in Figs. 5.126 – 5.141. Figure 4.2 compares the drag polars from XFOIL[4] analyses for the four airfoils at a Reynolds number of 200,000, while Fig. 4.3 shows the experimental results at the same Reynolds number. As seen from the two figures, the SA7035 and SA7036 airfoils are lower-lift versions of the SD7037 while the SA7038 is a higher-lift version. Based on these data, and the success of the SD7037, these four airfoils are highly recommended for use on R/C sailplanes.

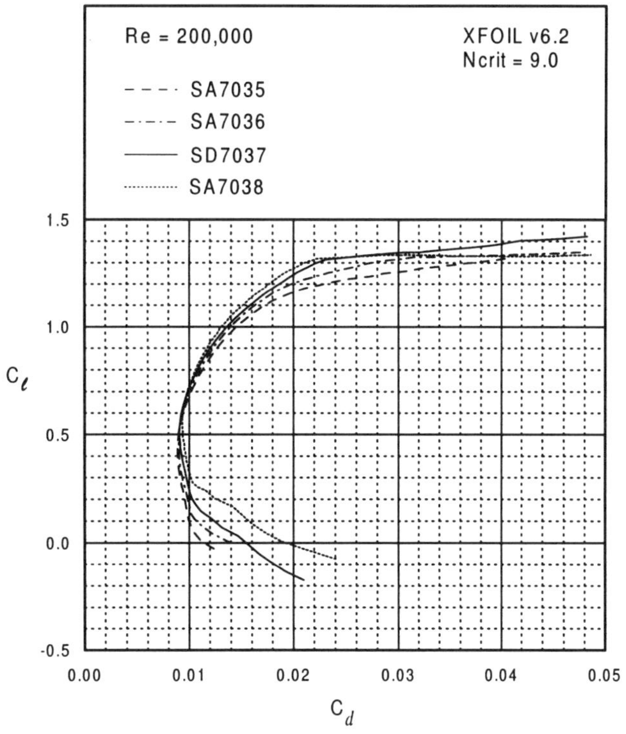

Fig. 4.2 Computational drag data comparison between the SD7037 and SA series airfoils at Re =200,000.

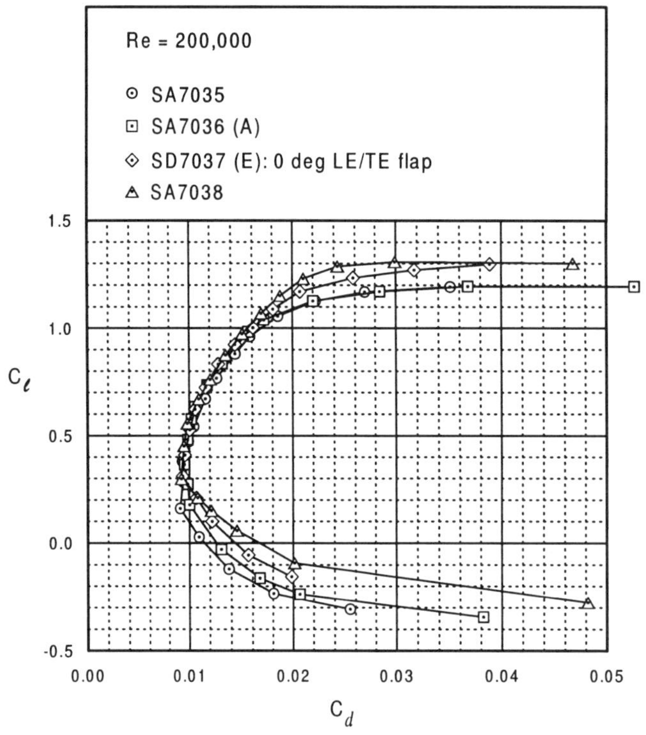

Fig. 4.3 Experimental drag data comparison between the SD7037 (E) and SA series airfoils at Re =200,000.

SD7080 The SD7080 is another popular airfoil for R/C soaring competition. This airfoil has better performance at lower C_l's than the SD7037 and has been used on faster sailplanes. For this reason, the SD7080 was used as a benchmark while designing the SA airfoils, in particular the SA7035. XFOIL[4] computations for the SA7035 and the SD7080 (not shown) indicate that the SA7035 achieves better performance than the SD7080 at the low-lift/high-speed and the high-lift/low-speed ends of the polar at the price of slightly higher drag through the middle range. For this reason, the SD7080 model from *SoarTech 8* was re-tested to provide a comparison with the SA7035 results. The comparison shows that, except in the middle range of the polar for Re =100,000, the SA7035 has better performance than the SD7080. A significant reason for this departure from the computational results is the inaccuracy of the SD7080 model, as shown in Fig. 5.176. This inaccuracy is primarily a result of the shrinkage and warping of the balsa sheeting used to cover the model. To provide a better comparison, a new model of the SD7080 should be built and tested.

S7075 (A) with trips The S7075 airfoil was based on a new design philosophy that integrates boundary-layer trips into the airfoil design process,[3] as opposed to considering boundary-layer trips as an afterthought on an otherwise poorly performing airfoil. As reported in Ref. 3: "The idea behind placing the trip so far aft can be explained as follows. When no trip is used, the bubble ramp on the upper surface is tailored so that the bubble starts far aft at low angles of attack and moves forward toward the leading edge until the airfoil stalls. Thus, the entire upper surface serves as a bubble ramp to promote transition. In the current philosophy, the bubble ramp and the trip assume more or less equal roles in promoting transition. At low angles of attack, the trip placed far aft on the airfoil acts to promote transition; whereas, for higher angles of attack the bubble ramp causes transition. Thus, the bubble ramp can be better tailored to operate over a narrower range while the trip promotes transition over the remainder of the operating range. These ideas were integrated into the S7075 during the design process."

In the first tests reporting on the S7075,[3] the trip was located at 68% chord. Since then, an exhaustive trip study was performed (see Chapter 6 and/or Ref. 12) which suggests that the trip on the S7075 was likely located too far aft. Figure 5.105 shows the trip configuration most recently tested. The trip was positioned at 57.5% chord as referenced to the leading edge of the trip (see Table 4.2). The zigzag trip itself was attached to 0.003-in. thick tape which is shown in the drawing. The 0.021-in. trip thickness includes both the trip and tape. Results shown in Figs. 5.106 and 5.108 for the clean and tripped cases, respectively, show that there is a substantial drag reduction at the lower C_l's as intended. For the new trip position, the performance rivals that of the SD7037.

The high-lift performance of the S7075 at Re = 60,000 is particularly intriguing. As previously described, the boundary-layer trip on the S7075 is most effective at low lift coefficients, while the bubble ramp is optimized for higher lift coefficients. When compared with the SD7037 (Fig. 5.152), the result of tailoring the bubble ramp for a narrow range of lift coefficients has lead to lower drag at Re = 60,000 for higher C_l's. This desirable high-lift/low-Re performance makes the tripped S7075 a strong candidate for R/C handlaunch competition.

RG15 (C) and S7012 (B) with flaps In *Volume 2*, data on the RG15 (C) and S7012 (B) was presented for 5° and 10° flap deflections. Since thermal soaring often requires smaller flap deflections, these airfoils were re-tested with flaps set at 2.5°. The results obtained are consistent with past data (*Volume 1* and *2*) in that the two airfoils compare well for similar flap settings. What is especially interesting, however, is the improvement in soaring performance for the 2.5° flap cases as opposed to the 5° cases. For reference, a 2.5° flap setting on a 10-in. chord with a 21% chord flap produces approximately a 3/32-in. deflection, and 5° flap setting gives a 3/16-in. deflection.

RG14 The Rolf Girsberger RG14, first reported on at the 1983 ISF-Seminar,[13] was designed to have slightly less pitching moment, thickness and maximum lift coefficient than the popular RG15. It consequently shares many of the desirable characteristics of the RG15.

S6063 & E231 The 7%-thick S6063 was designed as one in a series of airfoils for slope soaring (SD6060, S6061, S6062, and S6063). Although the wind-tunnel model is reflexed relative to the true airfoil, it still produces the lowest drag measured by the UIUC LSATs for *Re* = 300,000. When high speed is needed above all else, this airfoil in its modified form (Fig. 5.88) is applicable.

In contrast to the thin S6063, the 12.3%-thick Eppler E231 is a newly designed airfoil for R/C slope soaring where higher thickness is desirable. Also, it was designed to operate at higher *Re*'s than those considered in the design of the S6063. Thus, it is not surprising that the performance below *Re* = 200,000 suffers. However, it should be noted that the airfoil was designed to make use of trips located at 72% chord on the upper surface — just upstream of the flap hinge line at 80% chord. During the next test series, this airfoil will be tested with boundary layer trips guided by the results discussed in Chapter 6. For large models requiring thick wings, this airfoil may prove suitable.

4.2 Airfoils for Powered Aircraft

Table 4.4: Summary of Airfoil Data for Powered Aircraft

Airfoil	Configuration	Figures Begin
Symmetric Airfoils		
CG Ultimate	clean	Fig. 5.15, p. 74
DH4009	clean	Fig. 5.25, p. 88
E472	clean	Fig. 5.45, p. 114
Trainer 60	clean	Fig. 5.207, p. 314
Ultra-Sport 1000	clean	Fig. 5.211, p. 320
Semi-Symmetric Airfoils		
Avistar	clean	Fig. 5.5, p. 62
ESA	clean	Fig. 5.49, p. 120
Falcon 56 Mk II	clean	Fig. 5.53, p. 124
S8036	clean	Fig. 5.114, p. 190
S8037	clean	Fig. 5.118, p. 196
S8052	clean	Fig. 5.122, p. 202
Flat-Bottom Airfoils		
Clark-Y (B)	clean	Fig. 5.19, p. 80
PT-40 (A)	clean (sheeted)	Fig. 5.65, p. 138
PT-40 (B)	clean (open bay)	Fig. 5.69, p. 142

Powered R/C model builders typically group airfoils into one of three categories: symmetrical, semi-symmetrical, and flat-bottom. From an aerodynamics standpoint, the terms symmetrical and cambered are more proper descriptions, but for the sake of this discussion, cambered airfoils will be separated into the more specific semi-symmetrical and flat-bottom groupings.

Previous volumes of *Summary of Low-Speed Airfoil Data* have provided only limited data regarding power-plane airfoils. As a result, many readers may be unfamiliar with the characteristics of these airfoils. Some initial comments are therefore warranted to provide a better understanding of power-plane airfoil geometries and performance.

The most obvious characteristic of power-plane airfoils is their relatively large thickness. Frequent flight in high-g maneuvers in combination with conventional construction techniques (non-composite) require powered R/C aircraft to have a deep spar for adequate wing strength. By using a thick airfoil, adequate spar depth can be achieved. In addition to increasing the structural integrity of the wing, thick airfoils are also helpful when using retractable landing gear, allowing ample space for the retraction unit.

While thick airfoils may be structurally advantageous, aerodynamically they produce significantly more drag than their thin counterparts. High-drag airfoils, however, serve two purposes on powered R/C aircraft. First, they help to moderate flight speed during aerobatics or rapid descents, and second, they result in easier landings due to increased glide slopes. As one would expect, however, airfoil drag can sometimes be excessive, thereby producing more disadvantages than advantages. In such cases, it becomes beneficial for power-plane designers to seek out properly designed airfoils that are not only thick but also low drag.

Because of the large emphasis placed on the handling qualities of powered R/C aircraft, one of the largest performance drivers for power-plane airfoils is stall behavior. Excluding airfoils for use on high-performance aerobatic aircraft, an airfoil with a gentle stall is highly desirable. Some of the airfoils tested, however, have less than ideal stall characteristics partly owing to a lack of proper airfoil design techniques. By studying such airfoils, they serve as a first step in designing the next generation of high-performance airfoils for powered aircraft.

4.2.1 Symmetric Airfoils

Note: While the following airfoils are symmetric in design, some contour plots and velocity distributions are slightly asymmetrical owing to inaccuracies during the coordinate digitization process. See Table 4.1 for airfoils to which this may apply.

CG Ultimate This airfoil was taken from the highly aerobatic and popular Carl Goldberg ULTIMATE BIPE. Its broad region of constant drag followed by a sharp corner in the drag polar just prior to $C_{l,max}$ is typical of several of the power-plane airfoils tested. Focusing attention on the lift curves (Fig. 5.18), a non-linearity can be seen for $Re = 100{,}000$ and $C_l = 0$. Such non-linearities often cause handling problems when they occur near aircraft operational lift coefficients. For this airfoil, however, the non-linearity should have little impact on performance since the CG Ultimate airfoil typically operates at high lift coefficients when near $Re = 100{,}000$.

Being an aerobatic aircraft, the ULTIMATE BIPE often experiences periods of high lift coefficients when performing abrupt maneuvers (square loops, snap rolls, etc.). While the CG Ultimate airfoil does exhibit a fairly distinct break at stall allowing for crisp snap maneuvers, the break occurs at a relatively low $C_{l,max}$. If an airfoil was designed to increase this $C_{l,max}$ while maintaining the same crisp stall, an improvement in square maneuvers could be produced with little impact on snap behavior.

Once past the initial stall, this airfoil exhibits only a small reduction in lift. Interestingly, however, a critical point is reached near $\alpha = 16°$ after which the lift again begins to increase. With a skilled pilot, an aircraft using this airfoil can capitalize on this behavior and perform some rather unconventional maneuvers when flown in the post-stall regime.

DH4009 The DH4009 is a thinned NACA section with a maximum thickness of 9% occurring at $x/c = 40\%$. It has been used on the STORM aerobatic (pattern) aircraft and is typical of many of the airfoils currently in use on high-performance pattern aircraft. Owing to the sharp leading-edge radius, the DH4009 lift and drag performance is quite limited. Despite this limited performance range, minimum drag values are what one would expect from such a thin airfoil. Lift behavior is fairly typical except at low Reynolds numbers where a reversal in the lift-curve slope is seen around $\alpha = 0°$. This reversal has been seen on other symmetrical airfoils at low Reynolds numbers[14] and can usually be eliminated through the use of boundary-layer trips.[2,14]

E472 The E472, designed by Dr. Richard Eppler, was designed for use on full-scale aerobatic aircraft.[15] As part of the design requirements, Eppler identified four important factors of successful aerobatic airfoils: high lift coefficients both upright and inverted, a hard stall for flying snap maneuvers, high drag near zero lift to aid in speed management, and low drag at high C_l's to improve maneuvering performance.

At Reynolds numbers greater than 200,000, the E472 does produce a rather high $C_{l,max}$. Below $Re = 200{,}000$, however, a dramatic reduction in $C_{l,max}$ takes place which may be problematic for some model aircraft that land near a Reynolds number of 100,000. For this reason, it is recommended to only use this airfoil on larger R/C aircraft.

The E472 meets the criterion for hard stall, but only above $Re = 100{,}000$. One interesting aspect of this airfoil is its pronounced hysteresis even at the higher Reynolds numbers. Just as the post-stall behavior of the CG Ultimate may help when performing unusual maneuvers, the hysteresis produced by the E472 may prove beneficial as well.

The E472 does have high drag near $C_l = 0$, but again only at the higher Reynolds numbers. Around Re = 100,000, the drag is actually quite low for a low Reynolds number airfoil. This is a result of the short laminar regions on both the upper and lower surfaces near zero lift which limit the formation of laminar separation bubbles. At the higher lift coefficients, the short laminar region remains on the upper surface, thereby minimizing the possibility of laminar separation, while on the lower surface, an extensive favorable pressure gradient exists which generates a large amount of laminar flow with little chance of laminar separation. These two effects combine to produce fairly low drag coefficients at high C_l's, at least for a relatively thick, symmetric airfoil such as the E472.

Although this airfoil was not tested with flaps, Eppler did show computational results showing an extremely hard stall with positive flap deflections (trailing-edge down).[15] While this may be useful for snap maneuvers, it could prove less than desirable when maneuvering at low speeds during landing.

Trainer 60 This airfoil was taken from the Great Planes TRAINER 60. The unusual geometry at the trailing edge resulted from the inability of the coordinate smoothing code, AFSMO,[10] to accurately simulate flat, strip ailerons. Because it is unusual to use a symmetric airfoil on a trainer, this airfoil may find more appropriate use on aerobatic aircraft. An indication of how suited this airfoil is for aerobatic aircraft can be gained by comparing its performance with that of the E472.

Referring back to Eppler's four factors that constitute a successful aerobatic airfoil, the Trainer 60 airfoil adequately meets these requirements. Its $C_{l,max}$ of 1.2 for Re = 400,000 is only 0.1 less than the E472, yet it produces a noticeably larger amount of drag than the E472 at low C_l's. This high drag at low C_l's can be attributed to both its extreme thickness and large region of adverse pressure gradient near the trailing edge resulting in separated flow at all angles of attack in this region. Further comparison shows that drag values at the higher C_l's are comparable between the two airfoils. Perhaps the Trainer 60's only shortcoming, with respect to aerobatic flight, is its stall behavior. It is not as severe as the E472 and therefore may produce difficulty in executing crisp snap maneuvers.

Ultra-Sport 1000 The Ultra-Sport 1000 is used as the root airfoil on the Great Planes ULTRA-SPORT 1000, a popular intermediate-level pattern ship. The lift-curve results, which show an extremely soft stall, lend support to the praise garnered on the ULTRA-SPORT 1000 for its low-speed handling qualities. This soft stall, however, comes at the price of a low $C_{l,max}$, which fails to reach even 1.0. Comparison with the Trainer 60 shows the performance limitations of the Ultra-Sport 1000 quite clearly.

4.2.2 Semi-Symmetric Airfoils

Avistar The Avistar airfoil is used on the Hobbico AVISTAR—a design which in the opinion of this author (Lyon) has some of the best handling characteristics among contemporary R/C training aircraft. Such docile performance can almost entirely be attributed to the gentle stall of the Avistar airfoil. For example, at Re = 100,000 the initial effects of stall are first seen just after $\alpha = 6°$, yet the break does not occur until $\alpha = 15°$. Excluding wing planform effects, which may or may not be negligible, the early stall effects and the late break produce a 9° range over which the airfoil gives the pilot warning of an ensuing stall. While the actual break at stall is severe, the absence of hysteresis should make stall recovery of little concern. The excellent stall behavior of this airfoil is even more interesting when one considers that it was designed using none of the modern design techniques. For an example of the performance improvements possible by using these techniques, the reader is referred to the discussion of the S8037.

ESA The ESA (Evans Simitar Airfoil) was designed for use on Bill Evan's SIMITAR series of tailless aircraft. To obtain a positive pitching moment, the airfoil uses a deflected trailing-edge flap that results in a rather odd airfoil contour and an associated unusual velocity distribution near the trailing edge. Despite the unusual features of this airfoil, however, it has strikingly good performance when considered for use on aerobatic aircraft, which the SIMITAR series of aircraft are to a certain extent.

Upon comparison of the ESA with the E472 (an aerobatics airfoil by design), the ESA has nearly identical performance over the range of positive lift coefficients. The closeness in performance between the two airfoils, however, is not nearly as surprising when one considers the similarity of their upper-surface velocity distributions. Also evident from the velocity distributions is the reason for the higher low Reynolds number drag of the ESA near zero lift. The ESA's need for a positive pitching moment has resulted in a more adverse pressure gradient between $x/c = 20\%$ and 80% than that found on the E472. At lower Reynolds numbers, this provides more of an opportunity for a laminar separation bubble to form, thereby increasing drag.

Falcon 56 Mk II The Falcon airfoil was taken from plans for the Carl Goldberg FALCON 56 MK II, a time-tested intermediate aileron trainer with gentle handling qualities. From a performance standpoint, the Falcon shows evidence of a laminar separation bubble for all Reynolds numbers tested ($100,000 \leq Re \leq 400,000$). While pilots take little notice of a bubble's impact on handling qualities, this may not be true in the case of the FALCON 56 MK II. However, this is not because bubbles do not affect aircraft performance, but rather because a pilot typically flies at a combination of lift coefficients and Reynolds numbers where bubble effects are negligible. In the Falcon's case, however, a bubble remains even at the higher Reynolds numbers just below operational C_l's. Such bubble behavior may give the pilot a feeling of increased pitch sensitivity when diving or encountering gusts, which briefly decrease lift coefficients. While the performance of the Falcon airfoil is far from being poor, better performance can be achieved (see the S8037).

S8036 & S8037 The S8036 was designed as the root airfoil for the Top-Flite P-47 with two major constraints, a moderate thickness of 16% and a fairly gentle stall. As is common for many airfoils, the stall developed by the S8036 is rather sudden and sharp for $Re = 100{,}000$. While the sharp drop in C_l remains as the Reynolds number is increased, the effect is not as critical since the stall now develops at a much slower rate. When this slowly developing stall is combined with equipment limitations, it is not always possible to reach the angle of attack where the sharp drop in C_l occurs (if it occurs at all). In such cases, no hysteresis in the lift curve is seen. Interestingly, for cases when the angle of attack is large enough to result in a complete stall, a hysteresis loop is present which grows with increasing Reynolds number. Because of this unusual result, some explanation is warranted. As the Reynolds number increases, the largest angle of attack for which flow remains attached also increases, thereby increasing the stall angle of attack and the subsequent maximum lift coeffcient. However, for decreasing angles of attack, reattachment occurs at the same angle regardless of the Reynolds number. Therefore, the growing hysteresis loop is caused by an increase in the stall angle and not by some undesirable aerodynamic effect.

The S8037 was designed as the tip airfoil for the Top-Flite P-47 wing. As was the case with the S8036, stall behavior was a major design driver for this airfoil. Having been designed as a tip airfoil where the wing chord is typically the smallest, it was important to have gentle stall characteristics at Reynolds numbers lower than those experienced by the root airfoil (S8036). Comparisons between the S8036 and S8037 at $Re = 100{,}000$ show that this goal was achieved. The $C_{l,max}$ for the S8037 was also increased over the S8036 at the lower Reynolds numbers ($100,000 \leq Re \leq 200,000$), but this can be partly attributed to the higher camber of the S8037.

For aircraft that may tend to tip stall, the S8036/S8037 combination is a good choice. By using the S8036 at the wing root and the S8037 at the wing tip, the difference in low Reynolds number stall behavior between the two airfoils is different enough to moderate any tip-stall tendencies. On aircraft exhibiting a severe tip-stall problem, however, the difference in stall behavior between the S8036 and S8037 may not be large enough to completely eliminate the tip-stall problem. On the other hand, if an aircraft is such that there is no chance of tip stall occurring, the designer will be better served by using the S8037 over the entire span of the wing.

S8052 This airfoil was designed to meet Quickie 500 (Q-500) pylon racing requirements. Performance comparisons between the S8052 and another Q-500 airfoil, the R140,[2] show an increase in $C_{l,max}$ of 0.35 and a dramatic decrease in drag at the higher lift coefficients for the S8052. During a race, this higher $C_{l,max}$ and lower drag directly translate into shorter radius turns at higher speeds.

While race conditions can lead to Reynolds numbers reaching 1,000,000, such speeds are beyond the capability of the LSATs experimental apparatus. Because of this limited testing capability, data was only taken up to $Re = 500{,}000$. If the reader is interested in higher Reynolds numbers, they are referred to *Volume 1*, which provides limited computational results ($C_l - C_d$ data) for the S8052 at $Re =$ 300,000, 600,000, and 1,000,000.

4.2.3 Flat-Bottom Airfoils

Clark-Y (B) The Clark-Y, which has a flat bottom rearward of $x/c \approx 20\%$, has long been a member of the successful group of airfoils used on R/C aircraft. While this may largely be due to its partially flat lower surface and nearly 12% thickness helping to ease wing construction, its performance has proven to be a contest winner in situations requiring low drag at moderately high lift coefficients ($0.5 \leq C_l \leq 0.75$). Inspection of the velocity distributions over this range of lift coefficients shows an extended region of favorable pressure gradient followed by a fairly mild adverse gradient. When these two factors are combined with a nearly flat lower surface velocity distribution, low drag is the result over a large range of Reynolds numbers. For $Re = 60{,}000$ and 100,000, however, evidence of a laminar separation bubble can be seen in the drag polar centered around $C_l = 0.5$. Based on the velocity distributions, the nearly flat to slightly adverse pressure gradient present on the lower surface leads to the formation of a lower surface bubble at this C_l. For more discussion regarding the Clark-Y, the reader is referred to Section 4.3 — Airfoils for Small Wind Turbines.

PT-40 (A) & (B) The PT-40 airfoil was taken from the Great Planes PT-40 trainer. Two models of the PT-40 airfoil were tested, a fully-sheeted version (version A) and then one with no sheeting (open rib bays, version B) identical to that used on the actual aircraft (see Fig. 5.70). By testing these two versions, it was hoped to identify the effects of spar ridges and covering sag on airfoil performance.

The sheeted PT-40 (A) shows evidence of two distinct laminar separation bubbles for $Re = 100{,}000$ and 200,000 — one on the upper surface and one on the lower surface. The lower-surface bubble is present up to a C_l of approximately 0.65, after which point an upper-surface bubble appears and the lower-surface bubble becomes either insignificant or disappears entirely. From an aircraft handling standpoint, the lift curves do show a decidedly abrupt and sharp break at stall, particularly near the landing speed of the PT-40 ($Re = 100{,}000$). It is difficult, however, to make any comparisons between the stall characteristics of the airfoil and those of the aircraft. The reason for this difficulty lies in the incorporation of washout into the wing of the PT-40. By including washout, the impact of airfoil stall on aircraft stall is considerably reduced.

Owing to the large spanwise-drag variations expected for the open-bay PT-40 (B), drag values from six wake surveys spaced 3 in. apart were averaged instead of the usual four. The increased number of spanwise surveys should serve to adequately average the spanwise effects created by covering sag between ribs. The first thing one notices when comparing drag polars between the A and B models is the higher drag of the open-bay B model at the higher Reynolds numbers. While the presence of the spar ridge should serve as a trip and subsequently increase the amount of turbulent flow at the higher Reynolds numbers, evidence of a bubble around $C_l = 0.6$ makes this explanation seem at least partially incorrect. In other words, if the spar ridge had forced premature transition, then bubble formation would be unlikely.

For $C_l < 0.6$ the B model produces lower drag than the A model over a broad range of lift coefficients and Reynolds numbers. The low drag is particularly evident at $Re = 100{,}000$. The reason for the lower drag is unclear since the effects of spar ridges and covering sag are expected to be negligible on the lower surface of the model owing to its flat bottom. For a complete answer to the problem more data is required. As a final note, there is no significant difference in lift curves between the A and B models.

4.3 Airfoils for Small Wind Turbines

Table 4.5: Summary of Airfoil Data for Small Wind Turbines

Airfoil	Configuration	Figures Begin
A18	trip type C	Fig. 5.1, p. 58
BW-3	clean trip type C	Fig. 5.9, p. 66 Fig. 5.13, p. 71
Clark-Y (B)	trip type C	Fig. 5.23, p. 85
Goe 417a	clean	Fig. 5.57, p. 128
S7012 (B)	trip type C	Fig. 5.91, p. 164
SD7032 (D)	trip type C	Fig. 5.142, p. 226
SD7037 (B)	trip type C	Fig. 5.146, p. 230
SD7062 (B)	clean trip type C	Fig. 5.169, p. 256 Fig. 5.173, p. 261
SG6040	clean trip type C	Fig. 5.179, p. 270 Fig. 5.183, p. 275
SG6041	clean trip type C	Fig. 5.185, p. 280 Fig. 5.189, p. 285
SG6042	clean trip type C leading-edge tape	Fig. 5.191, p. 290 Fig. 5.195, p. 295 Fig. 5.199, p. 302
SG6043	clean trip type C	Fig. 5.201, p. 304 Fig. 5.205, p. 309
USNPS-4	clean	Fig. 5.215, p. 326

In support of the ongoing UIUC research effort regarding the design of small horizontal-axis wind turbines (HAWTs),[16] several airfoils for use on wind turbines were tested. From these airfoils, only the BW-3 and SG series of airfoils were designed specifically for wind turbine applications. The other airfoils are either: (1) airfoils used on existing wind turbines (SD7062 and USNPS-4), (2) benchmark airfoils (Clark-Y and Goe 417a), or (3) low Reynolds number airfoils designed for aircraft (A18, S7012, SD7032, and SD7037). Some of these airfoils were tested with zigzag trips in order to fix transition and thus simulate roughness effects experienced by wind turbine blades. In the majority of fixed transition cases, 0.023-in. thick trips were positioned at 2% chord and 5% chord on the upper and lower surface respectively. Exceptions to this will be explicitly stated.

A18 Originally designed for free-flight airplanes by Randy Archer,[2] this airfoil was tested with fixed transition to investigate the effects of leading-edge roughness on $C_{l,max}$. As Fig. 4.4 shows, $C_{l,max}$ is not affected when fixing transition at the leading edge. As a result, this airfoil is a good candidate for constant-speed, stall-regulated wind turbines (see glossary for definitions). Additionally, the good L/D characteristics of the A18 when clean also makes this airfoil suitable for variable-speed wind turbines.

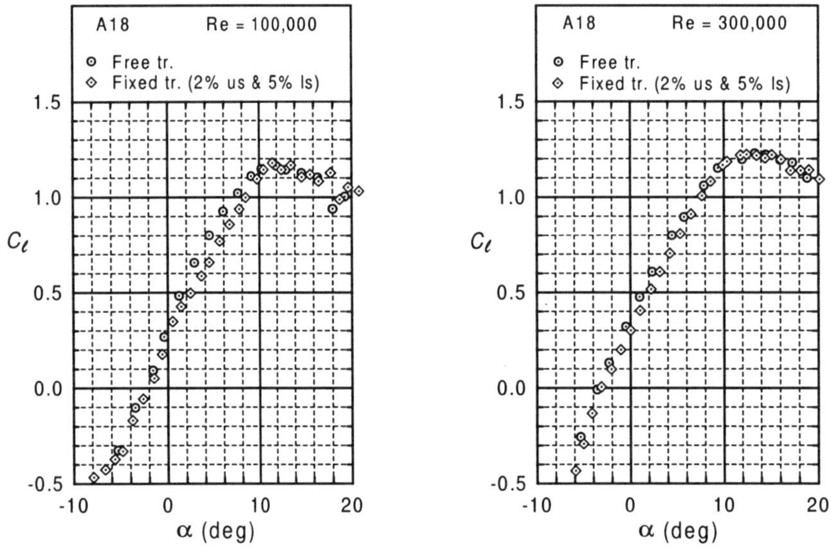

Fig. 4.4 Lift-curve comparsion for the A18 with free and fixed transition at Re = 100,000 and 300,000.

BW-3 This airfoil was designed by Bergey Windpower for use on their small wind turbine systems. An important design objective for the BW-3 was to have an easily pultrudable section which yielded a relative thickness of 5%. Note that the nominal coordinates of this airfoil were obtained by digitizing a section of the BWC Excel wind-turbine blade.

The drag polars for the BW-3 indicate that the low-drag bucket is virtually nonexistent. More specifically, there is a limited range of lift coefficients that result in minimum drag. As also seen from the drag polars, the BW-3 apparently does not experience the formation of large separation bubbles at Re =60,000 and 100,000 as usually seen with thicker airfoils. This observation is consistent with the absence of lift hysteresis for all Reynolds numbers considered. The overall drag of the BW-3, however, is relatively high compared with thicker low Reynolds number airfoils. Interestingly, the lift curves essentially overlap for all Reynolds numbers with a maximum lift coefficient of about 1.4, except for Re = 60,000. A high $C_{l,max}$, such as 1.4, is a good characteristic from an operational point of view since it provides a higher start-up torque that in turn produces a lower cut-in wind speed, thus improving overall energy capture.

Another distinguishing attribute of the BW-3 can be seen from the fixed transition results shown in Fig. 5.13. The lift characteristics of the BW-3 exhibit little difference when tested with or without trips. The loss in $C_{l,max}$ is on the order of 2.0% at Re =100,000 and 1.3% at 400,000. Therefore, the BW-3 is not only suitable for variable-speed/variable-pitch HAWTs but can also be applied to small stall-regulated wind turbines.

As a side note, drag independence of the BW-3 with respect to both Reynolds number and roughness (trips) is generated by the same effect: early transition for the clean airfoil. For changes in Reynolds number, drag variations typically result from changing amounts of laminar separation and turbulent flow. However, when natural transition occurs near the leading edge, as it does on the BW-3, these variations are small, resulting in relatively constant drag.

Clark-Y (B) The Clark-Y (B) is discussed here as a benchmark regarding the effects of fixing transition on airfoil performance. Note that the wind-tunnel model tested at UIUC is different than the Princeton model documented in *SoarTech 8*. While lift hysteresis is present for the clean model at Re = 200,000 and below, the addition of trips at the leading edge eliminated the hysteresis loop. This came at a price, however, as $C_{l,max}$ was reduced by an average of 15%. As expected, the trips also had a large detrimental effect on drag.

Goe 417a Like the Clark-Y (B), the Göttingen 417a was also used as a benchmark, this time for comparison with the BW-3 airfoil. Much like the BW-3 airfoil, the Goe 417a has a wedge-shaped drag polar and a $C_{l,max}$ of 1.4. Regarding maximum L/D's, the Goe 417a offers slightly better performance than the BW-3. However, from an applications standpoint, the larger thickness of the BW-3 may compensate for this small difference in L/D through structural benefits. While fixed transition data for the Goe 417a would have provided another means of comparison, these tests could not be performed owing to time constraints. It should be noted that data for Re = 400,000 is incomplete as a result of significant model vibrations and "bowing."

S7012 (B) The flapped model of the S7012 (B) was tested with fixed transition. (Results for the clean model can be found in Refs. 2 and 3.) It was expected that the performance of the S7012 would not degrade as much as other airfoils, such as the Clark-Y, since the transition point on the upper surface quickly moves towards the leading edge as $C_{l,max}$ approaches. This is indeed the case as seen in Fig. 4.5 where the the $C_{l,max}$ remained constant in both the free and fixed transition cases.

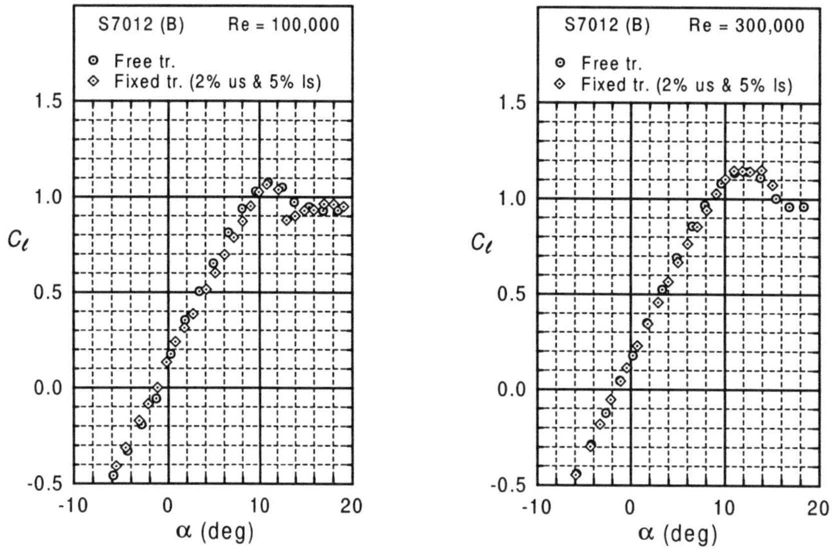

Fig. 4.5 Lift-curve comparsion for the S7012 (B) with free and fixed transition at Re = 100,000 and 300,000.

Consequently, the S7012 is applicable to variable and constant-speed wind turbines.

SD7032 (D) The SD7032 is another sailplane airfoil that was tested with simulated leading-edge roughness. As shown in Fig. 4.6, $C_{l,max}$ decreases by 5–8% with the addition of roughness. As a consequence, it is preferable to use the SD7032 on variable-speed wind turbines and not on stall-regulated machines ($C_{l,max}$ dependent).

SD7037 (B) Similar tests to those performed on the A18, S7012, and SD7032 were also performed on the SD7037. The results shown in Fig. 4.7 indicate that the potential use of the SD7037 can be extended beyond variable-speed wind turbines to small stall-regulated machines.

SD7062 (B) This airfoil has been used in the blade design of a 5-m/5-kW variable-speed HAWT. This design work was conducted by David Wood and his research group at the University of Newcastle in Australia.[17] The large relative thickness and the high-lift characteristics of the SD7062 were what first triggered the selection of this low Reynolds number airfoil for their wind turbine. Field tests of this wind turbine have indicated that the SD7062 provides good overall performance under clean conditions. It then seemed appropriate to test the SD7062 with fixed transition to simulate the performance of wind turbines under dirty-blade conditions. For comparison purposes, tests with free transition were also carried out. Note that the B model of the SD7062 tested at UIUC was a refurbished version of the original model built for the Princeton tests.[1]

As indicated by the drag polar for Re = 60,000, the flow seems to quickly separate from the upper surface, thus greatly limiting applications for the SD7062 when Re < 100,000. While lift hysteresis is present in the free-transition case for

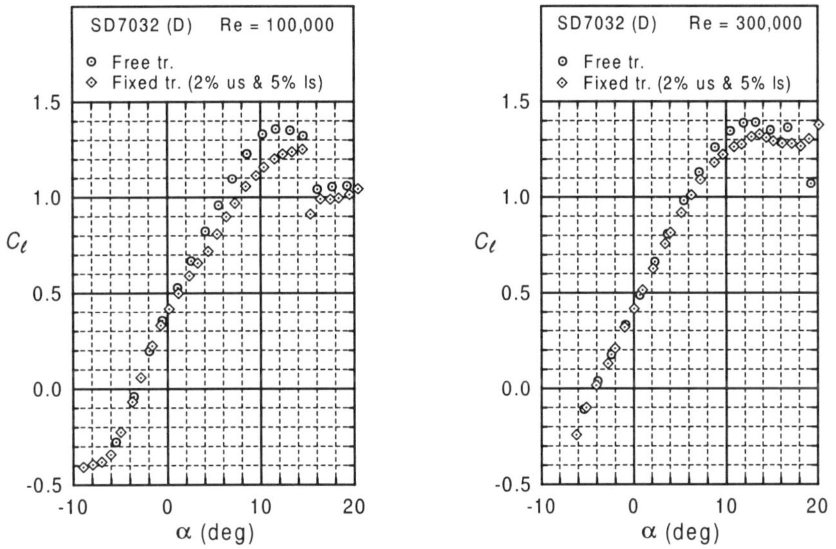

Fig. 4.6 Lift-curve comparsion for the SD7032 (D) with free and fixed transition at Re = 100,000 and 300,000.

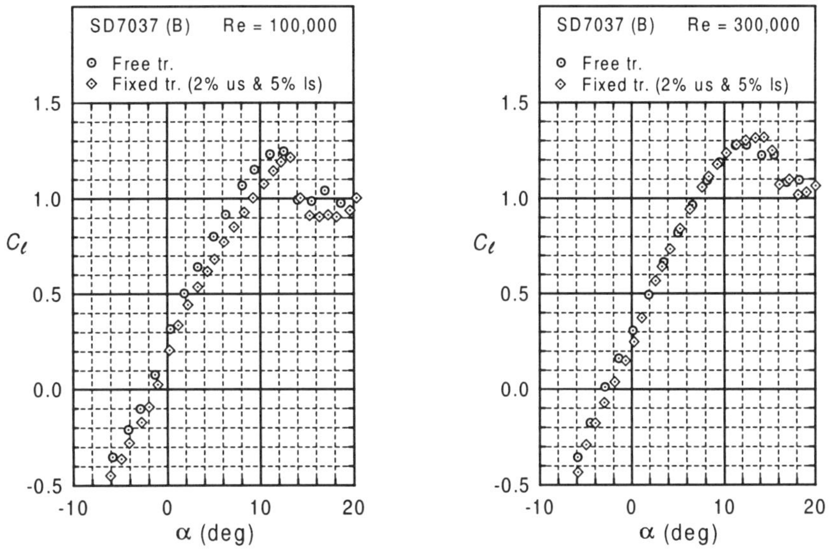

Fig. 4.7 Lift-curve comparsion for the SD7037 (B) with free and fixed transition at Re = 100,000 and 300,000.

Re = 100,000 and 200,000, hysteresis is not particularly alarming for variable-speed wind turbines as their blades normally operate below stall. The large drag buckets for the free-transition cases provide the good L/D characteristics essential to high-performance wind turbines. However, the ability of a wind turbine using the SD7062 to capture wind energy may be significantly degraded under dirty-blade conditions.

The fixed transition cases used to simulate dirty-blade conditions not only caused severe drag penalties but also decreased $C_{l,max}$. At Re = 100,000, the $C_{l,max}$ was reduced from 1.52 to 1.22 (a 19.7% decrease), and for Re = 400,000, the $C_{l,max}$ was reduced from 1.68 to 1.19 (a 29.2% decrease). Consequently, it seems clear that the SD7062 is not a good candidate for stall-regulated wind turbines. As a last comment, wind turbine blades that use the SD7062 airfoil should be given special attention to ensure they remain relatively clean.

SG6040, SG6041, SG6042, & SG6043 Over the last 12 years a considerable number of airfoils have been developed for HAWTs. For example, the advanced NREL airfoils such as the S822/S823 documented in *Volume 1* and *2* provide both aerodynamic and structural advantages as compared with the myriad of aircraft airfoils adapted for wind turbine use. There are, however, only a limited number of wind-turbine airfoils designed exclusively for wind turbines with small blades. To fill this niche, a total of four airfoils were designed for use on $1-5$ kW rated power wind turbines.[18]

During this design effort, the focus was to create a series of airfoils that could be used along the entire span of small variable-speed wind turbine blades. Considering the low operating Reynolds number and beneficial centrifugal stiffening effects of small HAWTs, the thickness of the primary airfoils (SG6041, SG6042, and SG6043) for use at the 75% blade station was fixed at 10%. Additionally, a 16% root airfoil (SG6040) was designed to accommodate possible large root bending moments and large blade-stiffness requirements. Several other design constraints were imposed on both the primary and root airfoils. The primary airfoils were designed to achieve as high an L/D as possible for $0.6 \leq C_l \leq 1.2$ with $250{,}000 \leq Re \leq 500{,}000$. The root airfoil was designed to operate at $C_l = 1.1$ for $Re = 200{,}000$ making it suitable for use over the inboard 30% of the blade. For all four of these airfoils, it was desired to keep drag near the design lift coefficient as low as possible. However, owing to the high design C_l's, the shallow upper-surface pressure gradients needed to produce low drag lead to aft-loaded airfoils. As a consequence, the SG6040 — SG6043 series of airfoils have fairly high pitching moments ($-0.08 < C_{m,c/4} < -0.14$).

Concentrating on baseline airfoil performance, penalties owing to off-design conditions are relatively small except for $Re \leq 100{,}000$ where laminar separation effects cause large increases in drag. Such performance penalties, however, are typical of most low Reynolds number airfoils. As indicated in Fig. 4.8, the new airfoils produce L/D's that equal or exceed those of previously existing low Reynolds number airfoils applicable to HAWTs.

Each airfoil was also tested with fixed transition (trip type C) at 2% upper surface and 5% lower surface. As expected, the higher the design C_l, the higher the performance loss. Furthermore, the C_l corresponding to the maximum L/D is also affected. However, the angle of attack at which the maximum L/D occurs remains unchanged. This constant angle of attack for maximum L/D allows a blade designed with the new airfoils to produce optimum performance even with leading-

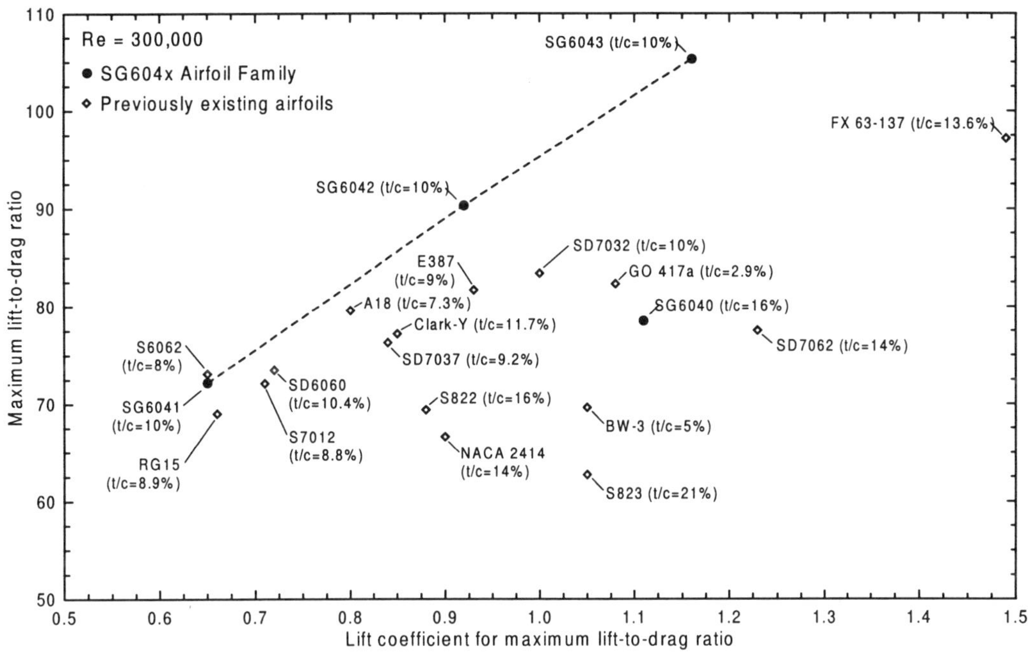

Fig. 4.8 Maximum L/D's versus corresponding C_l's for various HAWT airfoils at Re = 300,000.

edge roughness. Additional data taken on the SG6042 at Re = 300,000 for various trip heights (Fig. 5.197) indicates that trips thicker than $h/c = 0.04\%$ cause a rapid degradation in airfoil performance. Consequently, if leading-edge tape is used on blades utilizing these new airfoils, only a very thin layer should be applied. Drag data for such leading-edge tape is shown in Fig. 5.199.

USNPS-4 The USNPS-4 is a vintage airfoil that closely approximates the tip airfoil on the Jacobs 17.5-kW wind-turbine. The USNPS-4 shows a lift range comparable to the SD7062 but with much higher drag. It is likely that the performance of the USNPS-4 will significantly degrade with fixed transition.

4.4 Airfoils for Special Interest

Table 4.6: Summary of Airfoil Data for Special Interest

Airfoil	Configuration	Figures Begin
LRN1007 (B)	clean	Fig. 5.61, p. 134

LRN1007 (B) The LRN1007 airfoil is presented in the special interest category for two reasons. First, the airfoil is representative of a series of attempts at low Reynolds number airfoil design conducted by Pfenninger, Mangalam, and others during the mid-1980's using ISES and the Eppler code.[19,20,21] Second, as this airfoil approaches stall, the flow becomes very unsteady resulting in intense lift and drag fluctuations.

In light of the many fine low Reynolds number airfoils designed, tested, and published in *SoarTech 8* and *Summary of Low-Speed Airfoil Data*, the performance of the LRN1007 airfoil leaves much to be desired. This not withstanding, the original design and validation efforts for the LRN1007 should not go unrecognized. The drag polars and lift curves for Re = 60,000 and 100,000 show the performance degradations produced by large areas of upper-surface laminar separation without subsequent reattachment. The "jump" in lift near $\alpha = 8°$, and corresponding decrease in drag, illustrates what happens when the upper-surface flow changes from a large mid-chord bubble without reattachment to a small leading-edge bubble with turbulent reattachment.[22] As one might expect, the use of boundary-layer trips greatly enhances the airfoil performance at these Reynolds numbers in the mid-range of the polar.[21,23] For higher Reynolds numbers (Re = 200,000 and 300,000) the lift and drag characteristics show smoother behavior, indicating that the laminar separation on the upper surface occurs with turbulent reattachment and that the bubble located near mid-chord makes a more gradual transition to a leading-edge bubble as angle of attack is increased.[22]

While it is clear that the LRN1007 is not well suited for practical applications, its unsteady flow behavior at stall motivates further study of this airfoil. For angles of attack between 14.5° and 16.5°, and for a wide range of Reynolds numbers (Re = 40,000 to 1,400,000), the flow over the airfoil changes in a regular pattern between stalled and unstalled conditions.[24,25] Owing to the low frequency and large force fluctuations of this oscillation, a potentially destructive combination for aircraft wings, rotor blades, etc. is created. Recent lift measurements show very similar oscillations for the E374 and E387 airfoils. Such periodic variations in C_l are shown in Fig. 4.9 for the LRN1007 (B) and E374 (B) airfoils. These measurements, which were performed using much of the existing LSATs experimental apparatus with the exception of a specially configured lift balance, show the periodic nature and low-frequency of the oscillations. The total fluctuation in C_l is about 37% of the mean for the LRN1007 (B) and about 35% of the mean for the E374 (B). Presently, research is also being conducted on other airfoils to see if they exhibit similar results. Fortunately, a majority of airfoils do not show evidence of this peculiar unsteady-stall behavior.

Despite extensive research carried out for the LRN1007 airfoil, the cause of this low-frequency oscillation is not yet known. Flowfield velocity measurements on the upper surface of the airfoil show that the stalling and unstalling behavior is related to a leading-edge laminar separation bubble that grows with time during the oscillation. As this bubble grows, it then interacts with the separated turbulent boundary layer near the trailing edge causing the entire upper-surface boundary layer to separate, thereby stalling the airfoil. The flow then gradually reattaches from the leading edge and the airfoil unstalls.[26] The dependency of this oscillation on the leading-

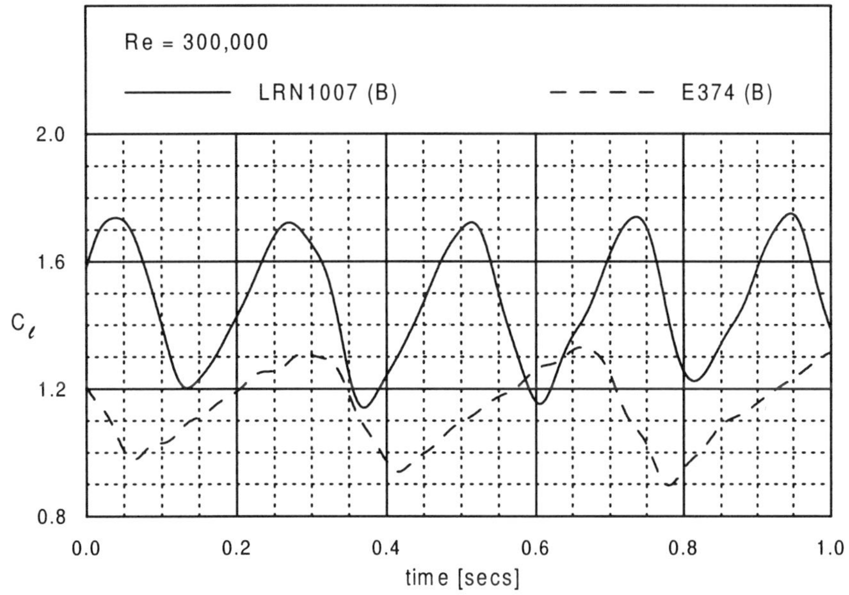

Fig. 4.9 Fluctuations in C_l's over a 1-sec. duration for the LRN1007 (B) and E374 (B) near stall at Re = 300,000.

edge separation bubble was verified when the oscillation ceased to occur after a zigzag trip was placed near the leading edge to eliminate the bubble.[25] While detailed measurements have not yet been performed on the E374 (B) airfoil, a leading-edge separation bubble very similar to that found on the LRN1007 (B) was observed prior to the onset of unsteady flow through the use of surface oil flow visualization. At present, it is not clear what flowfield characteristics must be present for the unsteadiness to occur. Consequently, this is the subject of ongoing investigations. More detailed information on this research can be found in Refs. 24–27, and it is expected that more information will be provided in future volumes of *Summary of Low-Speed Airfoil Data*.

Chapter 5

Airfoil Profiles and Performance Plots

In this chapter, the airfoil profiles and performance plots are presented. For quick reference, the airfoil names are listed in the margins. As a matter of record, a table listing all the data sets, associated figures, figure page numbers, as well as run numbers is included at the beginning of this chapter. In the table, a '**I**' indicates that the corresponding data is not plotted to save space, and an '*' indicates that the C_l's presented in the drag polars are values interpolated from corresponding lift runs. Also, the 'Avg. Difference' between the true and actual coordinates listed in the comparison plots is the average error in model construction. Finally, all models have a 12-in. chord except the 6-in. E387 (D).

Model (Builder) Designer	Configuration	V-dist. & Profile Fig.	V-dist. & Profile p.	Drag Data Fig.	Drag Data p.	Drag Data *Re*	Drag Data Run #	Lift & Moment Data Fig.	Lift & Moment Data p.	Lift & Moment Data *Re*	Lift & Moment Data Run #
A18	trip type C	5.1	58	5.3	59	100,000	PG02559	5.4	60	100,000	PG02558
(Cooney)	(see Fig. 5.2)	5.2				200,000	PG02561			200,000	PG02560
Archer						300,000	PG02563/PG02564		61	300,000	PG02562
Avistar	clean	5.5	62	5.7	63	100,000	CL01118	5.8	64	100,000	CL01117
(B. Champine)		5.6				200,000	CL01120			200,000	CL01119
Hobbico						300,000	CL01122		65	300,000	CL01121
						400,000	CL01124			400,000	CL01123
										500,000	CL01766
BW-3	clean	5.9	66	5.11	67	60,000	PG01723	5.12	68	60,000	PG01722
(M. Allen)		5.10				100,000	PG01515			100,000	PG01514
Bergey						200,000	PG01517		69	200,000	PG01516
Windpower						300,000	PG01519			300,000	PG01518
						400,000	PG01521		70	400,000	PG01520
	trip type C			5.13	71	100,000	PG01773	5.14	72	100,000	PG01772
	(see Fig. 5.10)					200,000	PG01775			200,000	PG01774
						300,000	PG01777		73	300,000	PG01776
						400,000	PG01779/PG01780			400,000	PG01778
CG Ultimate	clean	5.15	74	5.17	75	100,000	AB01671/AB01672	5.18	76	100,000	AB01670
(M. Levoe)		5.16				200,000	AB01674			200,000	AB01673
C. Goldberg						300,000	CL01676		77	300,000	AB01675
						400,000	CL01678			400,000	CL01677
						500,000	CL01760		78	500,000	CL01730
Clark-Y (B)	clean	5.19	80	5.21	81	60,000	PG01709/PG01710	5.22	82	60,000	PG01708
(J. Robertson)		5.20				100,000	PG01712/PG01713			100,000	PG01711
Clark						200,000	PG01715		83	200,000	PG01714
						300,000	PG01717			300,000	PG01716
(continues)						400,000	PG01719		84	400,000	PG01718

Clark-Y (B) (continued)	trip type C (see Fig. 5.20)			5.23	85	100,000	PG01752/PG01753	5.24	86	100,000	PG01751
						200,000	PG01755			200,000	PG01754
						300,000	PG01757		87	300,000	PG01756
						400,000	PG01759			400,000	PG01758
DH4009 (D. Stanley) D. House	clean	5.25 5.26	88	5.27	89	100,000	CL01082	5.28	90	100,000	CL01089
						200,000	CL01084			200,000	CL01083
						300,000	CL01086		91	300,000	CL01085
						400,000	CL01088			400,000	CL01087
						500,000	CL02072		92	500,000	CL02071
E231 (M. Lachowski) Eppler	clean	5.29 5.30	94	5.31	95	60,000	CL01252/CL01256	5.32	96	60,000	CL01251
						100,000	CL01254/CL01255			100,000	CL01253
						200,000	CL01258		97	200,000	CL01257
						300,000	CL01260			300,000	CL01259
						400,000	CL01262		98	400,000	CL01261
E374 (B) (M. Bame) Eppler	clean	5.33 5.34	100	5.35	101	60,000	CL01013	5.36	102	60,000	CL01012
						100,000	CL01015			100,000	CL01014
						200,000	CL01017		103	200,000	CL01016
						300,000	CL00983			300,000	CL01018
E387 (C) (J. Robertson) Eppler	clean	5.37 5.38	104	5.39	105	60,000	CL02130	5.40	106	60,000	CL02129
						100,000	CL02132			100,000	CL02131
						200,000	CL02134		107	200,000	CL02133
						300,000	CL02136			300,000	CL02135
						460,000	CL02138		108	460,000	CL02137
E387 (D) (Y. Tinel) Eppler	clean	5.41 5.42	110	5.43	111	60,000	CL02707*	5.44	112	60,000	CL02706
										200,000	CL02708
E472 (D. Stanley) Eppler	clean	5.45 5.46	114	5.47	115	100,000	CL01074	5.48	116	100,000	CL01073
						200,000	CL01076/CL01857			200,000	CL01075
						300,000	CL01078/CL01858		117	300,000	CL01077
						400,000	CL01080/CL01859			400,000	CL01079
						500,000	CL01856		118	500,000	CL01855

ESA	clean	5.49	120	5.51	121	100,000	AB01091	5.52	122	100,000	AB01090
(B. Halkett)		5.50				200,000	AB01093			200,000	AB01092
Evans						300,000	AB01095		123	300,000	AB01094
						400,000	AB01097/CL01107			400,000	AB01096
Falcon 56 Mk II	clean	5.53	124	5.55	125	100,000	CL01424/CL01451	5.56	126	100,000	CL01423
(B. Devaney)		5.54				200,000	CL01426/CL01427			200,000	CL01425
C. Goldberg						300,000	CL01447		127	300,000	CL01446
						400,000	CL01450			400,000	CL01449
Goe 417a	clean	5.57	128	5.59	129	60,000	PG01725	5.60	130	60,000	PG01724
(M. Fox)		5.58				100,000	PG01126			100,000	PG01125
Göttingen						200,000	CL01128		131	200,000	CL01127
University						300,000	CL01130			300,000	CL01129
						400,000	CL01132		132	400,000	CL01131
LRN1007 (B)	clean	5.61	134	5.63	135	60,000	AB01697	5.64	136	60,000	AB01696
(M. Allen)		5.62				100,000	AB01365/AB01698			100,000	AB01364
Mangalam/						200,000	AB01700		137	200,000	AB01699
Pfenniger						300,000	AB01363			300,000	AB01362
PT-40 (A)	clean	5.65	138	5.67	139	100,000	CL01021	5.68	140	100,000	CL01020
(T. Lampe)		5.66				200,000	CL01023			200,000	CL01022
GPMM						300,000	CL01025		141	300,000	CL01024
						400,000	AB01027			400,000	AB01026
PT-40 (B)	open bay	5.69	142	5.71	143	100,000	CL01066	5.72	144	100,000	CL01065
(T. Lampe)		5.70				200,000	CL01068			200,000	CL01067
GPMM						300,000	CL01070		145	300,000	CL01069
						400,000	CL01072			400,000	CL01071
RG14	clean	5.73	146	5.75	147	60,000	AB01058	5.76	148	60,000	AB01057
(R. Cavazos)		5.74				100,000	AB01060			100,000	AB01059
R. Girsberger						200,000	AB01062		149	200,000	AB01061
						300,000	CL01064			300,000	AB01063

RG15 (C) (B. Champine/ J. Robertson) R. Girsberger	clean 2.5 deg flap	5.77 5.78	150	5.79	151	60,000 100,000 200,000 300,000	CL02374 CL02376 CL02378 PG02380	5.80	152 153	60,000 100,000 200,000 300,000	CL02373 CL02375 CL02377 PG02379
	clean 15 deg flap	5.81	154					5.82	155	100,000 200,000	MS01738 MS01737
	clean 20 deg flap	5.83	156					5.84	157	100,000 200,000	MS01739 MS01740
	clean 25 deg flap	5.85	158					5.86	159	100,000 200,000	MS01741 MS01742
S6063 (R. Cavazos) Selig	clean	5.87 5.88	160	5.89	161	60,000 100,000 200,000 300,000	CL01689 AB01691 AB01693 AB01695	5.90	162 163	60,000 100,000 200,000 300,000	CL01688 AB01690 AB01692 AB01694
S7012 (B) (M. Lachowski) Selig	trip type C (see Fig. 5.92) 0 deg flap	5.91 5.92	164	5.93	165	100,000 200,000 300,000	PG01782 PG01784/PG01785 PG01787	5.94	166 167	100,000 200,000 300,000	PG01781 PG01783 PG01786
	clean 2.5 deg flap	5.95	168	5.96	169	60,000 100,000 200,000 300,000	CL02257* CL02259 CL02261 CL02263	5.97	170 171	60,000 100,000 200,000 300,000	CL02256 CL02258 CL02260 CL02262
	clean 15 deg flap	5.98	172					5.99	173	100,000 200,000	MS01731 MS01732
	clean 20 deg flap	5.100	174					5.101	175	100,000 200,000	MS01734 MS01733
	clean 25 deg flap	5.102	176					5.103	177	100,000 200,000	MS01735 MS01736
S7075 (A) (J. Robertson) Selig (continues)	clean 0 deg flap	5.104 5.105	178	5.106	179	60,000 100,000 200,000 300,000	CL01318 CL01320 CL01322 CL01324	5.107	180 181	60,000 100,000 200,000 300,000	CL01325 CL01319 CL01321 CL01323

S7075 (A) (continued)	trip type E (see Fig. 5.105) 0 deg flap			5.108	182	60,000	CL01806*	5.109	183	60,000	CL01805
						100,000	CL01808*			100,000	CL01807
						200,000	CL01810		184	200,000	CL01809
						300,000	CL01812			300,000	CL01811
S7075 (B) (J. Thomas) Selig	clean 0 deg flap	5.110 5.111	186	5.112	187	60,000	CL01109*	5.113	188	60,000	CL01111
						100,000	CL01112			100,000	CL01110
						200,000	CL01114		189	200,000	CL01113
						300,000	CL01116			300,000	CL01115
S8036 (T. Lampe) Selig	clean	5.114 5.115	190	5.116	191	100,000	CL01029/CL01038	5.117	192	100,000	CL01028
						200,000	CL01031			200,000	CL01030
						300,000	CL01033/CL01034/CL01035		193	300,000	CL01032
						400,000	CL01037			400,000	CL01036
						500,000	CL01771		194	500,000	CL01770
S8037 (T. Lampe) Selig	clean	5.118 5.119	196	5.120	197	100,000	CL01040	5.121	198	100,000	CL01039
						200,000	CL01042			200,000	CL01041
						300,000	CL01044		199	300,000	CL01043
						400,000	CL01046			400,000	CL01045
						500,000	CL01789		200	500,000	CL01788
S8052 (T. Lampe) Selig	clean	5.122 5.123	202	5.124	203	100,000	CL01413/CL01414	5.125	204	100,000	CL01412
						200,000	CL01416/CL01417			200,000	CL01415
						300,000	CL01420		205	300,000	CL01418
						400,000	CL01422			400,000	CL01421
						500,000	CL01769		206	500,000	CL01768
SA7035 (M. Allen) Selig/Ashok Gopalarathnam	clean	5.126 5.127	208	5.128	209	60,000	AG02573*	5.129	210	60,000	AG02583
						100,000	AG02576			100,000	AG02575
						150,000	AG02578		211	150,000	AG02577
						200,000	AG02580			200,000	AG02579
						300,000	AG02582		212	300,000	AG02581
SA7036 (A) (D. Thompson) Selig/Ashok Gopalarathnam	clean	5.130 5.131	214	5.132	215	100,000	AG02629	5.133	216	100,000	AG02628
						150,000	AG02635			150,000	AG02634
						200,000	AG02631		217	200,000	AG02630
						300,000	AG02633			300,000	AG02632

SA7036 (B) (T. Foster) Selig/Ashok Gopalarathnam	clean	5.134 5.135	218	5.136	219	100,000 150,000 200,000 300,000	AG02699 AG02701 AG02703 AG02705	5.137	220 221	100,000 150,000 200,000 300,000	AG02698 AG02700 AG02702 AG02704
SA7038 (T. Foster) Selig/Ashok Gopalarathnam	clean	5.138 5.139	222	5.140	223	100,000 150,000 200,000 300,000	AG02645 AG02647 AG02657/AG02658 AG02660*	5.141	224 225	100,000 150,000 200,000 300,000	AG02644 AG02646 AG02656 AG02659
SD7032 (D) (S. Watson) Selig/Donovan	trip type C (see Fig. 5.143)	5.142 5.143	226	5.144	227	100,000 200,000 300,000	PG02566 PG02568 PG02570/PG02571	5.145	228 229	100,000 200,000 300,000	PG02565 PG02567 PG02569
SD7037 (B) (D. Thompson) Selig/Donovan	trip type C (see Fig. 5.147)	5.146 5.147	230	5.148	231	100,000 200,000 300,000	PG02638* PG02640 PG02642	5.149	232 233	100,000 200,000 300,000	PG02637 PG02639 PG02641
SD7037 (D) (H. Stokely/ B. Williams) Selig/Donovan	clean	5.150 5.151	234	5.152	235	60,000 100,000 200,000 300,000	AB01100 AB01102 PG01104 PG01106	5.153	236 237	60,000 100,000 200,000 300,000	AB01099 AB01101 PG01103 PG01105
	w/ u.s. covering			5.154	238	200,000	CL02018	5.155	239	200,000	CL02017
SD7037 (E) (J. Robertson) Selig/Donovan	clean 0 deg LE flap 0 deg TE flap	5.156 5.157	240	5.158	241	100,000 200,000 300,000	CL02307 CL02309 CL02311	5.159	242 243	100,000 200,000 300,000	CL02306 CL02308 CL02310
	clean -3 deg LE flap -2.5 deg TE flap	5.160	244	5.161	245	100,000 200,000 300,000	CL02323 CL02315 CL02313	5.162	246 247	100,000 200,000 300,000	CL02316 CL02314 CL02312
	clean 2.5 deg LE flap 2.5 deg TE flap	5.163	248	5.164	249	100,000 200,000 300,000	CL02325 CL02327 CL02329	5.165	250 251	100,000 200,000 300,000	CL02324 CL02326 CL02328
	clean 0 deg LE flap 5 deg TE flap	5.166	252	5.167	253	100,000 200,000 300,000	AG02368 CL02370 CL02372	5.168	254 255	100,000 200,000 300,000	AG02367 CL02369 CL02371

SD7062 (B)	clean	5.169	256	5.171	257	60,000	PG01721	5.172	258	60,000	PG01720
(K. Widiner/		5.170				100,000	CF01155/PG01163			100,000	CF01154
M. Allen)						200,000	PG01157/PG01158		259	200,000	PG01156
Selig/Donovan						300,000	PG01160			300,000	PG01159
						400,000	PG01162		260	400,000	PG01161
	trip type C			5.173	261	100,000	PG01744	5.174	262	100,000	PG01743
	(see Fig. 5.170)					200,000	PG01746			200,000	PG01745
						300,000	PG01748		263	300,000	PG01747
						400,000	PG01750			400,000	PG01749
SD7080	clean	5.175	264	5.177	265			5.178	266	60,000	CL02584
(P. Illick)		5.176				100,000	AG02588			100,000	AG02586
Selig/Donovan						150,000	AG02590		267	150,000	AG02589
						200,000	AG02592			200,000	AG02591
						300,000	AG02594		268	300,000	AG02593
SG6040	clean	5.179	270	5.181	271	100,000	PG02399/PG02400	5.182	272	100,000	PG02398
(M. Allen)		5.180				150,000	CF02402/CF02403			150,000	CF02401
Selig/Giguère						200,000	CF02405/CF02406		273	200,000	CF02404
						300,000	CF02408			300,000	CF02407
						400,000	CF02411*		274	400,000	CF02409
						500,000	CF02413			500,000	CF02412
	trip type C			5.183	275			5.184	276	100,000	PG02507
	(see Fig. 5.180)					150,000	PG02505/PG02506			150,000	PG02504
									277	200,000	PG02508
						300,000	PG02510			300,000	PG02509
									278	400,000	CF02511
						500,000	CF02513			500,000	CF02512
SG6041	clean	5.185	280	5.187	281	100,000	AG02331	5.188	282	100,000	AG02330
(M. Allen)		5.186				150,000	PG02496			150,000	PG02503
Selig/Giguère						200,000	AG02333		283	200,000	AG02332
						300,000	AG02335			300,000	AG02334
						400,000	AG02337		284	400,000	PG02381
(continues)						500,000	AG02340			500,000	PG02382

SG6041 (continued)	trip type C (see Fig. 5.186)			5.189	285			5.190	286	100,000	PG02526
						150,000	PG02528			150,000	PG02527
									287	200,000	PG02529
						300,000	PG02531			300,000	PG02530
									288	400,000	PG02532
						500,000	PG02534			500,000	PG02533
SG6042 (M. Allen) Selig/Giguère	clean	5.191 5.192	290	5.193	291	100,000	AG02342/AG02343	5.194	292	100,000	AG02341
						150,000	PG02498			150,000	PG02502
						200,000	CL02345		293	200,000	CL02344
						300,000	CL02347			300,000	CL02346
						400,000	CL02349		294	400,000	CL02348
						500,000	CL02351/CL02352			500,000	CL02350
	trip type C (see Fig. 5.192)			5.195	295			5.196	296	100,000	PG02547
						150,000	PG02549			150,000	PG02548
									297	200,000	PG02550
						300,000	PG02552/PG02553			300,000	PG02551
									298	400,000	PG02554
						500,000	PG02556/PG02557			500,000	PG02555
	trip type C			5.197	299			5.198			
	$h/c = 0.04\%$					300,000	PG02649		300	300,000	PG02648
	$h/c = 0.08\%$					300,000	PG02651			300,000	PG02650
	$h/c = 0.12\%$					300,000	PG02653		301	300,000	PG02652
	$h/c = 0.15\%$					300,000	PG02655			300,000	PG02654
	(see Fig. 5.192)										
	LE tape (see Fig. 5.192)			5.199	302	300,000	PG02695	5.200	303	300,000	PG02694
SG6043 (M. Allen) Selig/Giguère	clean	5.201 5.202	304	5.203	305	100,000	PG02385	5.204	306	100,000	PG02384
						150,000	PG02500			150,000	PG02501
						200,000	PG02387/PG02388		307	200,000	PG02386
						300,000	PG02390/PG02391			300,000	PG02389
						400,000	PG02393/PG02394		308	400,000	PG02392
(continues)						500,000	PG02397			500,000	PG02396

SG6043	trip type C			5.205	309			5.206	310	100,000	PG02535
(continued)	(see Fig. 5.202)					150,000	PG02537			150,000	PG02536
									311	200,000	PG02538
						300,000	PG02540/PG02541			300,000	PG02539
									312	400,000	PG02542
						500,000	PG02545/PG02546			500,000	PG02544
Trainer 60	clean	5.207	314	5.209	315	100,000	CL01048	5.210	316	100,000	CL01047
(J. Thomas)		5.208				200,000	CL01050/CL01053			200,000	CL01049
J. Bridi						300,000	CL01052		317	300,000	CL01051
						400,000	AB01055/AB01056			400,000	AB01054
						500,000	CL02074		318	500,000	CL02073
Ultra-Sport 1000	clean	5.211	320	5.213	321	100,000	CL01680	5.214	322	100,000	CL01679
(D. Brengman)		5.212				200,000	CL01682			200,000	CL01681
GPMM						300,000	CL01684		323	300,000	CL01683
						400,000	CL01686			400,000	CL01685
						500,000	CL01861		324	500,000	CL01860
USNPS-4	clean	5.215	326	5.217	327	100,000	PG01166	5.218	328	100,000	PG01165
(J. Raley)		5.216				200,000	PG01507			200,000	PG01167
unknown						300,000	PG01511		329	300,000	PG01510
						400,000	PG01513			400,000	PG01512

A18

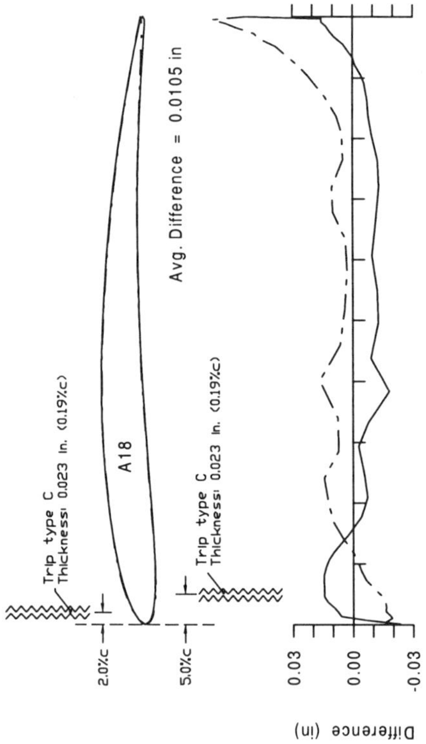
A18
Trip type C
Thickness: 0.023 in. (0.19%c)
Trip type C
Thickness: 0.023 in. (0.19%c)
2.0%c
5.0%c
Avg. Difference = 0.0105 in
0.03
0.00
-0.03
Difference (in)

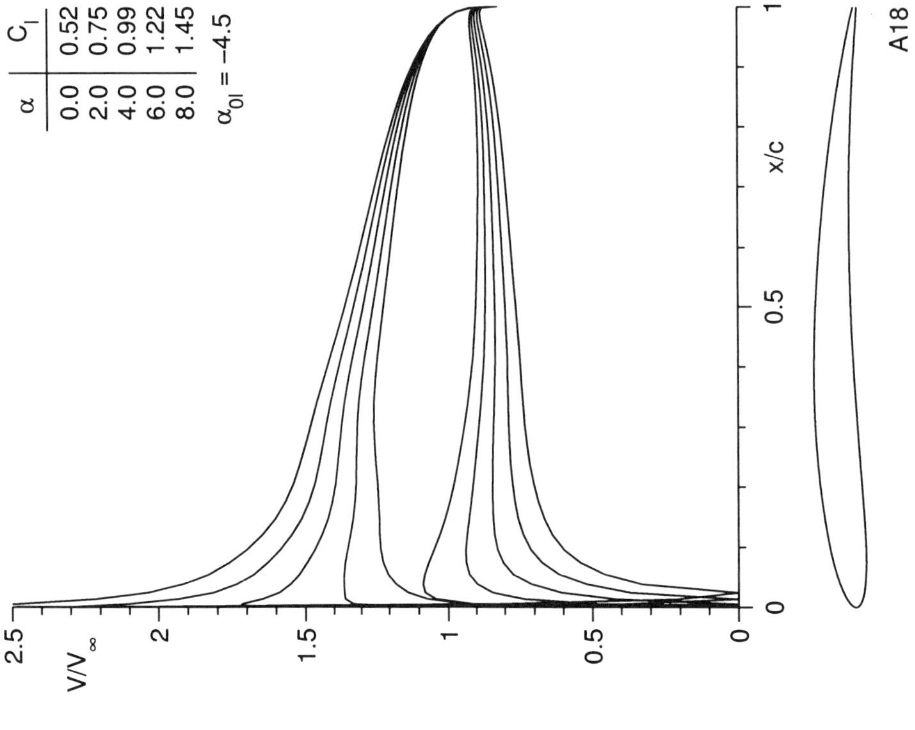
α
0.0
2.0
4.0
6.0
8.0
C_l
0.52
0.75
0.99
1.22
1.45
α_{0l} = −4.5
V/V_∞
2.5
2
1.5
1
0.5
0
x/c
0
0.5
1
A18

A18

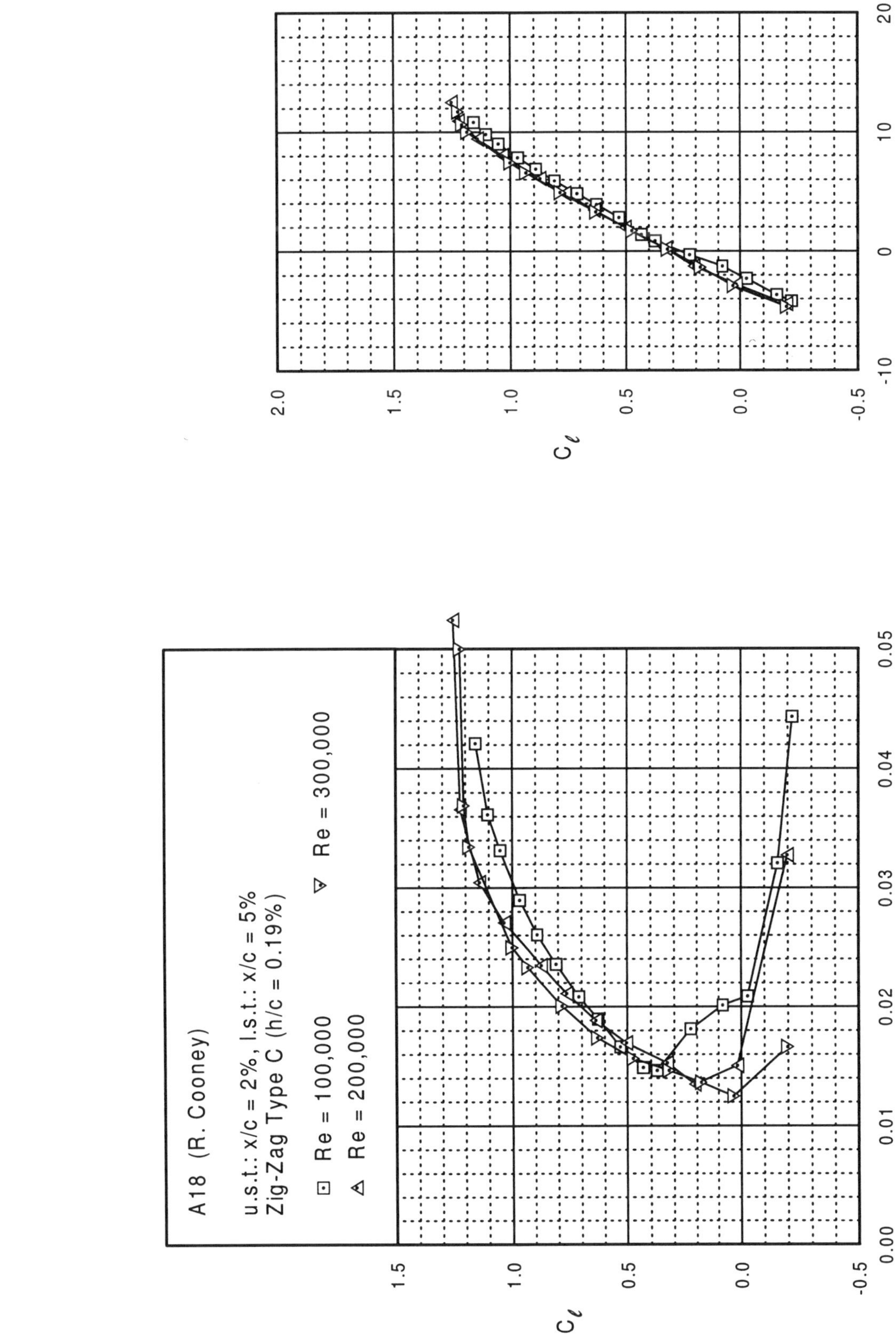
A18 (R. Cooney)
u.s.t.: x/c = 2%, l.s.t.: x/c = 5%
Zig-Zag Type C (h/c = 0.19%)
Re = 100,000
Re = 200,000
Re = 300,000
C_d
C_ℓ
α (deg)

A18

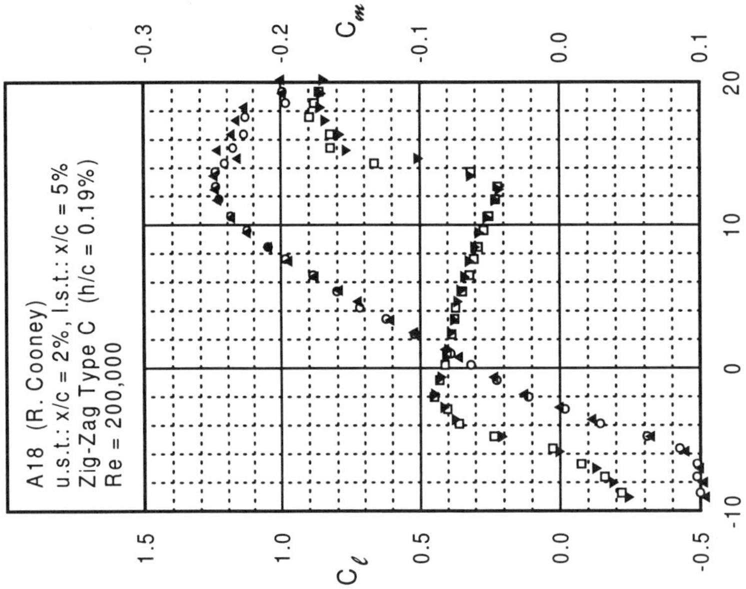
A18 (R. Cooney)
u.s.t.: x/c = 2%, l.s.t.: x/c = 5%
Zig-Zag Type C (h/c = 0.19%)
Re = 200,000
C_ℓ
C_m
α (deg)

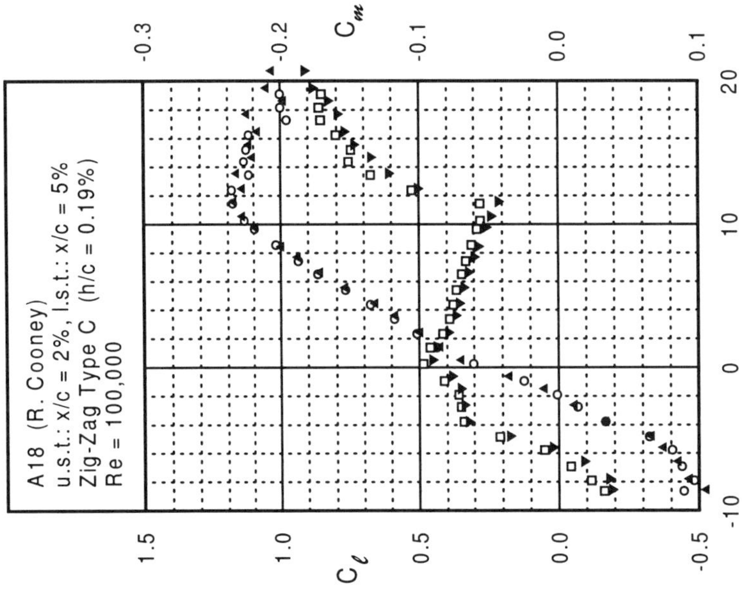
A18 (R. Cooney)
u.s.t.: x/c = 2%, l.s.t.: x/c = 5%
Zig-Zag Type C (h/c = 0.19%)
Re = 100,000
C_ℓ
C_m
α (deg)

A18

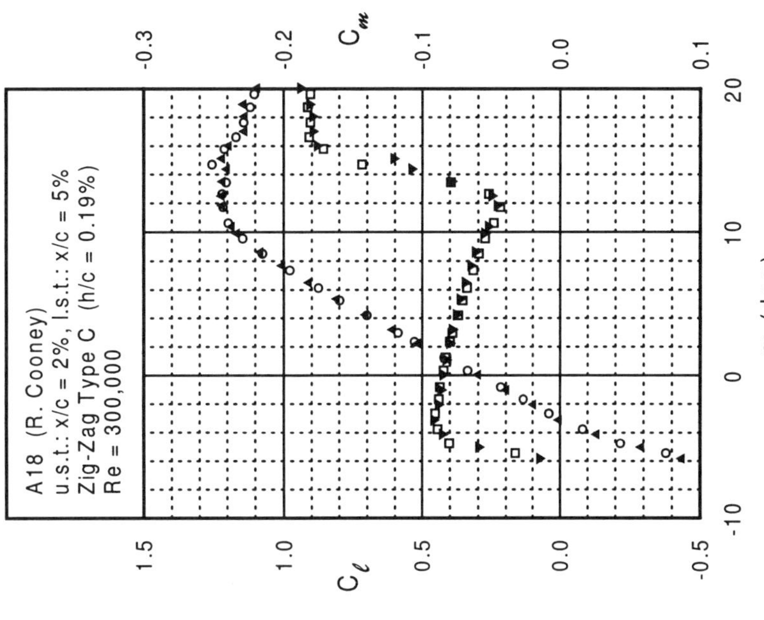
A18 (R. Cooney)
u.s.t.: x/c = 2%, l.s.t.: x/c = 5%
Zig-Zag Type C (h/c = 0.19%)
Re = 300,000
C_ℓ
1.5
1.0
0.5
0.0
-0.5
C_m
-0.3
-0.2
-0.1
0.0
0.1
-10
0
10
20
α (deg)

Avistar

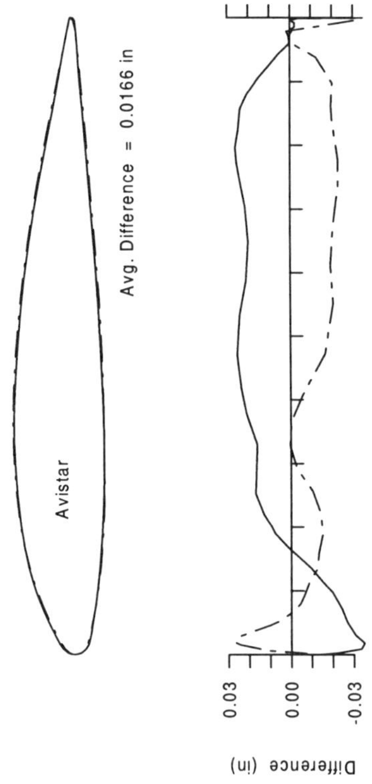
Avistar
Avg. Difference = 0.0166 in
0.03
0.00
-0.03
Difference (in)

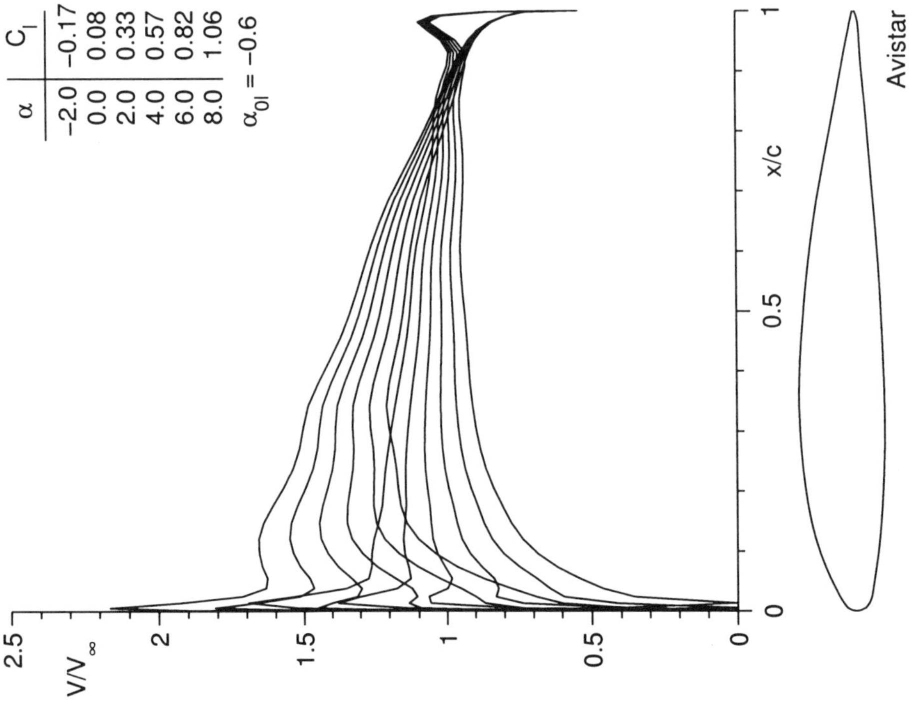
α | C_l
−2.0 | −0.17
0.0 | 0.08
2.0 | 0.33
4.0 | 0.57
6.0 | 0.82
8.0 | 1.06
α_{0l} = −0.6
2.5
V/V_∞
2
1.5
1
0.5
0
0
0.5
x/c
1
Avistar

Avistar

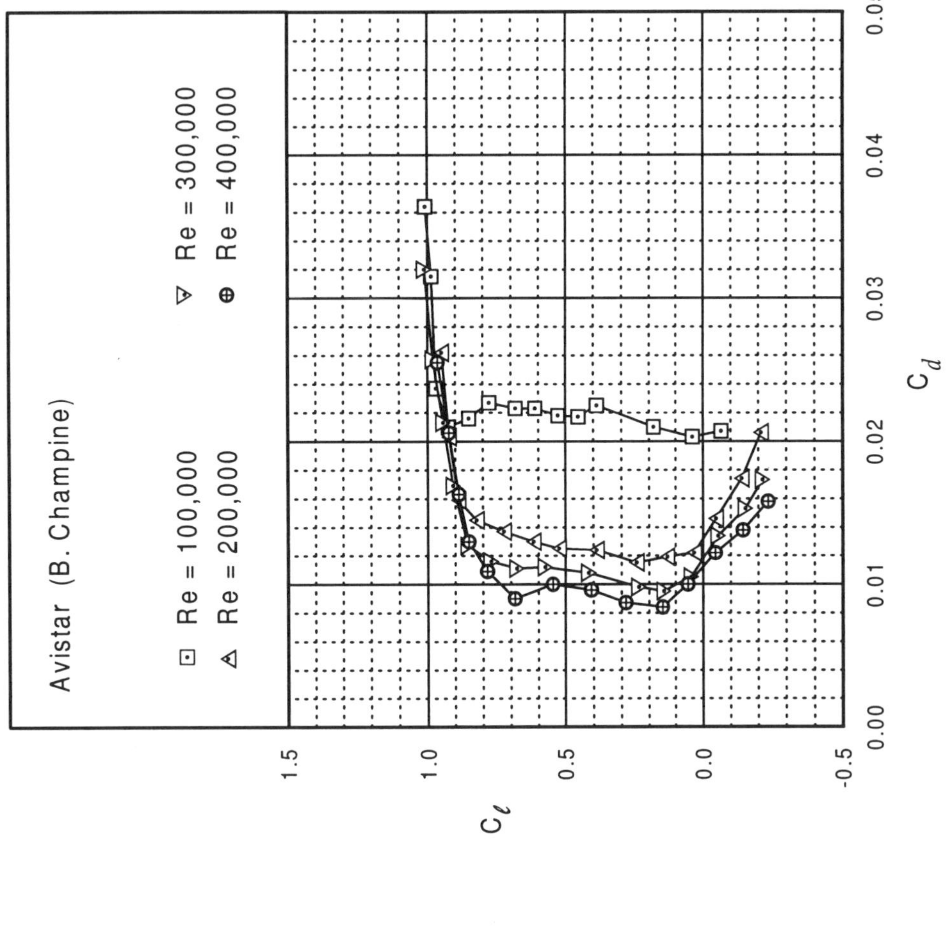

Avistar (B. Champine)
Re = 100,000
Re = 200,000
Re = 300,000
Re = 400,000
C_ℓ
C_d
1.5
1.0
0.5
0.0
-0.5
0.00
0.01
0.02
0.03
0.04
0.05

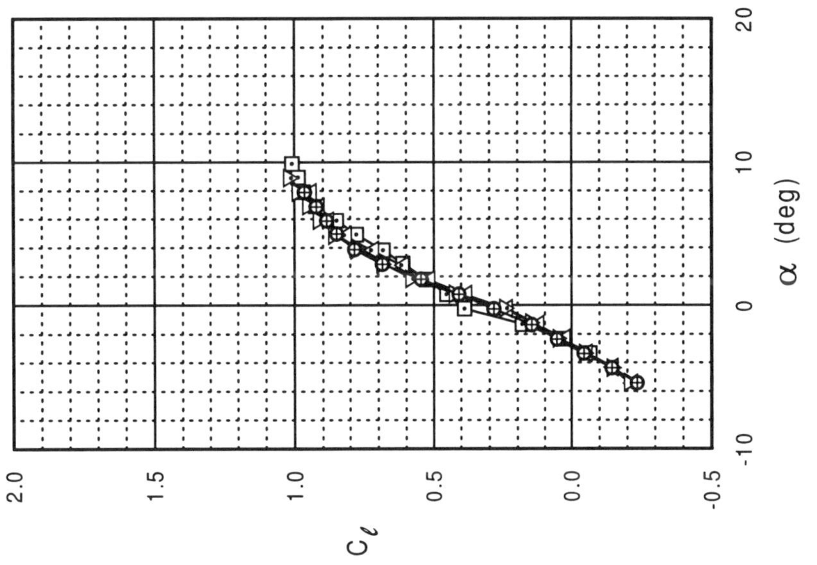

C_ℓ
α (deg)
2.0
1.5
1.0
0.5
0.0
-0.5
-10
0
10
20

Avistar

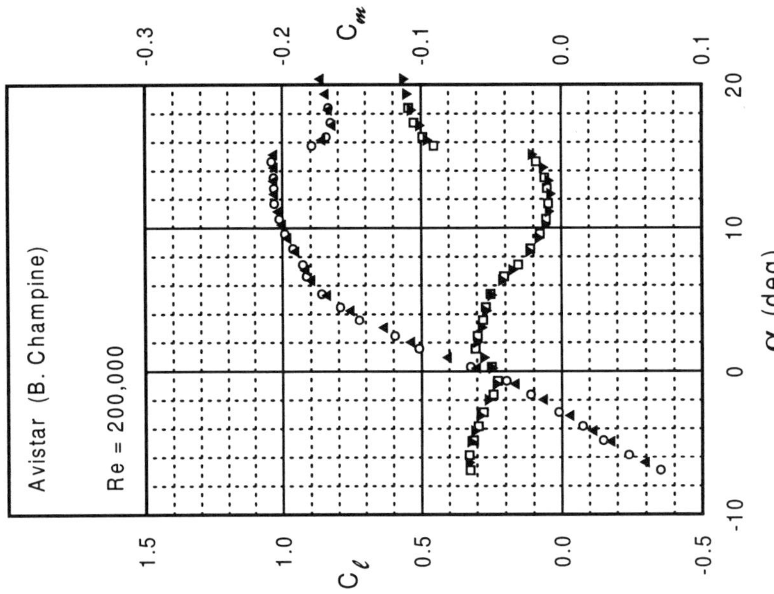
Avistar (B. Champine)
Re = 200,000
α (deg)
-10
0
10
20
C_l
1.5
1.0
0.5
0.0
-0.5
C_m
-0.3
-0.2
-0.1
0.0
0.1

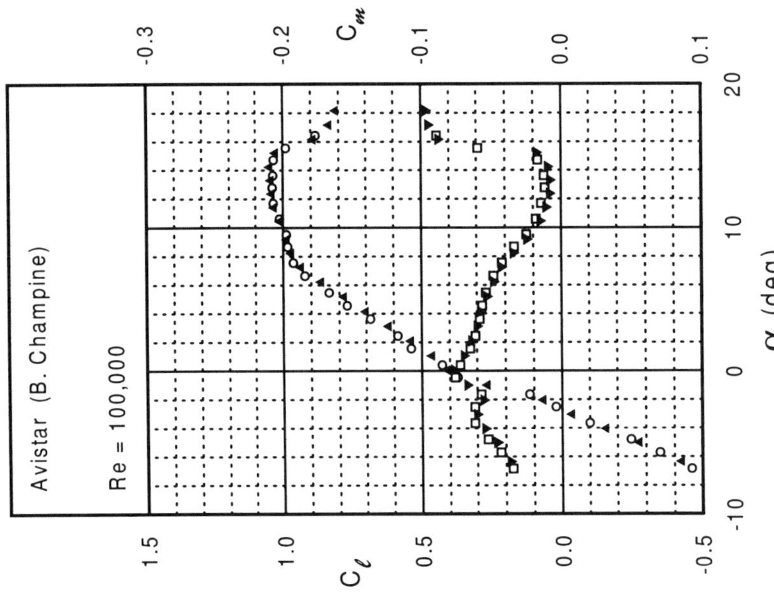
Avistar (B. Champine)
Re = 100,000
α (deg)
-10
0
10
20
C_l
1.5
1.0
0.5
0.0
-0.5
C_m
-0.3
-0.2
-0.1
0.0
0.1

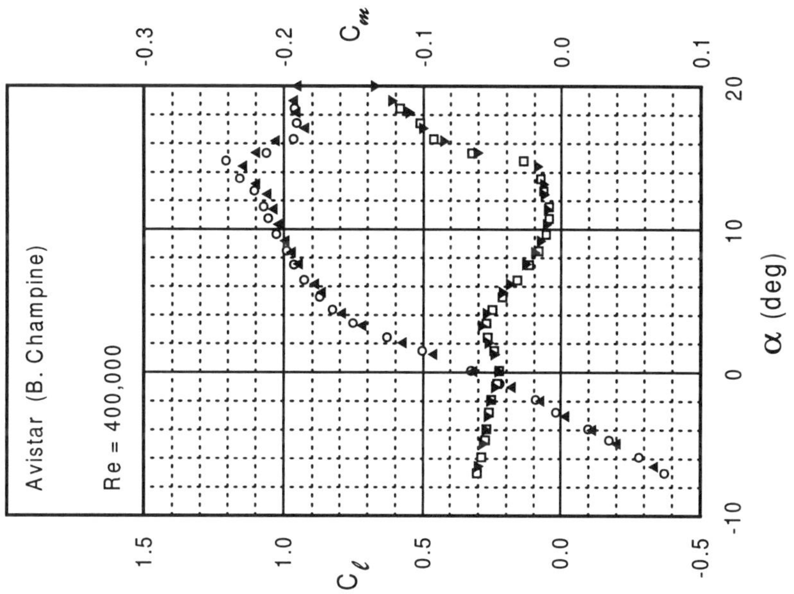
Avistar (B. Champine)
Re = 400,000
C_ℓ
C_m
α (deg)
1.5
1.0
0.5
0.0
-0.5
-0.3
-0.2
-0.1
0.0
0.1
-10
0
10
20

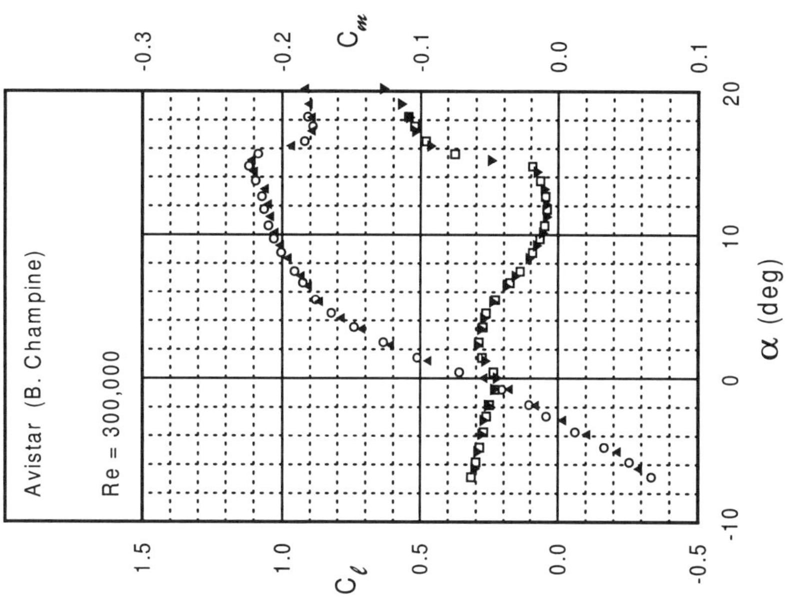
Avistar (B. Champine)
Re = 300,000
C_ℓ
C_m
α (deg)
1.5
1.0
0.5
0.0
-0.5
-0.3
-0.2
-0.1
0.0
0.1
-10
0
10
20

BW-3

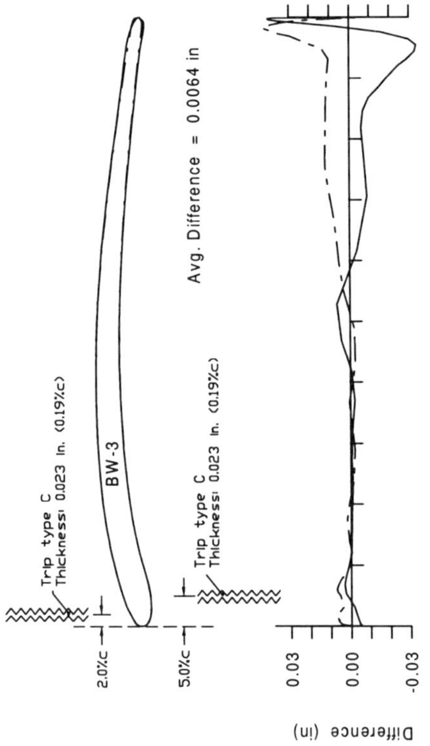
Trip type C
Thickness: 0.023 in. (0.19%c)
BW-3
Avg. Difference = 0.0064 in
2.0%c
5.0%c
0.03
0.00
-0.03
Difference (in)

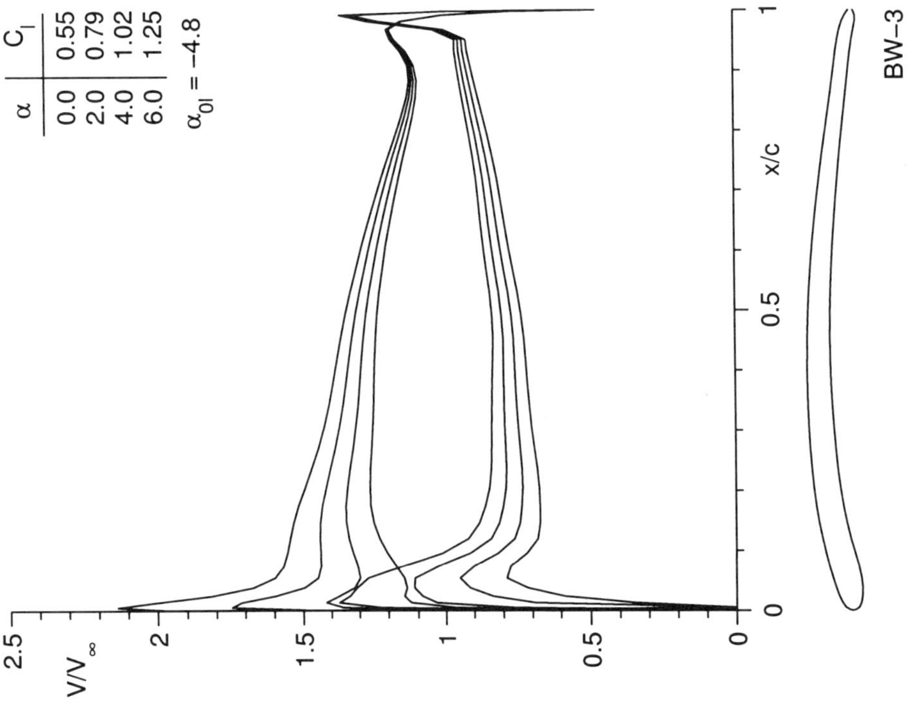
α | C_l
0.0 | 0.55
2.0 | 0.79
4.0 | 1.02
6.0 | 1.25
α_0l = −4.8
V/V∞
2.5
2
1.5
1
0.5
0
x/c
0.5
1
BW-3

BW-3

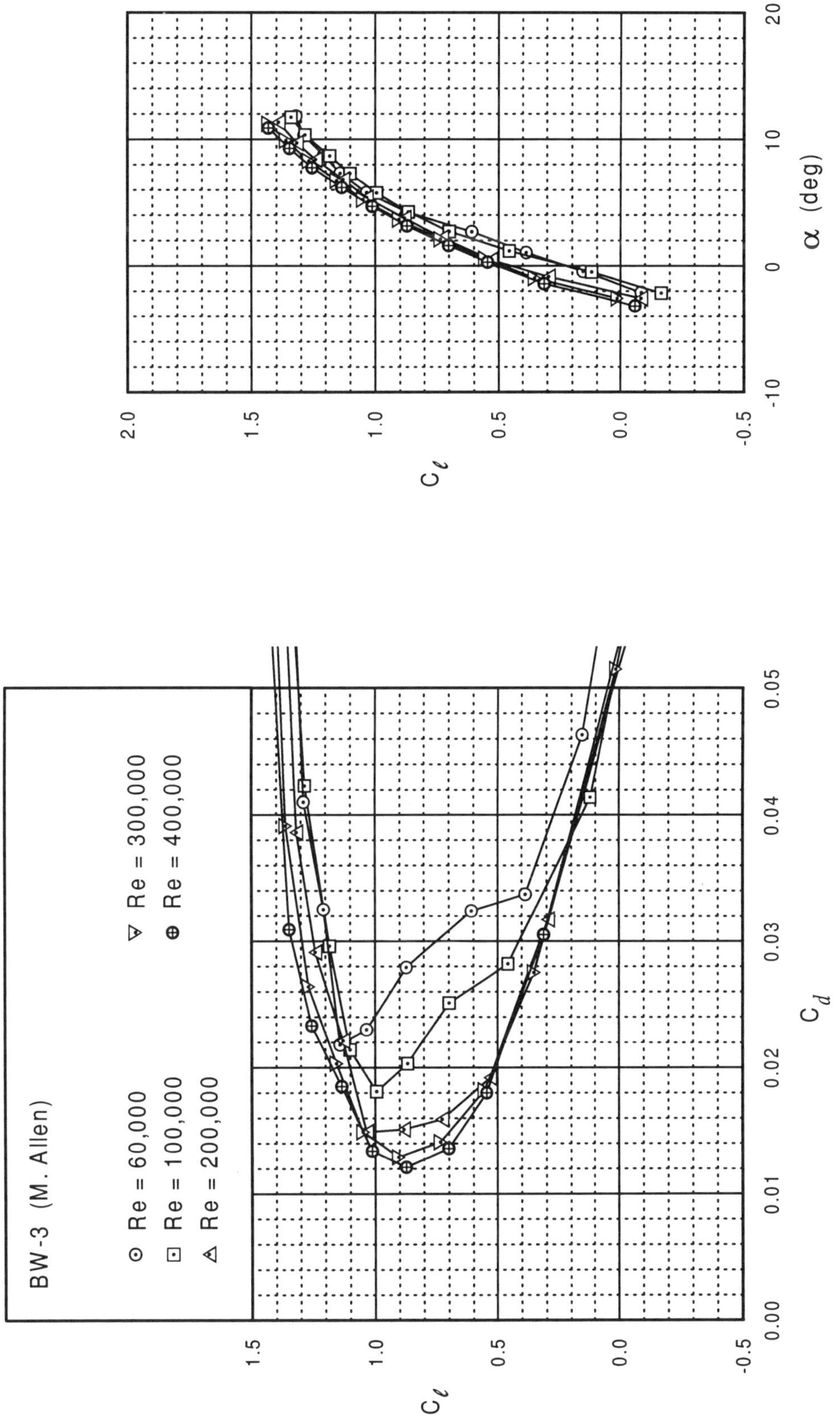
BW-3 (M. Allen)
Re = 60,000
Re = 100,000
Re = 200,000
Re = 300,000
Re = 400,000
C_ℓ
C_d
α (deg)
1.5
1.0
0.5
0.0
-0.5
2.0
0.00
0.01
0.02
0.03
0.04
0.05
-10
0
10
20

BW-3

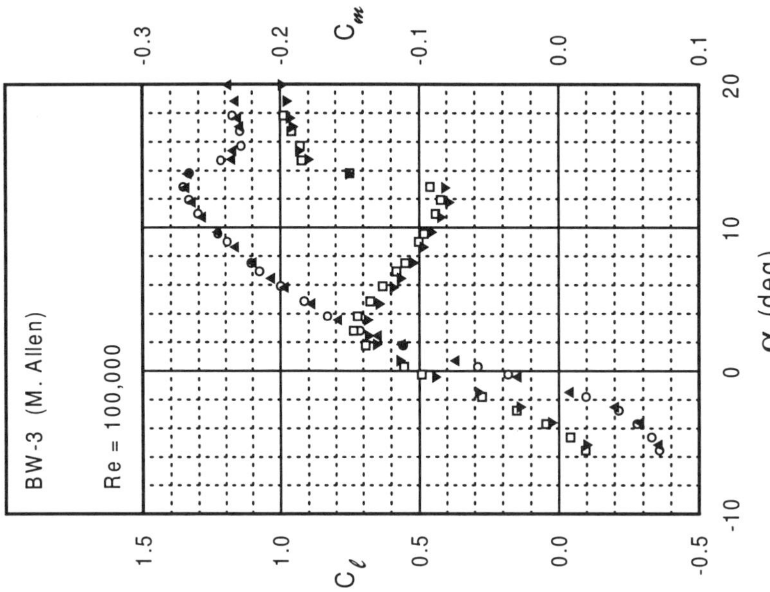
BW-3 (M. Allen)
Re = 100,000
C_ℓ
C_m
α (deg)

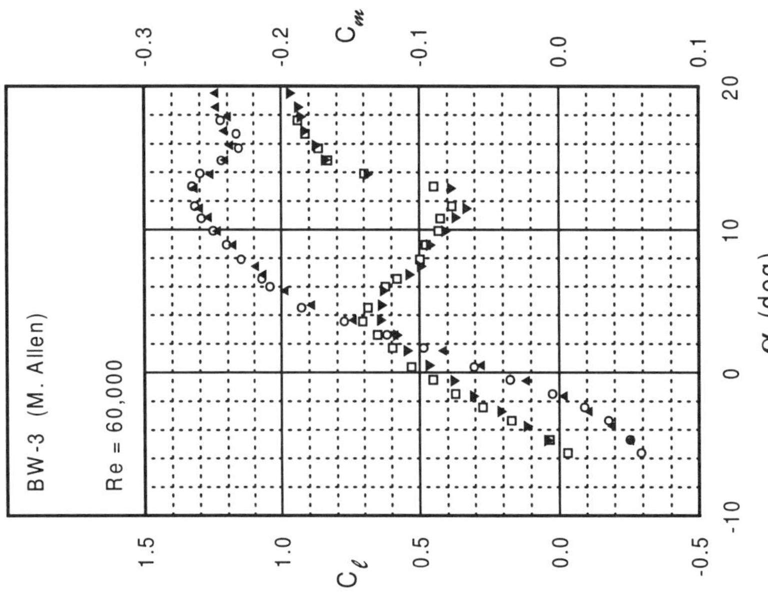
BW-3 (M. Allen)
Re = 60,000
C_ℓ
C_m
α (deg)

BW-3

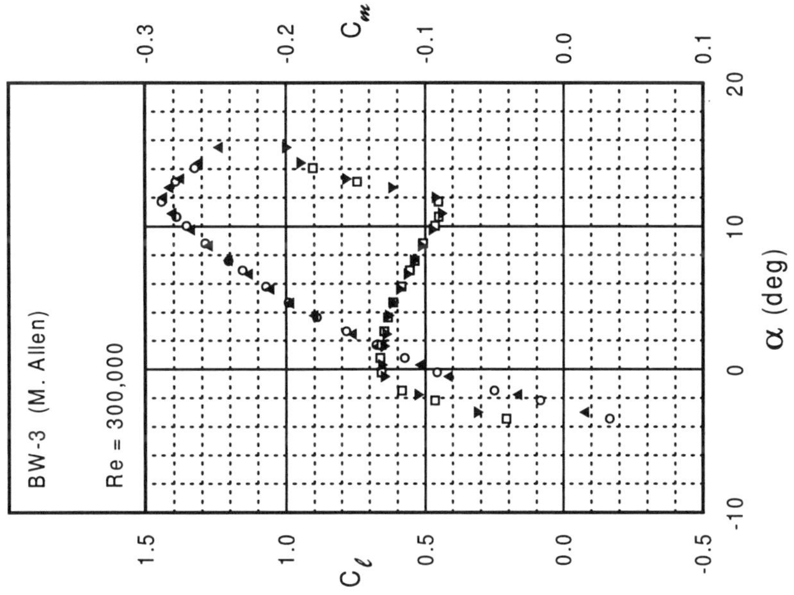
BW-3 (M. Allen)
Re = 300,000
C_l
C_m
α (deg)

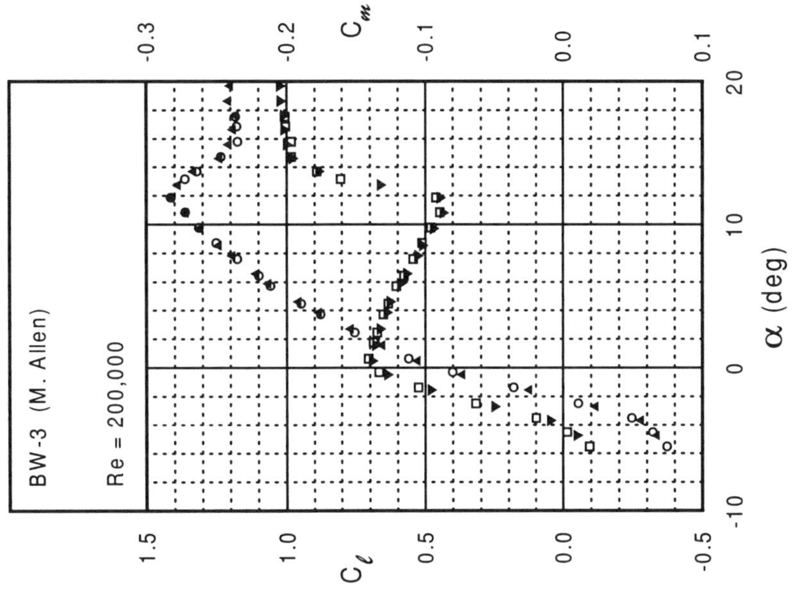
BW-3 (M. Allen)
Re = 200,000
C_l
C_m
α (deg)

BW-3

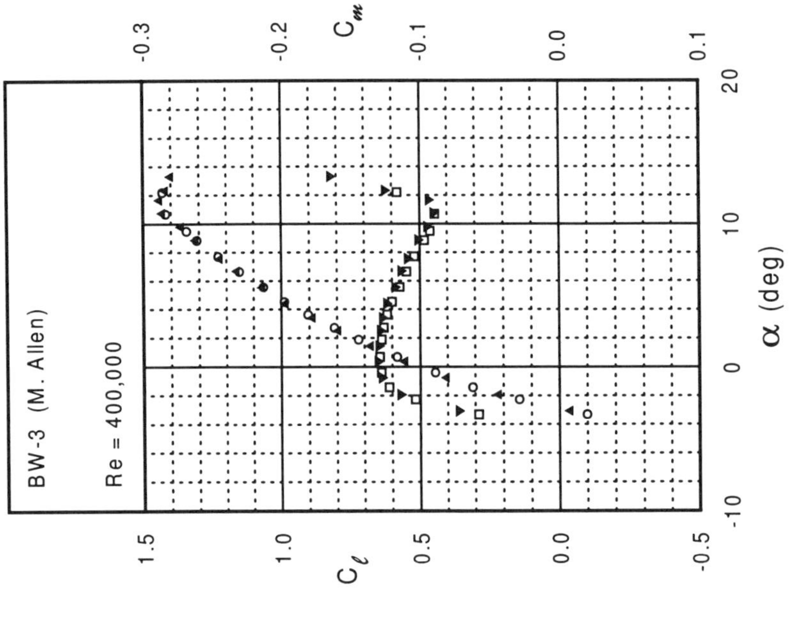
BW-3 (M. Allen)
Re = 400,000
C_ℓ
C_m
α (deg)
-10
0
10
20
1.5
1.0
0.5
0.0
-0.5
-0.3
-0.2
-0.1
0.0
0.1

BW-3

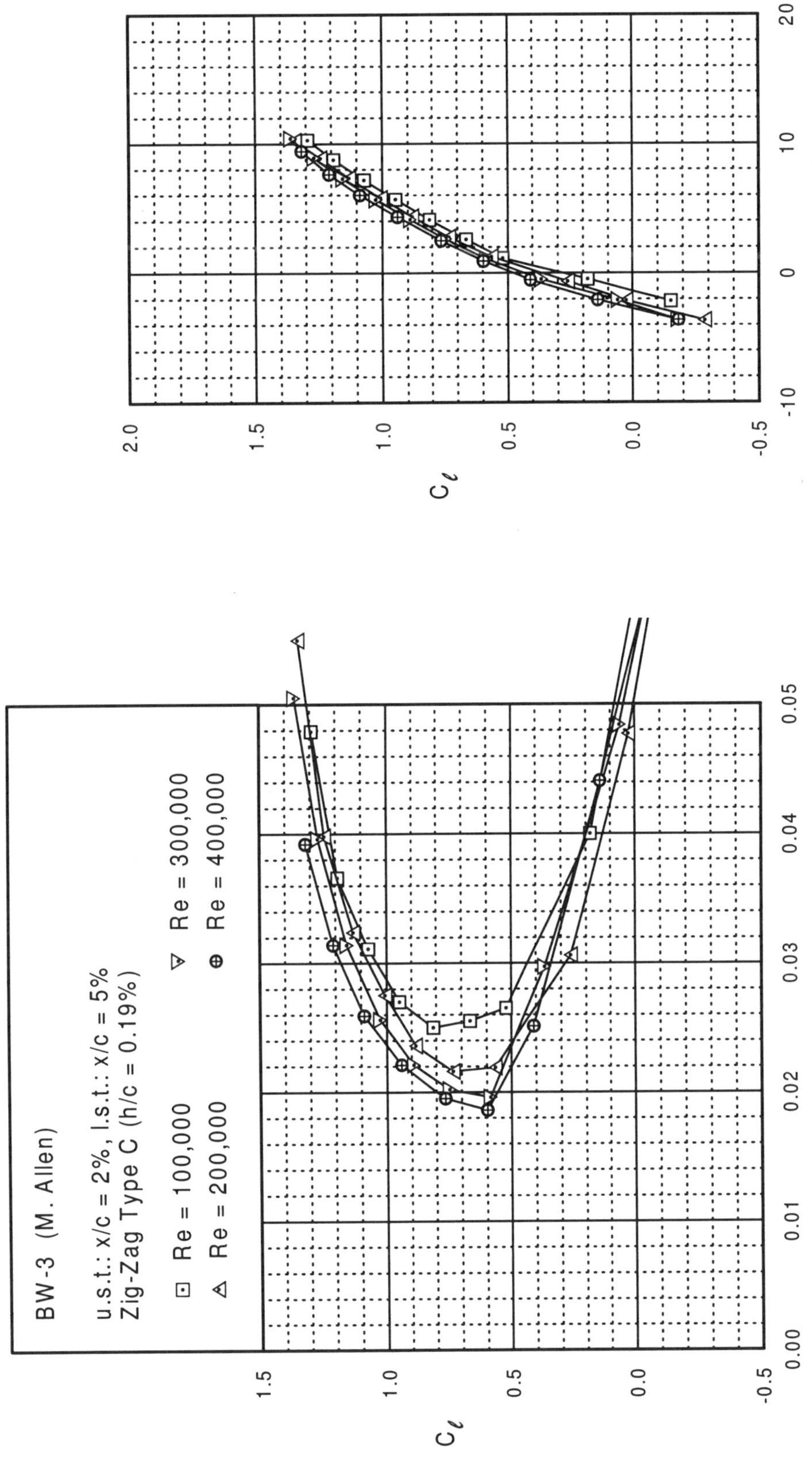
BW-3 (M. Allen)
u.s.t.: x/c = 2%, l.s.t.: x/c = 5%
Zig-Zag Type C (h/c = 0.19%)
Re = 100,000
Re = 200,000
Re = 300,000
Re = 400,000
C_ℓ
C_d
α (deg)
-0.5
0.0
0.5
1.0
1.5
2.0
0.00
0.01
0.02
0.03
0.04
0.05
-10
0
10
20

BW-3

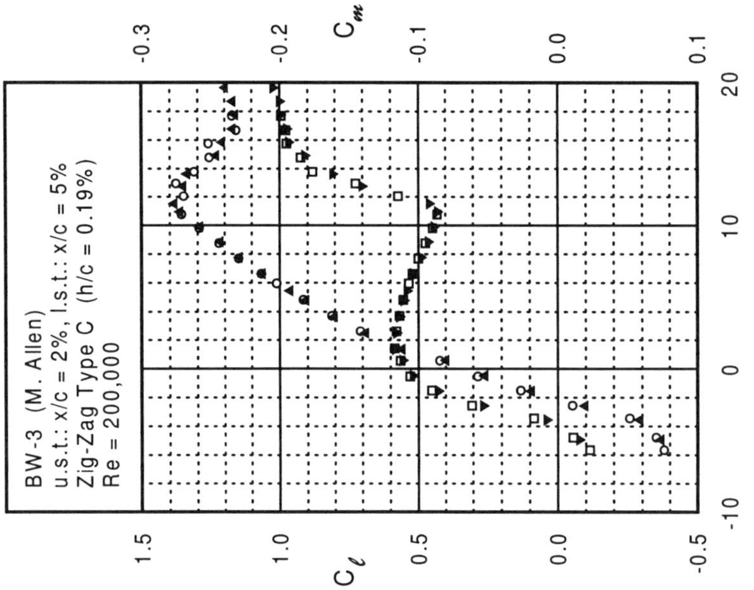
BW-3 (M. Allen)
u.s.t.: x/c = 2%, l.s.t.: x/c = 5%
Zig-Zag Type C (h/c = 0.19%)
Re = 200,000
C_m
-0.3
-0.2
-0.1
0.0
0.1
C_l
1.5
1.0
0.5
0.0
-0.5
α (deg)
-10
0
10
20

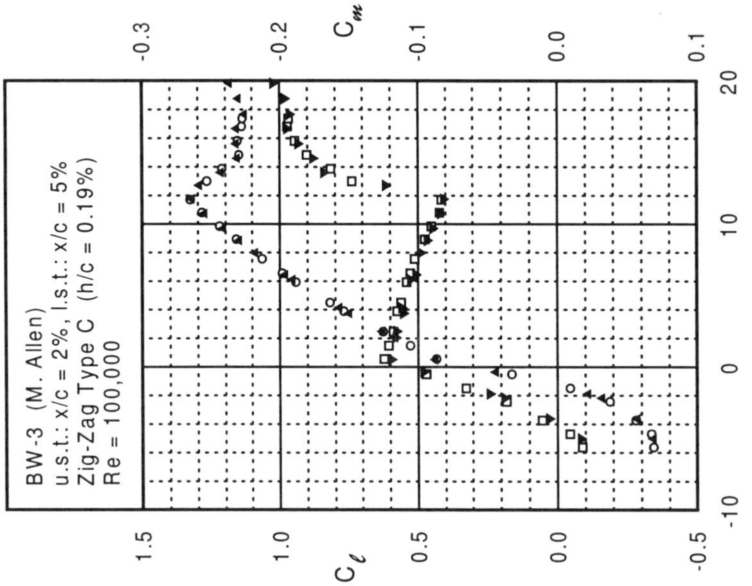
BW-3 (M. Allen)
u.s.t.: x/c = 2%, l.s.t.: x/c = 5%
Zig-Zag Type C (h/c = 0.19%)
Re = 100,000
C_m
-0.3
-0.2
-0.1
0.0
0.1
C_l
1.5
1.0
0.5
0.0
-0.5
α (deg)
-10
0
10
20

BW-3

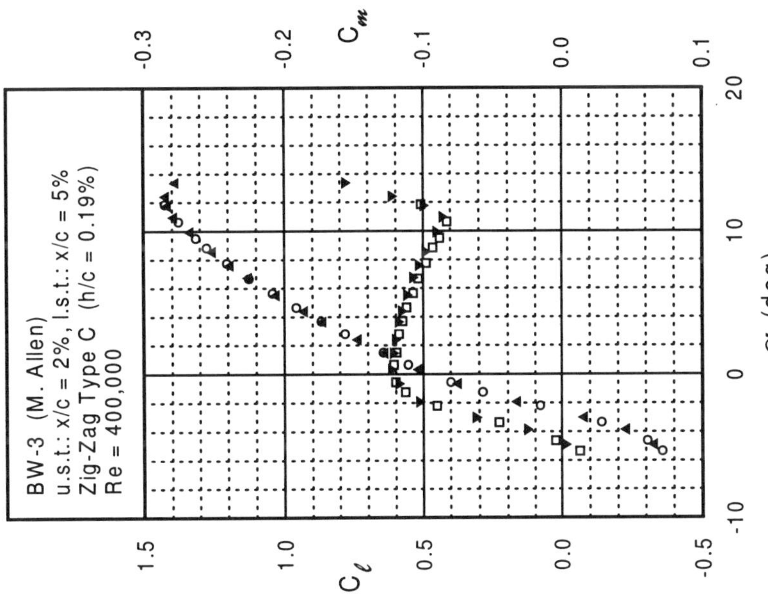
BW-3 (M. Allen)
u.s.t.: x/c = 2%, l.s.t.: x/c = 5%
Zig-Zag Type C (h/c = 0.19%)
Re = 400,000
α (deg)
C_ℓ
C_m

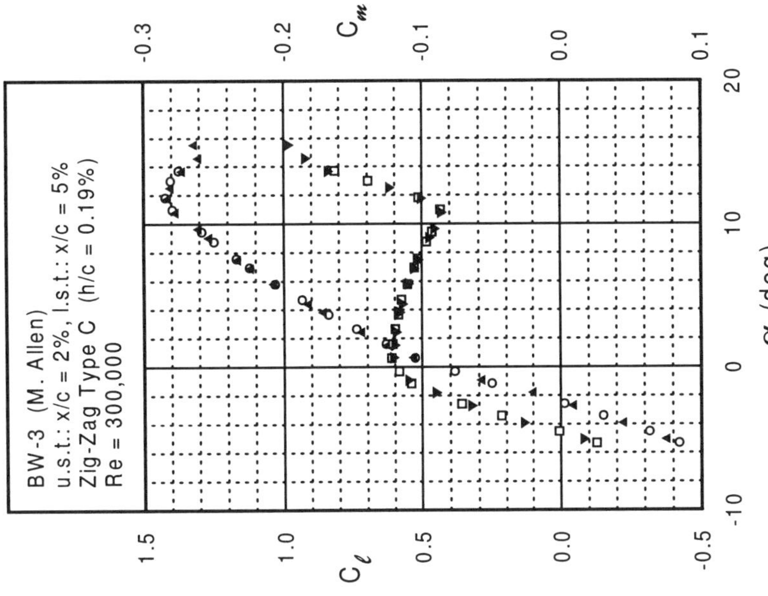
BW-3 (M. Allen)
u.s.t.: x/c = 2%, l.s.t.: x/c = 5%
Zig-Zag Type C (h/c = 0.19%)
Re = 300,000
α (deg)
C_ℓ
C_m

CG Ultimate

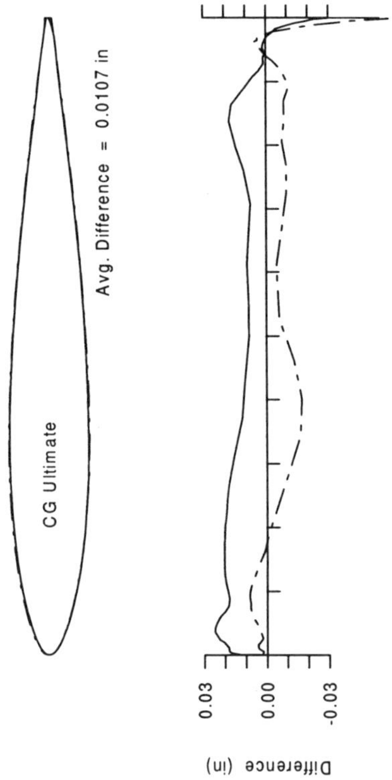
CG Ultimate
Avg. Difference = 0.0107 in
0.03
0.00
-0.03
Difference (in)

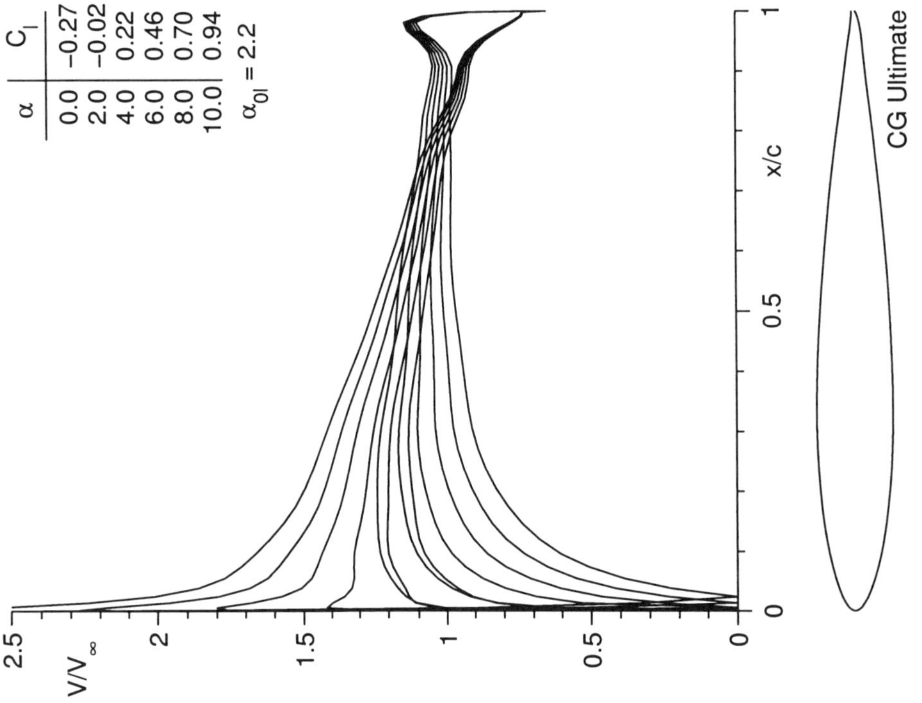
α | C_l
0.0 | −0.27
2.0 | −0.02
4.0 | 0.22
6.0 | 0.46
8.0 | 0.70
10.0 | 0.94
α_{0l} = 2.2
V/V_∞
2.5
2
1.5
1
0.5
0
x/c
0
0.5
1
CG Ultimate

CG Ultimate

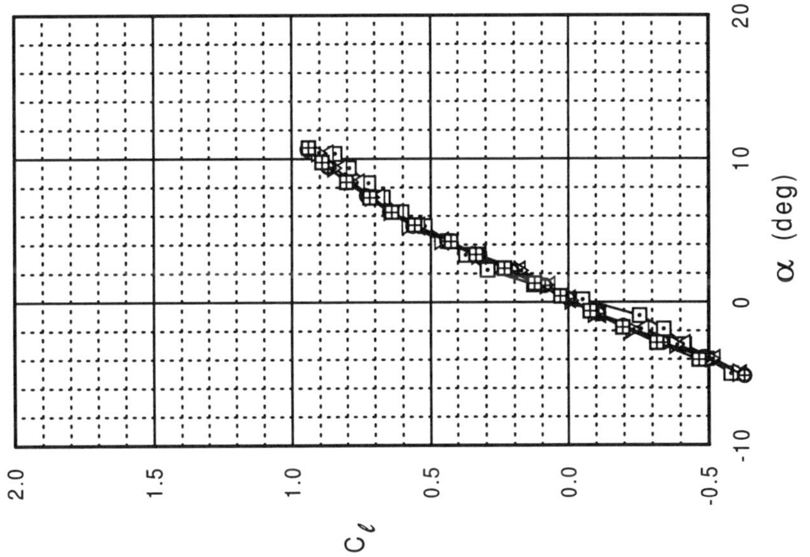
2.0
1.5
1.0
0.5
0.0
-0.5
C_ℓ
-10
0
10
20
α (deg)

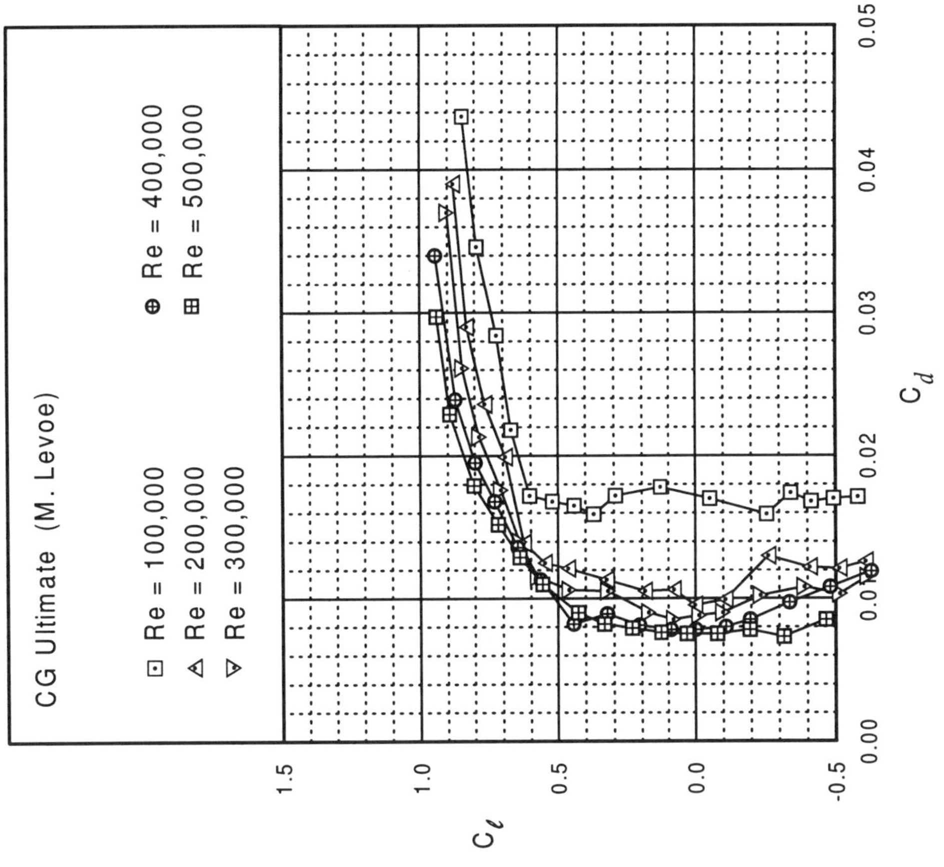
CG Ultimate (M. Levoe)
Re = 100,000
Re = 200,000
Re = 300,000
Re = 400,000
Re = 500,000
1.5
1.0
0.5
0.0
-0.5
C_ℓ
0.00
0.01
0.02
0.03
0.04
0.05
C_d

CG Ultimate

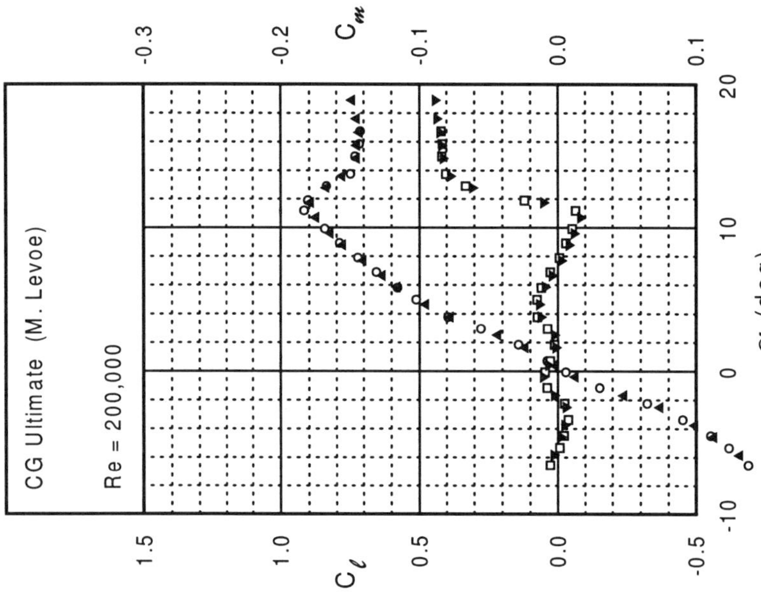
CG Ultimate (M. Levoe)
Re = 200,000
C_ℓ
C_m
α (deg)

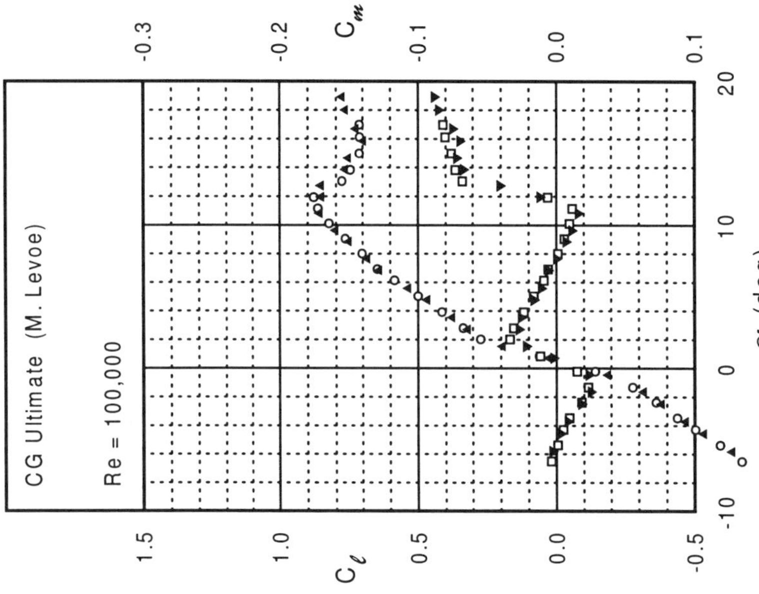
CG Ultimate (M. Levoe)
Re = 100,000
C_ℓ
C_m
α (deg)

CG Ultimate

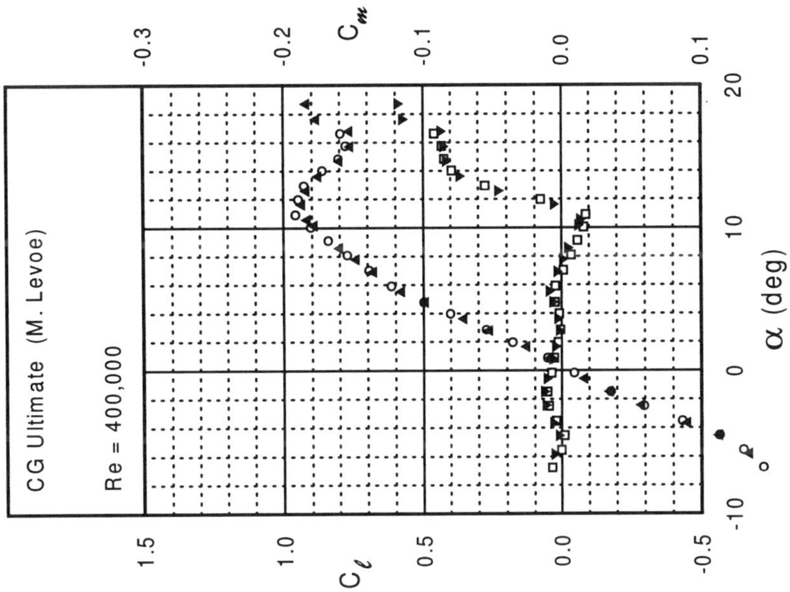
CG Ultimate (M. Levoe)
Re = 400,000
C_ℓ
1.5
1.0
0.5
0.0
-0.5
C_m
-0.3
-0.2
-0.1
0.0
0.1
-10
0
10
20
α (deg)

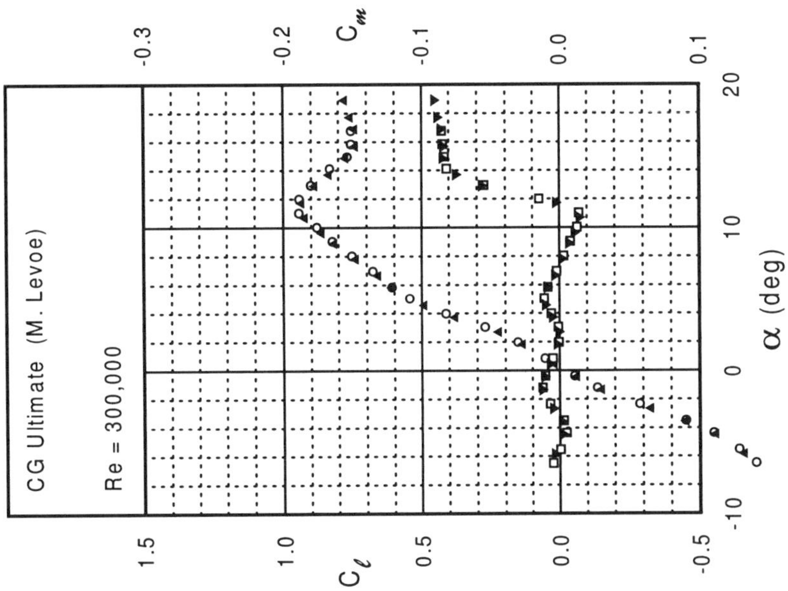
CG Ultimate (M. Levoe)
Re = 300,000
C_ℓ
1.5
1.0
0.5
0.0
-0.5
C_m
-0.3
-0.2
-0.1
0.0
0.1
-10
0
10
20
α (deg)

CG Ultimate

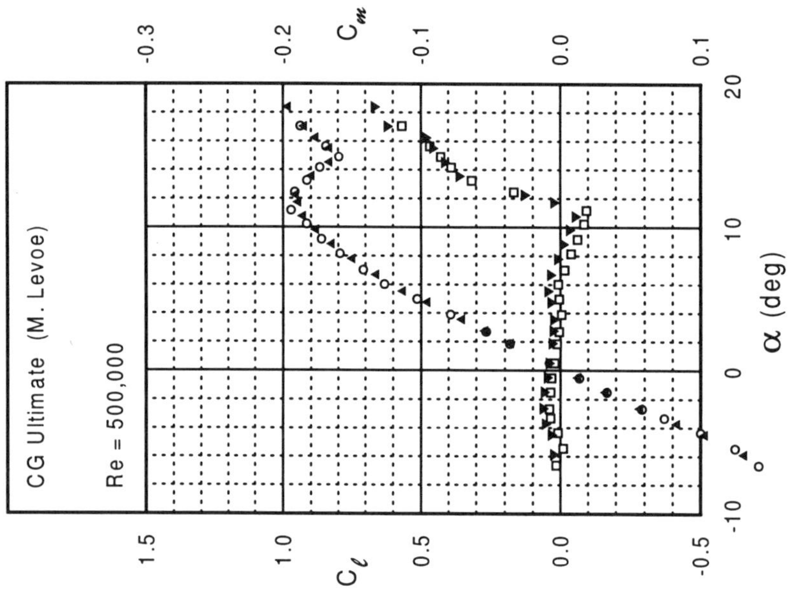
CG Ultimate (M. Levoe)
Re = 500,000
C_ℓ
1.5
1.0
0.5
0.0
-0.5
C_m
-0.3
-0.2
-0.1
0.0
0.1
-10
0
10
20
α (deg)

Clark-Y (B)

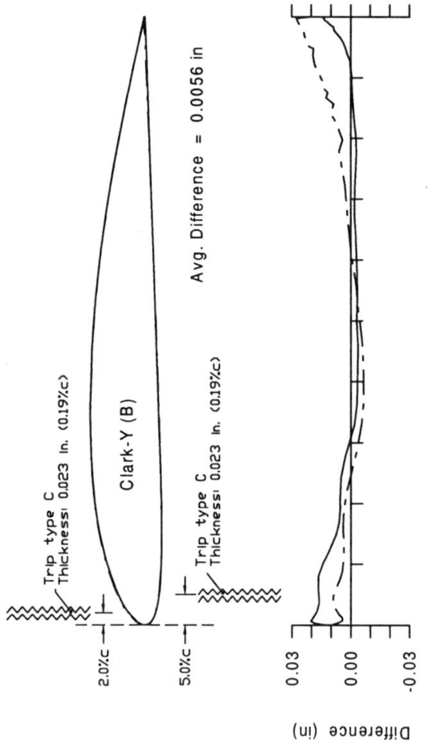
Trip type C
Thickness: 0.023 in. (0.19%c)
Clark-Y (B)
Trip type C
Thickness: 0.023 in. (0.19%c)
2.0%c
5.0%c
Avg. Difference = 0.0056 in
0.03
0.00
-0.03
Difference (in)

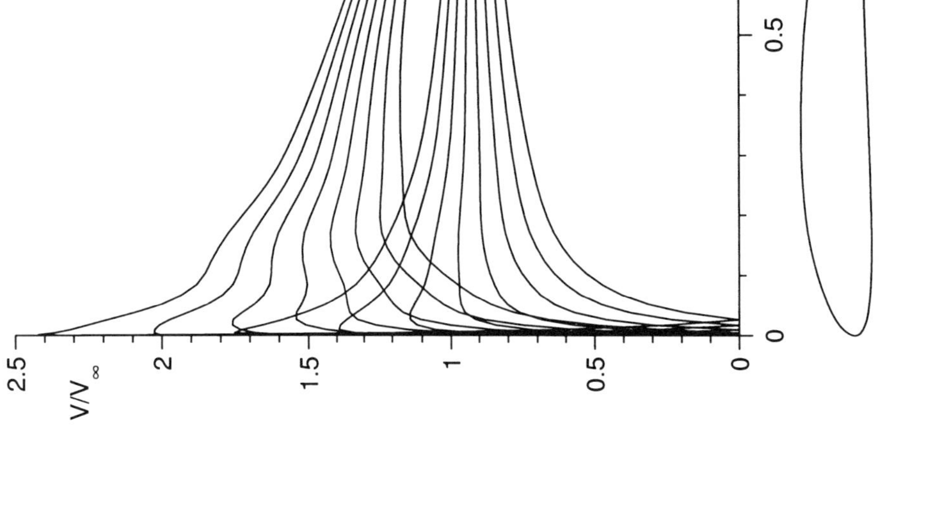
α C_l
-4.0 -0.05
-2.0 0.20
0.0 0.44
2.0 0.68
4.0 0.92
6.0 1.16
8.0 1.40
10.0 1.63
α_0l = -3.6
2.5
V/V∞
2
1.5
1
0.5
0
0
0.5
x/c
1
Clark-Y

Clark-Y (B)

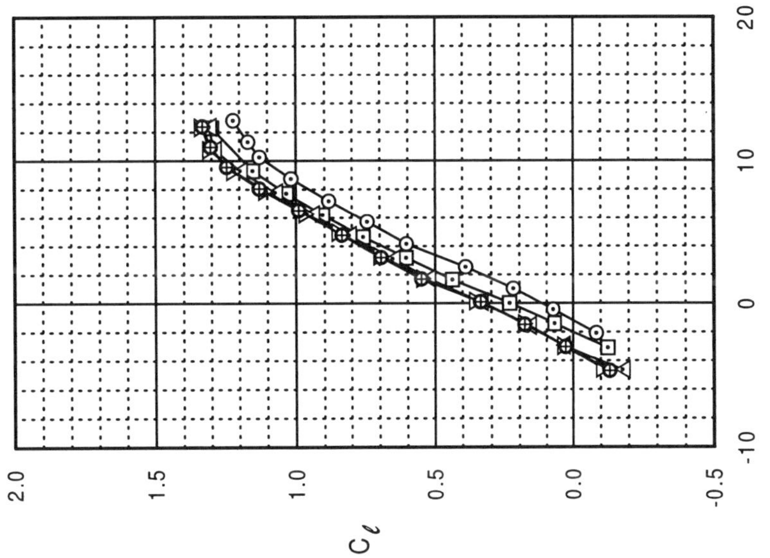
2.0
1.5
1.0
0.5
0.0
-0.5
C_ℓ
-10
0
10
20
α (deg)

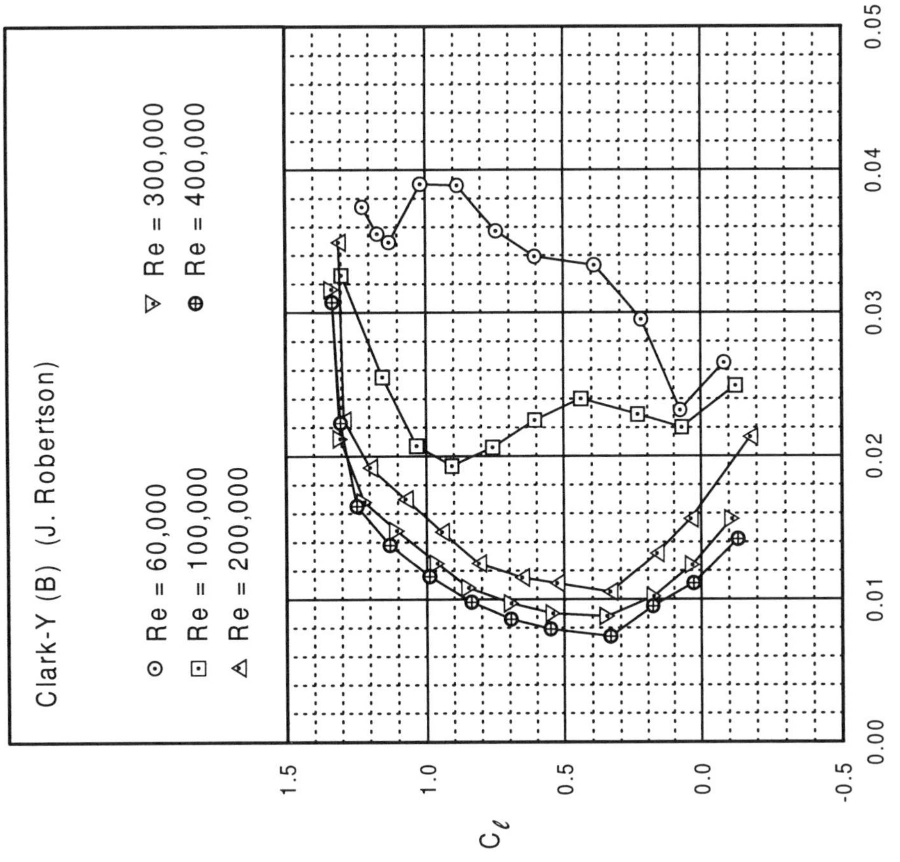
Clark-Y (B) (J. Robertson)
Re = 60,000
Re = 100,000
Re = 200,000
Re = 300,000
Re = 400,000
1.5
1.0
0.5
0.0
-0.5
C_ℓ
0.00
0.01
0.02
0.03
0.04
0.05
C_d

Clark-Y (B)

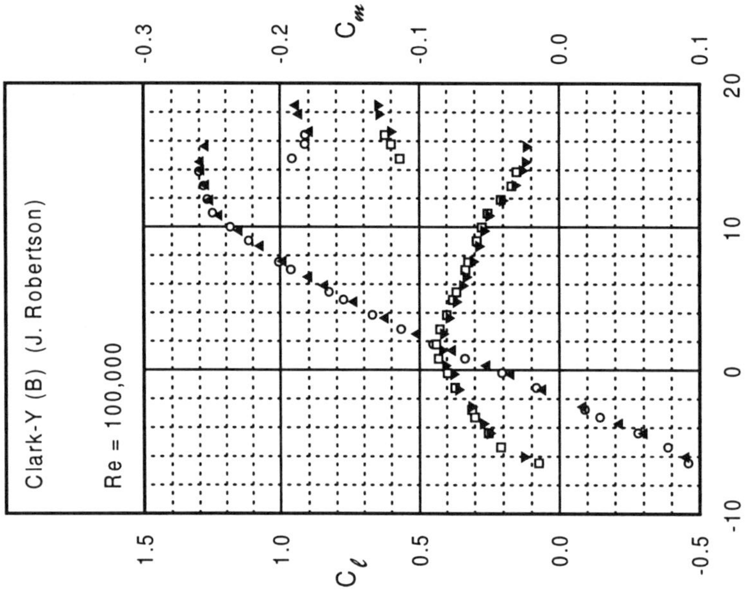
Clark-Y (B) (J. Robertson)
Re = 100,000
C_ℓ
C_m
α (deg)

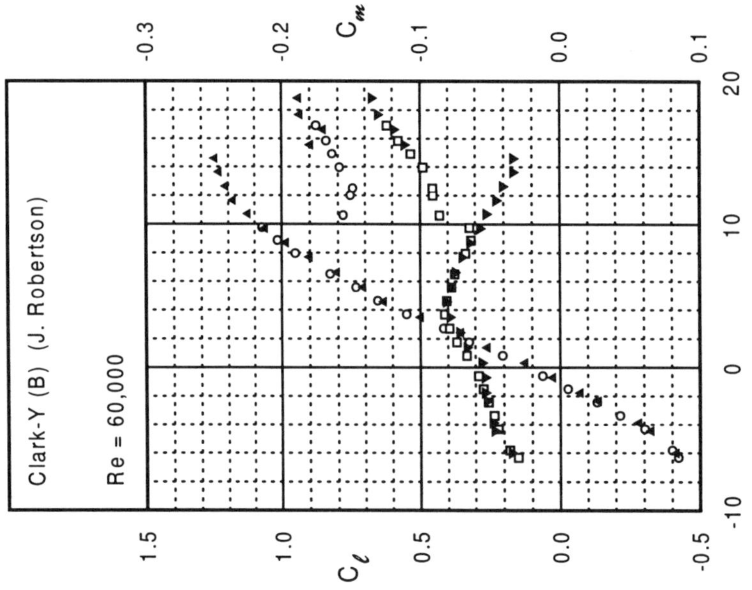
Clark-Y (B) (J. Robertson)
Re = 60,000
C_ℓ
C_m
α (deg)

Clark-Y (B)

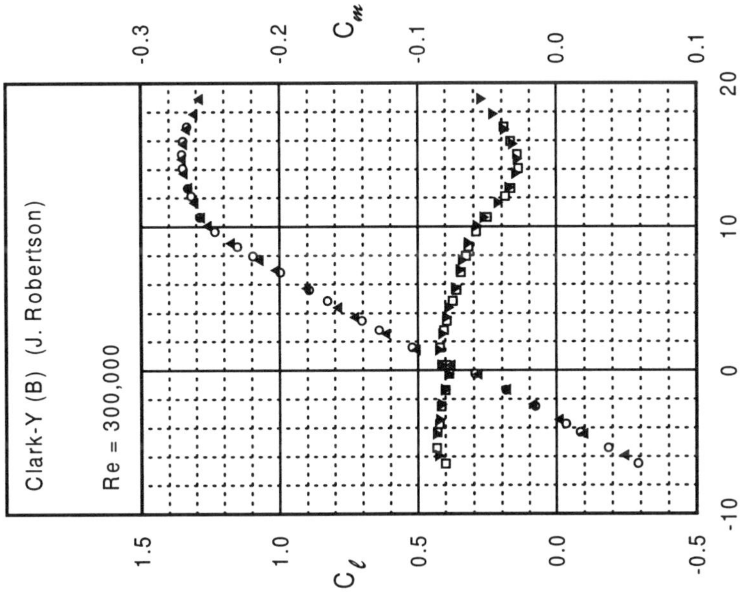
Clark-Y (B) (J. Robertson)
Re = 300,000
C_ℓ
C_m
α (deg)

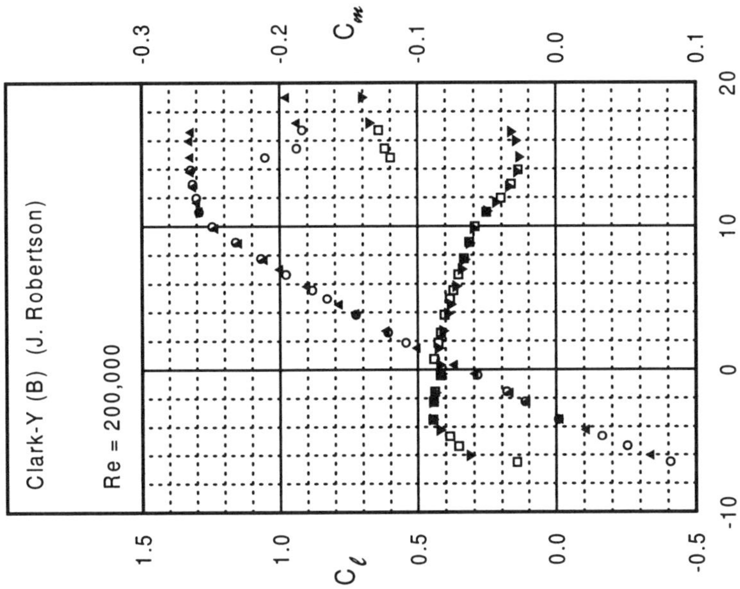
Clark-Y (B) (J. Robertson)
Re = 200,000
C_ℓ
C_m
α (deg)

Clark-Y (B)

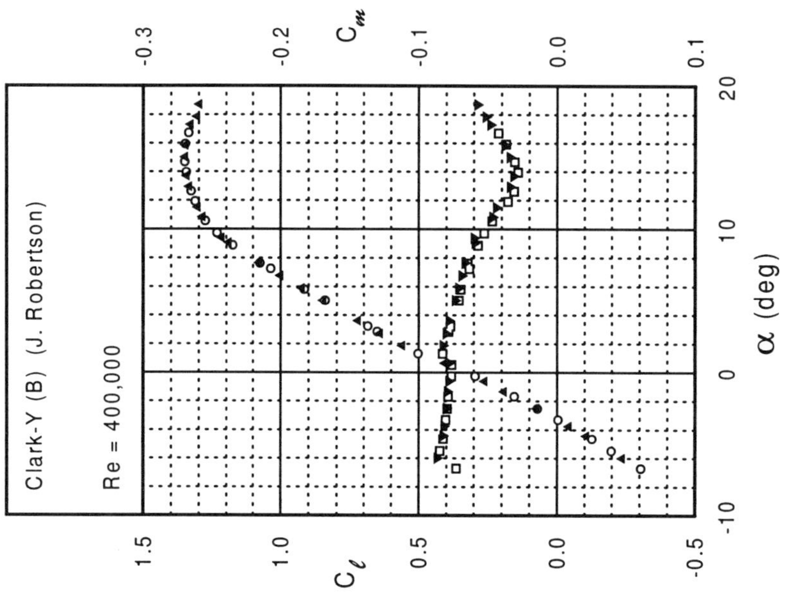
Clark-Y (B) (J. Robertson)
Re = 400,000
C_ℓ
1.5
1.0
0.5
0.0
-0.5
C_m
-0.3
-0.2
-0.1
0.0
0.1
α (deg)
-10
0
10
20

Clark-Y (B)

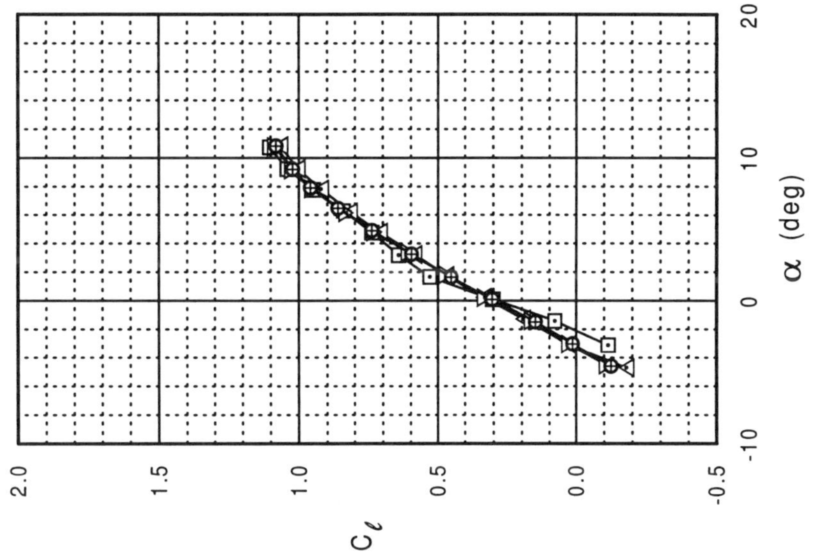

C_ℓ
α (deg)
2.0
1.5
1.0
0.5
0.0
-0.5
-10
0
10
20

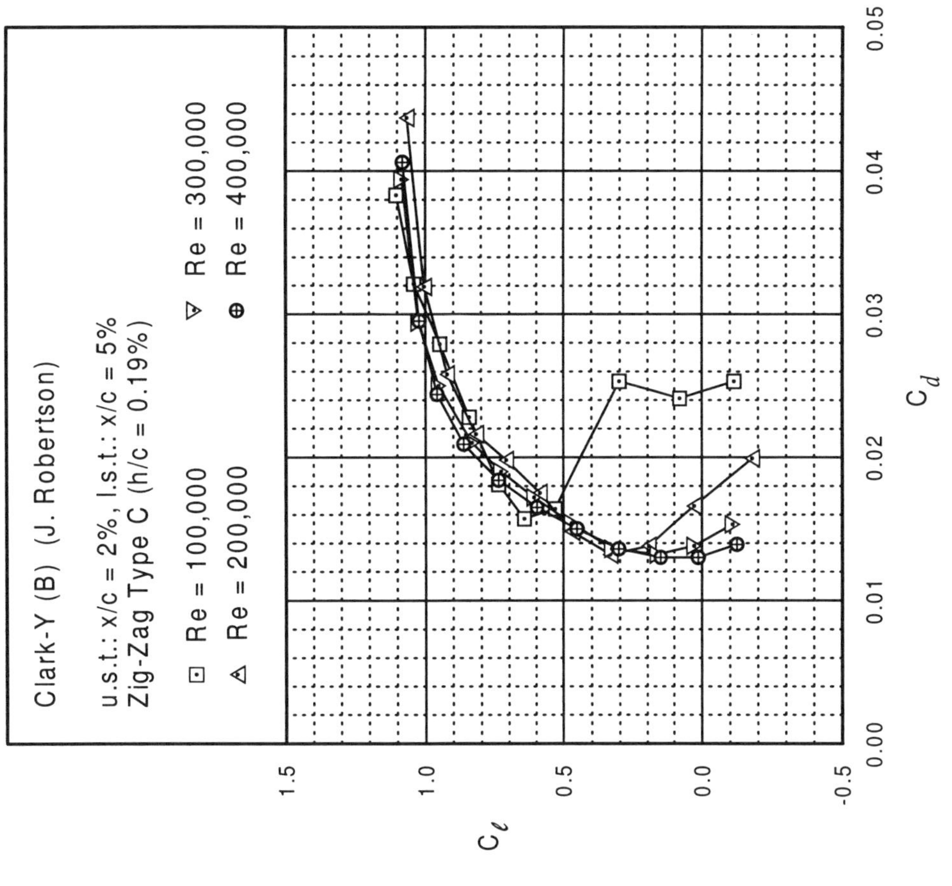

Clark-Y (B) (J. Robertson)
u.s.t.: x/c = 2%, l.s.t.: x/c = 5%
Zig-Zag Type C (h/c = 0.19%)
Re = 100,000
Re = 200,000
Re = 300,000
Re = 400,000
C_ℓ
C_d
1.5
1.0
0.5
0.0
-0.5
0.00
0.01
0.02
0.03
0.04
0.05

Clark-Y (B)

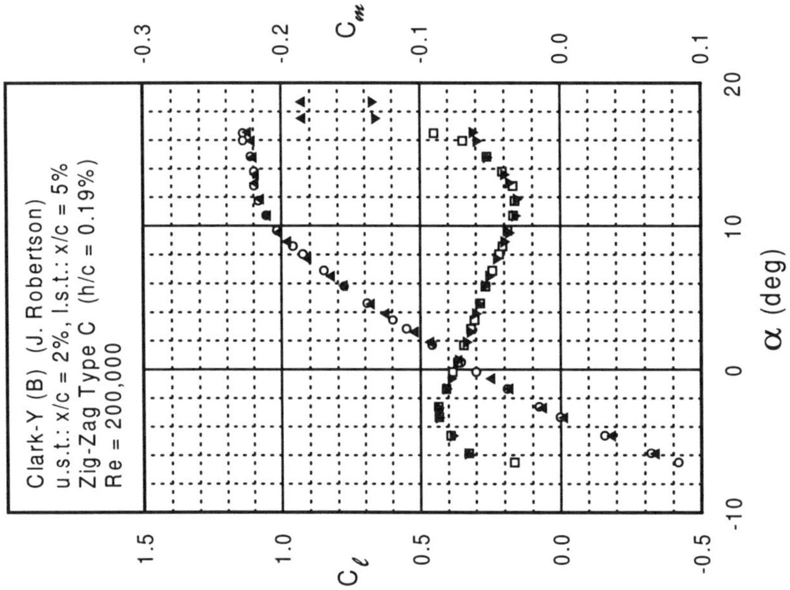
Clark-Y (B) (J. Robertson)
u.s.t.: x/c = 2%, l.s.t.: x/c = 5%
Zig-Zag Type C (h/c = 0.19%)
Re = 200,000
C_m
C_ℓ
α (deg)

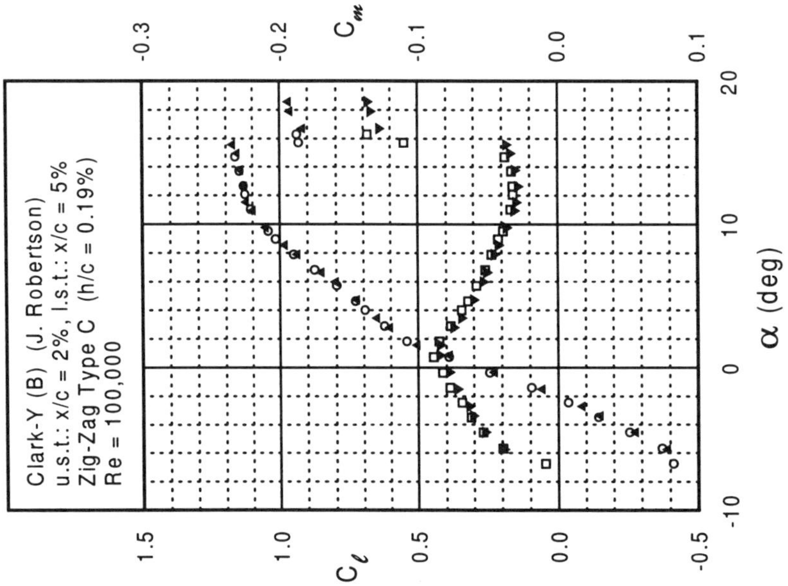
Clark-Y (B) (J. Robertson)
u.s.t.: x/c = 2%, l.s.t.: x/c = 5%
Zig-Zag Type C (h/c = 0.19%)
Re = 100,000
C_m
C_ℓ
α (deg)

Clark-Y (B)

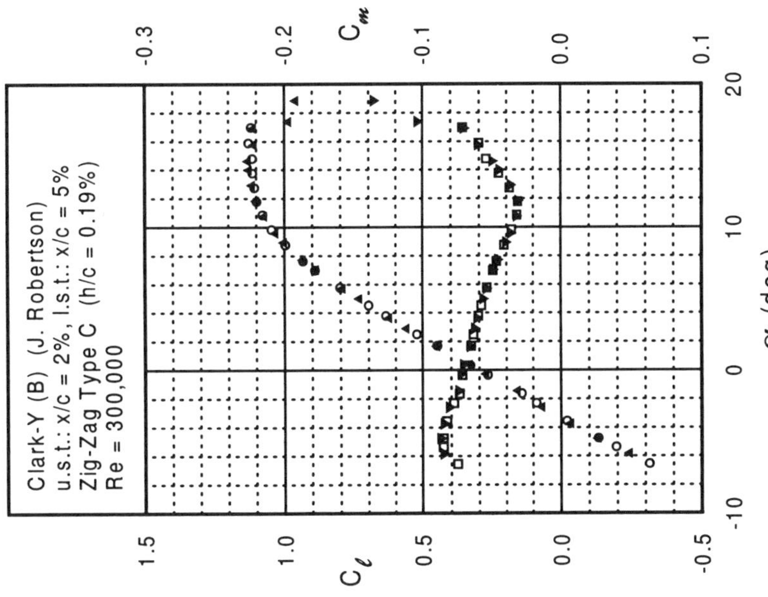
Clark-Y (B) (J. Robertson)
u.s.t.: x/c = 2%, l.s.t.: x/c = 5%
Zig-Zag Type C (h/c = 0.19%)
Re = 300,000
C_ℓ
C_m
α (deg)
1.5
1.0
0.5
0.0
-0.5
-0.3
-0.2
-0.1
0.0
0.1
-10
0
10
20

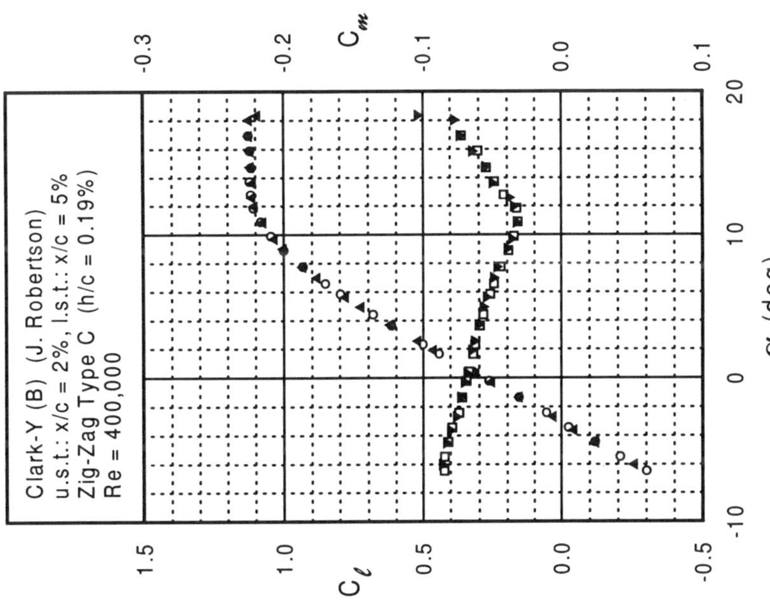
Clark-Y (B) (J. Robertson)
u.s.t.: x/c = 2%, l.s.t.: x/c = 5%
Zig-Zag Type C (h/c = 0.19%)
Re = 400,000
C_ℓ
C_m
α (deg)
1.5
1.0
0.5
0.0
-0.5
-0.3
-0.2
-0.1
0.0
0.1
-10
0
10
20

DH4009

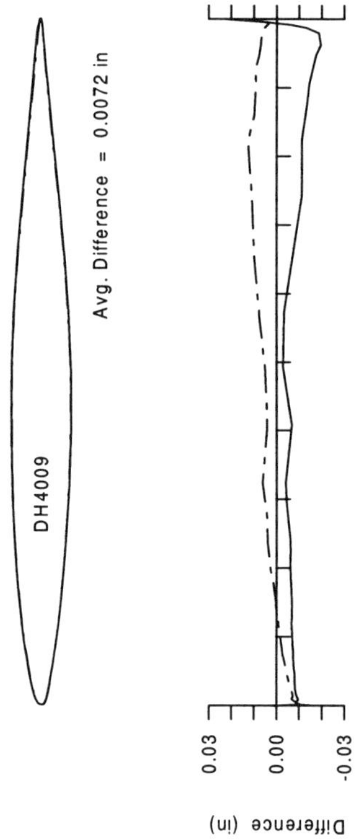
DH4009
Avg. Difference = 0.0072 in
0.03
0.00
-0.03
Difference (in)

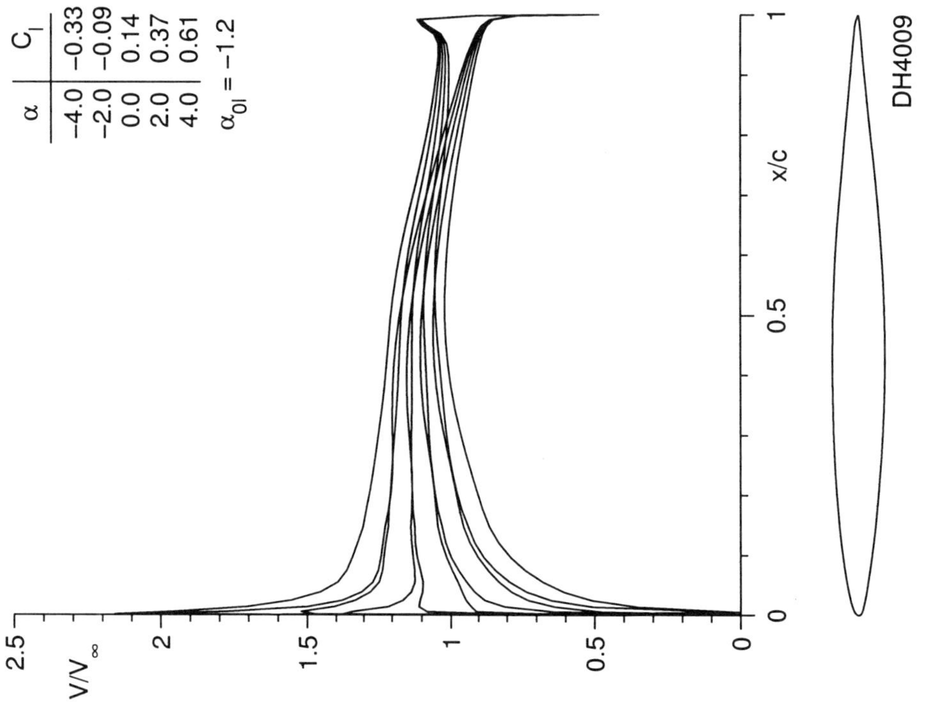
α | C_l
-4.0 | -0.33
-2.0 | -0.09
0.0 | 0.14
2.0 | 0.37
4.0 | 0.61
$\alpha_{0l} = -1.2$
V/V_∞
2.5
2
1.5
1
0.5
0
x/c
0.5
1
DH4009

DH4009

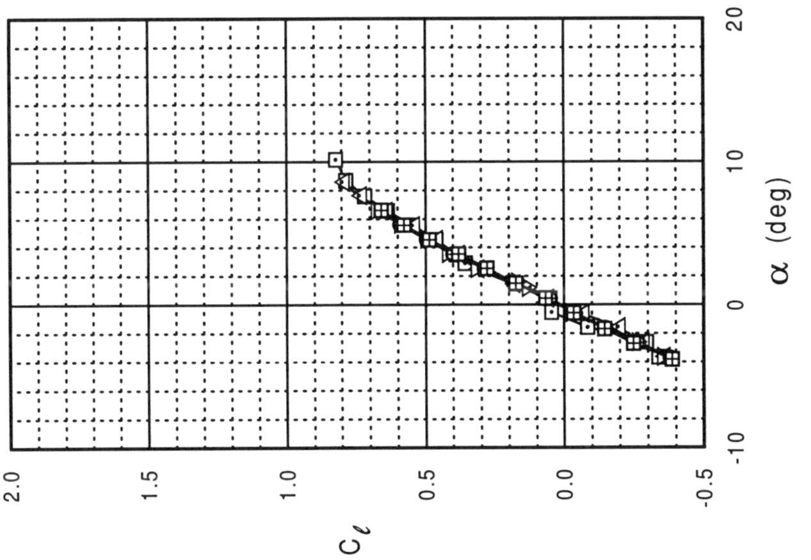

2.0
1.5
1.0
0.5
0.0
-0.5
C_ℓ
-10
0
10
20
α (deg)

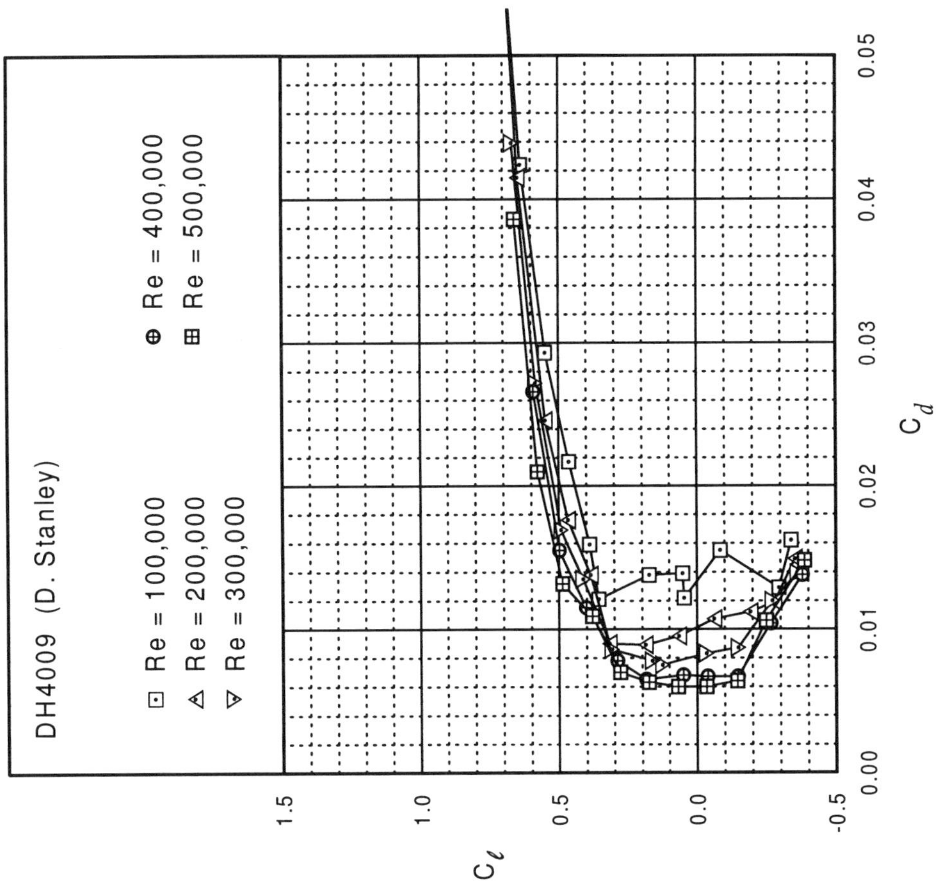

DH4009 (D. Stanley)
Re = 100,000
Re = 200,000
Re = 300,000
Re = 400,000
Re = 500,000
1.5
1.0
0.5
0.0
-0.5
C_ℓ
0.00
0.01
0.02
0.03
0.04
0.05
C_d

DH4009

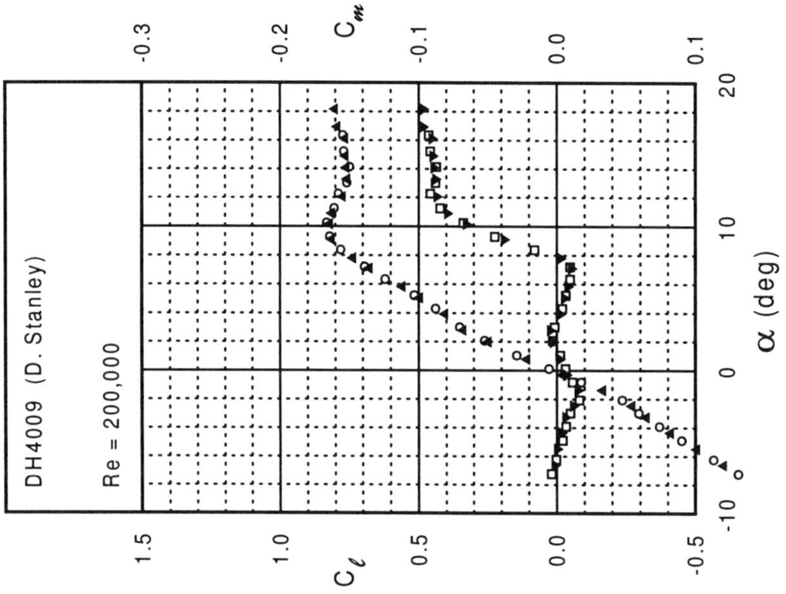
DH4009 (D. Stanley)
Re = 200,000
α (deg)
C_l
C_m

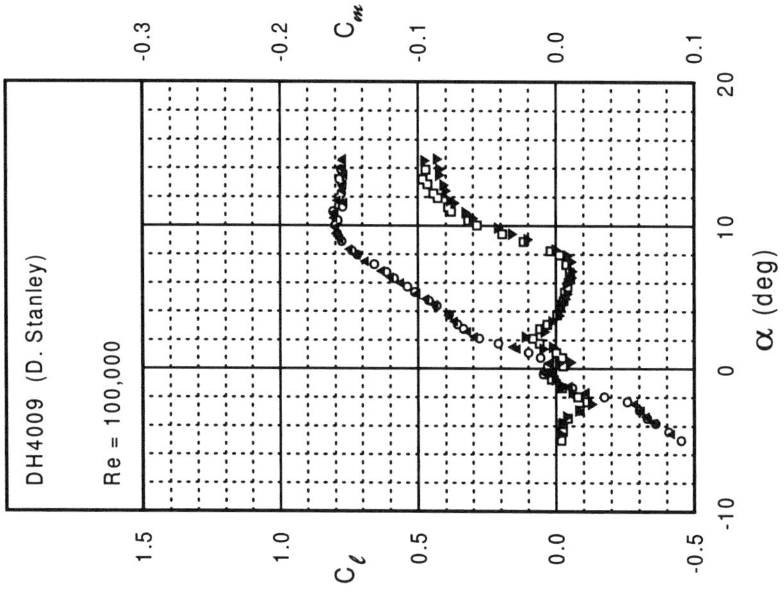
DH4009 (D. Stanley)
Re = 100,000
α (deg)
C_l
C_m

DH4009

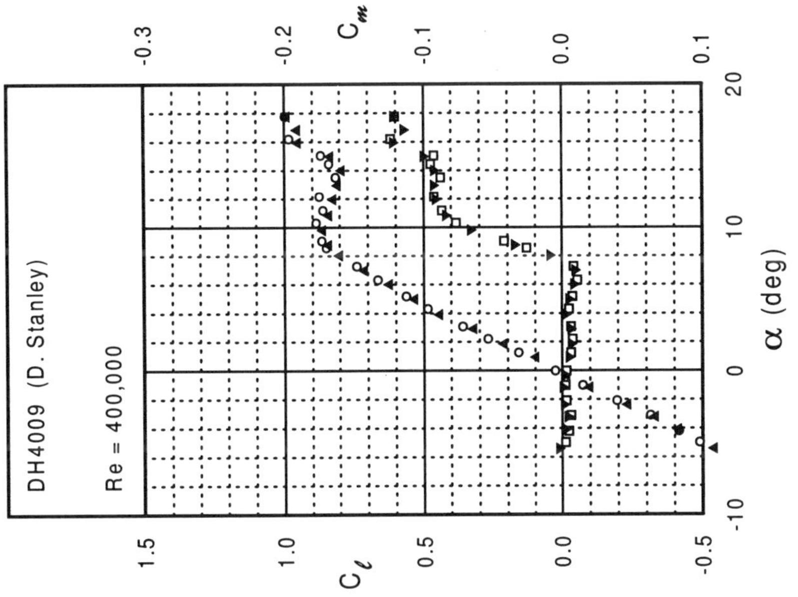
DH4009 (D. Stanley)
Re = 400,000
Cℓ
Cm
α (deg)
1.5
1.0
0.5
0.0
-0.5
-0.3
-0.2
-0.1
0.0
0.1
-10
0
10
20

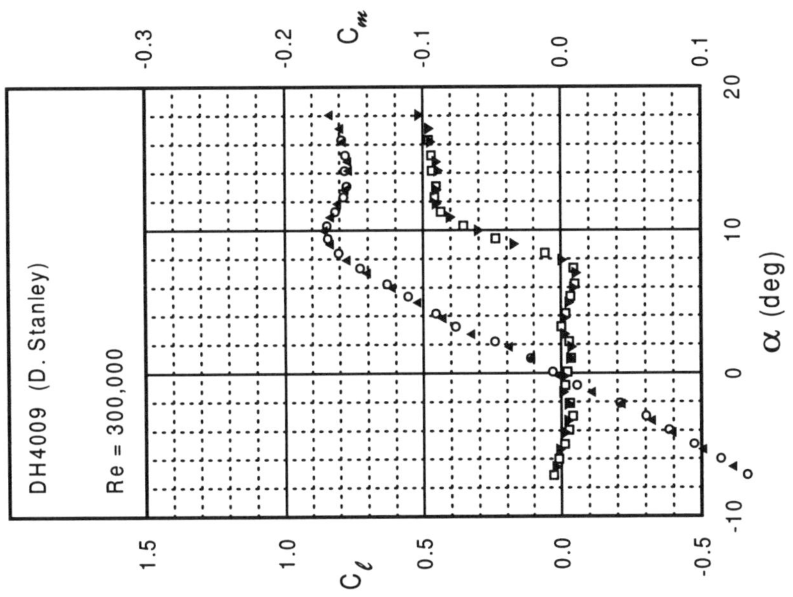
DH4009 (D. Stanley)
Re = 300,000
Cℓ
Cm
α (deg)
1.5
1.0
0.5
0.0
-0.5
-0.3
-0.2
-0.1
0.0
0.1
-10
0
10
20

DH4009

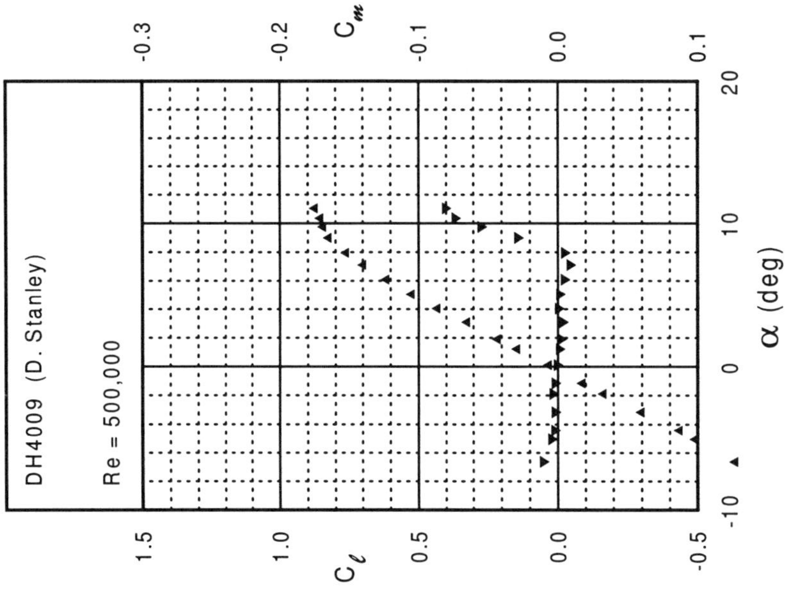
DH4009 (D. Stanley)
Re = 500,000
-0.3
-0.2
-0.1
0.0
0.1
C_m
1.5
1.0
0.5
0.0
-0.5
C_ℓ
-10
0
10
20
α (deg)

E231

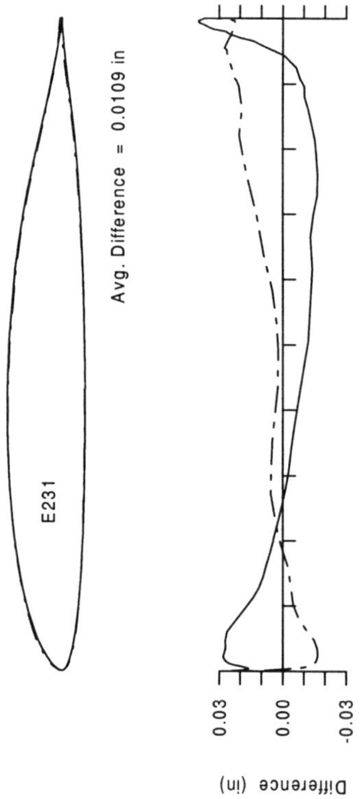
E231
Avg. Difference = 0.0109 in
0.03
0.00
-0.03
Difference (in)

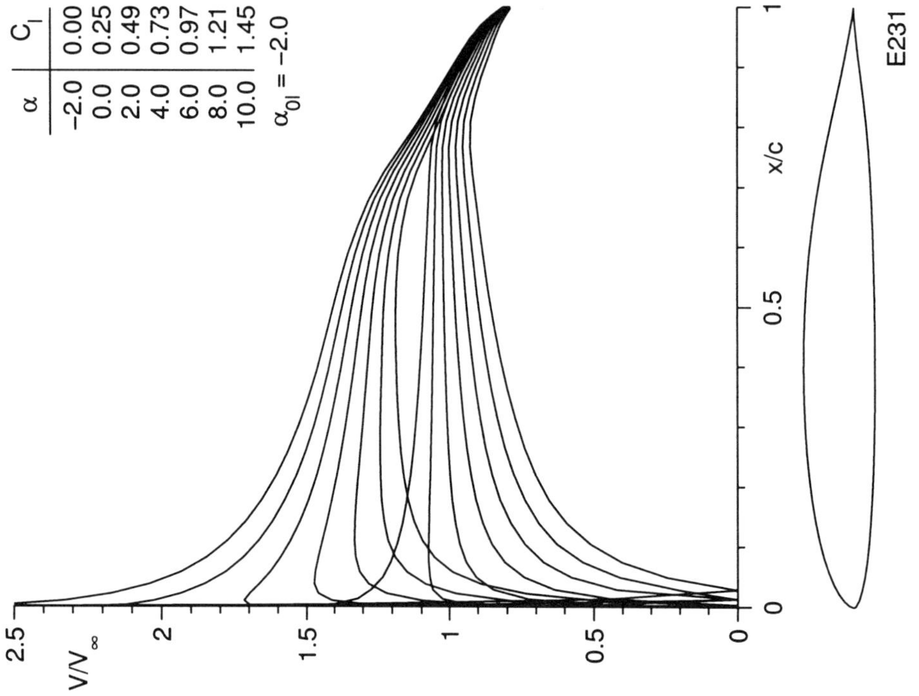
α
−2.0
0.0
2.0
4.0
6.0
8.0
10.0
C_l
0.00
0.25
0.49
0.73
0.97
1.21
1.45
α_{0l} = −2.0
V/V_∞
2.5
2
1.5
1
0.5
0
x/c
0
0.5
1
E231

E231

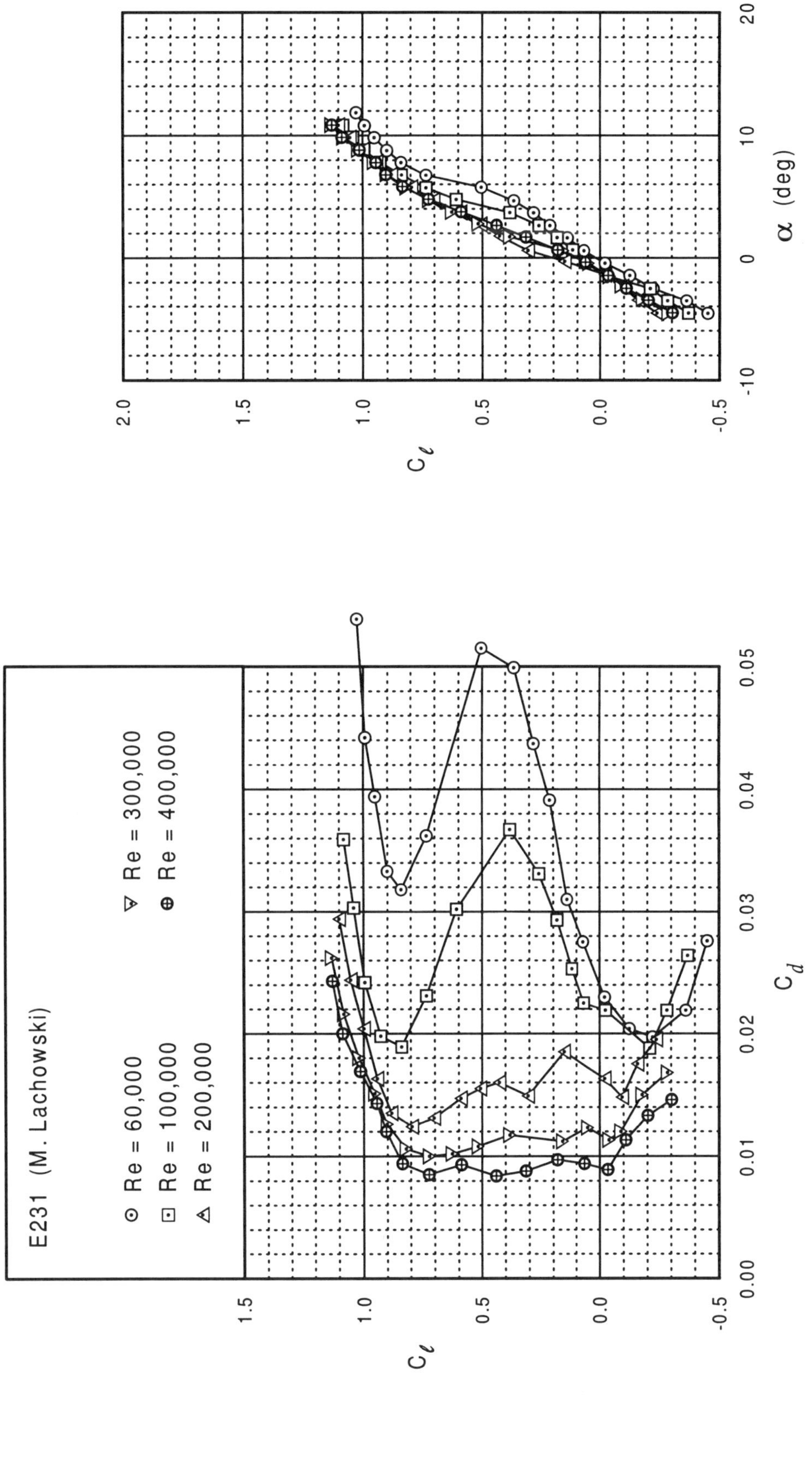

E231 (M. Lachowski)
Re = 60,000
Re = 100,000
Re = 200,000
Re = 300,000
Re = 400,000
C_l
C_d
α (deg)

E231

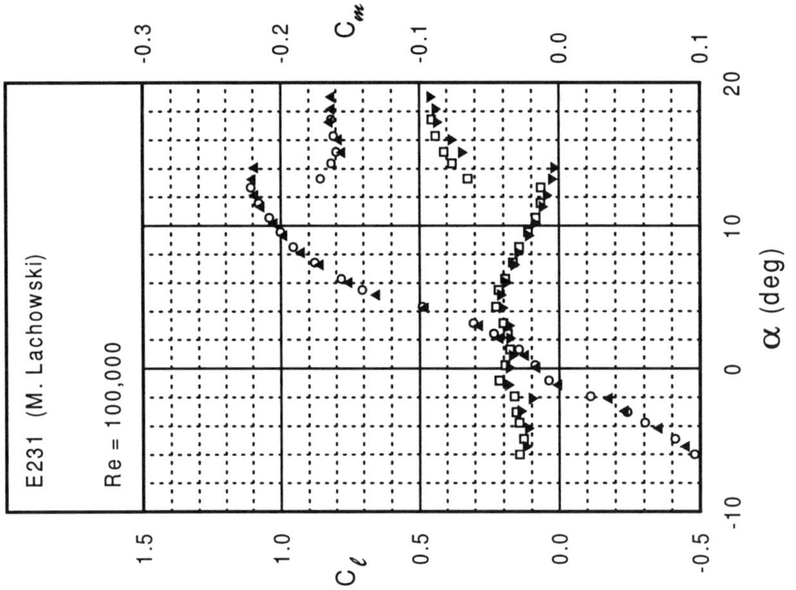
E231 (M. Lachowski)
Re = 100,000
C_m
-0.3
-0.2
-0.1
0.0
0.1
C_ℓ
1.5
1.0
0.5
0.0
-0.5
-10
0
10
20
α (deg)

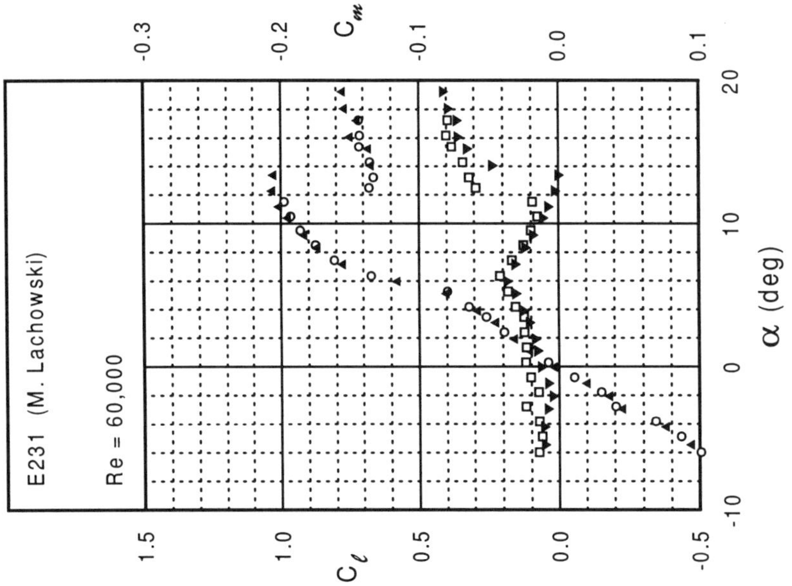
E231 (M. Lachowski)
Re = 60,000
C_m
-0.3
-0.2
-0.1
0.0
0.1
C_ℓ
1.5
1.0
0.5
0.0
-0.5
-10
0
10
20
α (deg)

E231

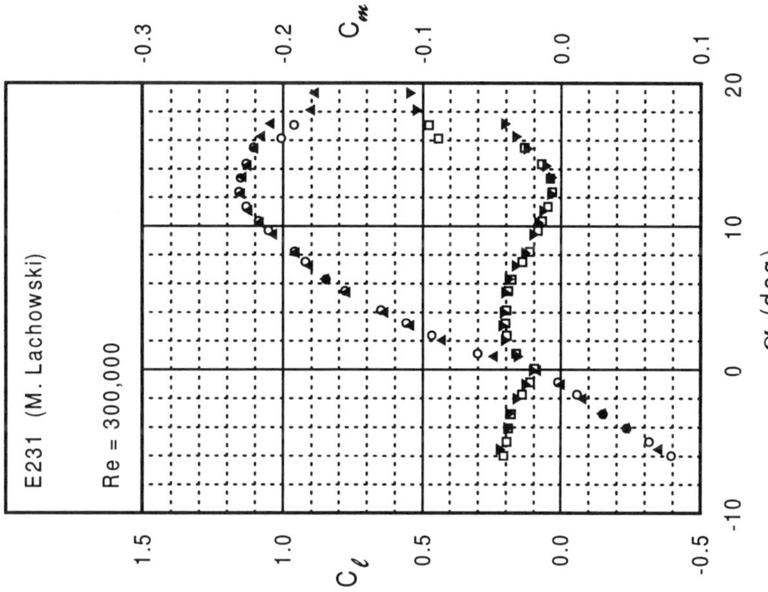
E231 (M. Lachowski)
Re = 300,000
C_ℓ
C_m
α (deg)

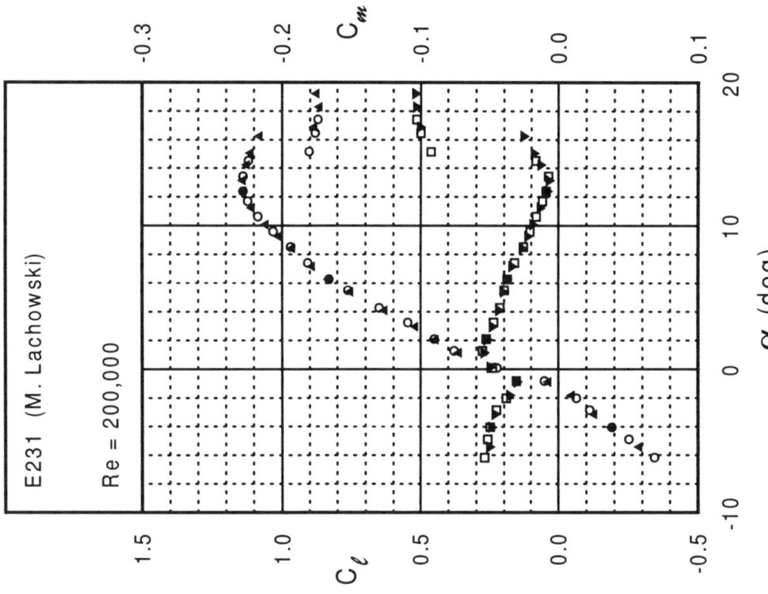
E231 (M. Lachowski)
Re = 200,000
C_ℓ
C_m
α (deg)

E231

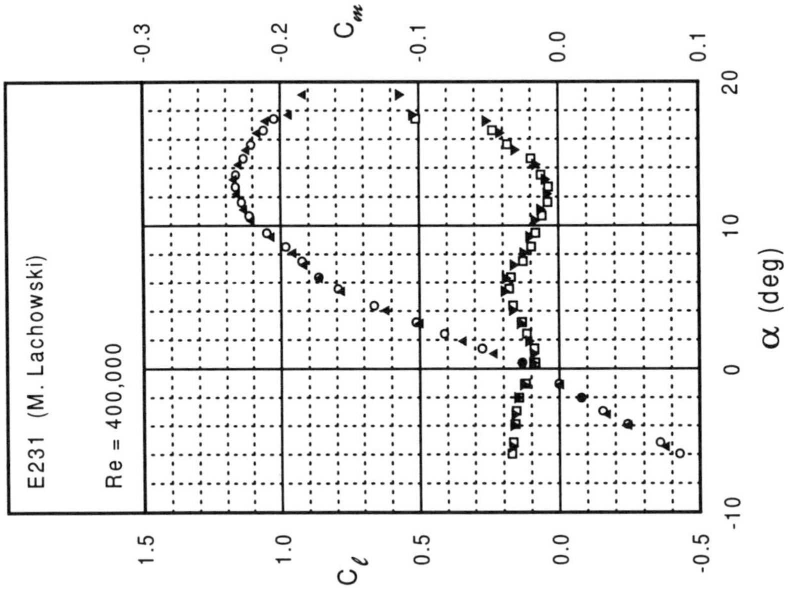

E374 (B)

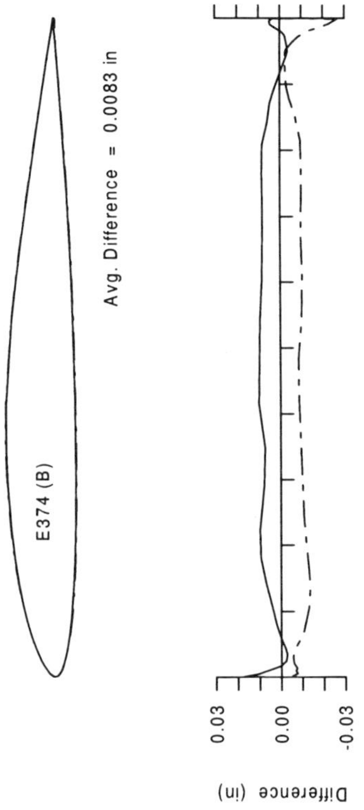
E374 (B)
Avg. Difference = 0.0083 in
0.03
0.00
-0.03
Difference (in)

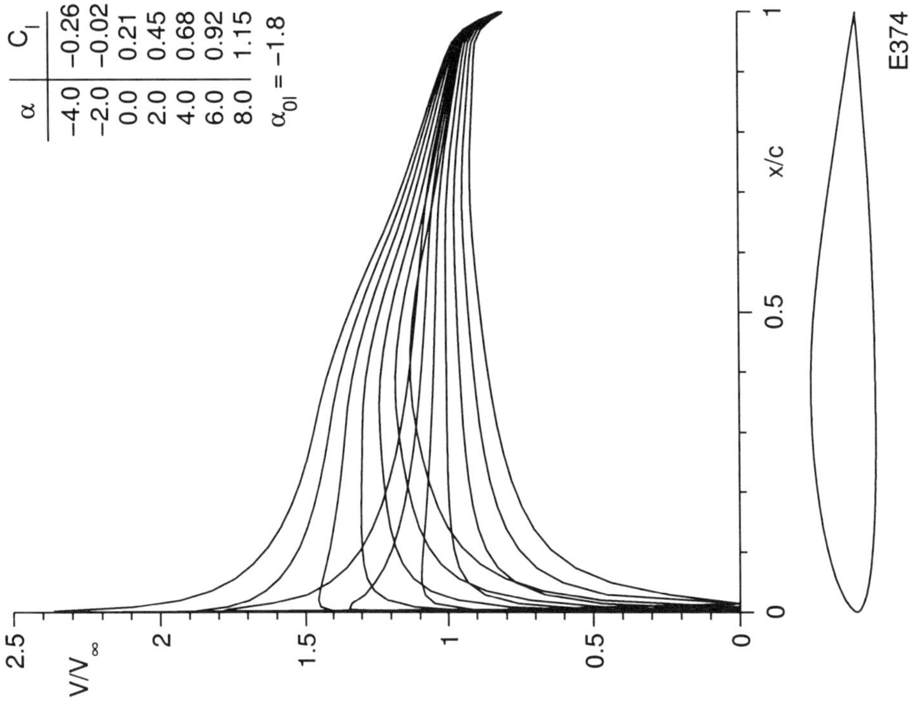
α | C_l
-4.0 | -0.26
-2.0 | -0.02
0.0 | 0.21
2.0 | 0.45
4.0 | 0.68
6.0 | 0.92
8.0 | 1.15
α_{0l} = -1.8
V/V_∞
2.5
2
1.5
1
0.5
0
x/c
0.5
1
E374

E374 (B)

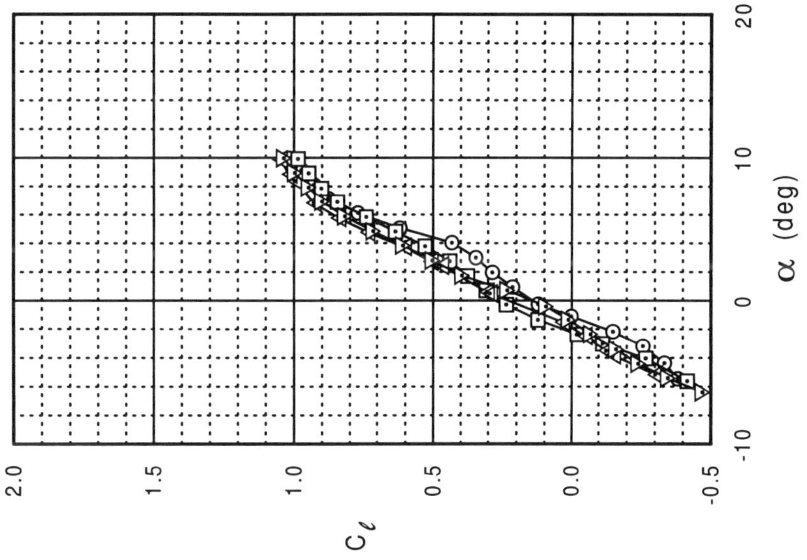
C_l
α (deg)

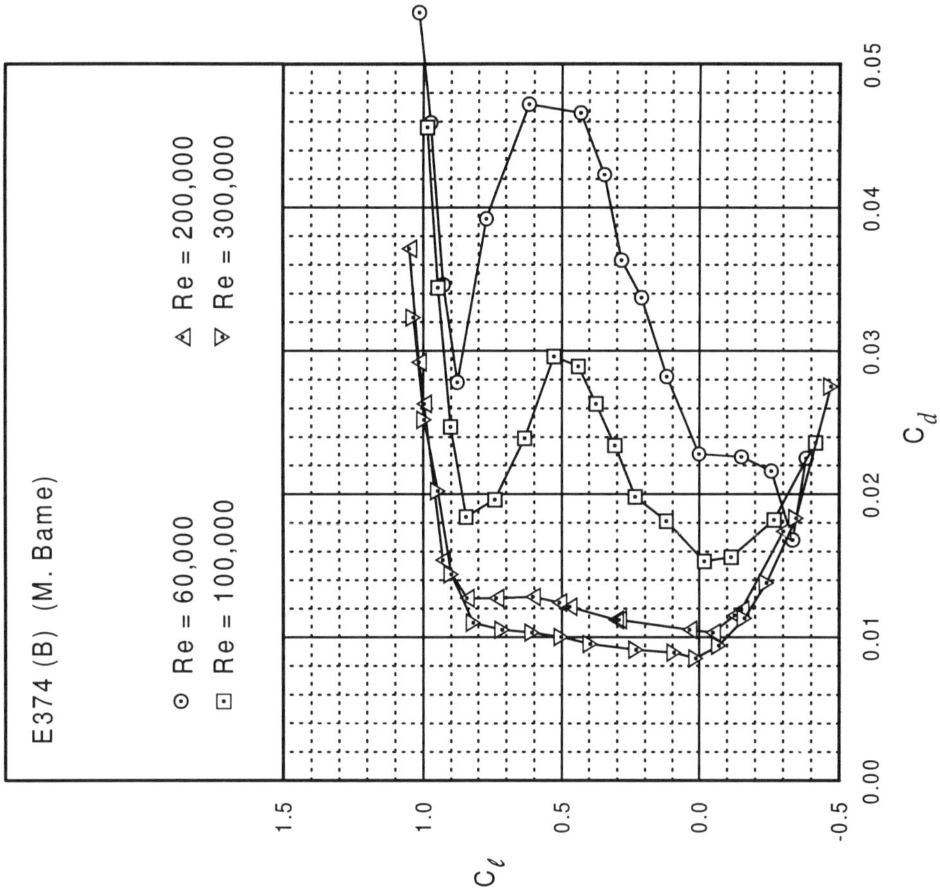
E374 (B) (M. Bame)
Re = 60,000
Re = 100,000
Re = 200,000
Re = 300,000
C_l
C_d

E374 (B)

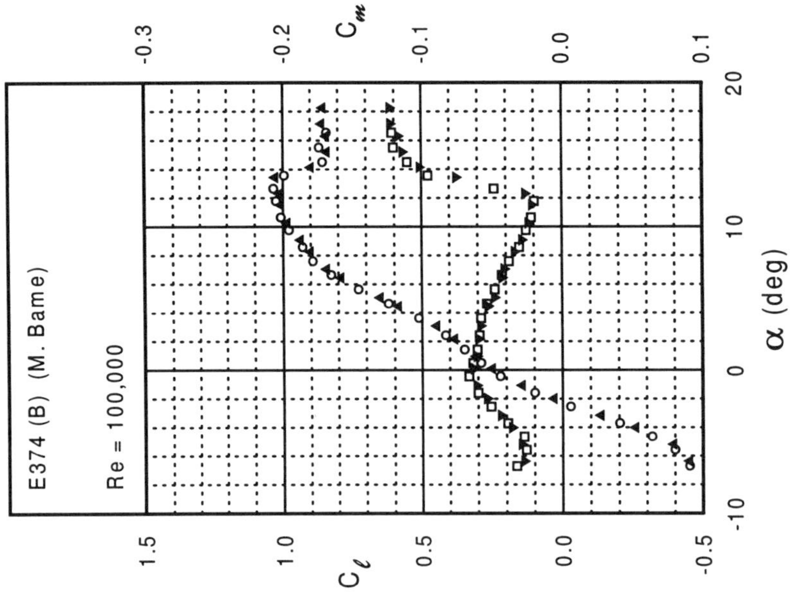
E374 (B) (M. Bame)
Re = 100,000
C_m
-0.3
-0.2
-0.1
0.0
0.1
C_ℓ
1.5
1.0
0.5
0.0
-0.5
-10
0
10
20
α (deg)

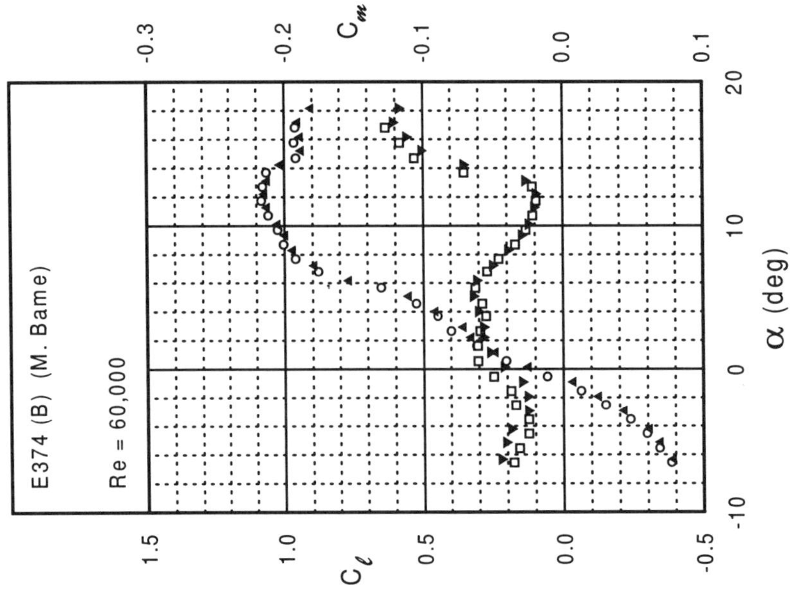
E374 (B) (M. Bame)
Re = 60,000
C_m
-0.3
-0.2
-0.1
0.0
0.1
C_ℓ
1.5
1.0
0.5
0.0
-0.5
-10
0
10
20
α (deg)

E374 (B)

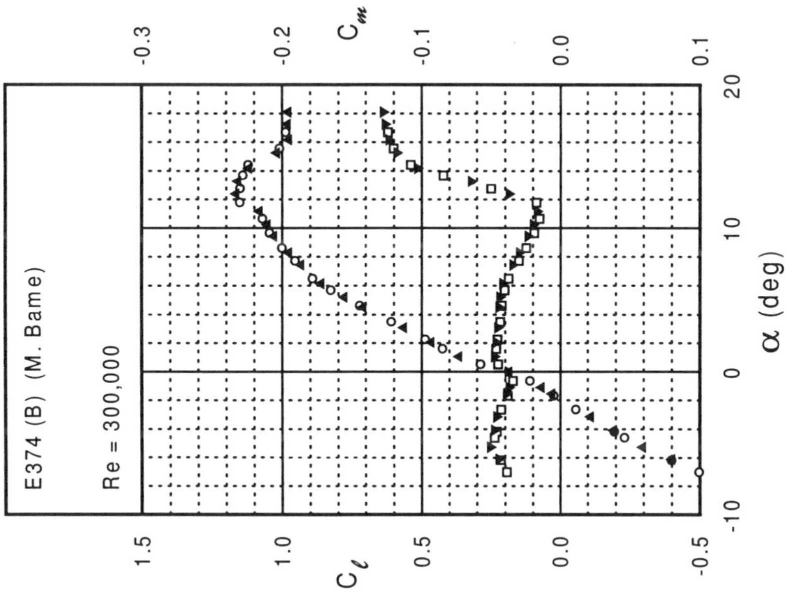
E374 (B) (M. Bame)
Re = 300,000
C_l
C_m
α (deg)
-0.3
-0.2
-0.1
0.0
0.1
1.5
1.0
0.5
0.0
-0.5
-10
0
10
20

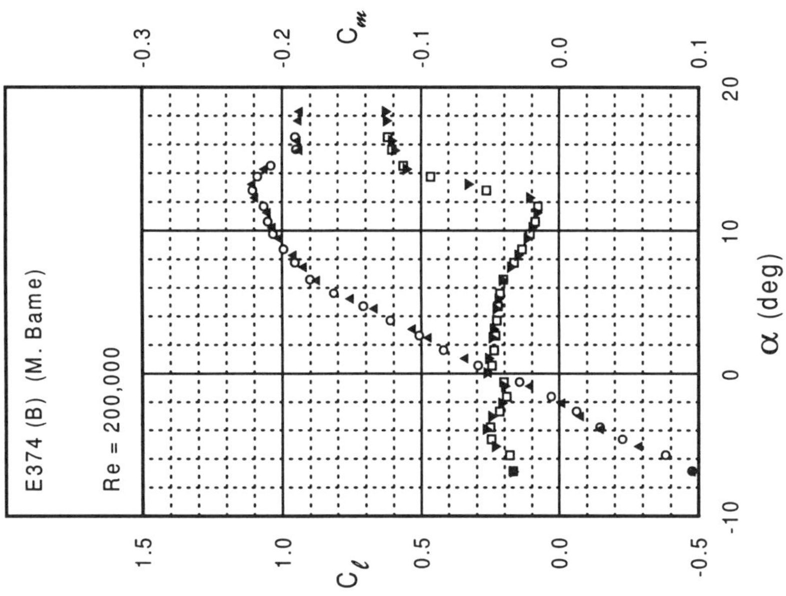
E374 (B) (M. Bame)
Re = 200,000
C_l
C_m
α (deg)
-0.3
-0.2
-0.1
0.0
0.1
1.5
1.0
0.5
0.0
-0.5
-10
0
10
20

E387 (C)

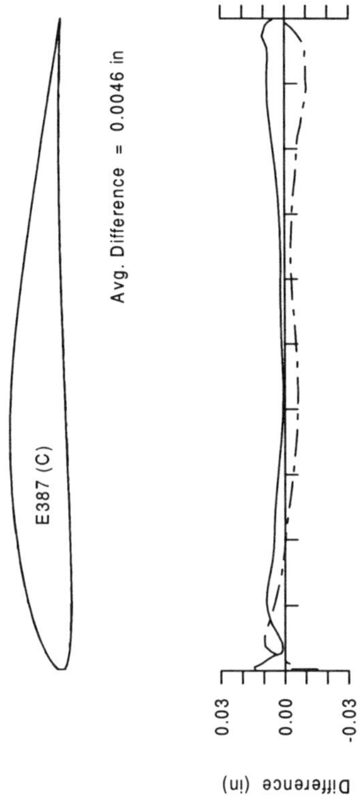
E387 (C)
Avg. Difference = 0.0046 in
0.03
0.00
-0.03
Difference (in)

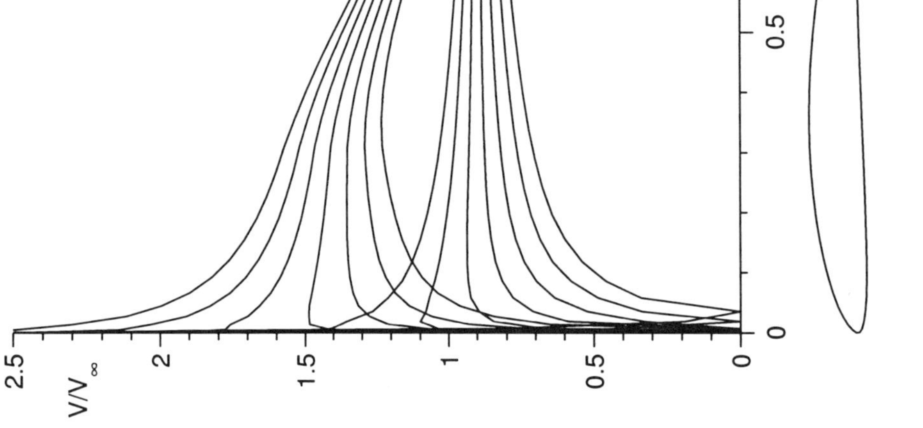
α | C_l
−2.0 | 0.19
0.0 | 0.42
2.0 | 0.65
4.0 | 0.88
6.0 | 1.11
8.0 | 1.34
10.0 | 1.57
$\alpha_{0l} = -3.7$
V/V_∞
2.5
2
1.5
1
0.5
0
x/c
0.5
1
E387

E387 (C)

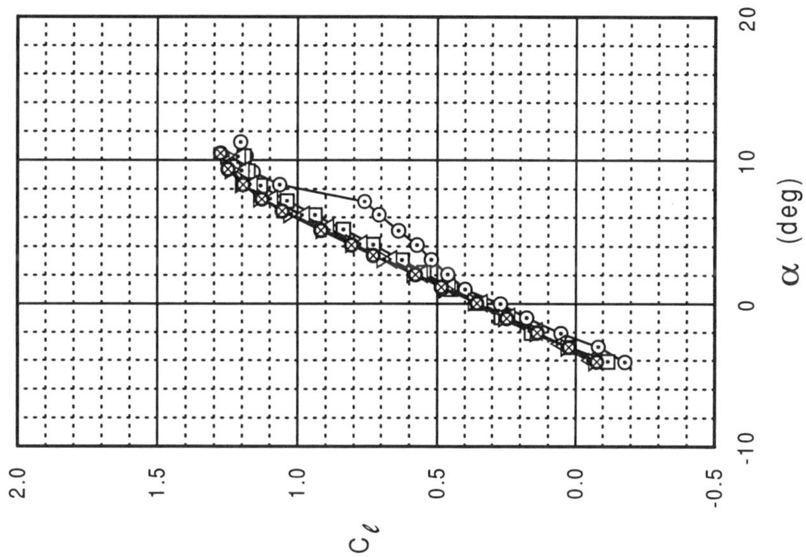
2.0
1.5
1.0
0.5
0.0
-0.5
C_ℓ
-10
0
10
20
α (deg)

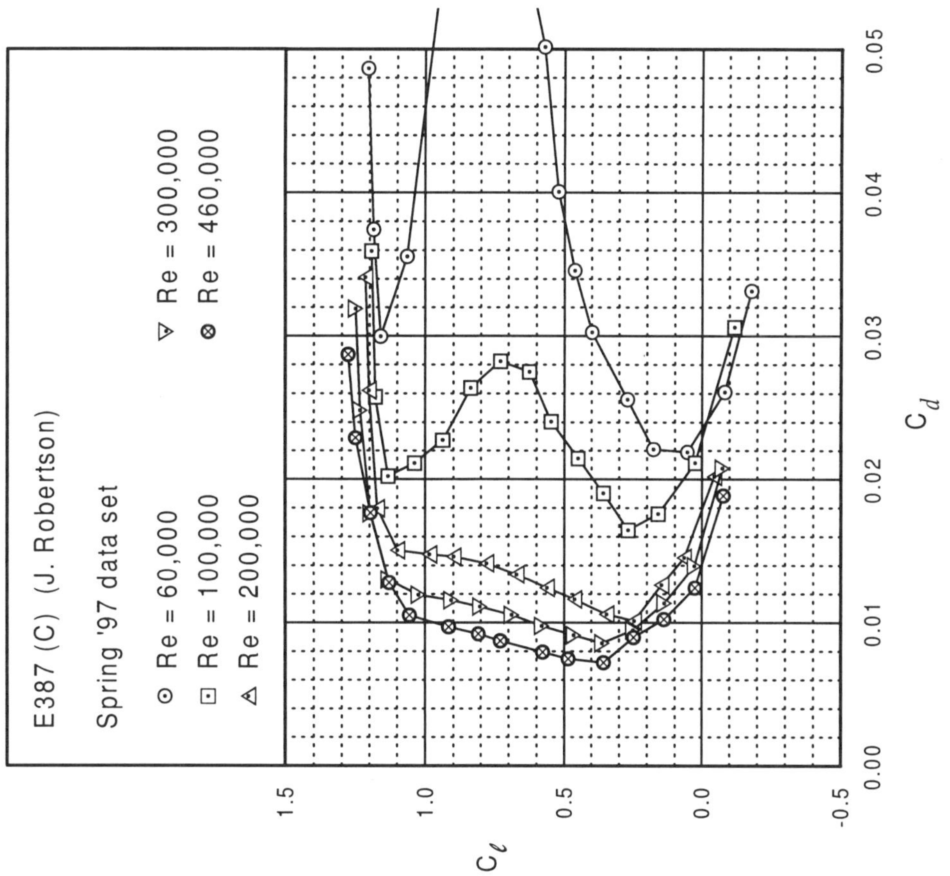
E387 (C) (J. Robertson)
Spring '97 data set
Re = 60,000
Re = 100,000
Re = 200,000
Re = 300,000
Re = 460,000
1.5
1.0
0.5
0.0
-0.5
C_ℓ
0.00
0.01
0.02
0.03
0.04
0.05
C_d

E387 (C)

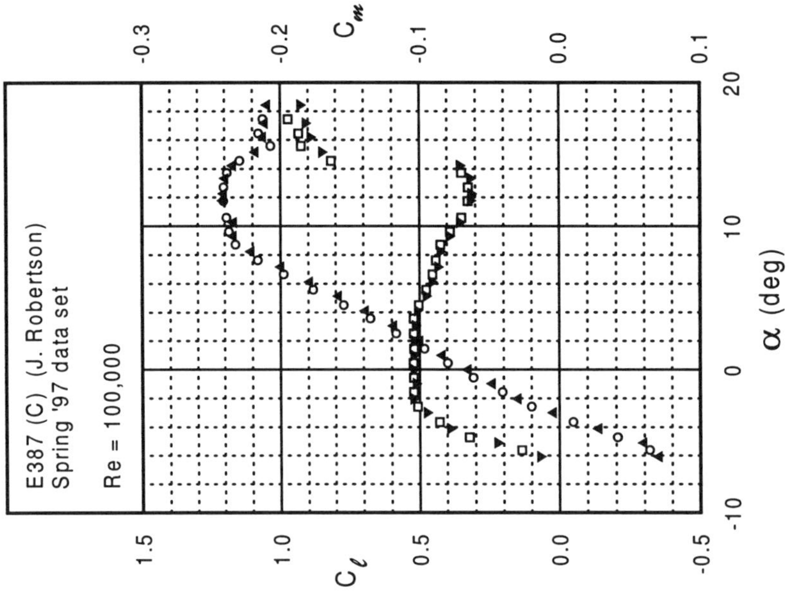
E387 (C) (J. Robertson)
Spring '97 data set
Re = 100,000
C_l
C_m
α (deg)
-10
0
10
20
1.5
1.0
0.5
0.0
-0.5
-0.3
-0.2
-0.1
0.0
0.1

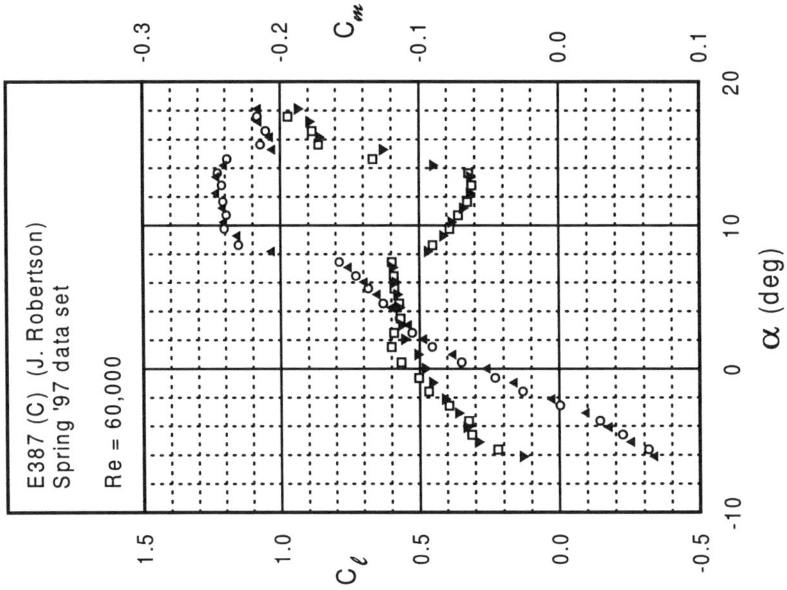
E387 (C) (J. Robertson)
Spring '97 data set
Re = 60,000
C_l
C_m
α (deg)
-10
0
10
20
1.5
1.0
0.5
0.0
-0.5
-0.3
-0.2
-0.1
0.0
0.1

E387 (C)

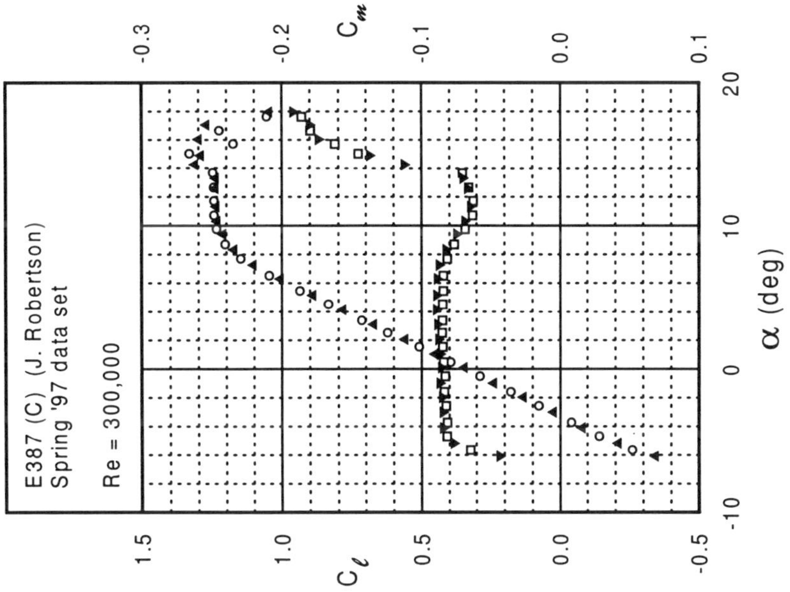
E387 (C) (J. Robertson)
Spring '97 data set
Re = 300,000
C_ℓ
C_m
α (deg)
-10
0
10
20
1.5
1.0
0.5
0.0
-0.5
-0.3
-0.2
-0.1
0.0
0.1

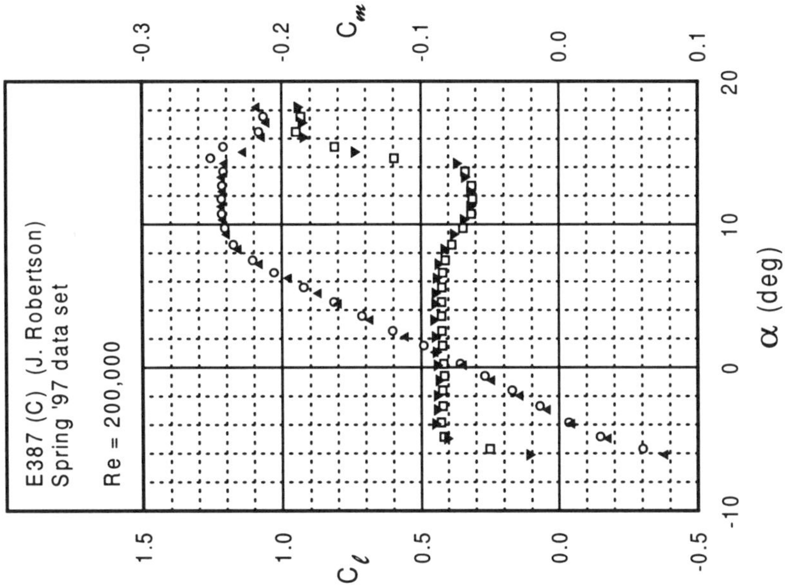
E387 (C) (J. Robertson)
Spring '97 data set
Re = 200,000
C_ℓ
C_m
α (deg)
-10
0
10
20
1.5
1.0
0.5
0.0
-0.5
-0.3
-0.2
-0.1
0.0
0.1

E387 (C)

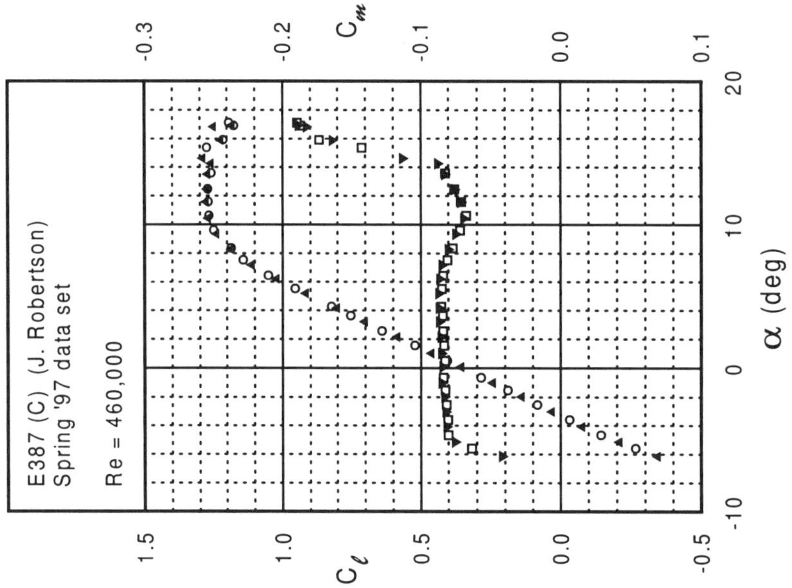
E387 (C) (J. Robertson)
Spring '97 data set
Re = 460,000
C_ℓ
1.5
1.0
0.5
0.0
-0.5
C_m
-0.3
-0.2
-0.1
0.0
0.1
-10
0
10
20
α (deg)

E387 (D)

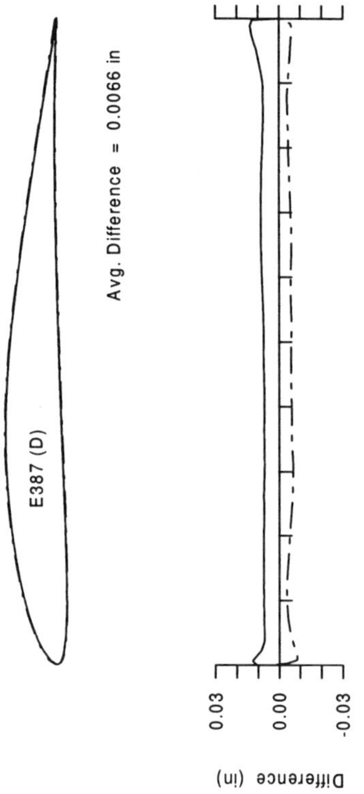
E387 (D)
Avg. Difference = 0.0066 in
0.03
0.00
-0.03
Difference (in)

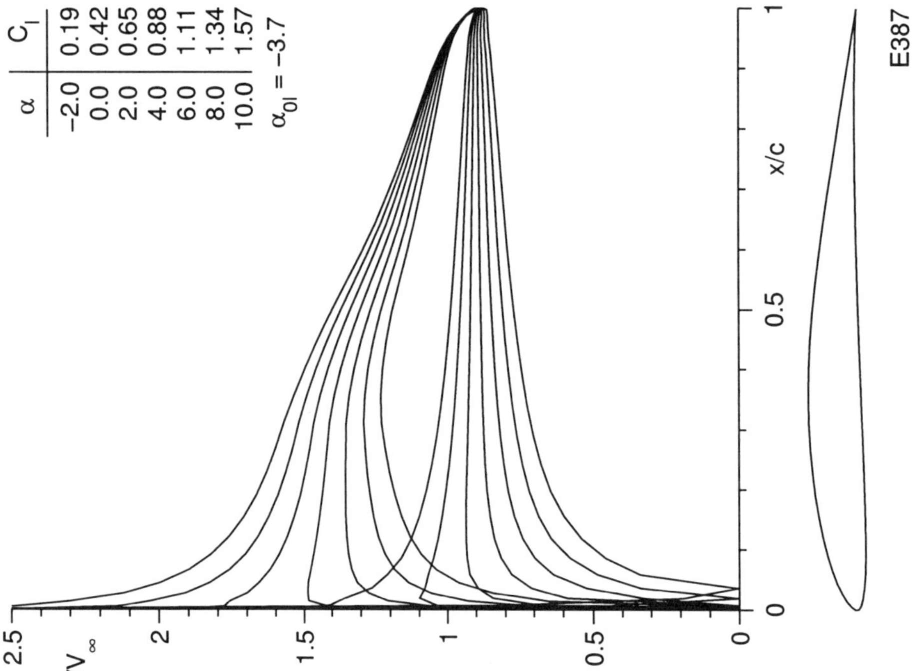
α | C_l
−2.0 | 0.19
0.0 | 0.42
2.0 | 0.65
4.0 | 0.88
6.0 | 1.11
8.0 | 1.34
10.0 | 1.57
α_{0l} = −3.7
V/V_∞
2.5
2
1.5
1
0.5
0
0.5
1
x/c
E387

E387 (D)

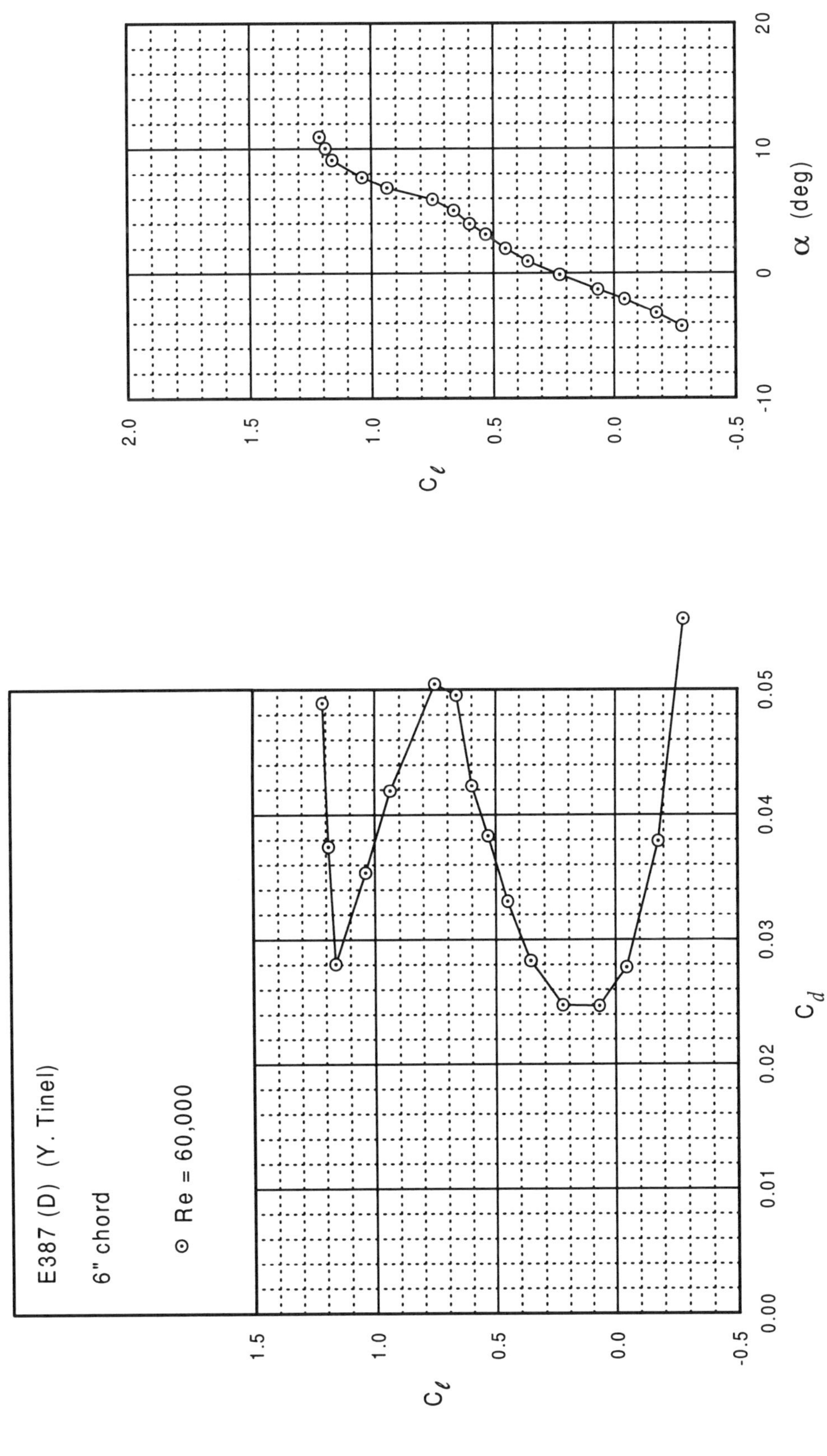
E387 (D) (Y. Tinel)
6" chord
⊙ Re = 60,000
C_ℓ
C_d
α (deg)
2.0
1.5
1.0
0.5
0.0
-0.5
0.00
0.01
0.02
0.03
0.04
0.05
-10
0
10
20

E387 (D)

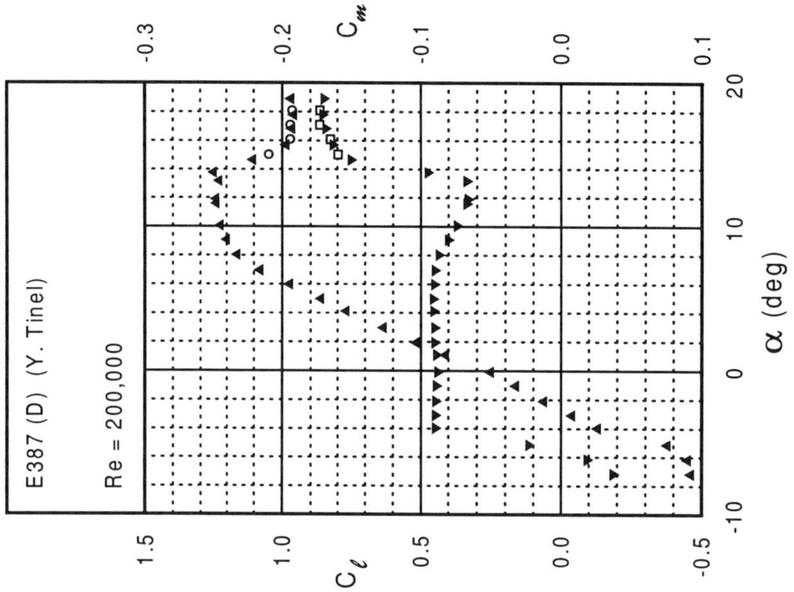
E387 (D) (Y. Tinel)
Re = 200,000
C_ℓ
C_m
α (deg)

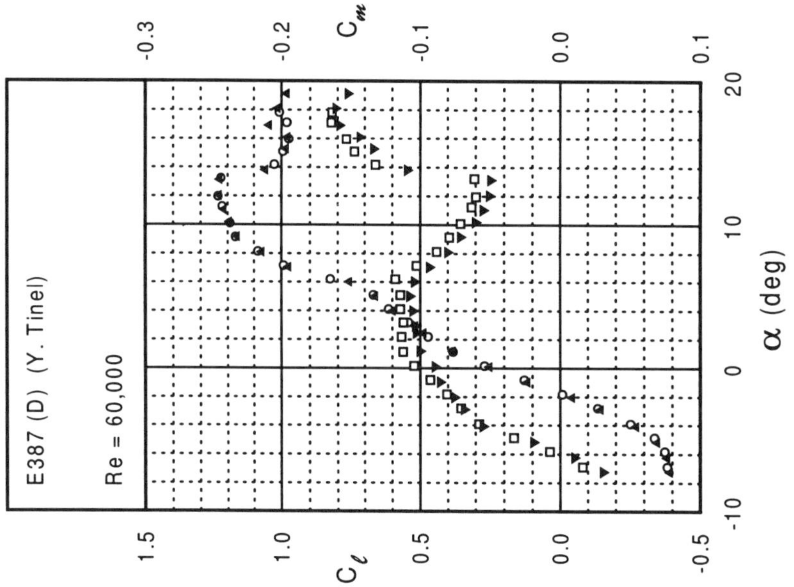
E387 (D) (Y. Tinel)
Re = 60,000
C_ℓ
C_m
α (deg)

E472

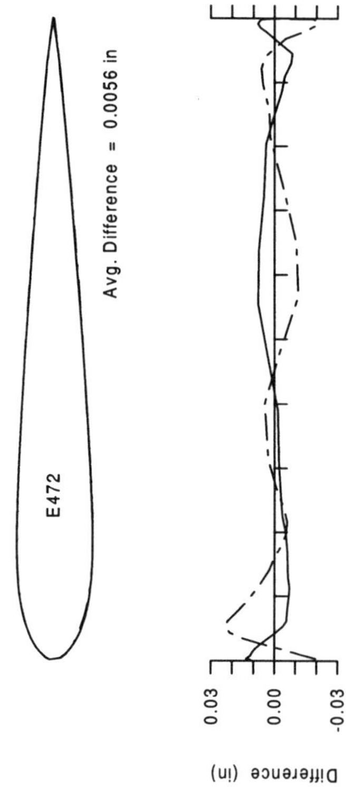
E472
Avg. Difference = 0.0056 in
0.03
0.00
-0.03
Difference (in)

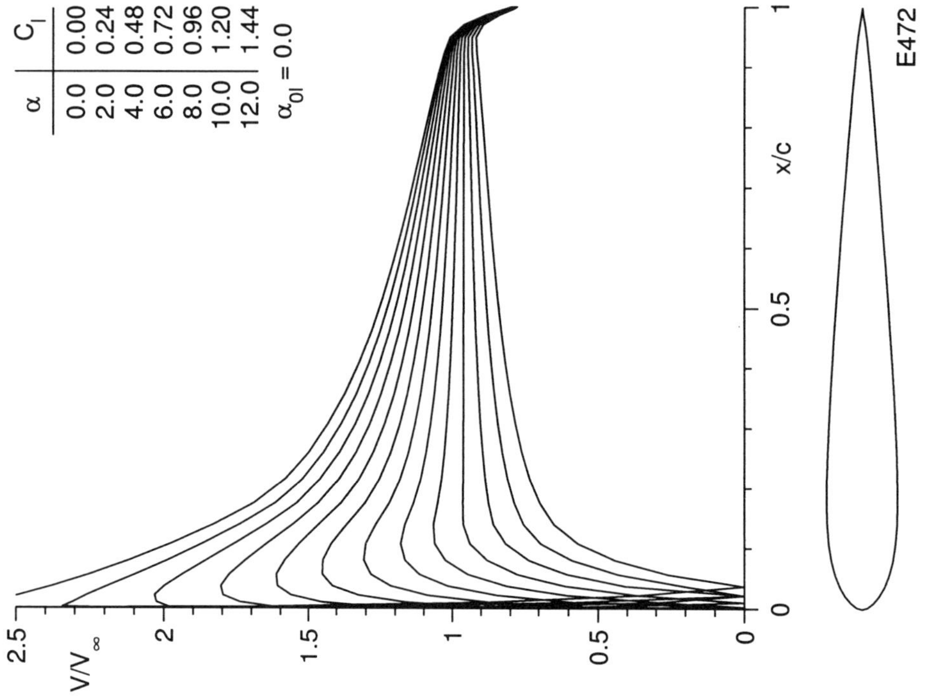
α | C_l
0.0 | 0.00
2.0 | 0.24
4.0 | 0.48
6.0 | 0.72
8.0 | 0.96
10.0 | 1.20
12.0 | 1.44
α_{0l} = 0.0
V/V_∞
2.5
2
1.5
1
0.5
0
x/c
0.5
1
E472

E472

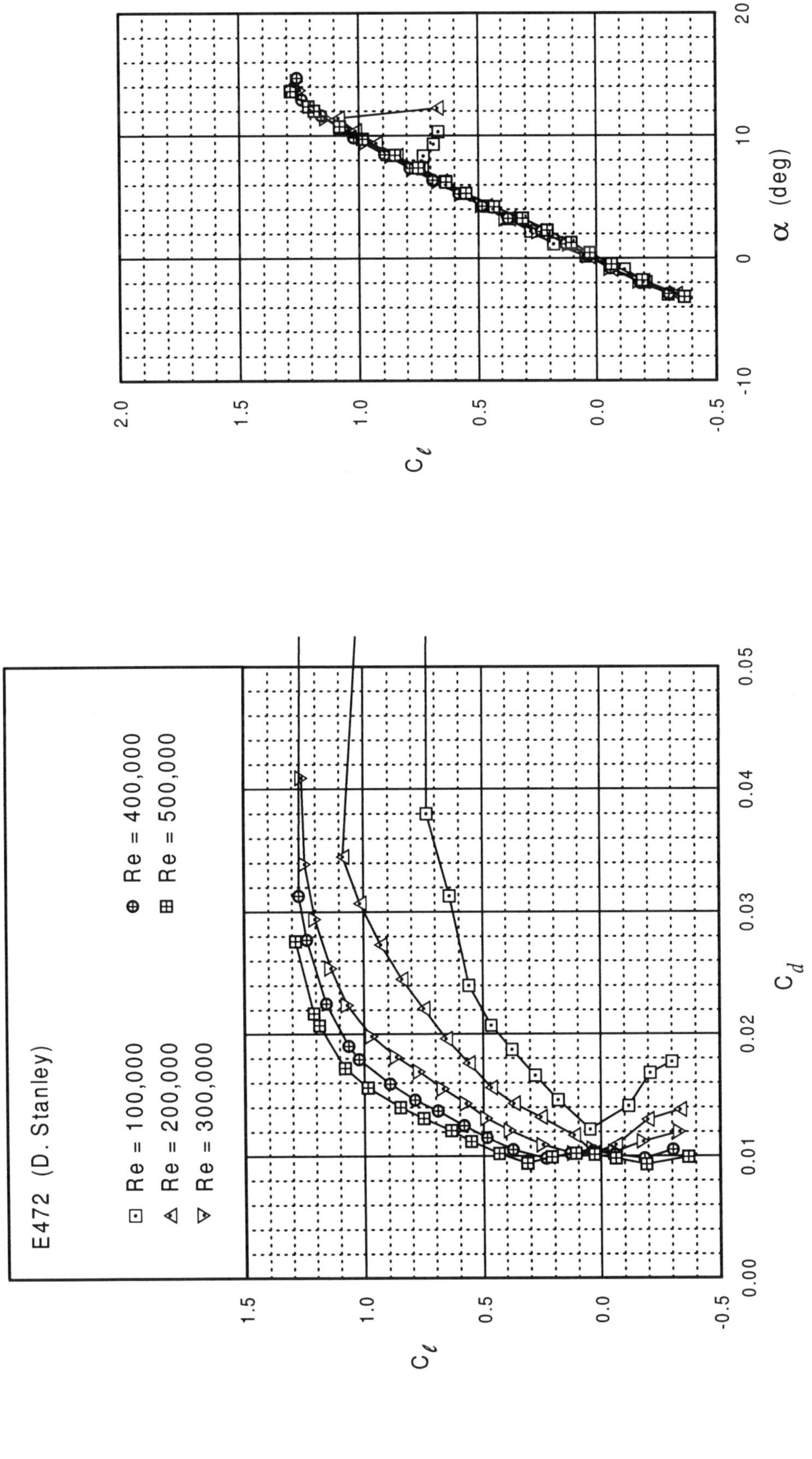
E472 (D. Stanley)
Re = 100,000
Re = 200,000
Re = 300,000
Re = 400,000
Re = 500,000
C_ℓ
C_d
α (deg)

E472

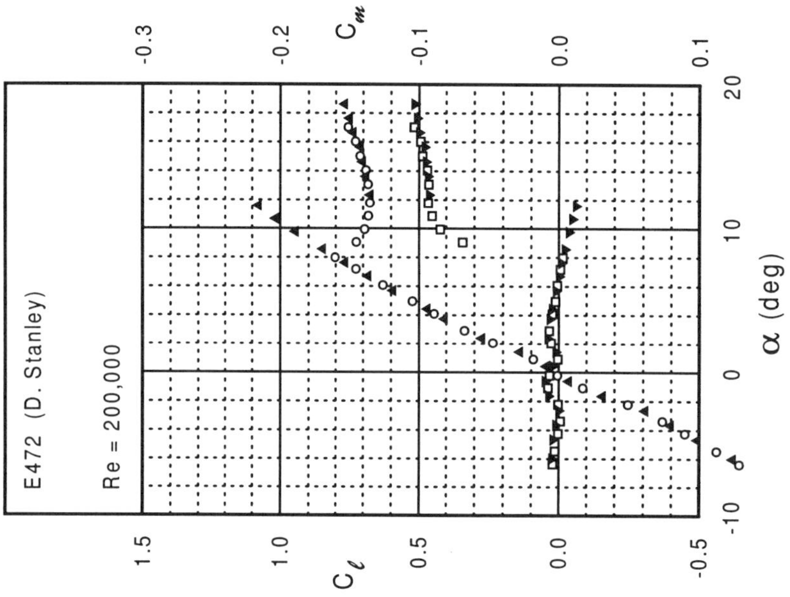
E472 (D. Stanley)
Re = 200,000
C_ℓ
C_m
α (deg)

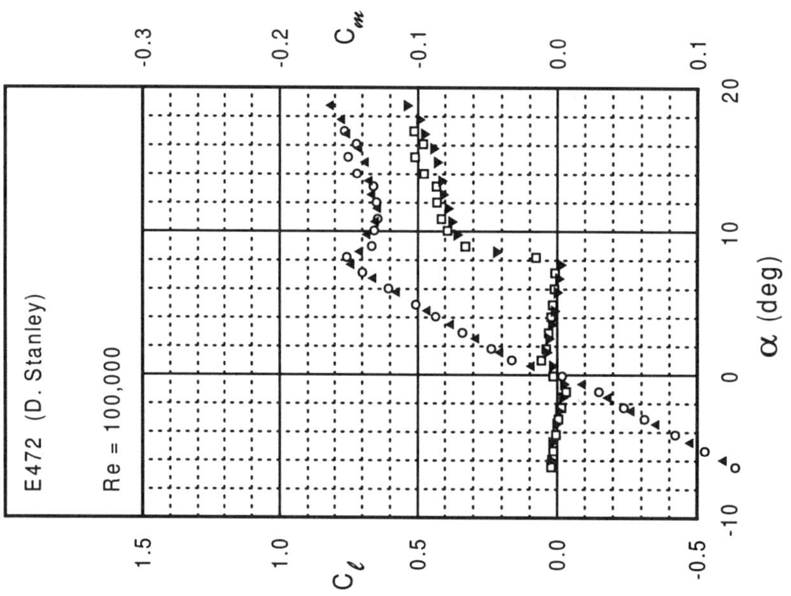
E472 (D. Stanley)
Re = 100,000
C_ℓ
C_m
α (deg)

E472

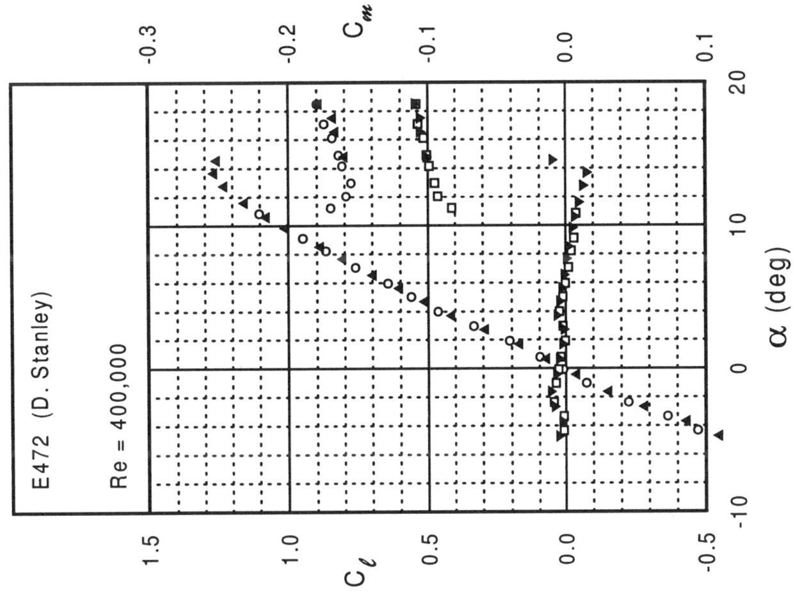
E472 (D. Stanley)
Re = 400,000
C_ℓ
1.5
1.0
0.5
0.0
-0.5
C_m
-0.3
-0.2
-0.1
0.0
0.1
-10
0
10
20
α (deg)

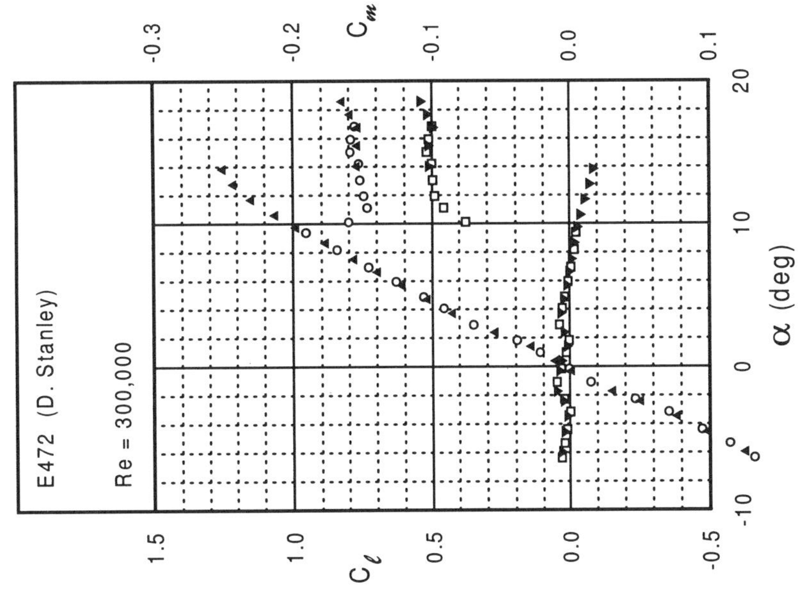
E472 (D. Stanley)
Re = 300,000
C_ℓ
1.5
1.0
0.5
0.0
-0.5
C_m
-0.3
-0.2
-0.1
0.0
0.1
-10
0
10
20
α (deg)

E472

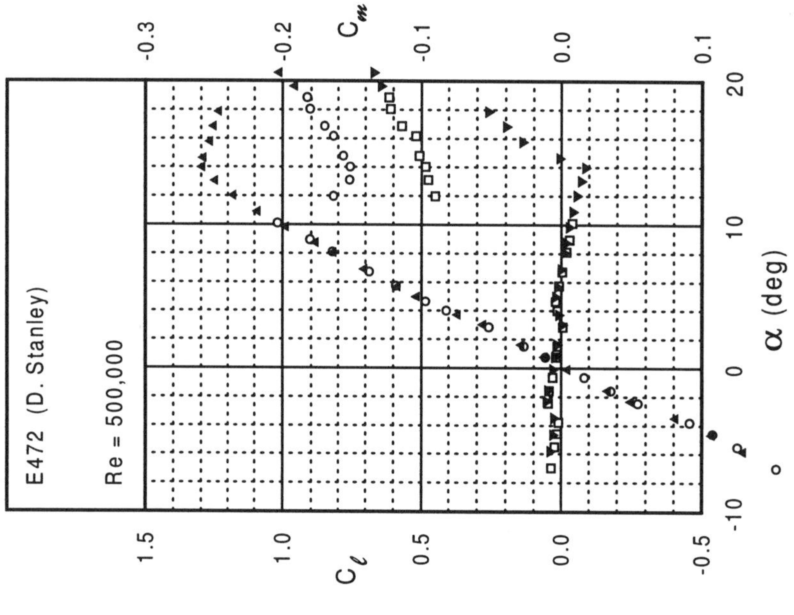
E472 (D. Stanley)
Re = 500,000
C_ℓ
C_m
α (deg)
-10
0
10
20
1.5
1.0
0.5
0.0
-0.5
-0.3
-0.2
-0.1
0.0
0.1

ESA

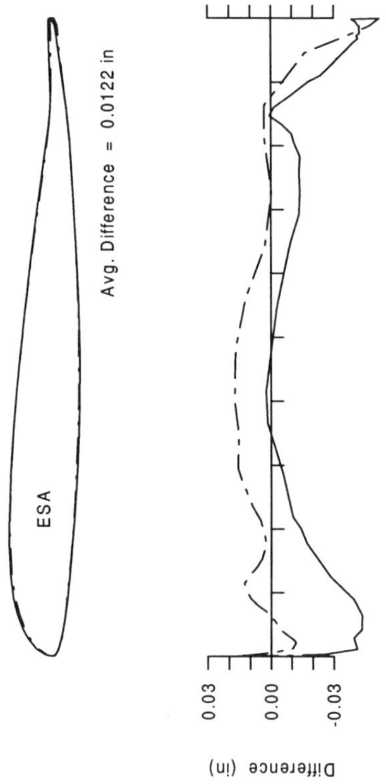
ESA
Avg. Difference = 0.0122 in
0.03
0.00
-0.03
Difference (in)

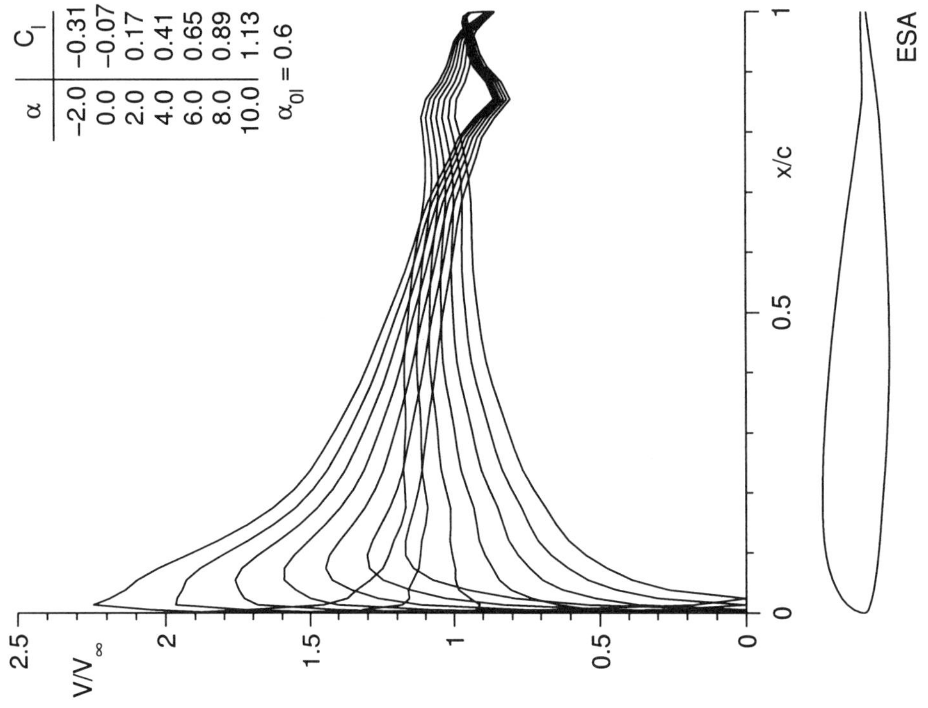
α | C_l
−2.0 | −0.31
0.0 | −0.07
2.0 | 0.17
4.0 | 0.41
6.0 | 0.65
8.0 | 0.89
10.0 | 1.13
$\alpha_{0l} = 0.6$
V/V_∞
2.5
2
1.5
1
0.5
0
0.5
1
x/c
ESA

ESA

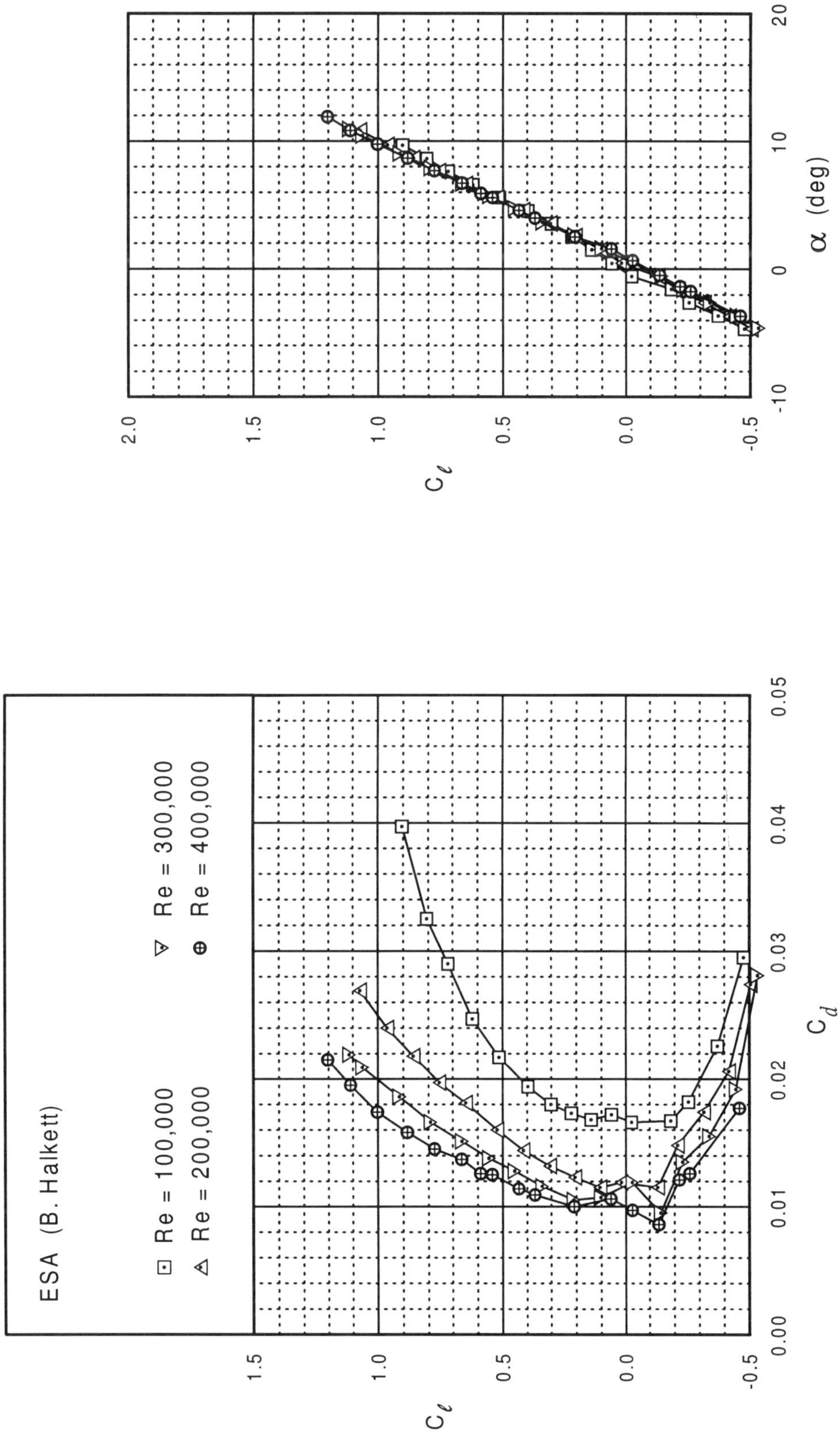
ESA (B. Halkett)
Re = 100,000
Re = 200,000
Re = 300,000
Re = 400,000
C_ℓ
C_d
α (deg)
-0.5
0.0
0.5
1.0
1.5
2.0
0.00
0.01
0.02
0.03
0.04
0.05
-10
0
10
20

ESA

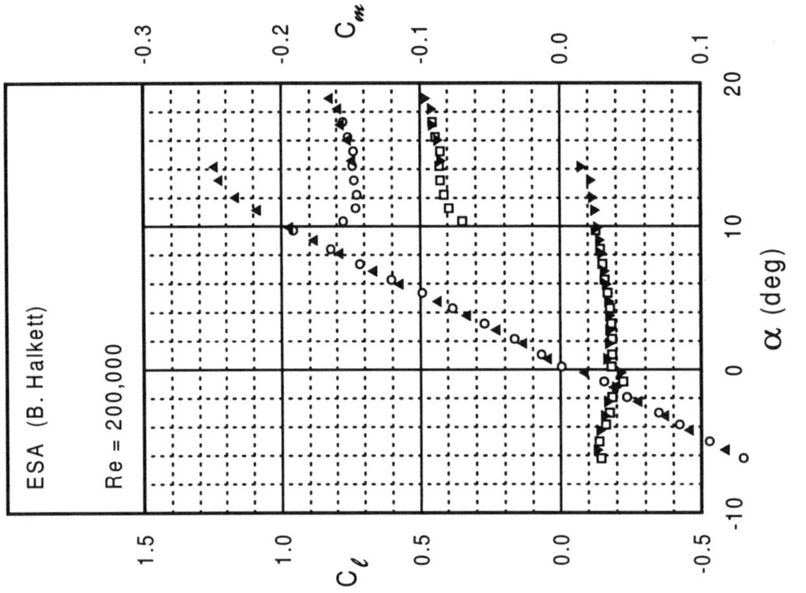
ESA (B. Halkett)
Re = 200,000
C_ℓ
C_m
α (deg)

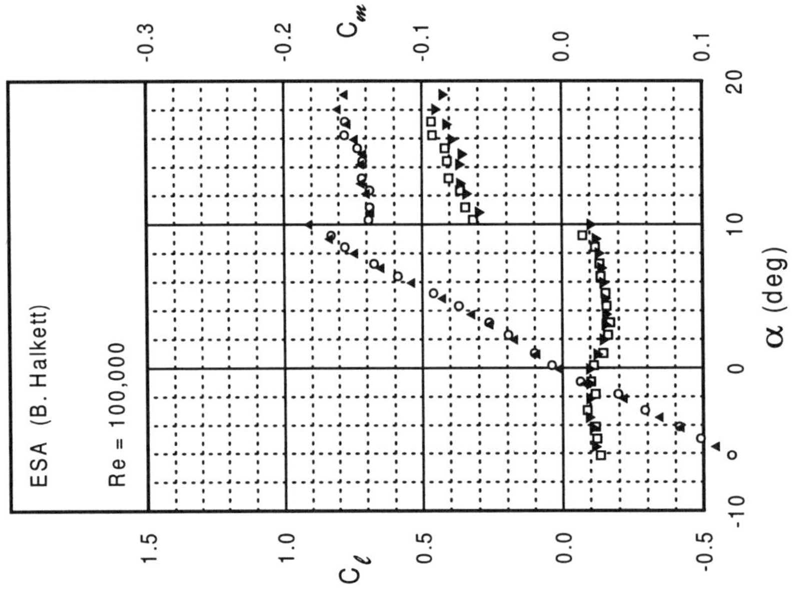
ESA (B. Halkett)
Re = 100,000
C_ℓ
C_m
α (deg)

ESA

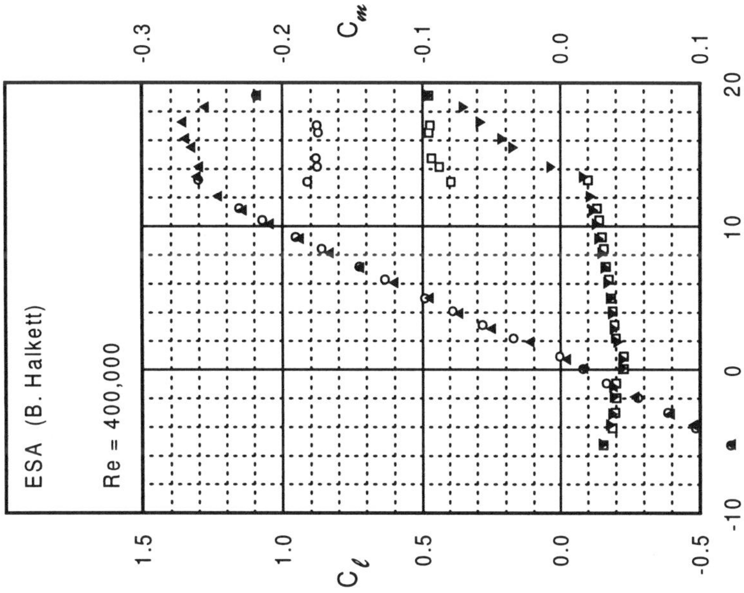
ESA (B. Halkett)
Re = 400,000
C_ℓ
C_m
α (deg)
1.5
1.0
0.5
0.0
-0.5
-0.3
-0.2
-0.1
0.0
0.1
-10
0
10
20

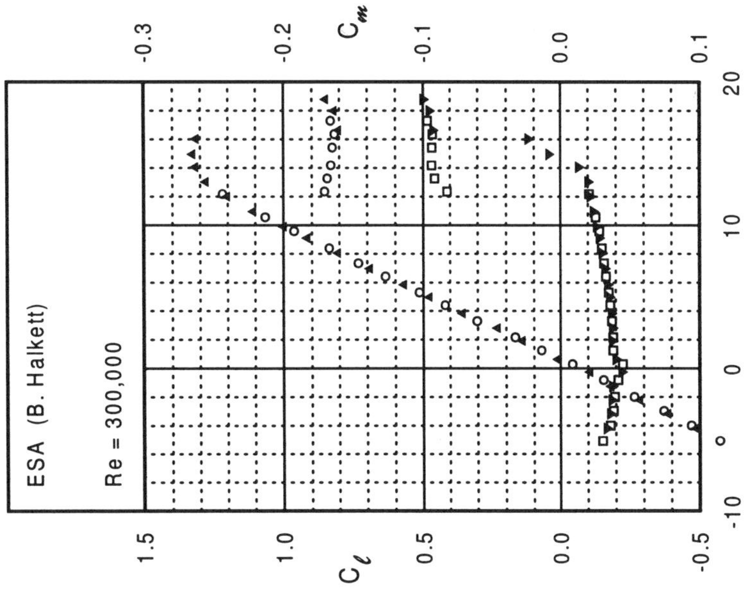
ESA (B. Halkett)
Re = 300,000
C_ℓ
C_m
α (deg)
1.5
1.0
0.5
0.0
-0.5
-0.3
-0.2
-0.1
0.0
0.1
-10
0
10
20

Falcon 56 Mk II

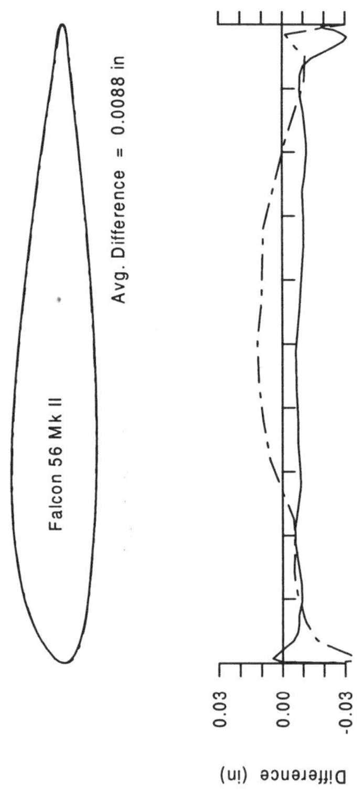

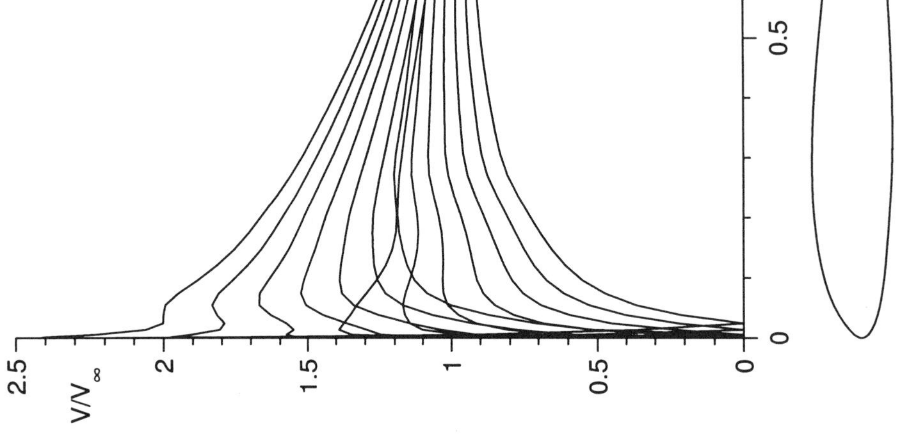

Falcon 56 Mk II

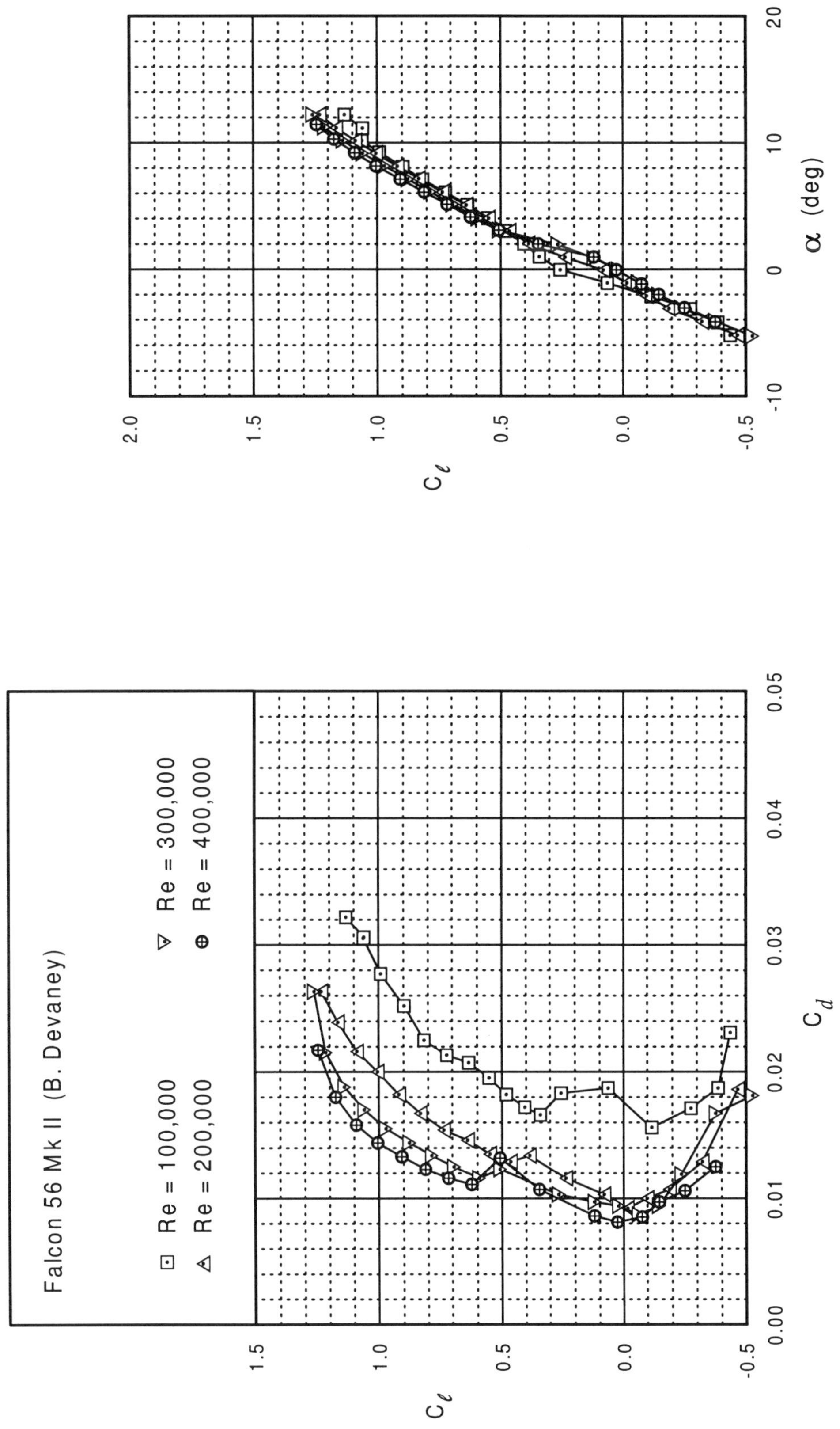

Falcon 56 Mk II

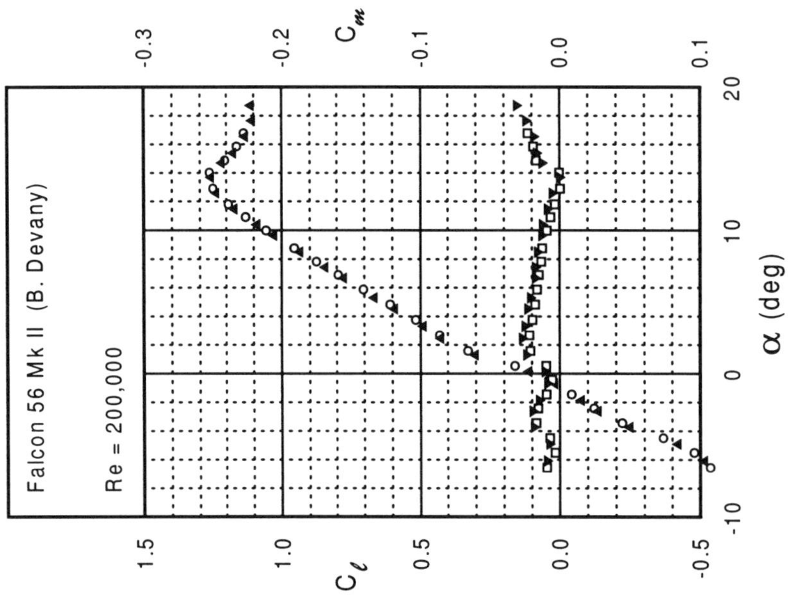

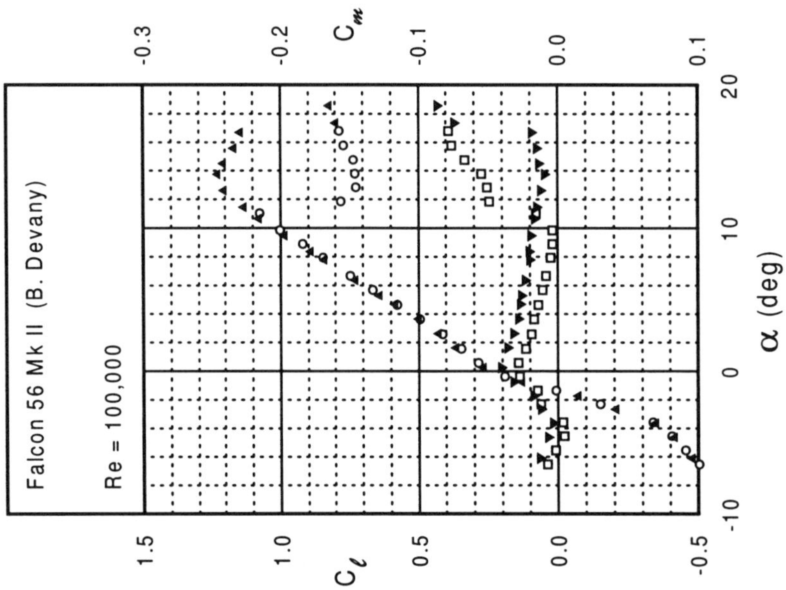

Falcon 56 Mk II

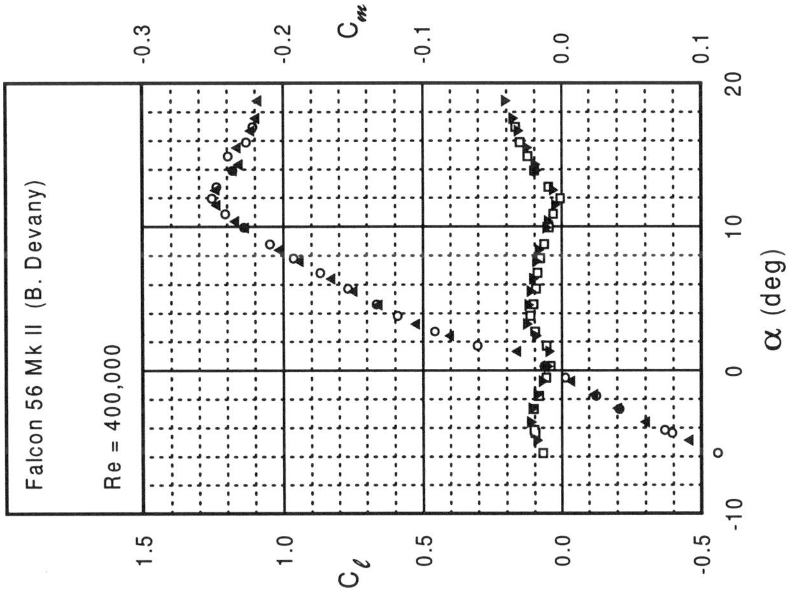

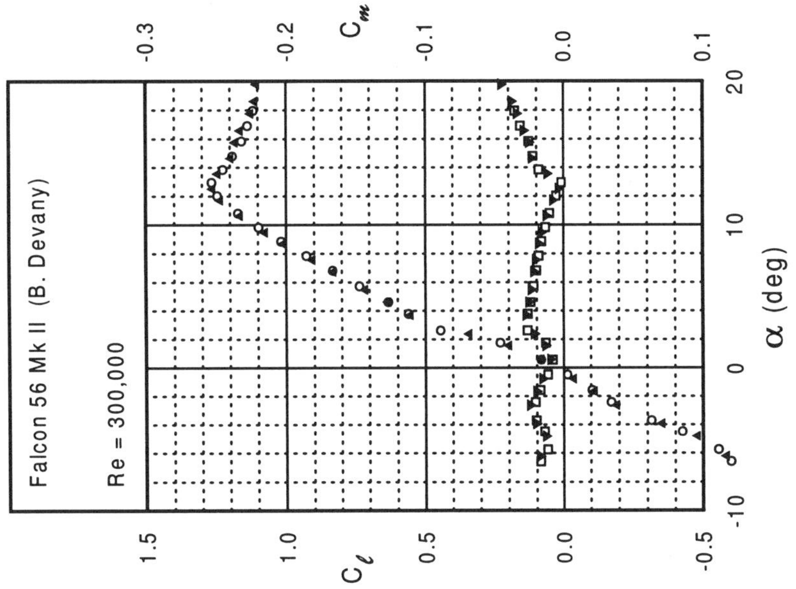

Goe 417a

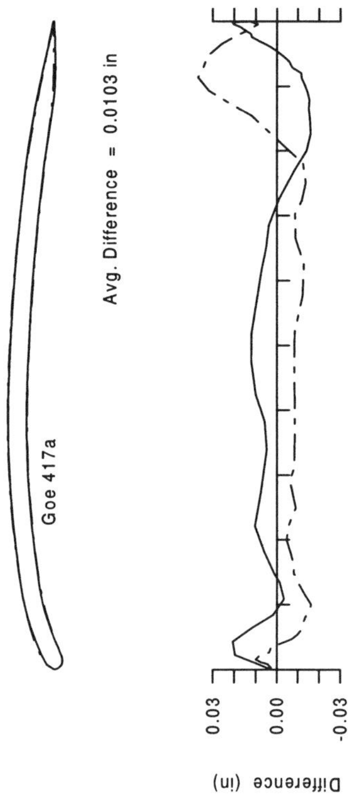
Goe 417a
Avg. Difference = 0.0103 in
0.03
0.00
-0.03
Difference (in)

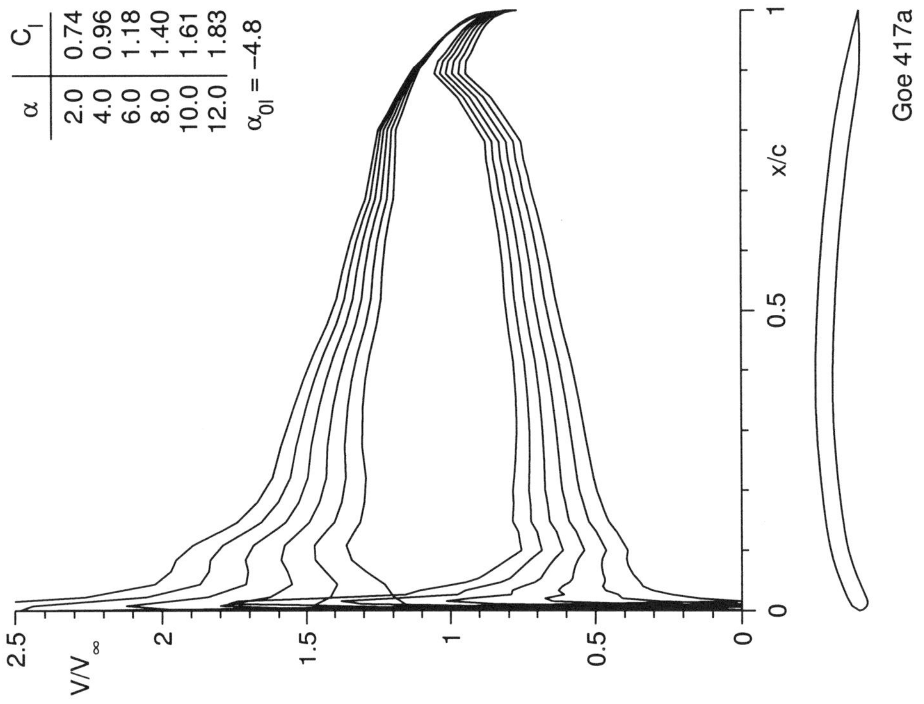
α | C_l
2.0 | 0.74
4.0 | 0.96
6.0 | 1.18
8.0 | 1.40
10.0 | 1.61
12.0 | 1.83
α_0l = −4.8
V/V∞
2.5
2
1.5
1
0.5
0
0.5
1
x/c
Goe 417a

Goe 417a

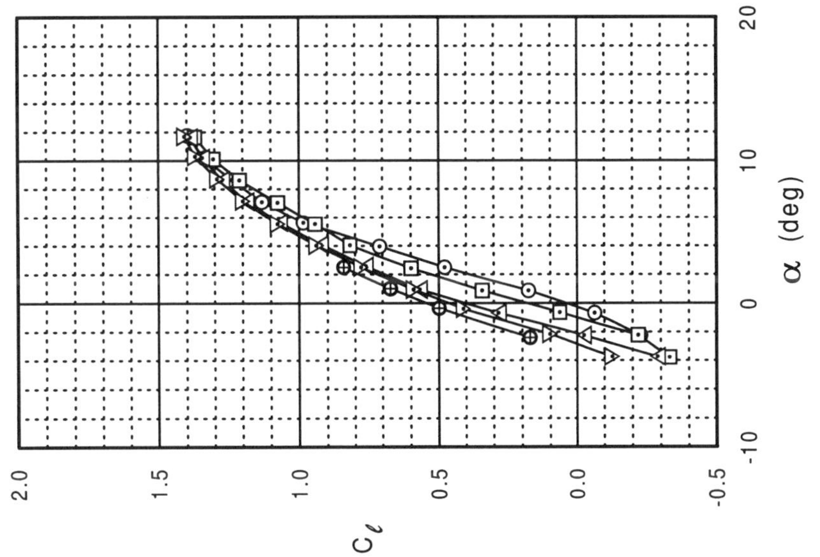

C_ℓ
α (deg)

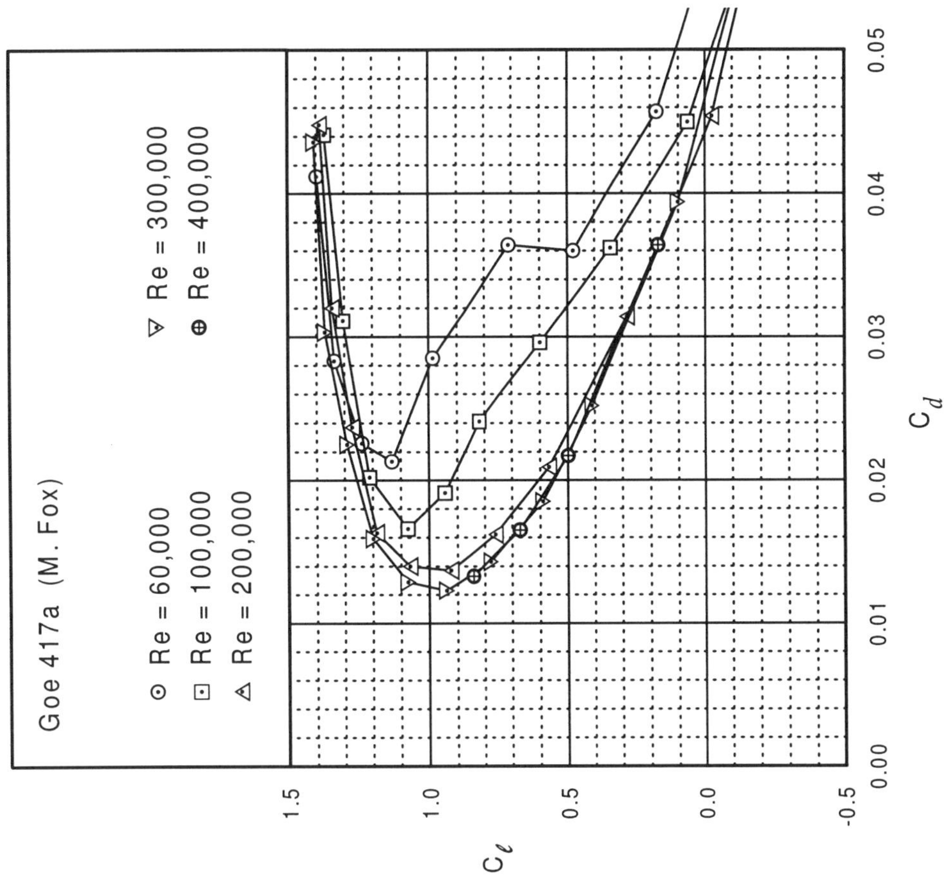

Goe 417a (M. Fox)
Re = 60,000
Re = 100,000
Re = 200,000
Re = 300,000
Re = 400,000
C_ℓ
C_d

Goe 417a

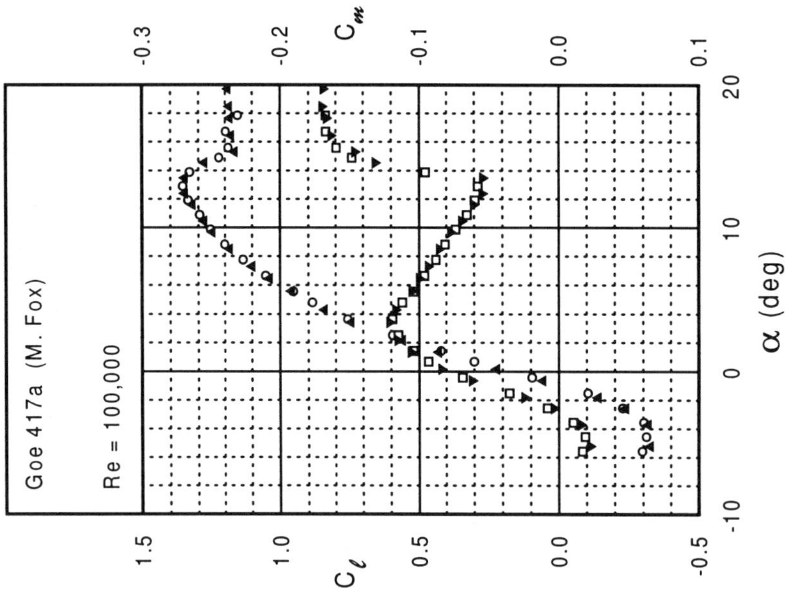
Goe 417a (M. Fox)
Re = 100,000
C_ℓ
C_m
α (deg)

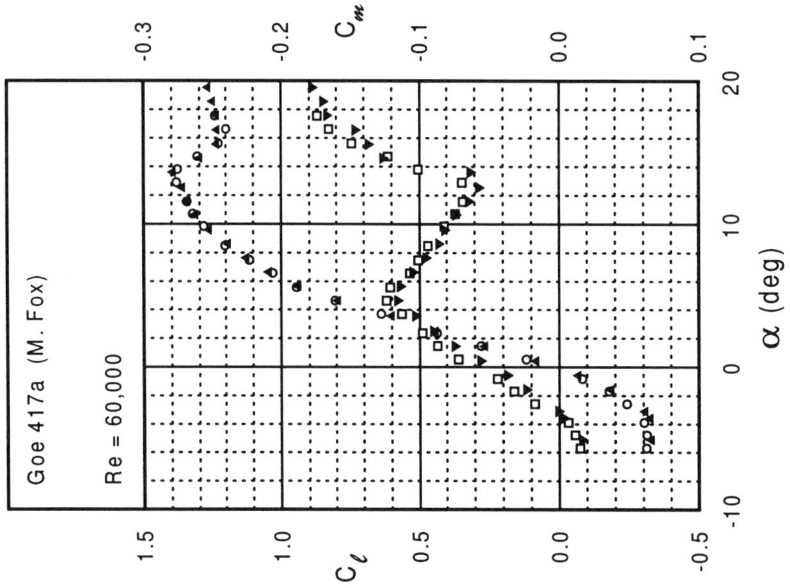
Goe 417a (M. Fox)
Re = 60,000
C_ℓ
C_m
α (deg)

Goe 417a

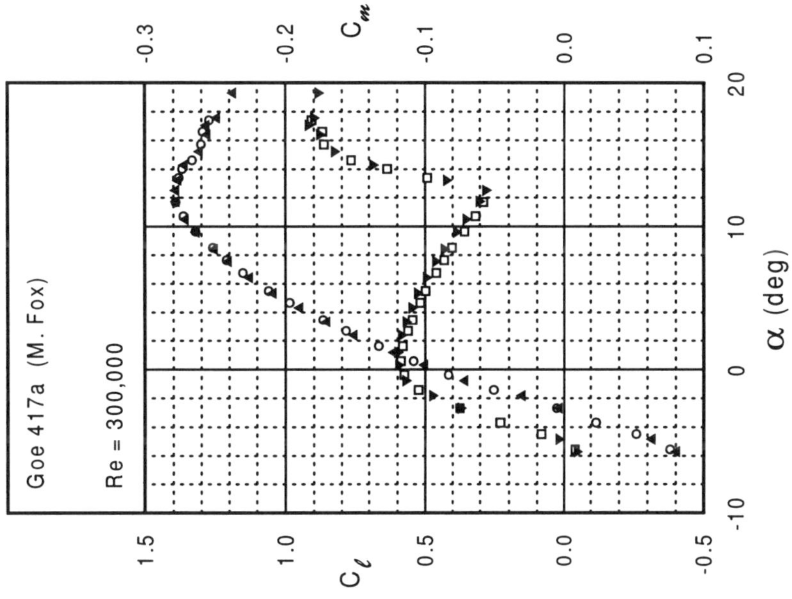
Goe 417a (M. Fox)
Re = 300,000
C_ℓ
C_m
α (deg)
1.5
1.0
0.5
0.0
-0.5
-0.3
-0.2
-0.1
0.0
0.1
-10
0
10
20

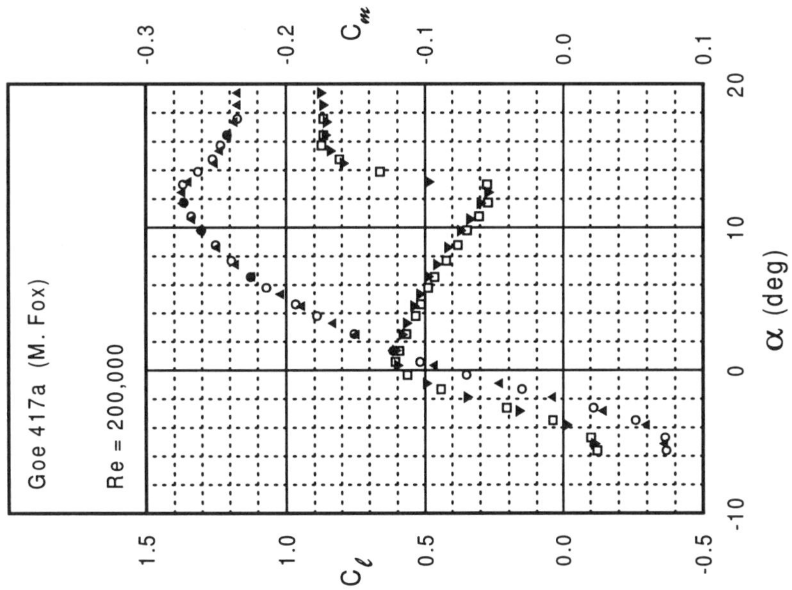
Goe 417a (M. Fox)
Re = 200,000
C_ℓ
C_m
α (deg)
1.5
1.0
0.5
0.0
-0.5
-0.3
-0.2
-0.1
0.0
0.1
-10
0
10
20

Goe 417a

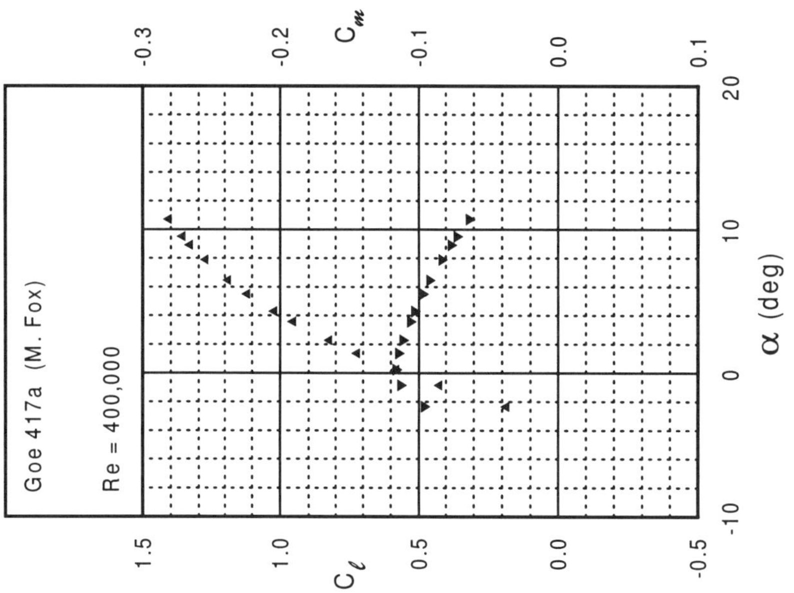
Goe 417a (M. Fox)
Re = 400,000
-0.3
-0.2
-0.1
0.0
0.1
C_m
1.5
1.0
0.5
0.0
-0.5
C_ℓ
-10
0
10
20
α (deg)

LRN1007 (B)

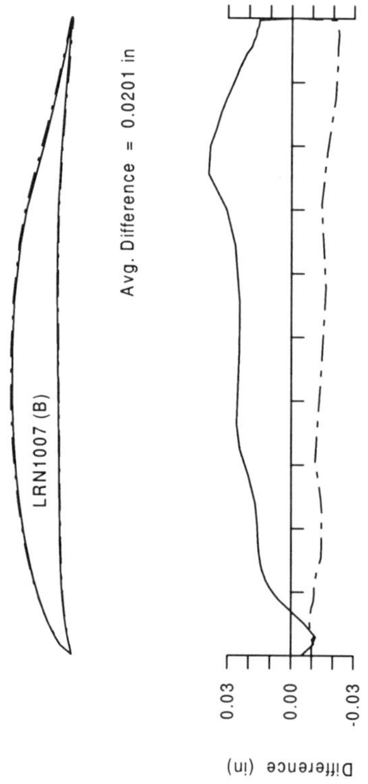
LRN1007 (B)
Avg. Difference = 0.0201 in
0.03
0.00
-0.03
Difference (in)

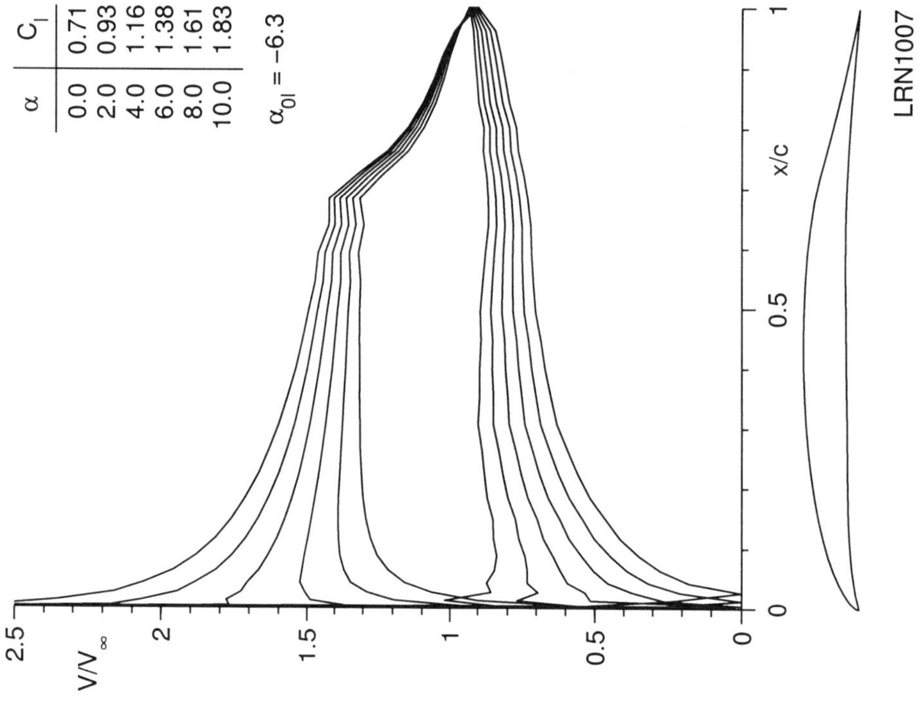
α | C_l
0.0 | 0.71
2.0 | 0.93
4.0 | 1.16
6.0 | 1.38
8.0 | 1.61
10.0 | 1.83
α_{0l} = −6.3
V/V_∞
2.5
2
1.5
1
0.5
0
x/c
0.5
LRN1007

LRN1007 (B)

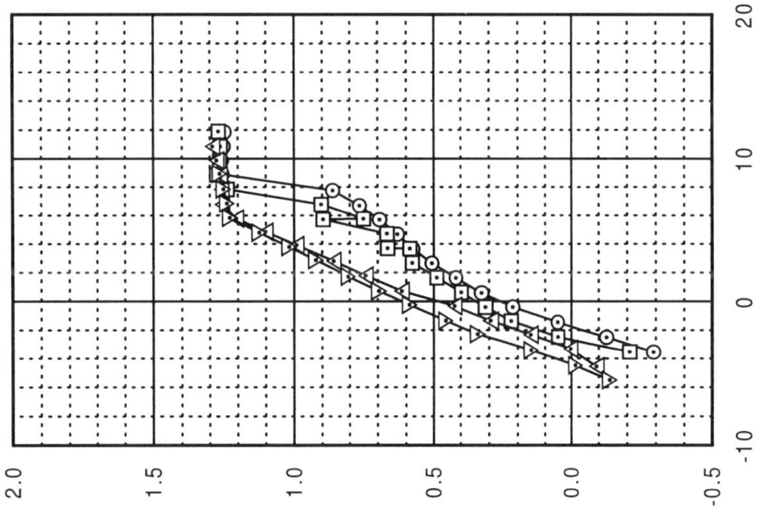

C_ℓ
α (deg)

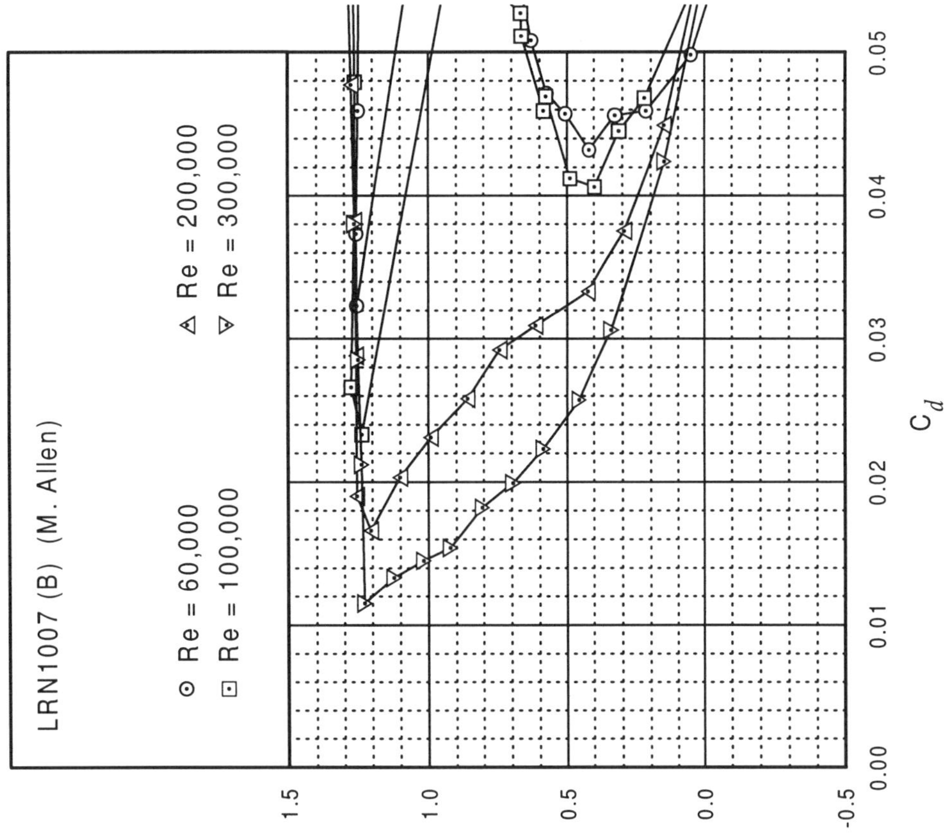

LRN1007 (B) (M. Allen)
Re = 60,000
Re = 100,000
Re = 200,000
Re = 300,000
C_ℓ
C_d

LRN1007 (B)

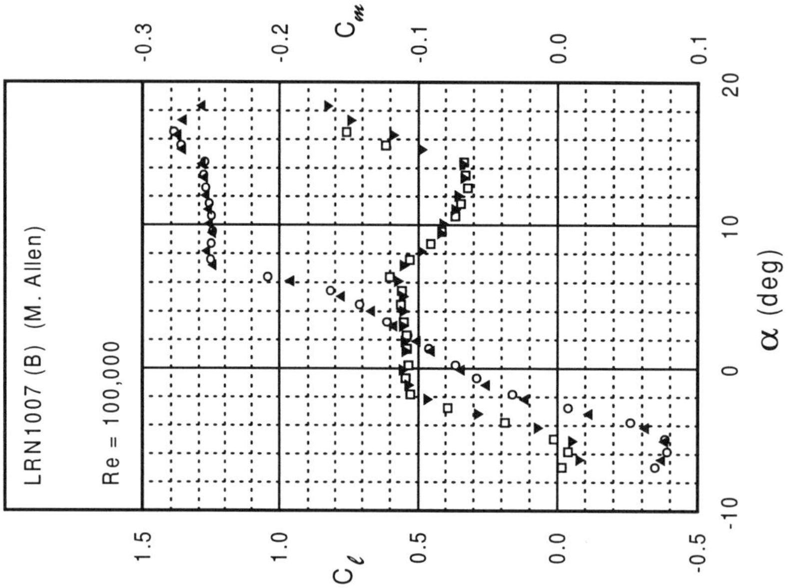
LRN1007 (B) (M. Allen)
Re = 100,000
C_m
C_ℓ
α (deg)

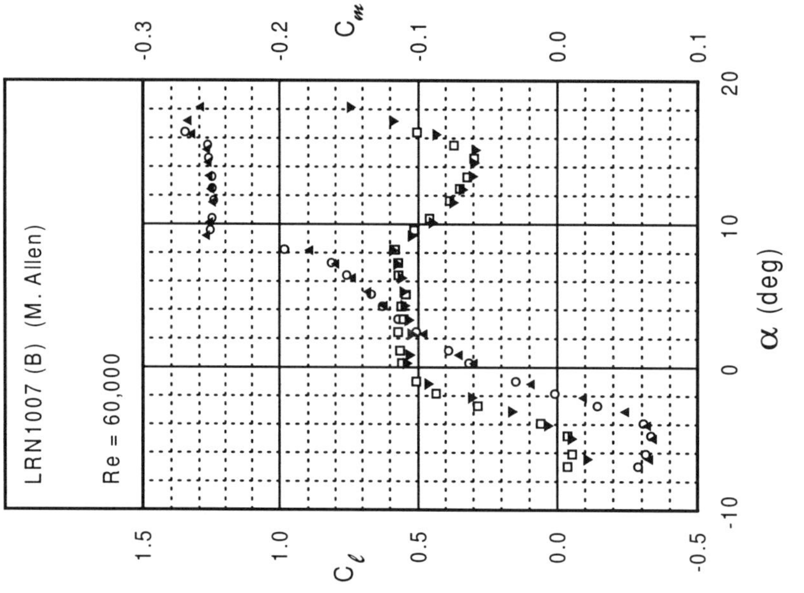
LRN1007 (B) (M. Allen)
Re = 60,000
C_m
C_ℓ
α (deg)

LRN1007 (B)

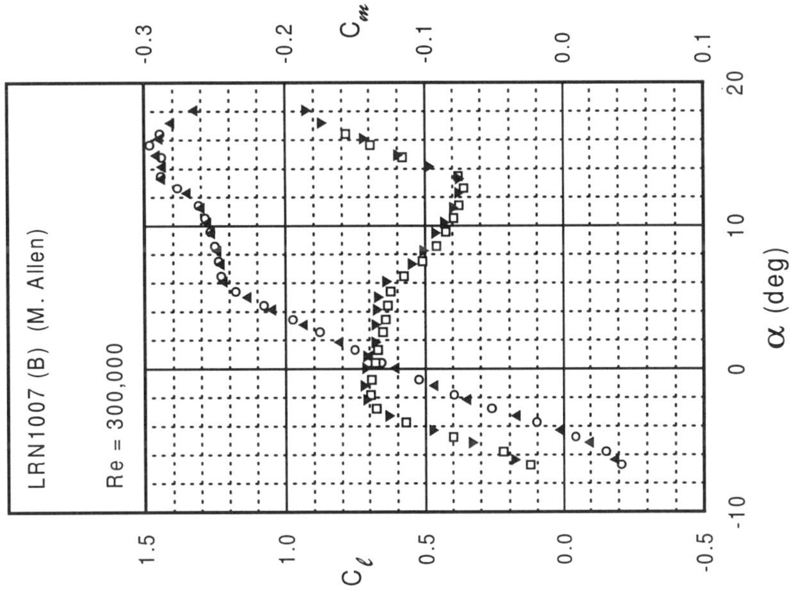
LRN1007 (B) (M. Allen)
Re = 300,000
C_l
1.5
1.0
0.5
0.0
-0.5
C_m
-0.3
-0.2
-0.1
0.0
0.1
-10
0
10
20
α (deg)

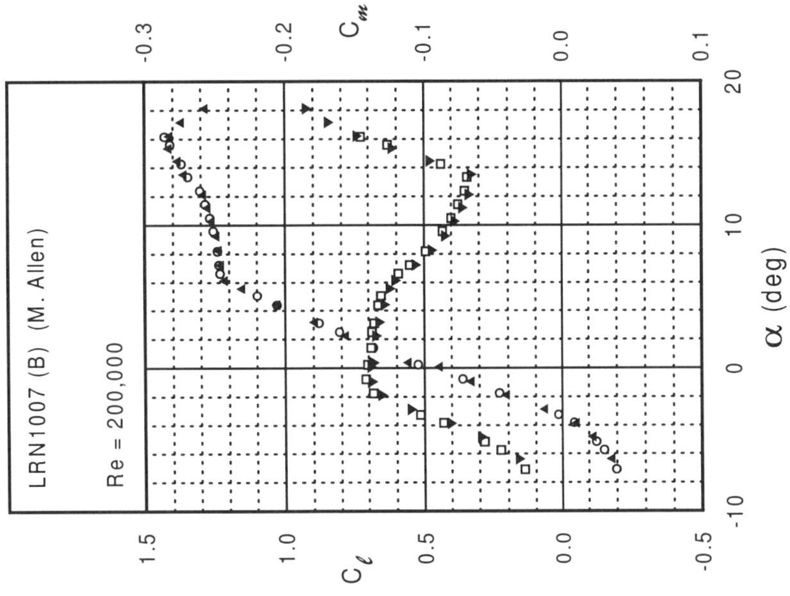
LRN1007 (B) (M. Allen)
Re = 200,000
C_l
1.5
1.0
0.5
0.0
-0.5
C_m
-0.3
-0.2
-0.1
0.0
0.1
-10
0
10
20
α (deg)

PT-40 (A)

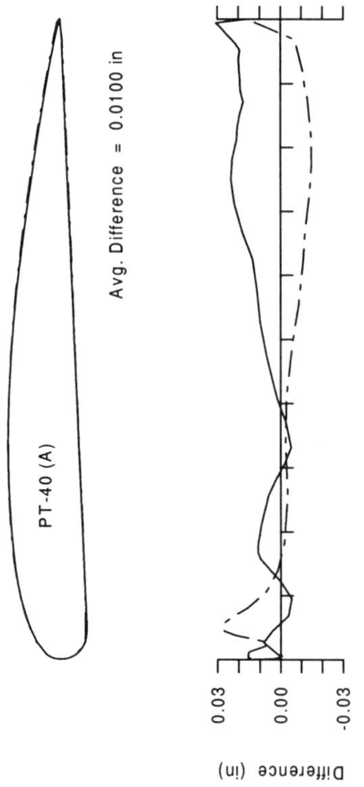
PT-40 (A)
Avg. Difference = 0.0100 in
0.03
0.00
-0.03
Difference (in)

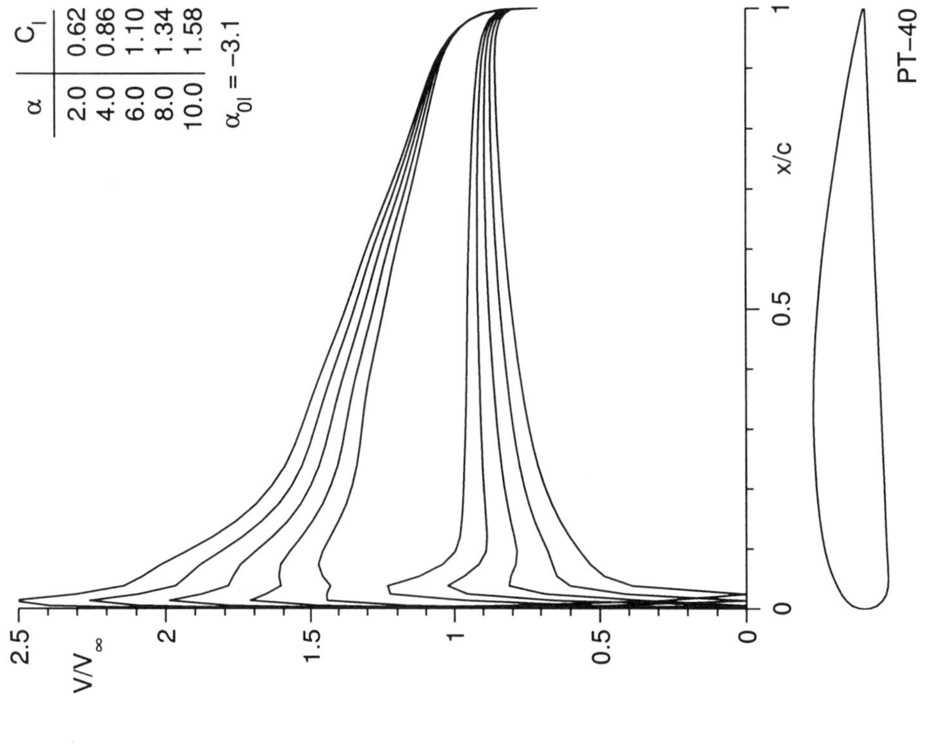
α | C_l
2.0 0.62
4.0 0.86
6.0 1.10
8.0 1.34
10.0 1.58
α_{0l} = −3.1
V/V_∞
2.5
2
1.5
1
0.5
0
x/c
0.5
1
PT–40

PT-40 (A)

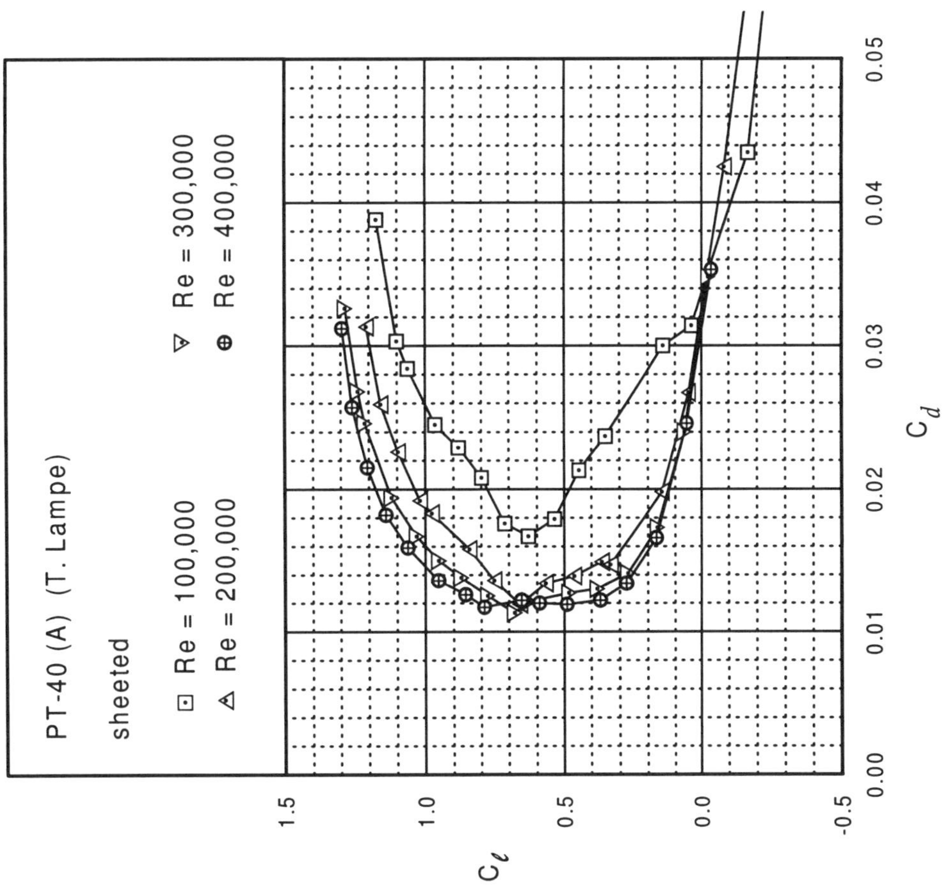
PT-40 (A) (T. Lampe)
sheeted
Re = 100,000
Re = 200,000
Re = 300,000
Re = 400,000
C_ℓ
C_d
1.5
1.0
0.5
0.0
-0.5
0.00
0.01
0.02
0.03
0.04
0.05

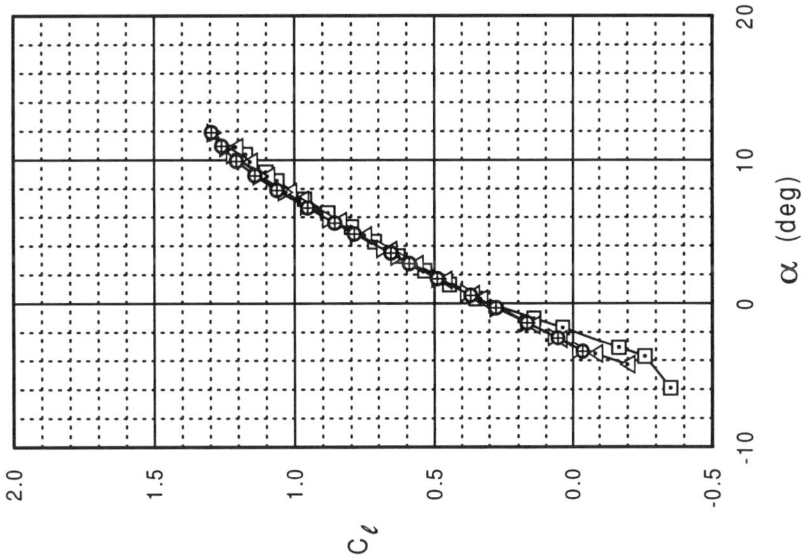
C_ℓ
α (deg)
2.0
1.5
1.0
0.5
0.0
-0.5
-10
0
10
20

PT-40 (A)

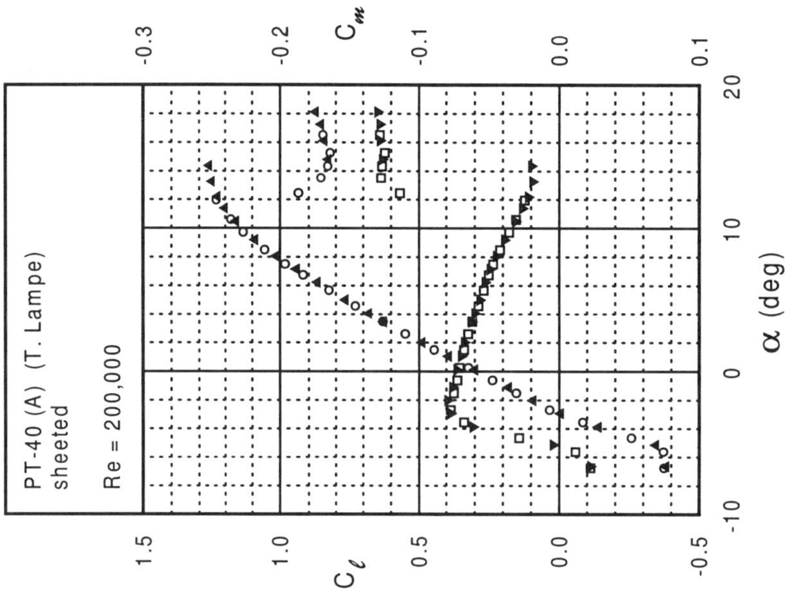
PT-40 (A) (T. Lampe)
sheeted
Re = 200,000
α (deg)
Cl
Cm

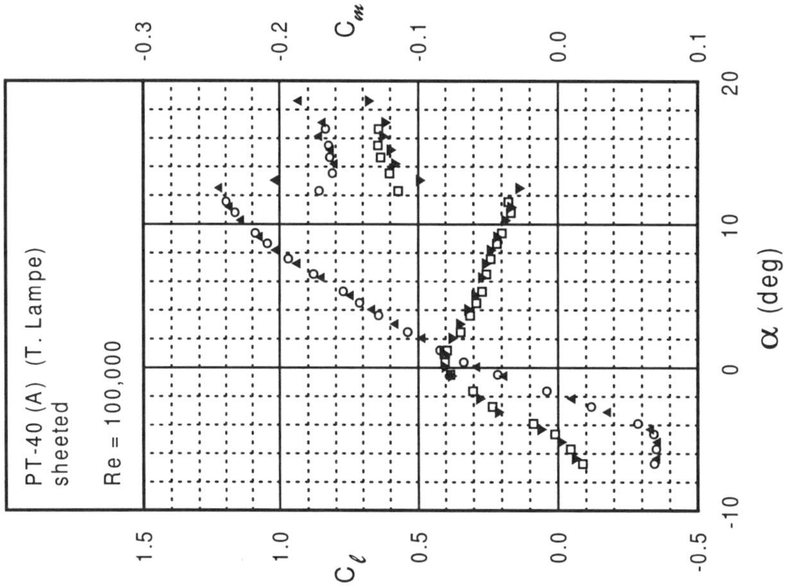
PT-40 (A) (T. Lampe)
sheeted
Re = 100,000
α (deg)
Cl
Cm

PT-40 (A)

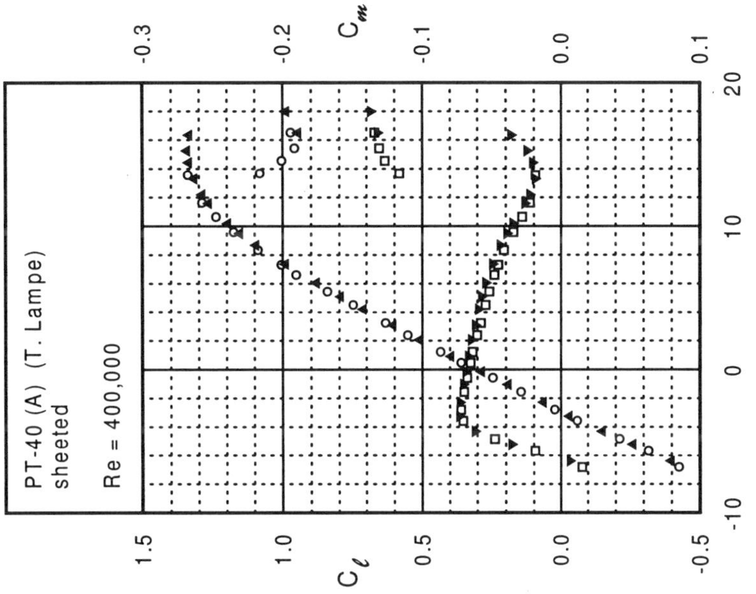
PT-40 (A) (T. Lampe)
sheeted
Re = 400,000
C_m
-0.3
-0.2
-0.1
0.0
0.1
C_ℓ
1.5
1.0
0.5
0.0
-0.5
-10
0
10
20
α (deg)

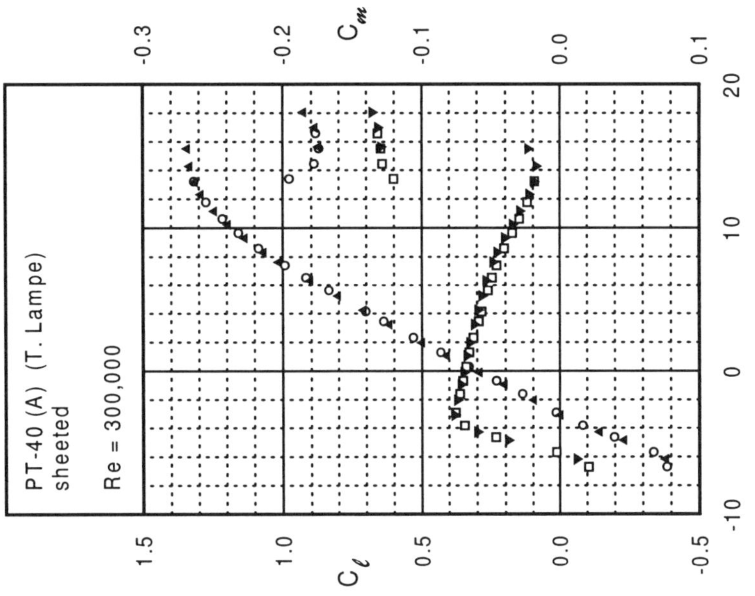
PT-40 (A) (T. Lampe)
sheeted
Re = 300,000
C_m
-0.3
-0.2
-0.1
0.0
0.1
C_ℓ
1.5
1.0
0.5
0.0
-0.5
-10
0
10
20
α (deg)

PT-40 (B)

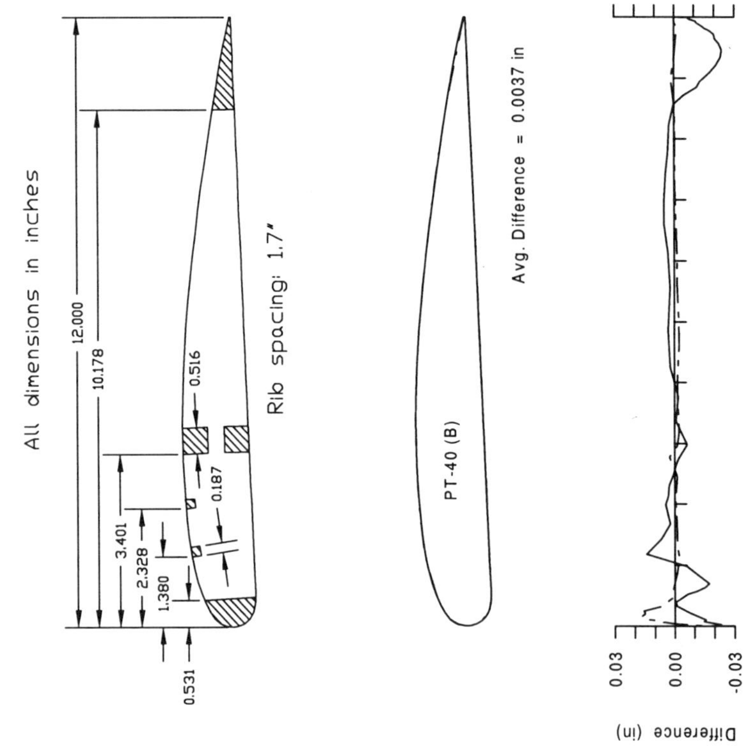
All dimensions in inches
12.000
10.178
0.516
Rib spacing: 1.7″
3.401
2.328
1.380
0.187
0.531
PT-40 (B)
Avg. Difference = 0.0037 in
0.03
0.00
-0.03
Difference (in)

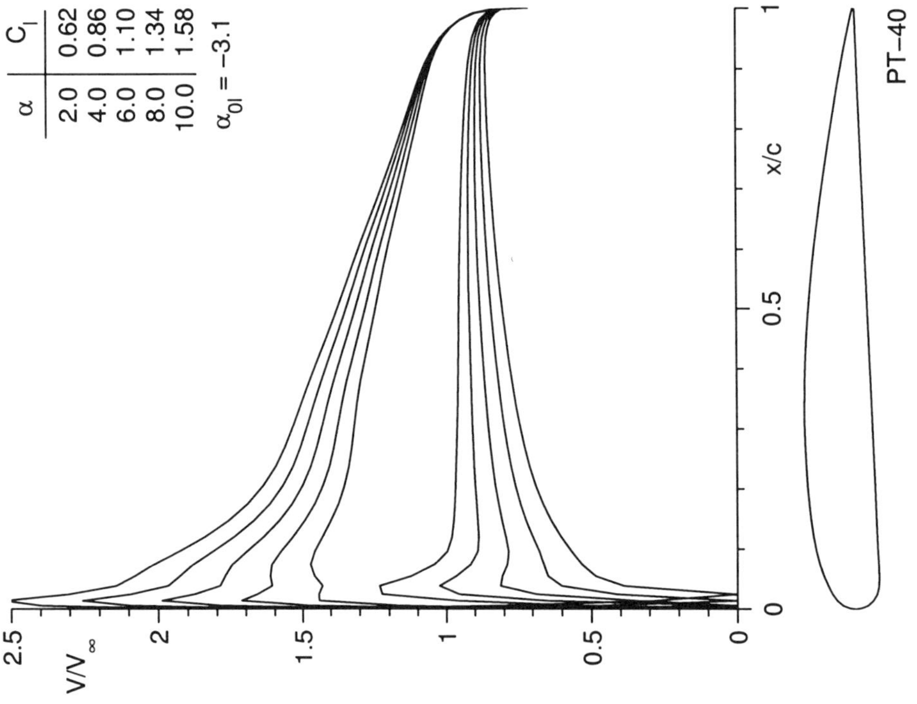
α
2.0
4.0
6.0
8.0
10.0
C_l
0.62
0.86
1.10
1.34
1.58
α_{0l} = −3.1
V/V_∞
2.5
2
1.5
1
0.5
0
x/c
0
0.5
1
PT–40

PT-40 (B)

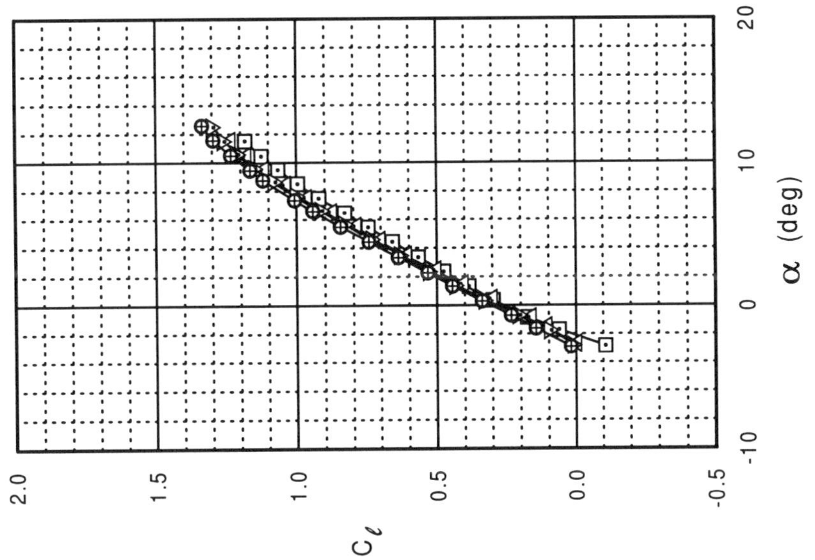
α (deg)
C_ℓ
2.0
1.5
1.0
0.5
0.0
-0.5
-10
0
10
20

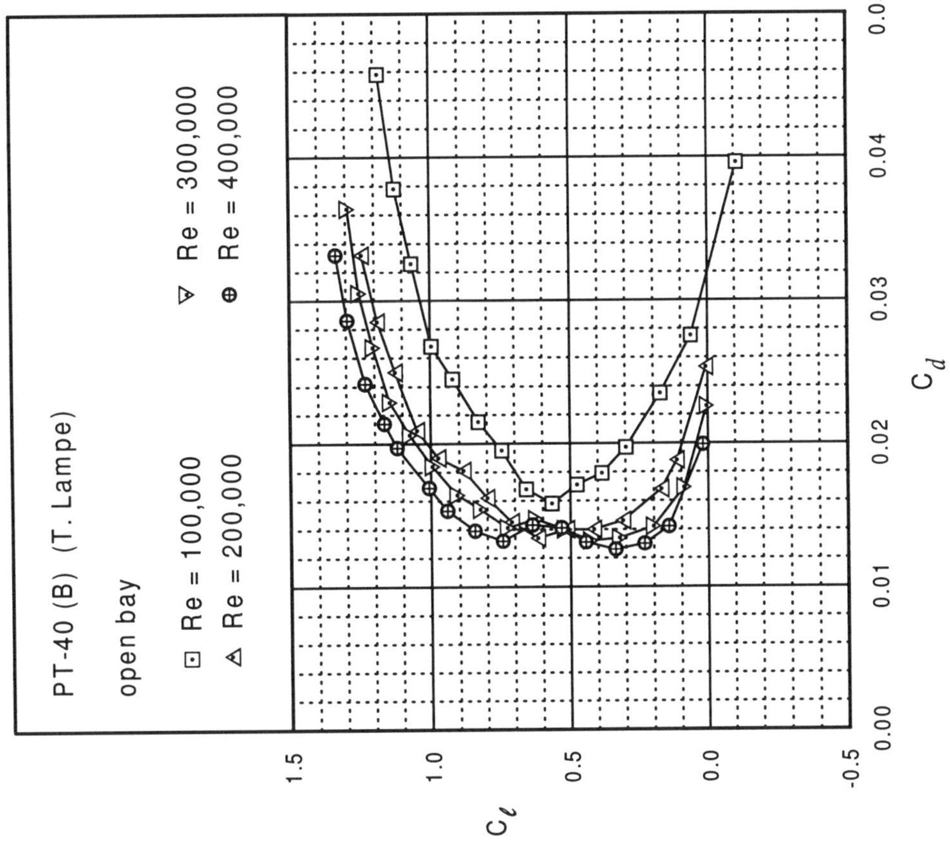
PT-40 (B) (T. Lampe)
open bay
Re = 100,000
Re = 200,000
Re = 300,000
Re = 400,000
C_d
C_ℓ
1.5
1.0
0.5
0.0
-0.5
0.00
0.01
0.02
0.03
0.04
0.05

PT-40 (B)

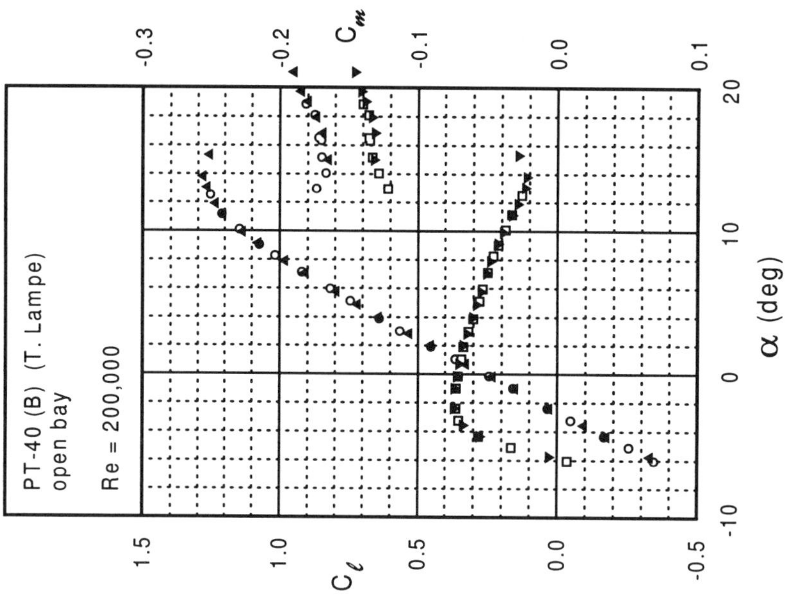
PT-40 (B) (T. Lampe)
open bay
Re = 200,000
C_ℓ
C_m
α (deg)
-0.3
-0.2
-0.1
0.0
0.1
1.5
1.0
0.5
0.0
-0.5
-10
0
10
20

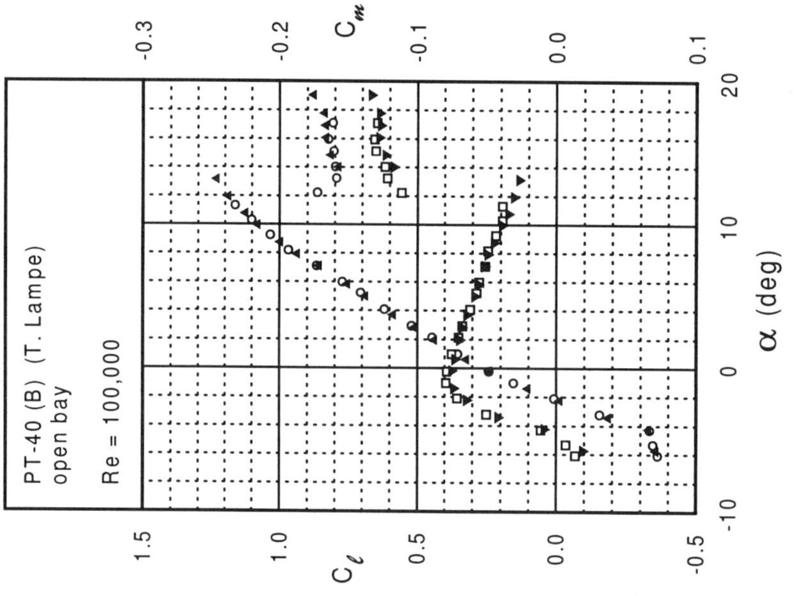
PT-40 (B) (T. Lampe)
open bay
Re = 100,000
C_ℓ
C_m
α (deg)
-0.3
-0.2
-0.1
0.0
0.1
1.5
1.0
0.5
0.0
-0.5
-10
0
10
20

PT-40 (B)

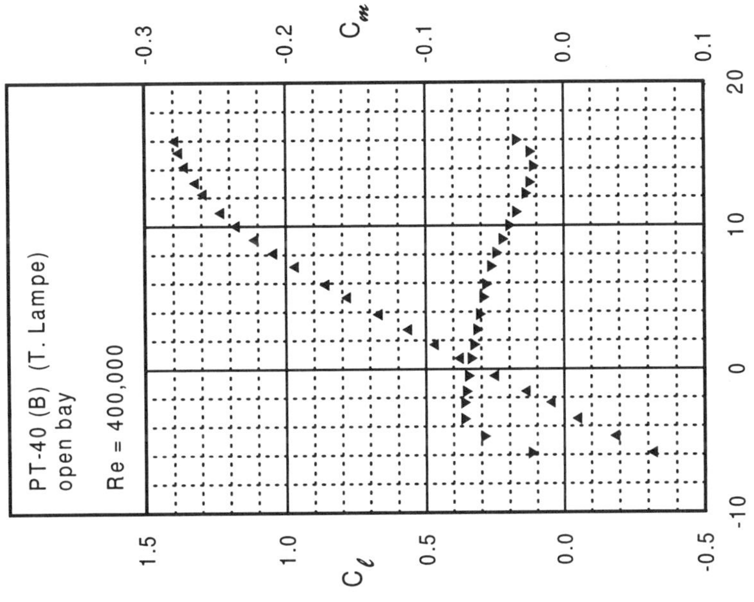
PT-40 (B) (T. Lampe)
open bay
Re = 400,000
C_ℓ
C_m
α (deg)

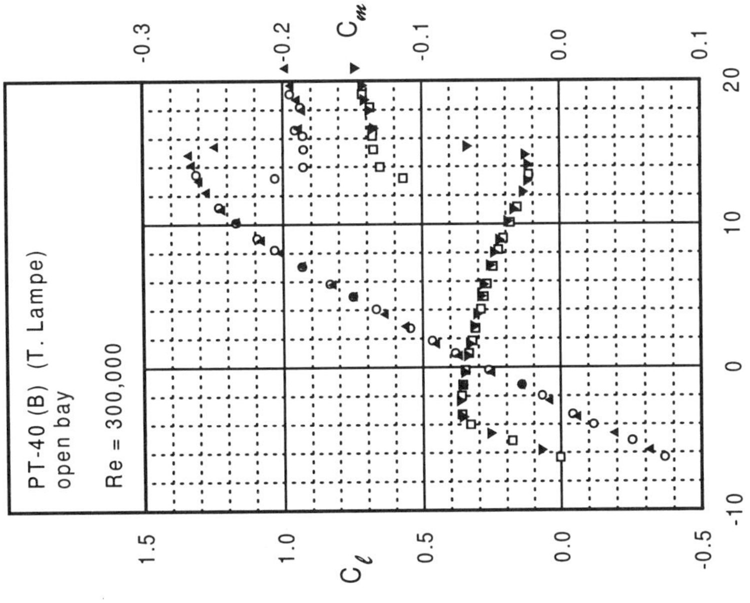
PT-40 (B) (T. Lampe)
open bay
Re = 300,000
C_ℓ
C_m
α (deg)

RG14

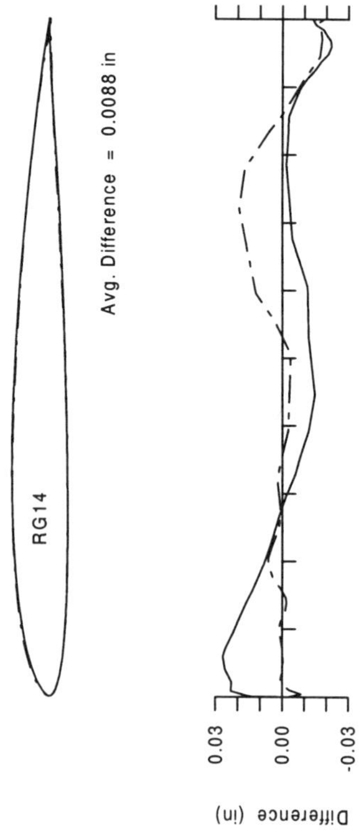

RG14
Avg. Difference = 0.0088 in
Difference (in)
0.03
0.00
-0.03

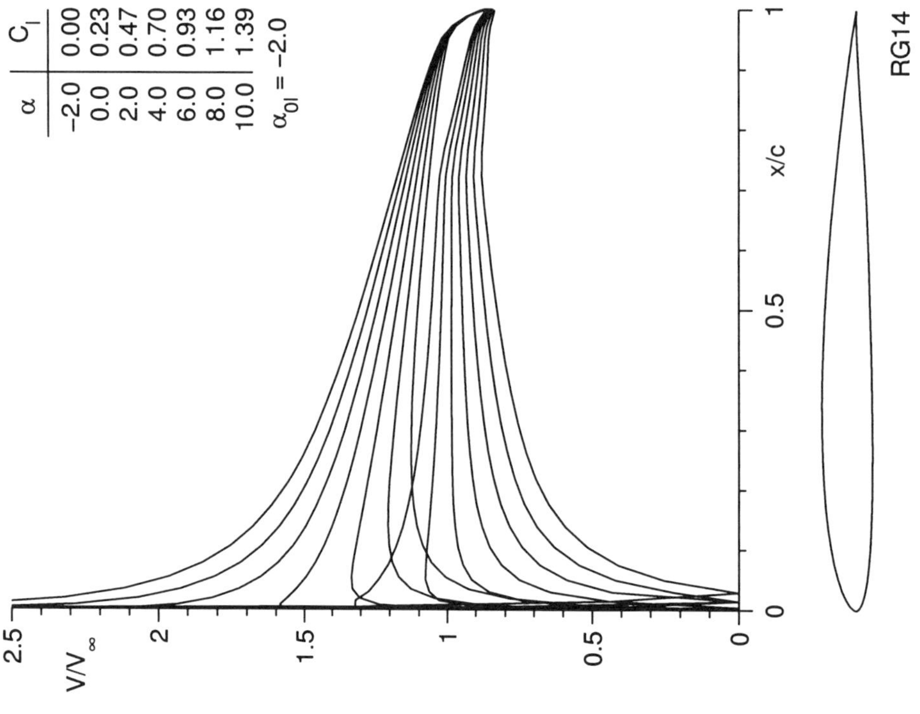

α
-2.0
0.0
2.0
4.0
6.0
8.0
10.0
C_l
0.00
0.23
0.47
0.70
0.93
1.16
1.39
α_{0l} = -2.0
V/V_∞
2.5
2
1.5
1
0.5
0
x/c
0.5
RG14

RG14

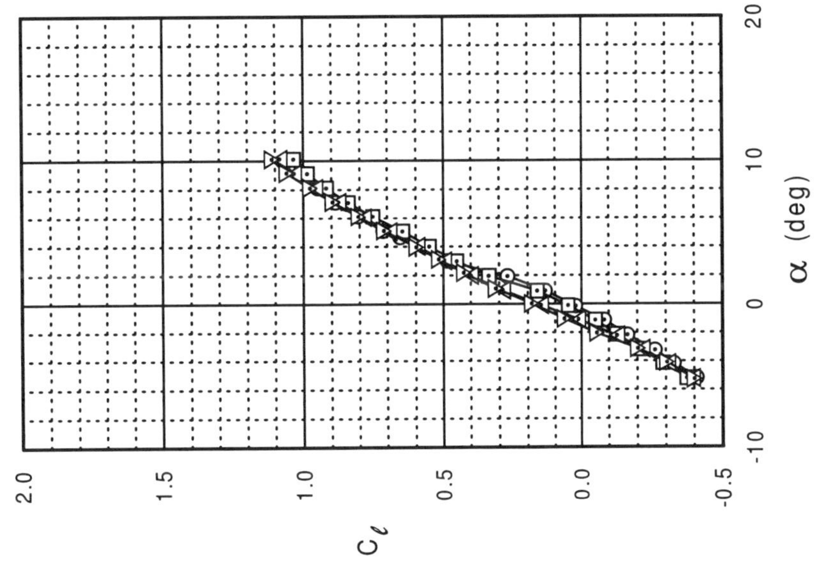
2.0
1.5
1.0
0.5
0.0
-0.5
-10
0
10
20
α (deg)

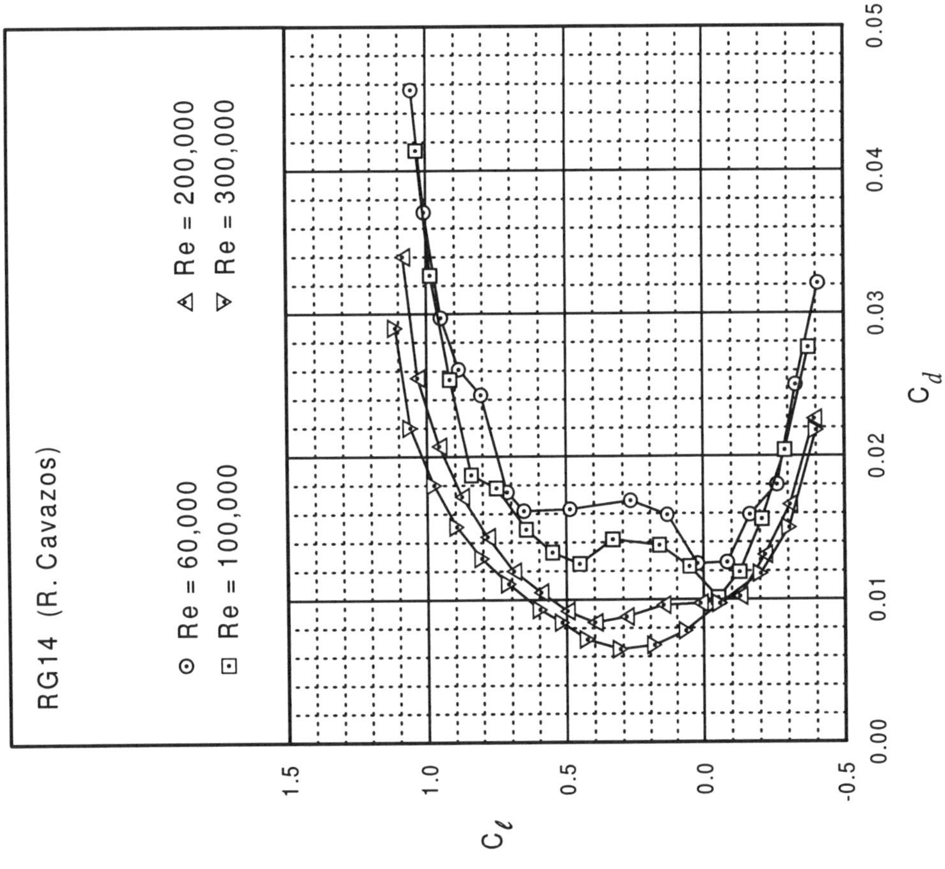
RG14 (R. Cavazos)
Re = 60,000
Re = 100,000
Re = 200,000
Re = 300,000
1.5
1.0
0.5
0.0
-0.5
0.00
0.01
0.02
0.03
0.04
0.05

RG14

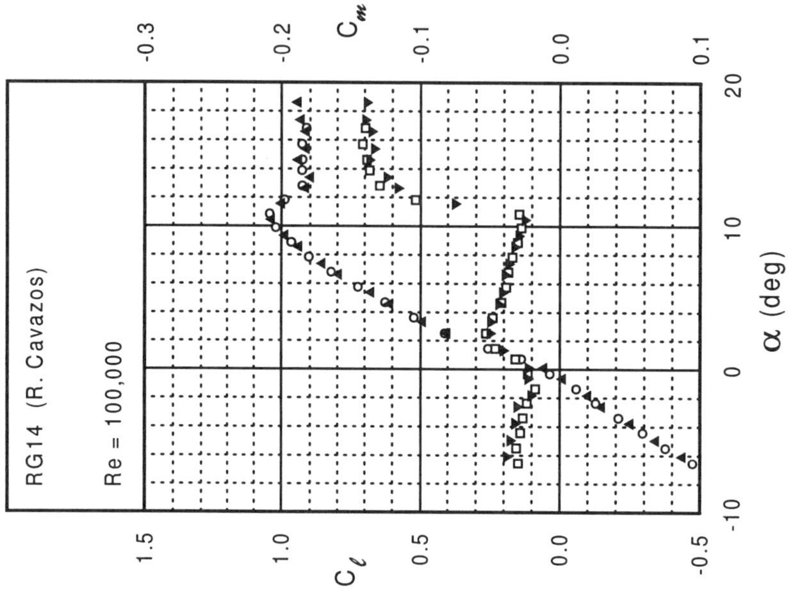
RG14 (R. Cavazos)
Re = 100,000
C_ℓ
1.5
1.0
0.5
0.0
-0.5
C_m
-0.3
-0.2
-0.1
0.0
0.1
-10
0
10
20
α (deg)

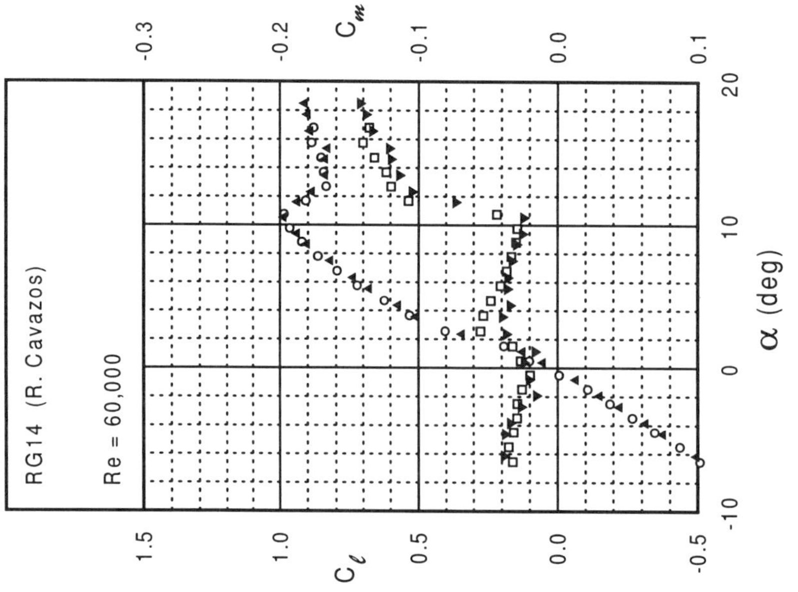
RG14 (R. Cavazos)
Re = 60,000
C_ℓ
1.5
1.0
0.5
0.0
-0.5
C_m
-0.3
-0.2
-0.1
0.0
0.1
-10
0
10
20
α (deg)

RG14

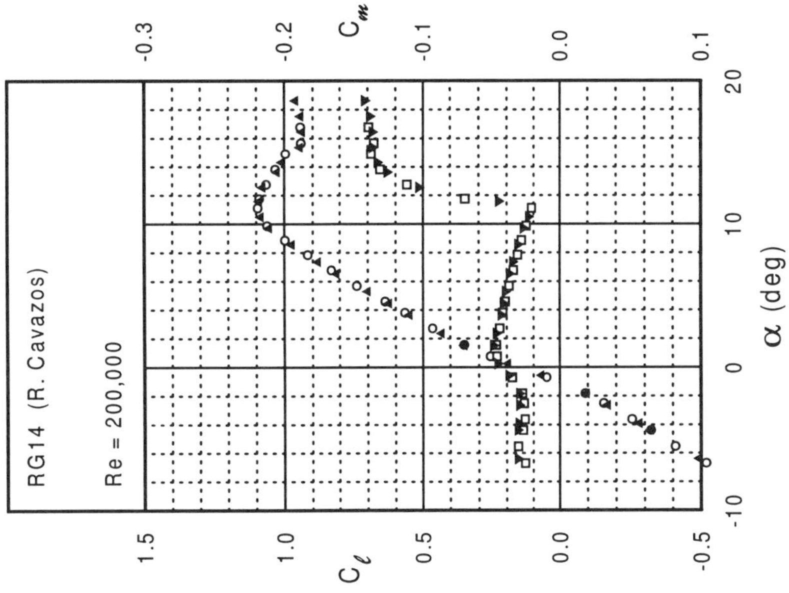
RG14 (R. Cavazos)
Re = 200,000
C_ℓ
C_m
α (deg)

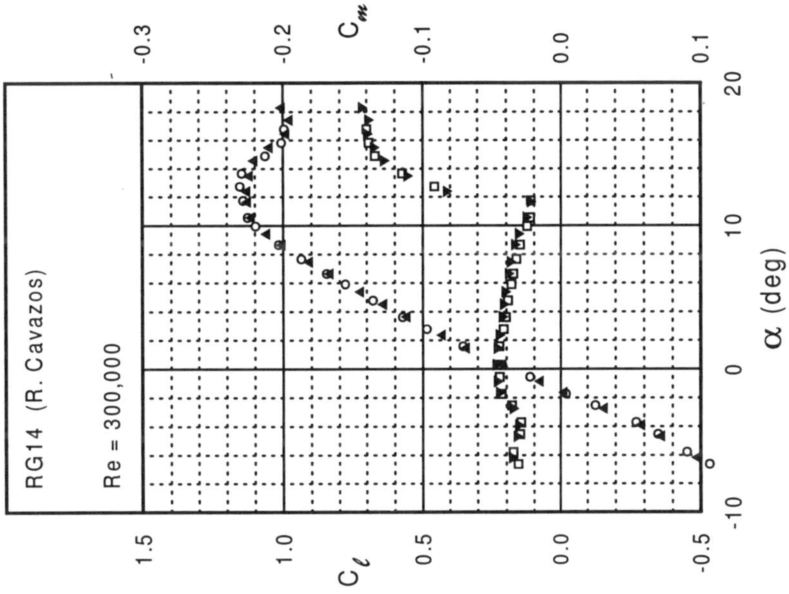
RG14 (R. Cavazos)
Re = 300,000
C_ℓ
C_m
α (deg)

RG15 (C)

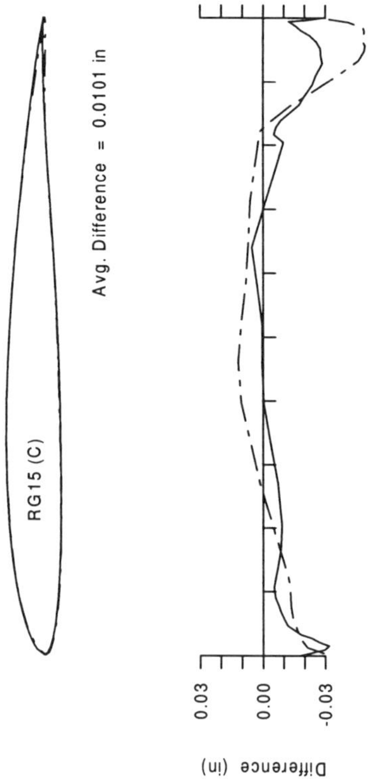
RG15 (C)
Avg. Difference = 0.0101 in
0.03
0.00
-0.03
Difference (in)

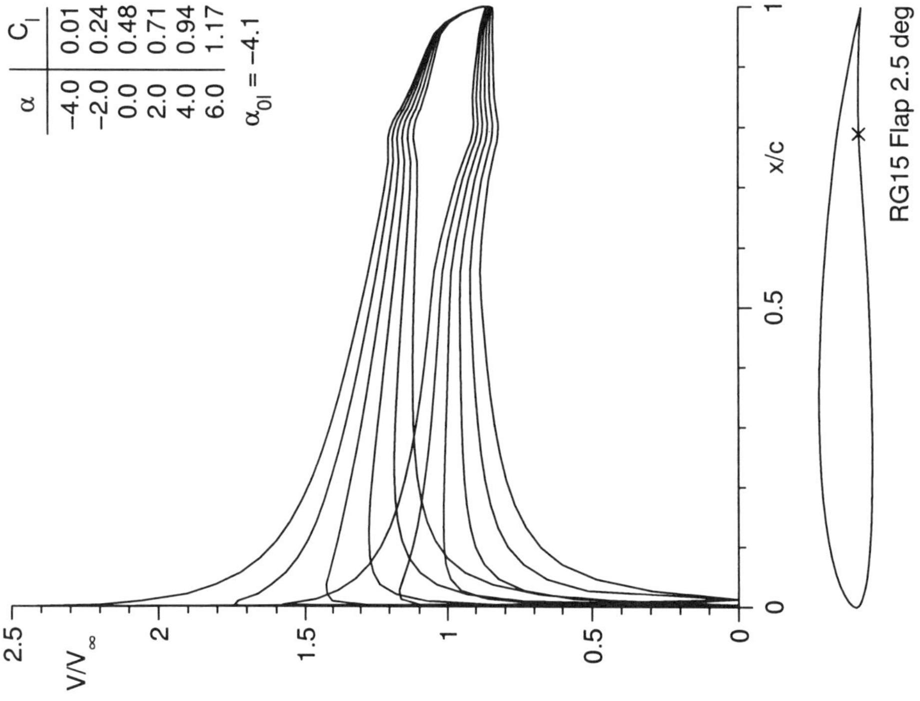
α Cl
-4.0 0.01
-2.0 0.24
0.0 0.48
2.0 0.71
4.0 0.94
6.0 1.17
α0l = -4.1
V/V∞
2.5
2
1.5
1
0.5
0
0
0.5
1
x/c
RG15 Flap 2.5 deg

RG15 (C)

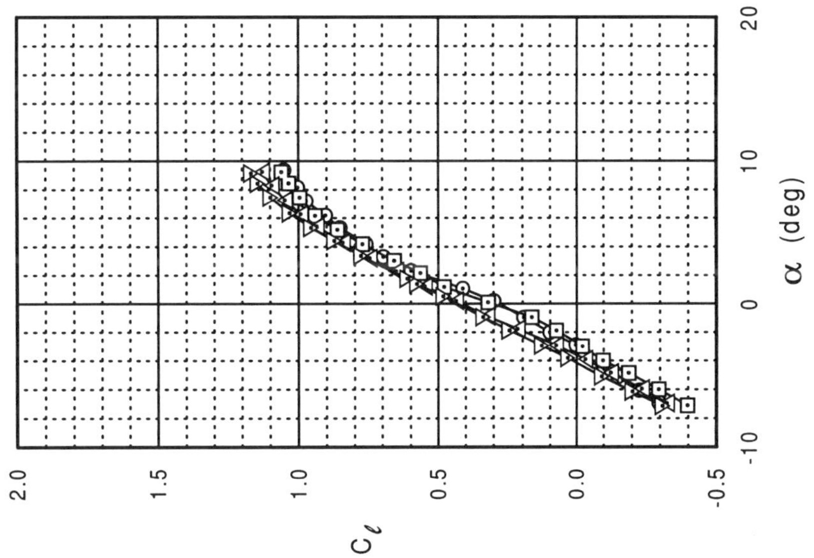
2.0
1.5
1.0
0.5
0.0
-0.5
C_ℓ
-10
0
10
20
α (deg)

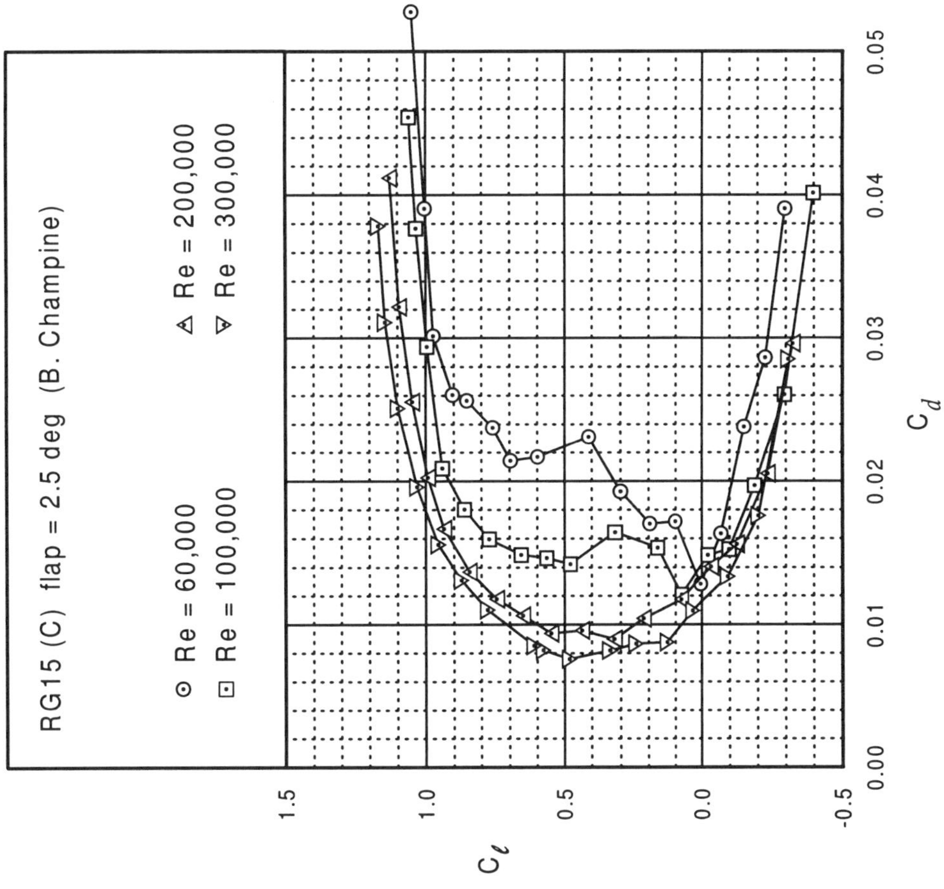
RG15 (C) flap = 2.5 deg (B. Champine)
⊙ Re = 60,000
⊡ Re = 100,000
△ Re = 200,000
▽ Re = 300,000
1.5
1.0
0.5
0.0
-0.5
C_ℓ
0.00
0.01
0.02
0.03
0.04
0.05
C_d

RG15 (C)

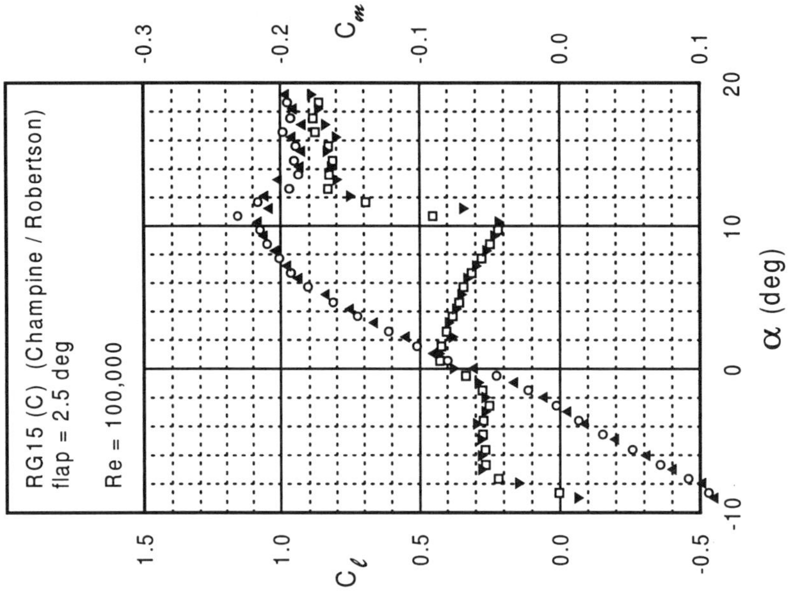
RG15 (C) (Champine / Robertson)
flap = 2.5 deg
Re = 100,000
C_ℓ
1.5
1.0
0.5
0.0
-0.5
C_m
-0.3
-0.2
-0.1
0.0
0.1
-10
0
10
20
α (deg)

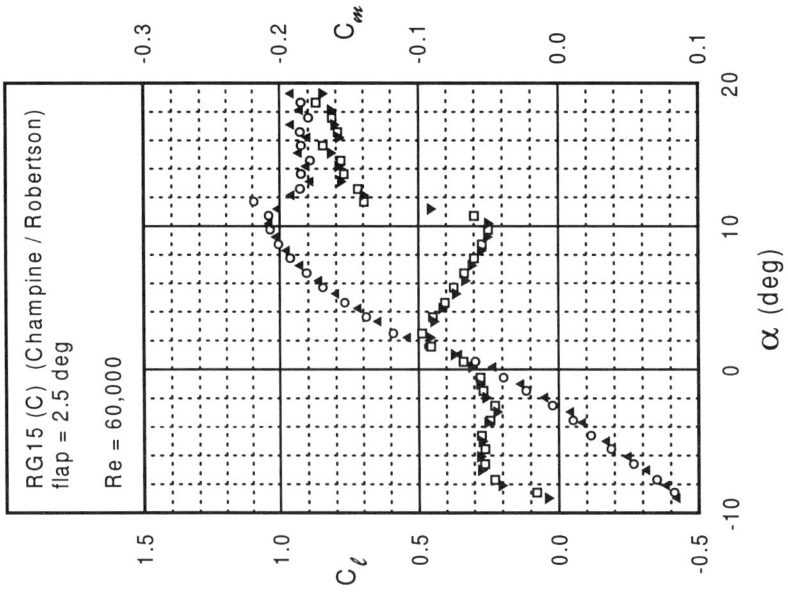
RG15 (C) (Champine / Robertson)
flap = 2.5 deg
Re = 60,000
C_ℓ
1.5
1.0
0.5
0.0
-0.5
C_m
-0.3
-0.2
-0.1
0.0
0.1
-10
0
10
20
α (deg)

RG15 (C)

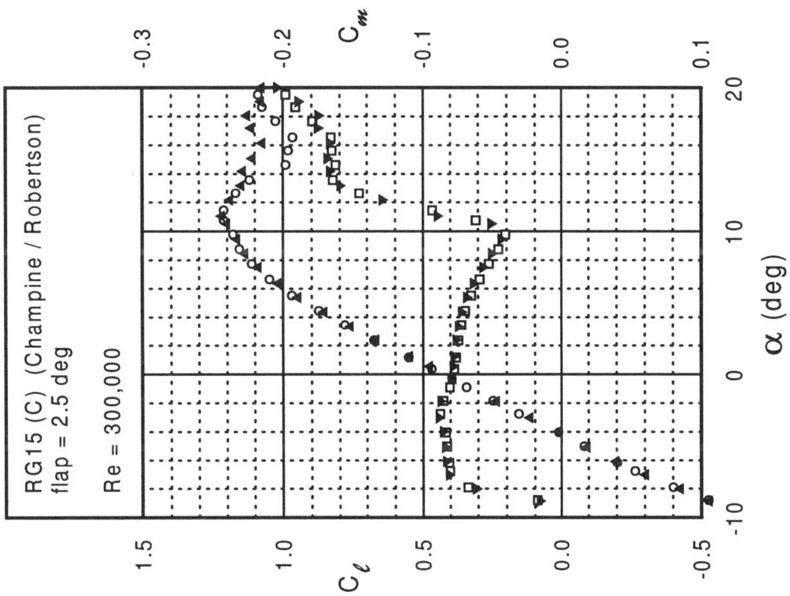
RG15 (C) (Champine / Robertson)
flap = 2.5 deg
Re = 300,000
C_l
C_m
α (deg)

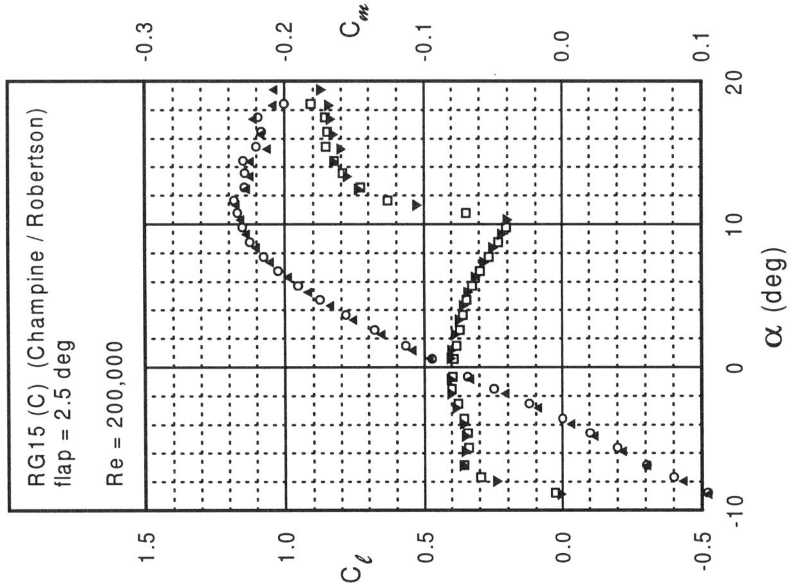
RG15 (C) (Champine / Robertson)
flap = 2.5 deg
Re = 200,000
C_l
C_m
α (deg)

RG15 (C)

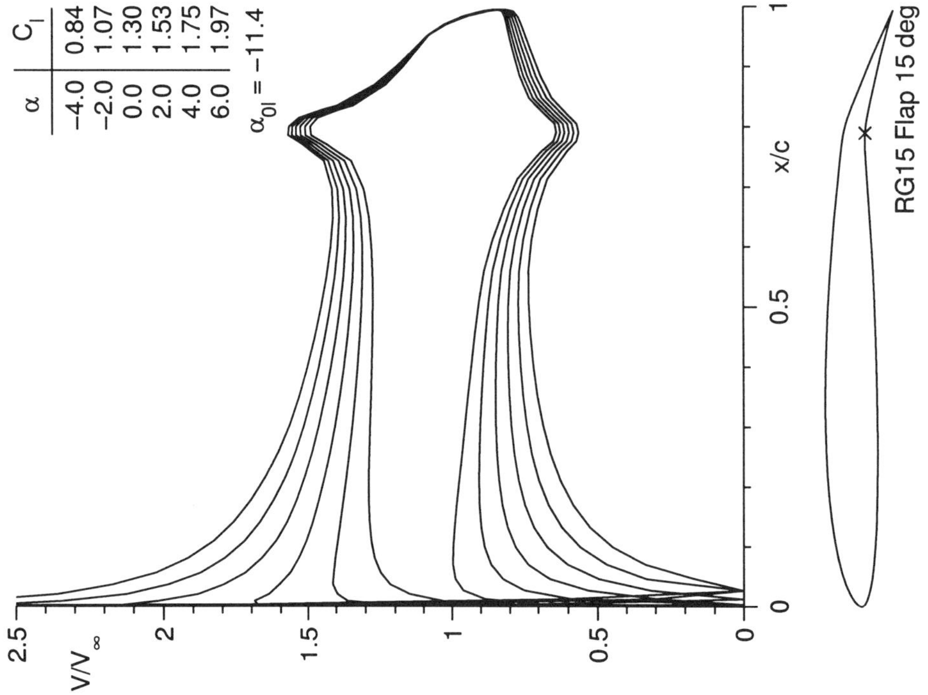
α | C_l
−4.0 | 0.84
−2.0 | 1.07
0.0 | 1.30
2.0 | 1.53
4.0 | 1.75
6.0 | 1.97
α_{0l} = −11.4
V/V_∞
2.5
2
1.5
1
0.5
0
x/c
0.5
RG15 Flap 15 deg

RG15 (C)

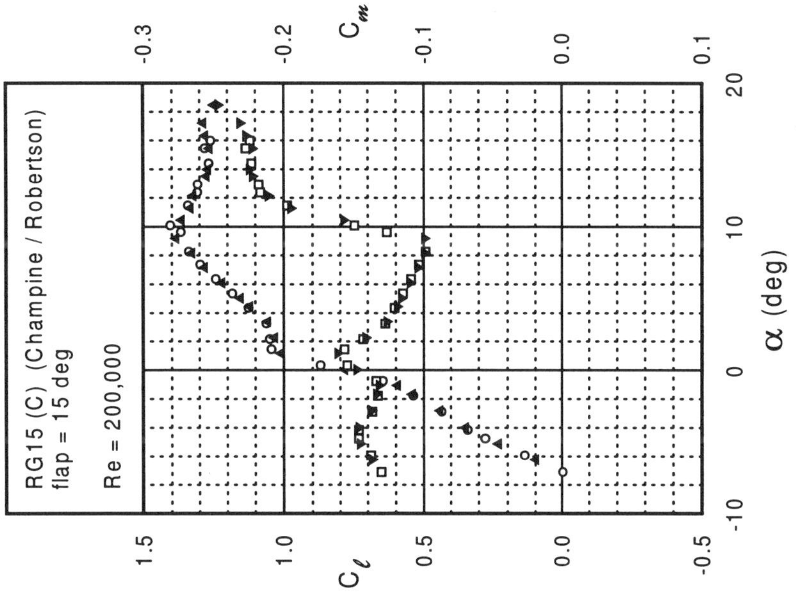
RG15 (C) (Champine / Robertson)
flap = 15 deg
Re = 200,000
C_m
-0.3
-0.2
-0.1
0.0
0.1
C_ℓ
1.5
1.0
0.5
0.0
-0.5
-10
0
10
20
α (deg)

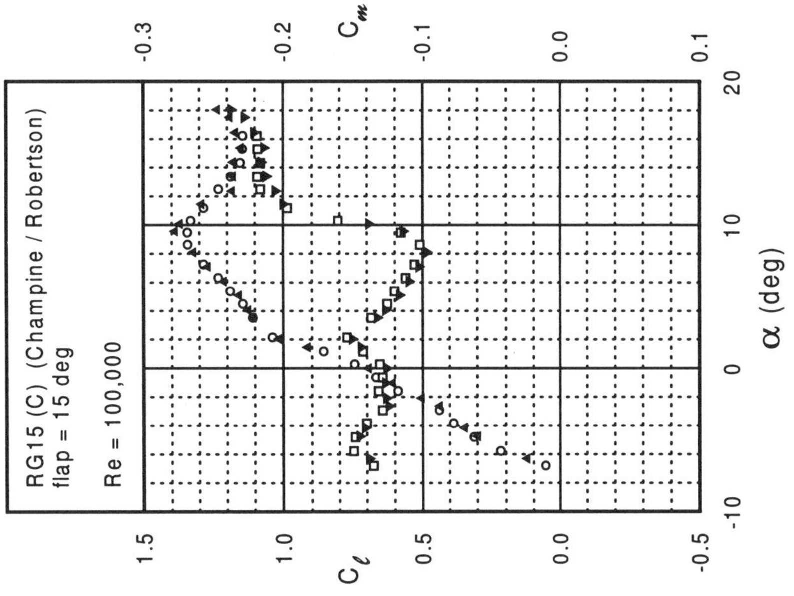
RG15 (C) (Champine / Robertson)
flap = 15 deg
Re = 100,000
C_m
-0.3
-0.2
-0.1
0.0
0.1
C_ℓ
1.5
1.0
0.5
0.0
-0.5
-10
0
10
20
α (deg)

RG15 (C)

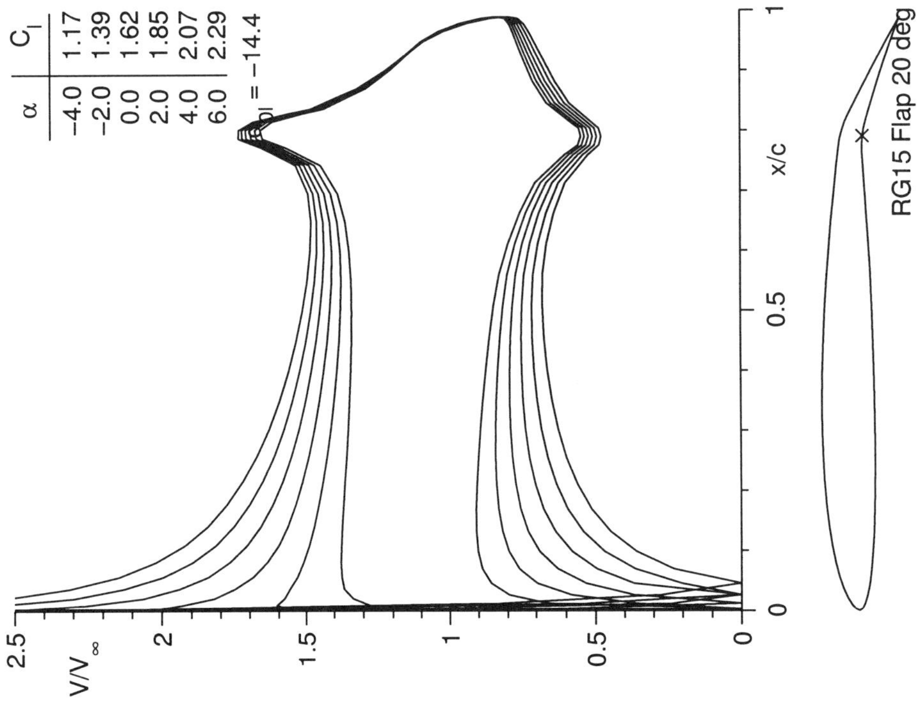

α | Cl
−4.0 | 1.17
−2.0 | 1.39
0.0 | 1.62
2.0 | 1.85
4.0 | 2.07
6.0 | 2.29
α0l = −14.4
V/V∞
x/c
RG15 Flap 20 deg

RG15 (C)

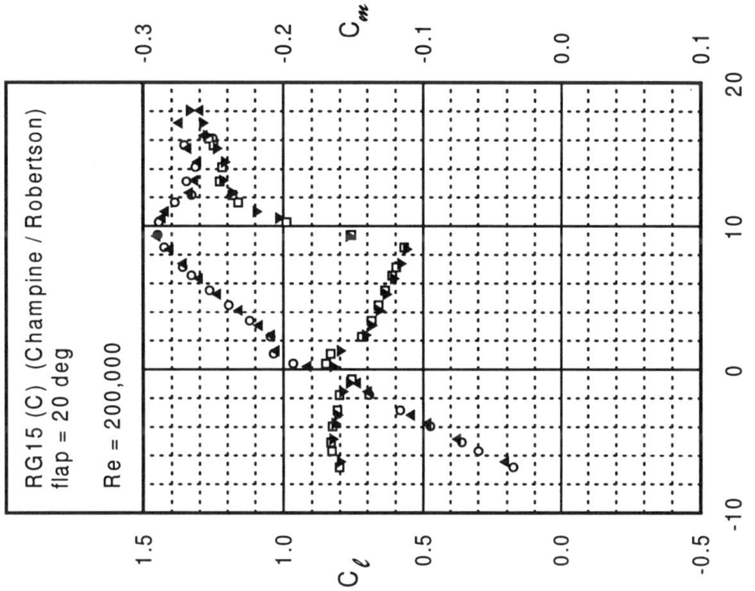
RG15 (C) (Champine / Robertson)
flap = 20 deg
Re = 200,000
C_ℓ
C_m
α (deg)

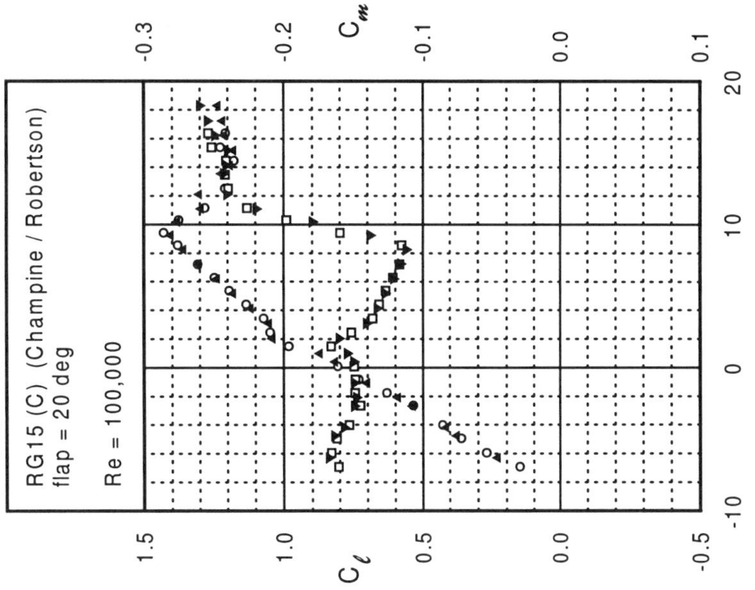
RG15 (C) (Champine / Robertson)
flap = 20 deg
Re = 100,000
C_ℓ
C_m
α (deg)

RG15 (C)

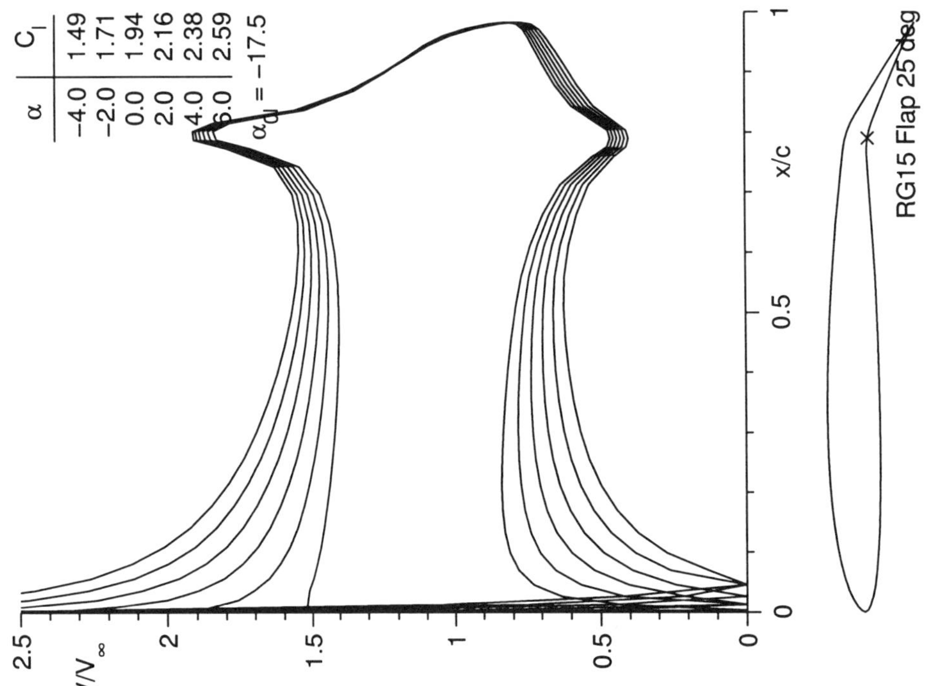
α C_l
−4.0 1.49
−2.0 1.71
0.0 1.94
2.0 2.16
4.0 2.38
6.0 2.59
α_0l = −17.5
V/V∞
2.5
2
1.5
1
0.5
0
x/c
0
0.5
1
RG15 Flap 25 deg

RG15 (C)

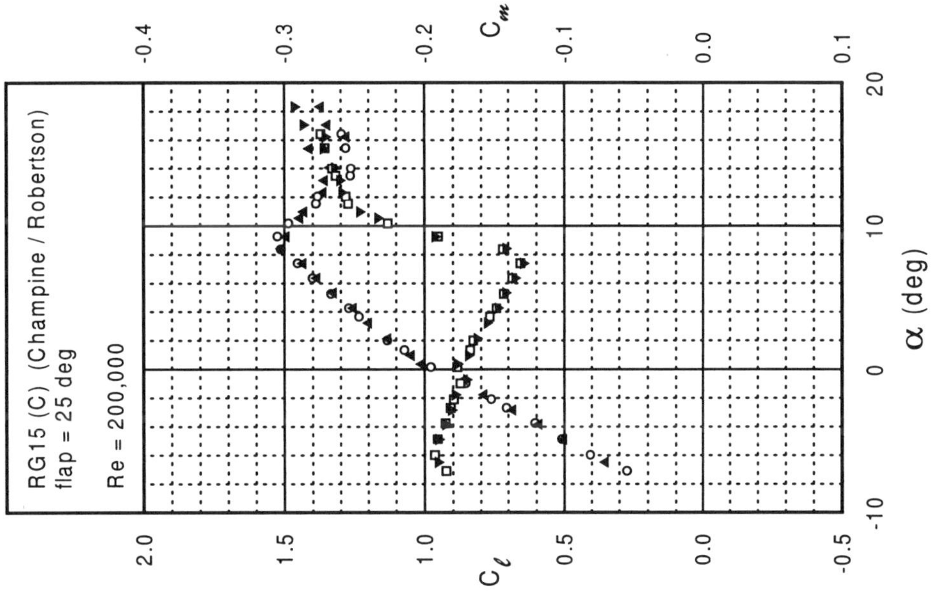
RG15 (C) (Champine / Robertson)
flap = 25 deg
Re = 200,000
α (deg)
C_ℓ
C_m

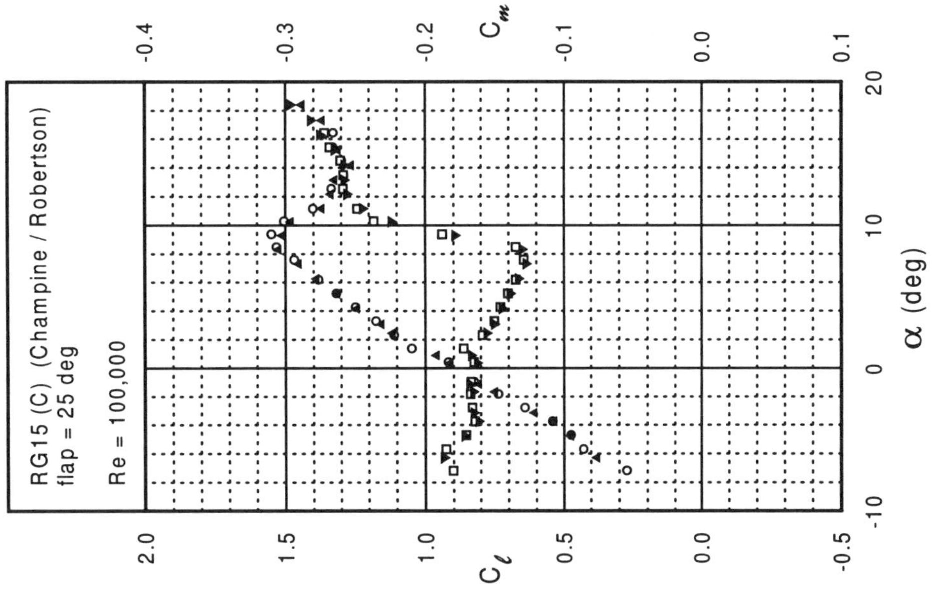
RG15 (C) (Champine / Robertson)
flap = 25 deg
Re = 100,000
α (deg)
C_ℓ
C_m

S6063

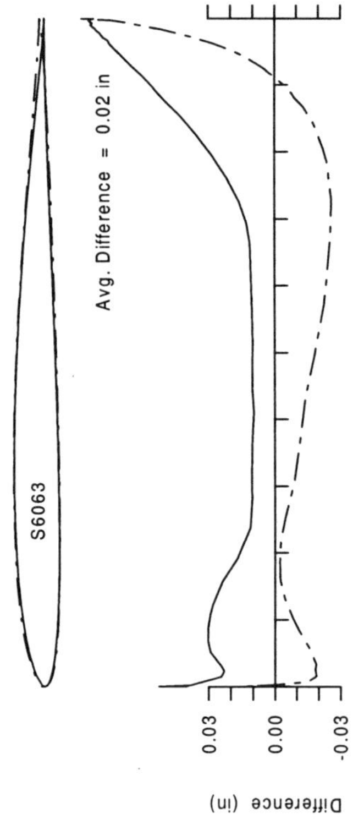
S6063
Avg. Difference = 0.02 in
0.03
0.00
-0.03
Difference (in)

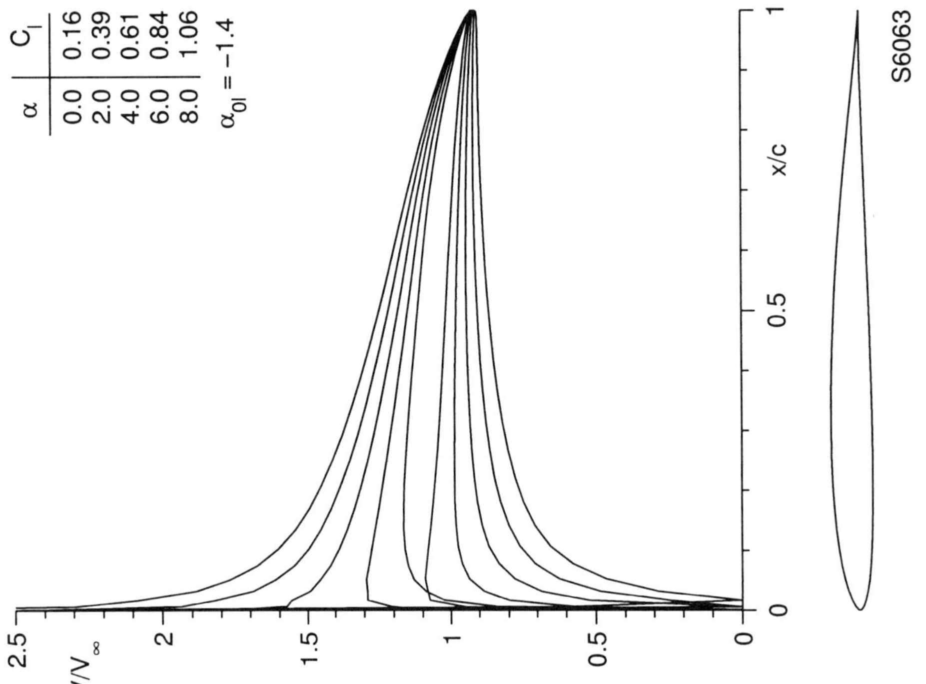
α | C_l
0.0 | 0.16
2.0 | 0.39
4.0 | 0.61
6.0 | 0.84
8.0 | 1.06
$\alpha_{0l} = -1.4$
V/V_∞
2.5
2
1.5
1
0.5
0
0
0.5
1
x/c
S6063

S6063

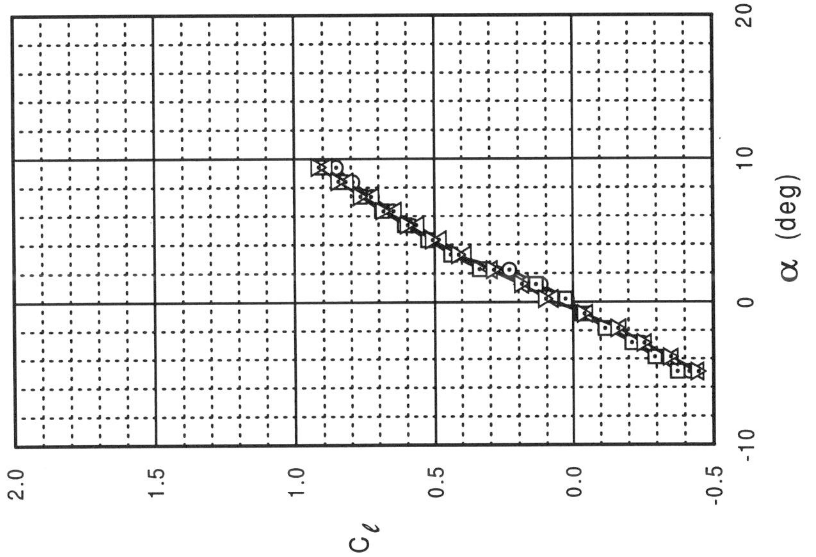

2.0
1.5
1.0
0.5
0.0
-0.5
C_ℓ
-10
0
10
20
α (deg)

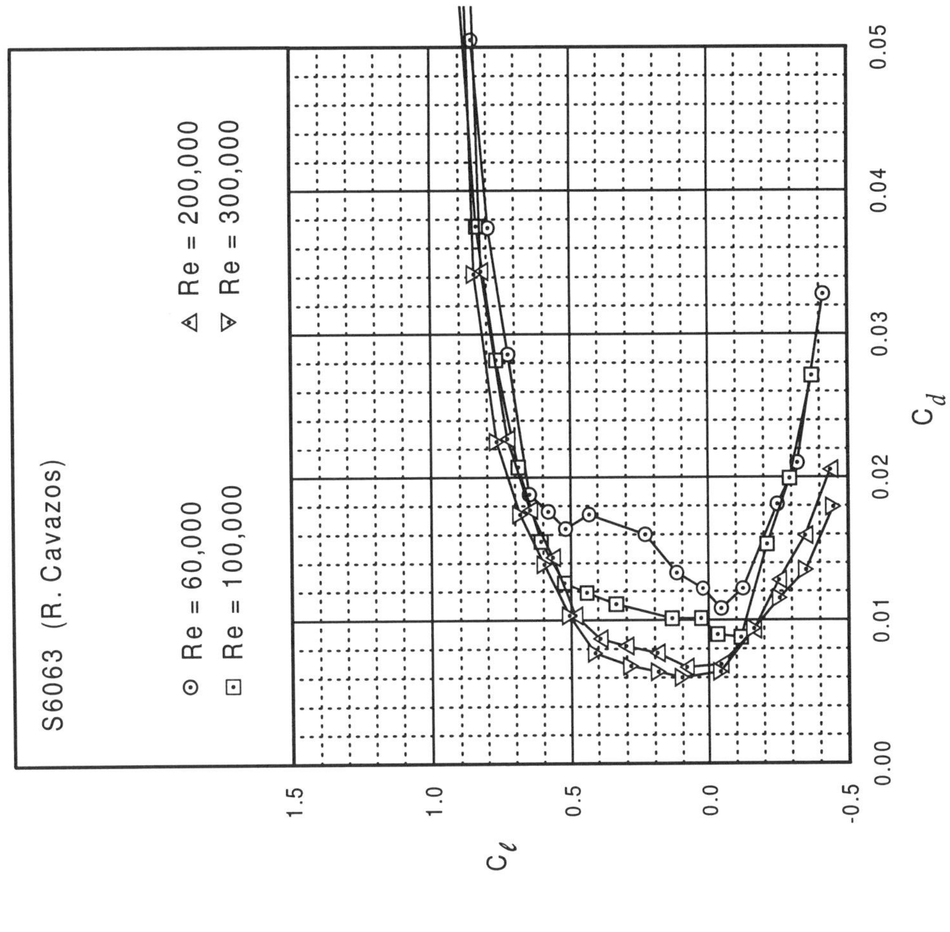

S6063 (R. Cavazos)
⊙ Re = 60,000
⊡ Re = 100,000
△ Re = 200,000
▽ Re = 300,000
1.5
1.0
0.5
0.0
-0.5
C_ℓ
0.00
0.01
0.02
0.03
0.04
0.05
C_d

S6063

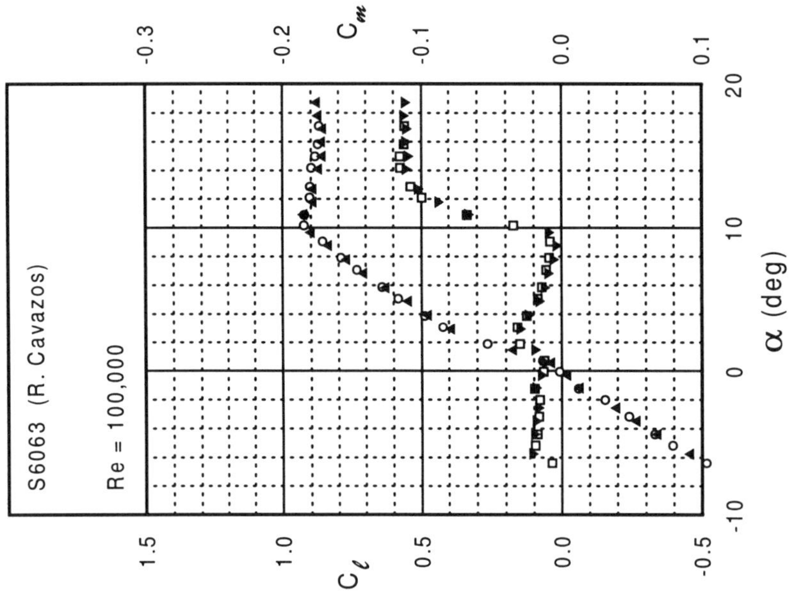

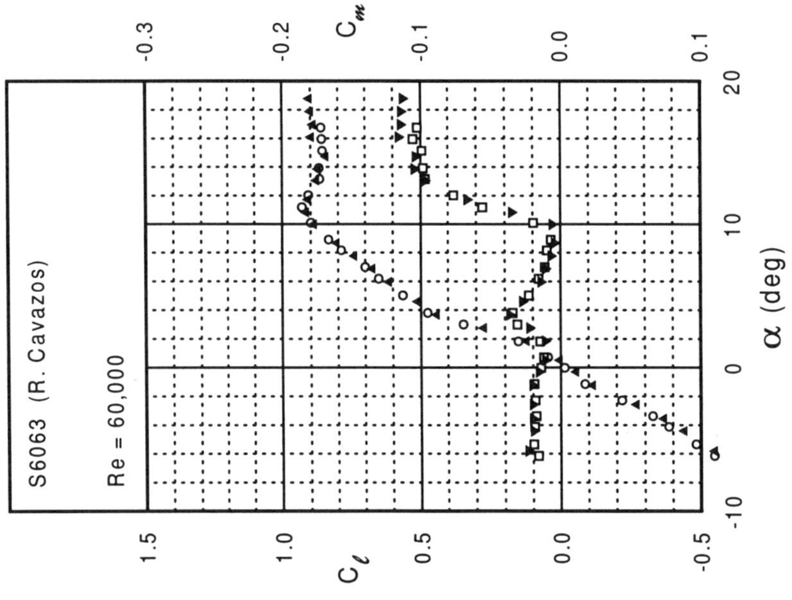

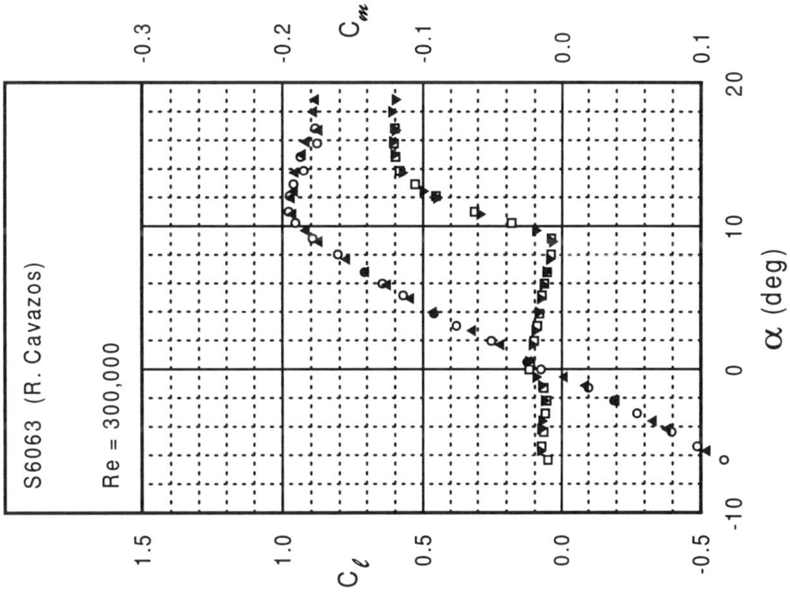
S6063 (R. Cavazos)
Re = 300,000
C_ℓ
1.5
1.0
0.5
0.0
-0.5
C_m
-0.3
-0.2
-0.1
0.0
0.1
-10
0
10
20
α (deg)

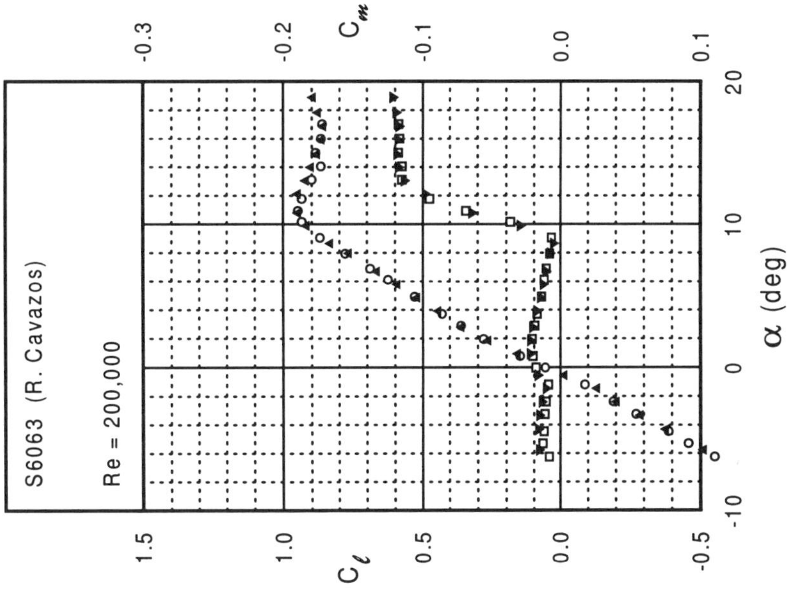
S6063 (R. Cavazos)
Re = 200,000
C_ℓ
1.5
1.0
0.5
0.0
-0.5
C_m
-0.3
-0.2
-0.1
0.0
0.1
-10
0
10
20
α (deg)

S7012 (B)

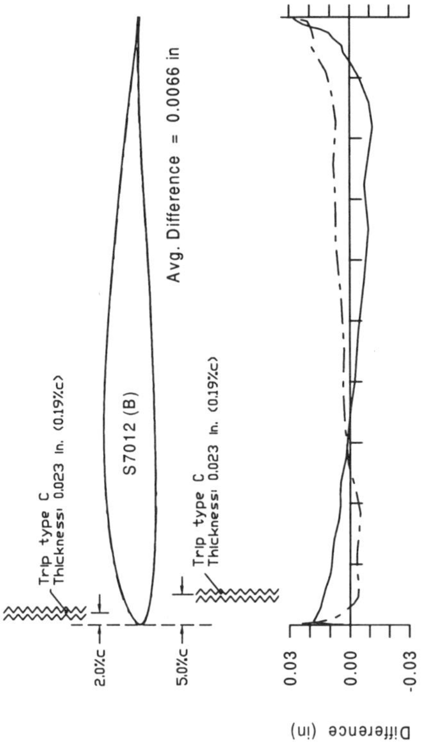

Trip type C
Thickness: 0.023 in. (0.19%c)
S7012 (B)
Trip type C
Thickness: 0.023 in. (0.19%c)
Avg. Difference = 0.0066 in
2.0%c
5.0%c
0.03
0.00
-0.03
Difference (in)

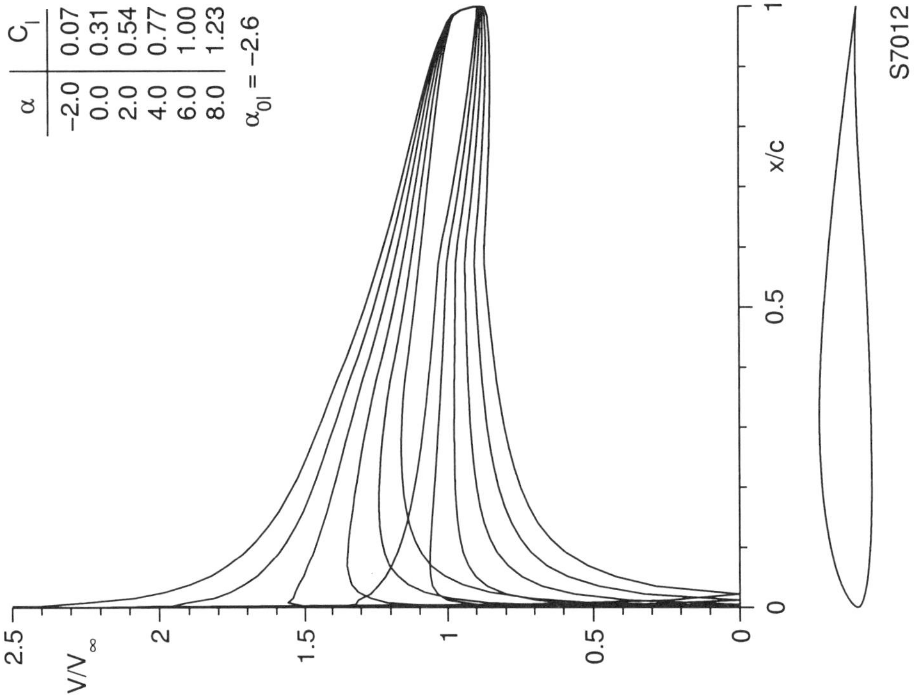

α
C_l
-2.0 0.07
0.0 0.31
2.0 0.54
4.0 0.77
6.0 1.00
8.0 1.23
α_0l = -2.6
V/V_∞
2.5
2
1.5
1
0.5
0
x/c
0
0.5
1
S7012

S7012 (B)

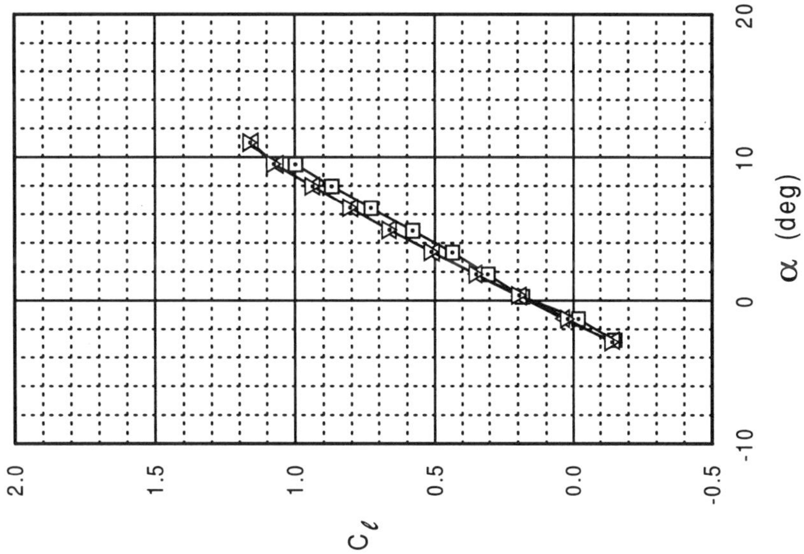
2.0
1.5
1.0
0.5
0.0
-0.5
C_ℓ
-10
0
10
20
α (deg)

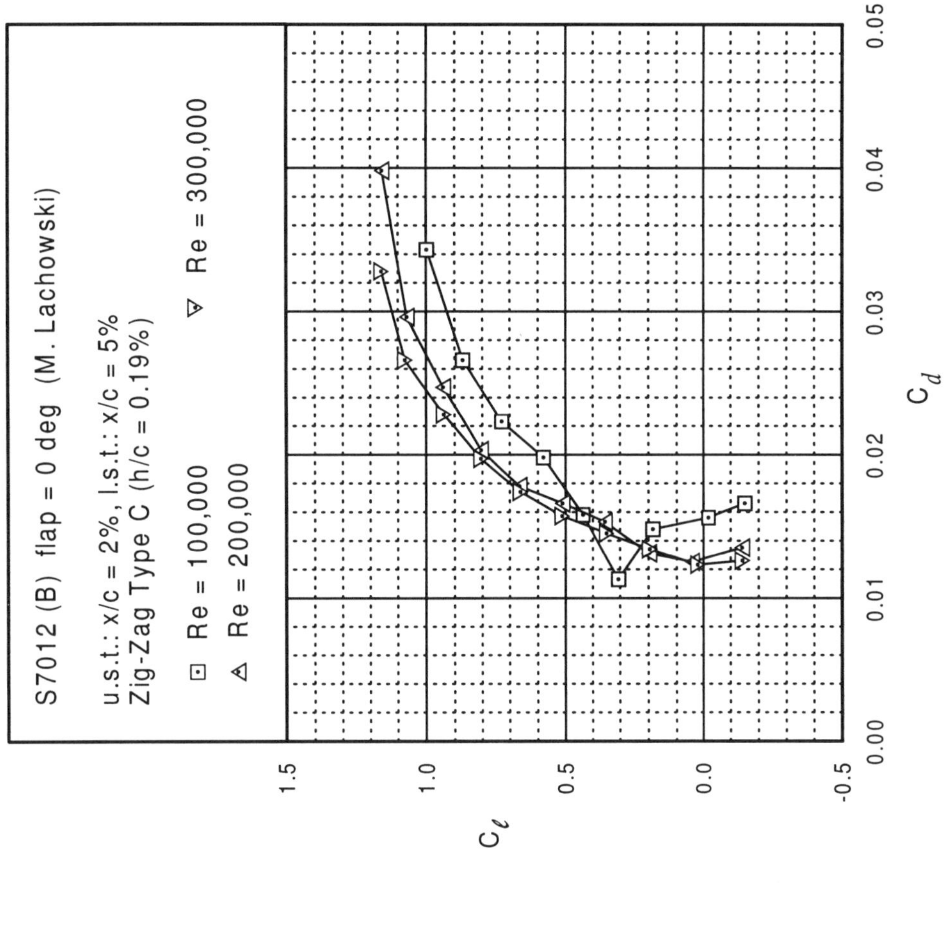
S7012 (B) flap = 0 deg (M. Lachowski)
u.s.t.: x/c = 2%, l.s.t.: x/c = 5%
Zig-Zag Type C (h/c = 0.19%)
Re = 100,000
Re = 200,000
Re = 300,000
1.5
1.0
0.5
0.0
-0.5
C_ℓ
0.00
0.01
0.02
0.03
0.04
0.05
C_d

S7012 (B)

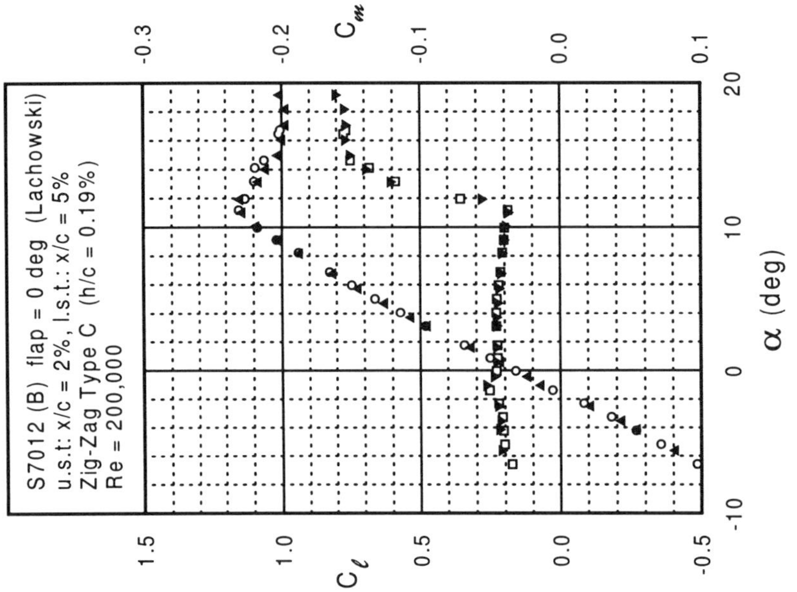
S7012 (B) flap = 0 deg (Lachowski)
u.s.t.: x/c = 2%, l.s.t.: x/c = 5%
Zig-Zag Type C (h/c = 0.19%)
Re = 200,000
C_ℓ
1.5
1.0
0.5
0.0
-0.5
C_m
-0.3
-0.2
-0.1
0.0
0.1
-10
0
10
20
α (deg)

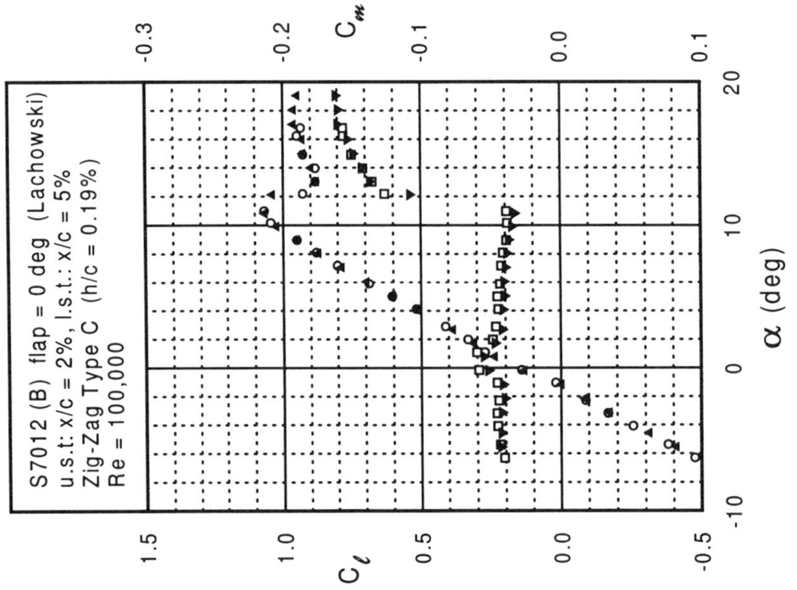
S7012 (B) flap = 0 deg (Lachowski)
u.s.t.: x/c = 2%, l.s.t.: x/c = 5%
Zig-Zag Type C (h/c = 0.19%)
Re = 100,000
C_ℓ
1.5
1.0
0.5
0.0
-0.5
C_m
-0.3
-0.2
-0.1
0.0
0.1
-10
0
10
20
α (deg)

S7012 (B)

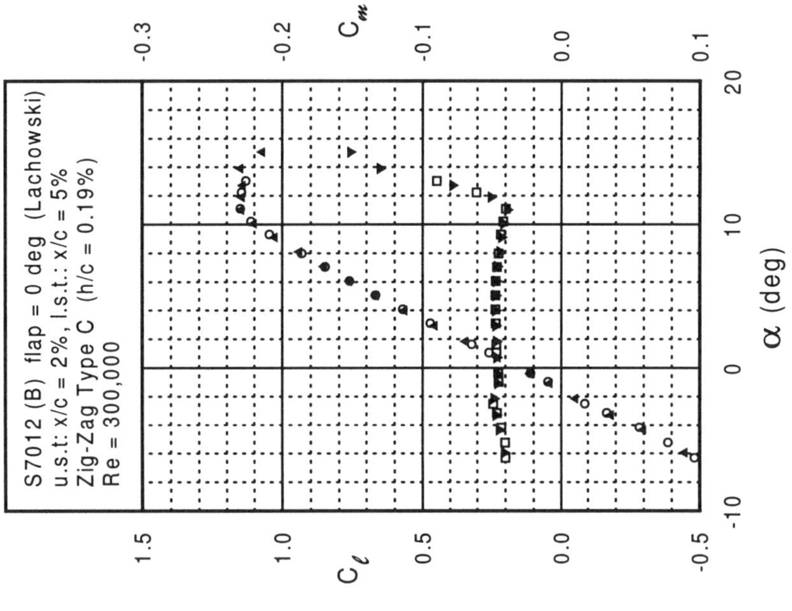
S7012 (B) flap = 0 deg (Lachowski)
u.s.t: x/c = 2%, l.s.t.: x/c = 5%
Zig-Zag Type C (h/c = 0.19%)
Re = 300,000
Cℓ
Cm
α (deg)

S7012 (B)

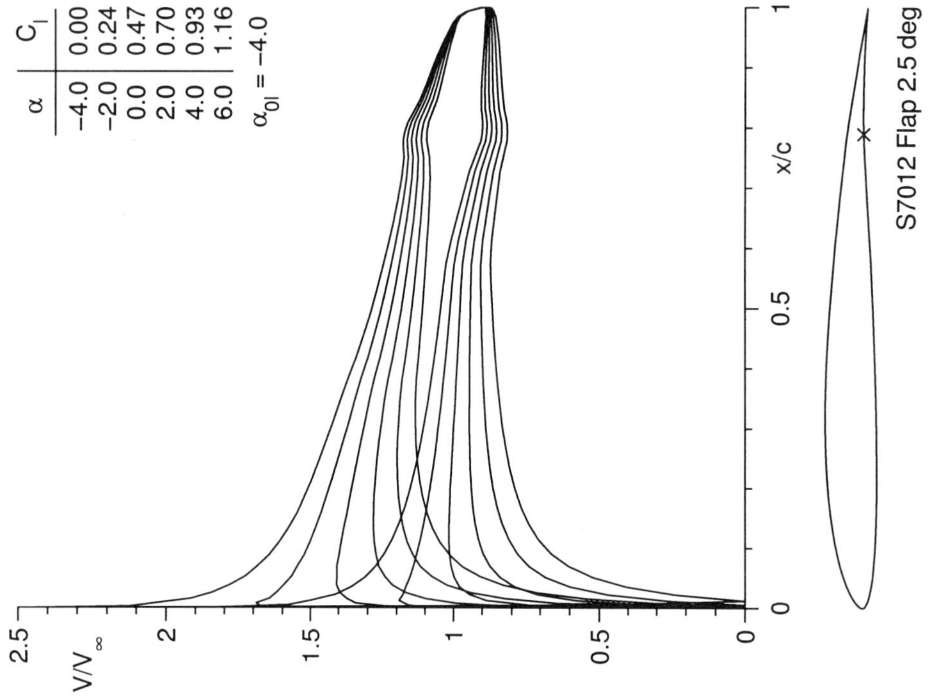
α | C_l
-4.0 | 0.00
-2.0 | 0.24
0.0 | 0.47
2.0 | 0.70
4.0 | 0.93
6.0 | 1.16
α_0l = -4.0
V/V∞
2.5
2
1.5
1
0.5
0
x/c
S7012 Flap 2.5 deg

S7012 (B)

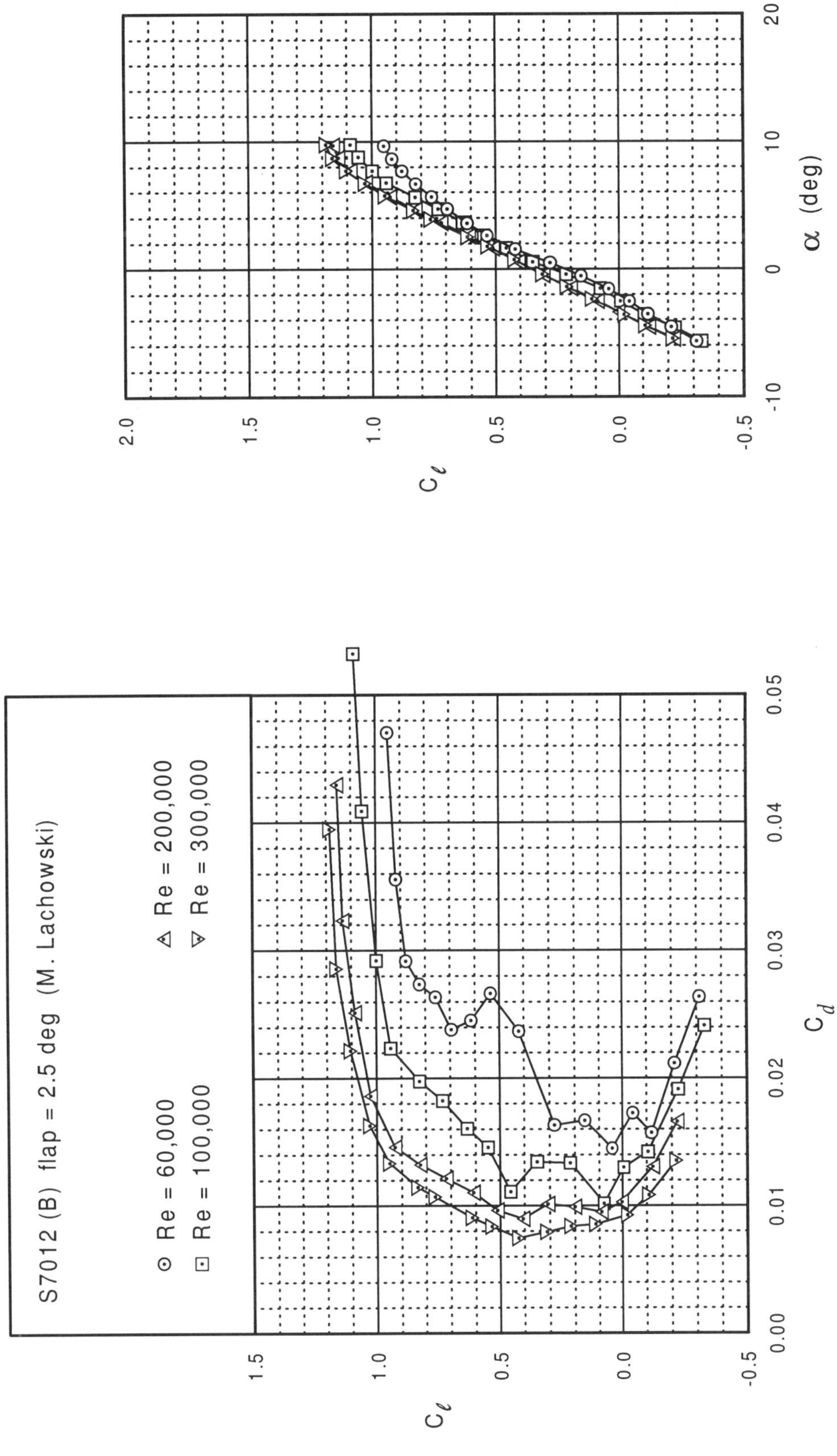
S7012 (B) flap = 2.5 deg (M. Lachowski)
Re = 60,000
Re = 100,000
Re = 200,000
Re = 300,000
C_ℓ
C_d
α (deg)

S7012 (B)

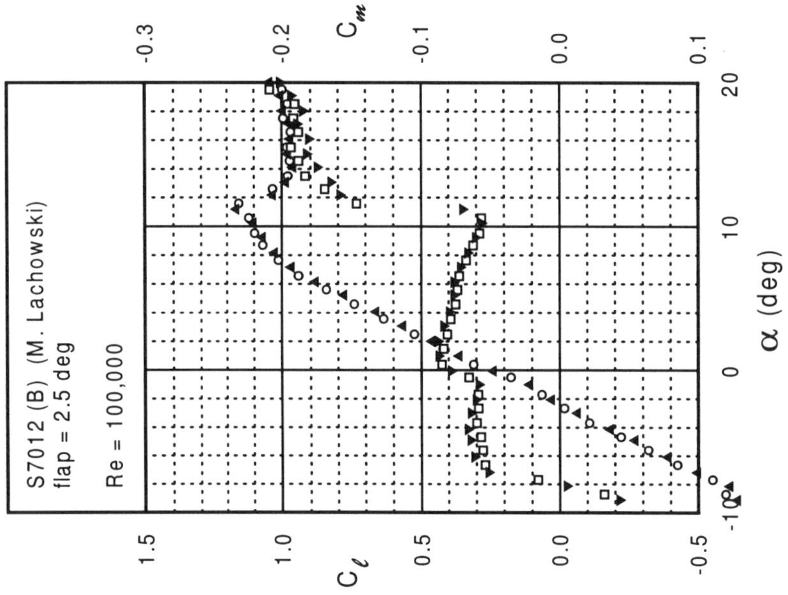
S7012 (B) (M. Lachowski)
flap = 2.5 deg
Re = 100,000
C_m
-0.3
-0.2
-0.1
0.0
0.1
C_ℓ
1.5
1.0
0.5
0.0
-0.5
-10
0
10
20
α (deg)

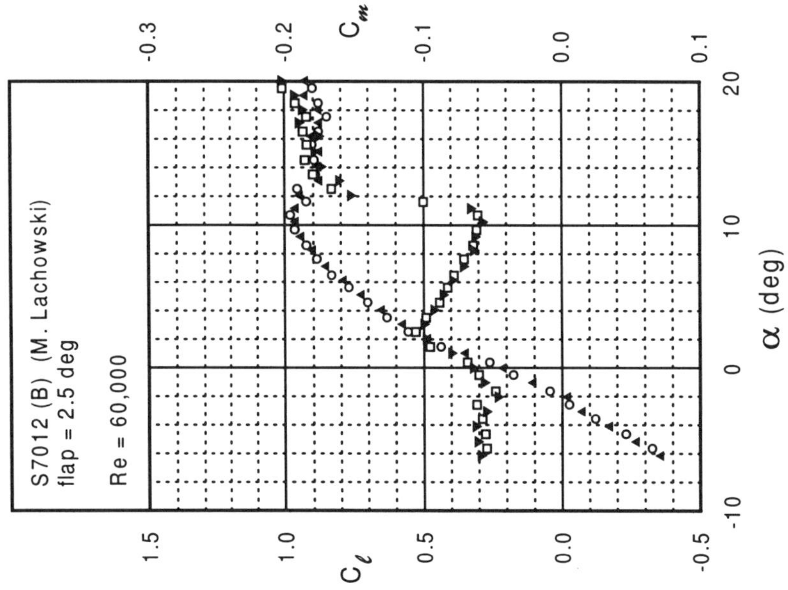
S7012 (B) (M. Lachowski)
flap = 2.5 deg
Re = 60,000
C_m
-0.3
-0.2
-0.1
0.0
0.1
C_ℓ
1.5
1.0
0.5
0.0
-0.5
-10
0
10
20
α (deg)

S7012 (B)

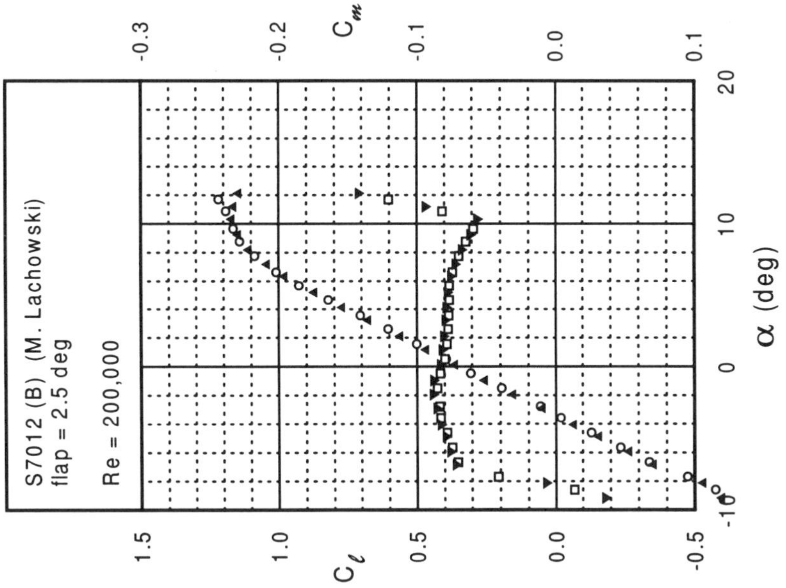
S7012 (B) (M. Lachowski)
flap = 2.5 deg
Re = 200,000
C_ℓ
C_m
α (deg)

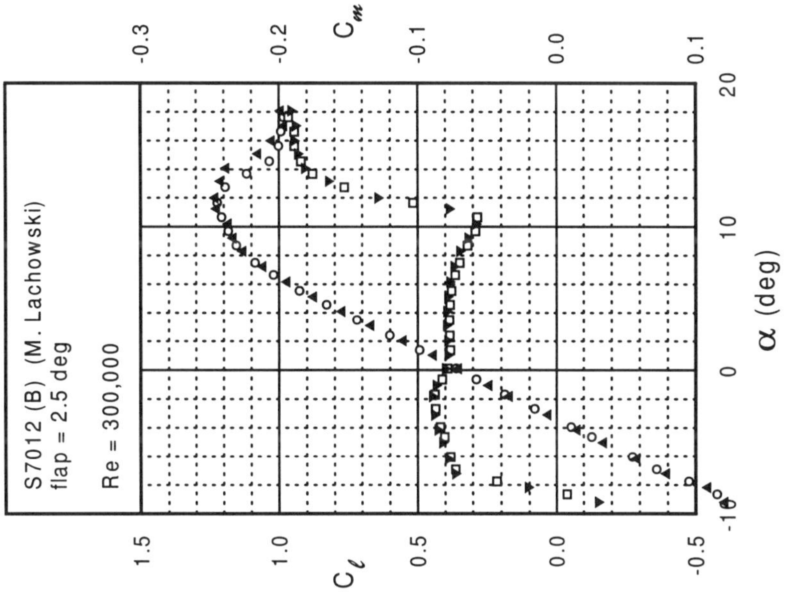
S7012 (B) (M. Lachowski)
flap = 2.5 deg
Re = 300,000
C_ℓ
C_m
α (deg)

S7012 (B)

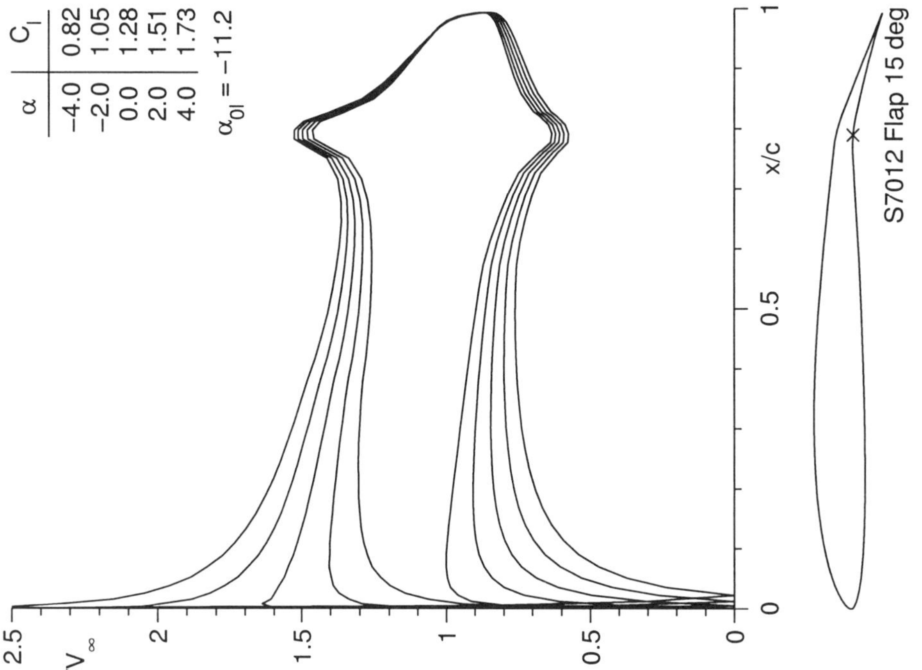
α | C_l
−4.0 | 0.82
−2.0 | 1.05
0.0 | 1.28
2.0 | 1.51
4.0 | 1.73
α_{0l} = −11.2
V/V_∞
2.5
2
1.5
1
0.5
0
x/c
0.5
1
S7012 Flap 15 deg

S7012 (B)

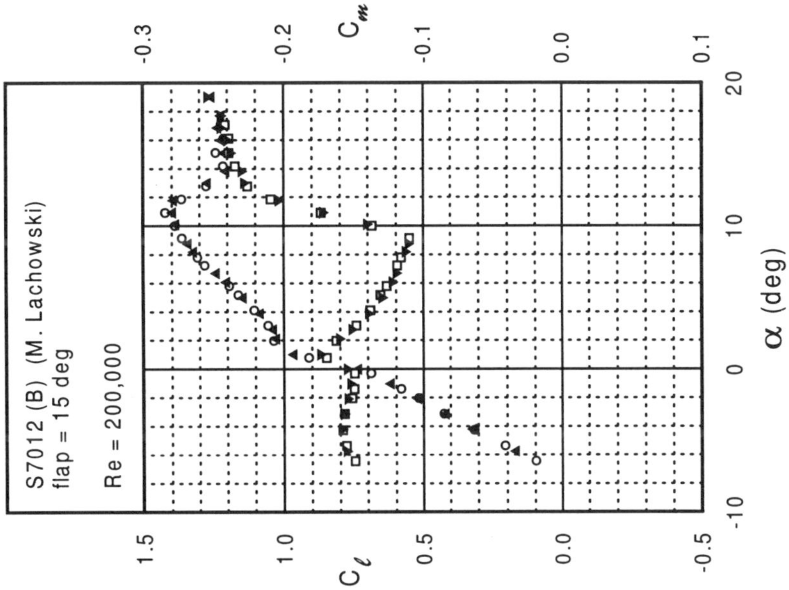
S7012 (B) (M. Lachowski)
flap = 15 deg
Re = 200,000
C_m
C_l
α (deg)

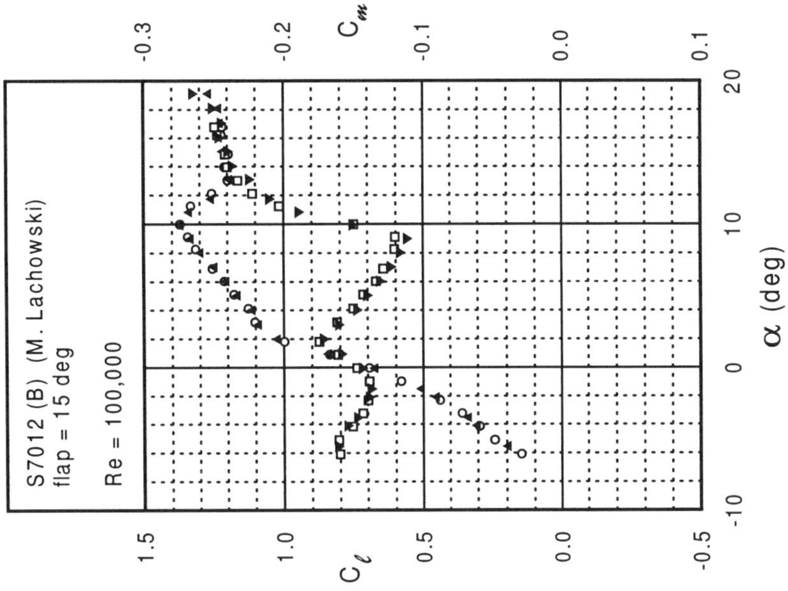
S7012 (B) (M. Lachowski)
flap = 15 deg
Re = 100,000
C_m
C_l
α (deg)

S7012 (B)

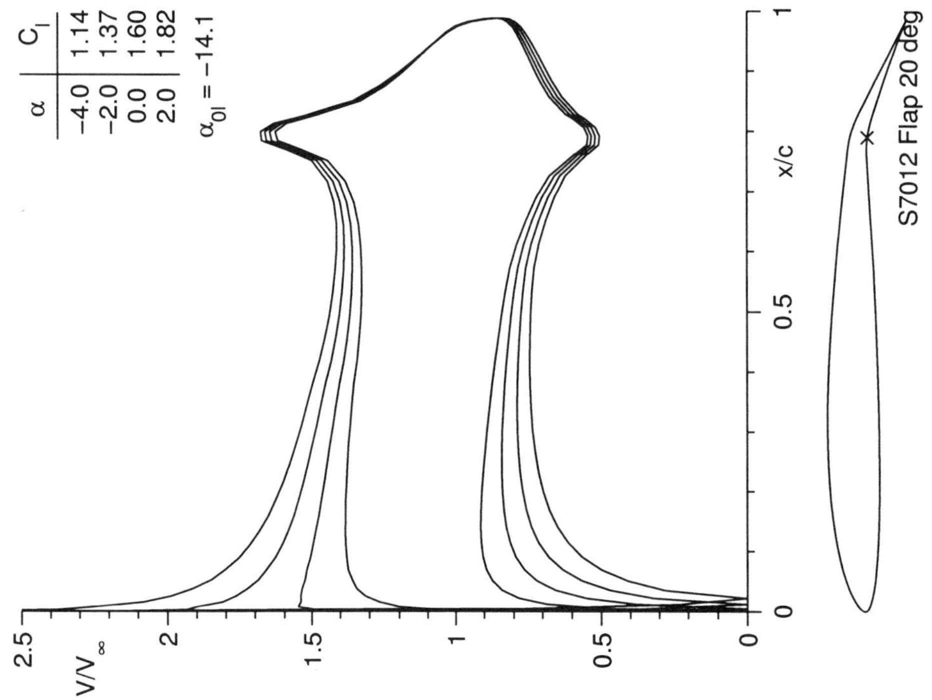

α | C_l
-4.0 | 1.14
-2.0 | 1.37
0.0 | 1.60
2.0 | 1.82
α_{0l} = −14.1
V/V∞
2.5
2
1.5
1
0.5
0
x/c
0.5
1
S7012 Flap 20 deg

S7012 (B)

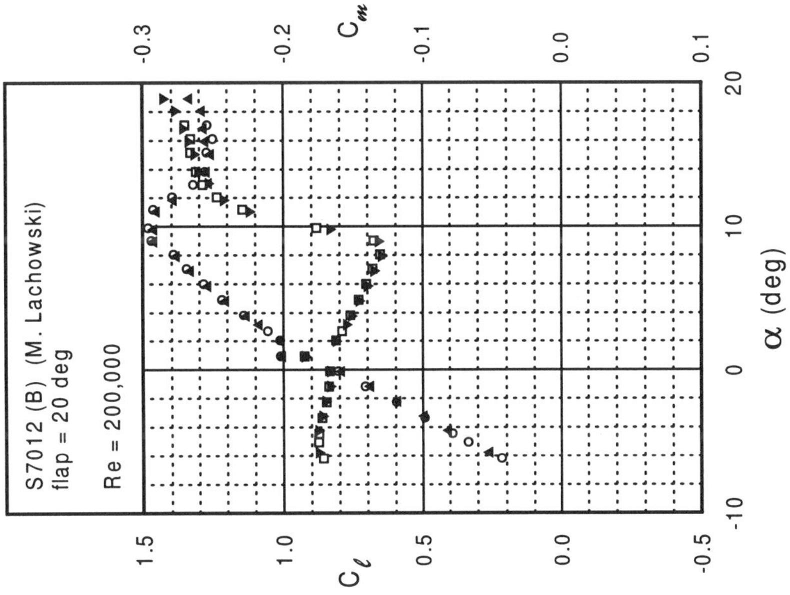
S7012 (B) (M. Lachowski)
flap = 20 deg
Re = 200,000
C_ℓ
C_m
α (deg)
1.5
1.0
0.5
0.0
-0.5
-0.3
-0.2
-0.1
0.0
0.1
-10
0
10
20

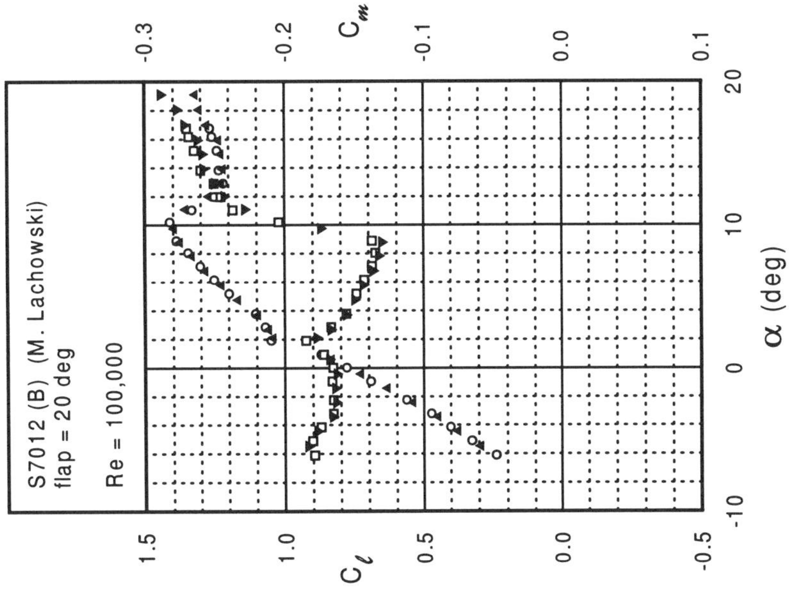
S7012 (B) (M. Lachowski)
flap = 20 deg
Re = 100,000
C_ℓ
C_m
α (deg)
1.5
1.0
0.5
0.0
-0.5
-0.3
-0.2
-0.1
0.0
0.1
-10
0
10
20

S7012 (B)

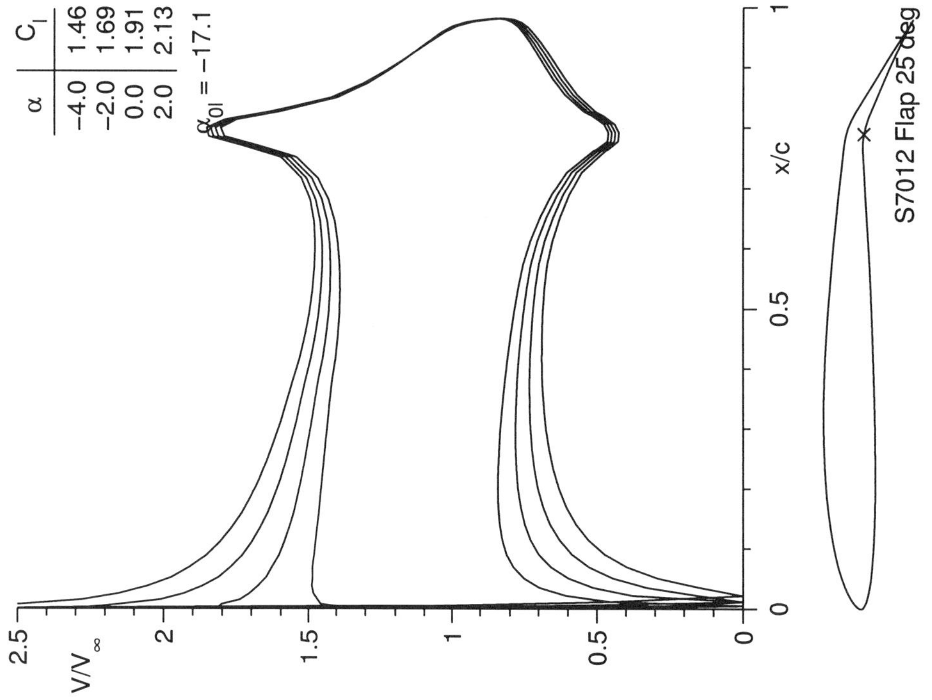
α | C_l
−4.0 | 1.46
−2.0 | 1.69
0.0 | 1.91
2.0 | 2.13
α_0l = −17.1
V/V∞
2.5
2
1.5
1
0.5
0
x/c
0.5
1
S7012 Flap 25 deg

S7012 (B)

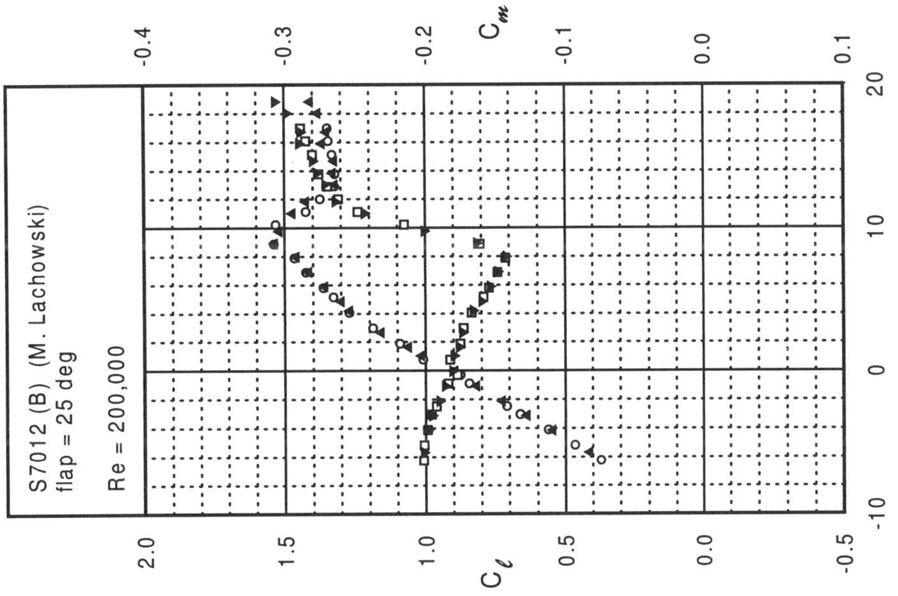
S7012 (B) (M. Lachowski)
flap = 25 deg
Re = 200,000
C_m
-0.4
-0.3
-0.2
-0.1
0.0
0.1
C_l
2.0
1.5
1.0
0.5
0.0
-0.5
-10
0
10
20
α (deg)

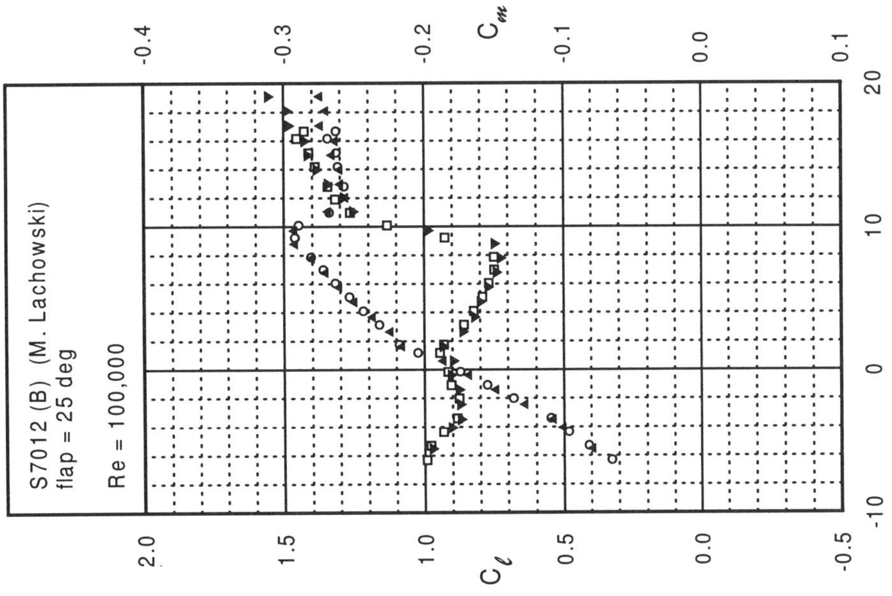
S7012 (B) (M. Lachowski)
flap = 25 deg
Re = 100,000
C_m
-0.4
-0.3
-0.2
-0.1
0.0
0.1
C_l
2.0
1.5
1.0
0.5
0.0
-0.5
-10
0
10
20
α (deg)

S7075 (A)

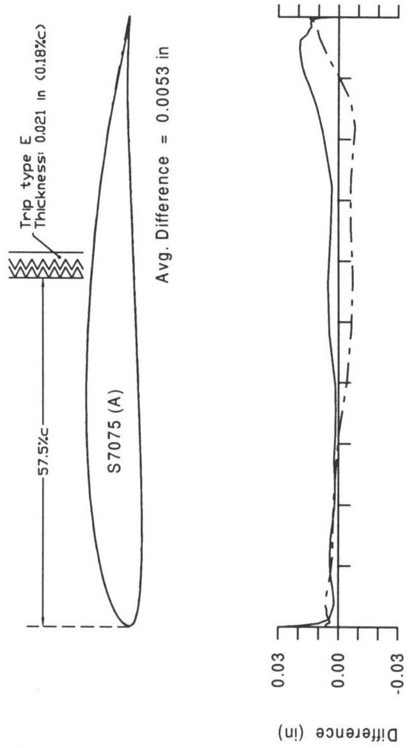
Trip type E
Thickness: 0.021 in (0.18%c)
57.5%c
S7075 (A)
Avg. Difference = 0.0053 in
0.03
0.00
-0.03
Difference (in)

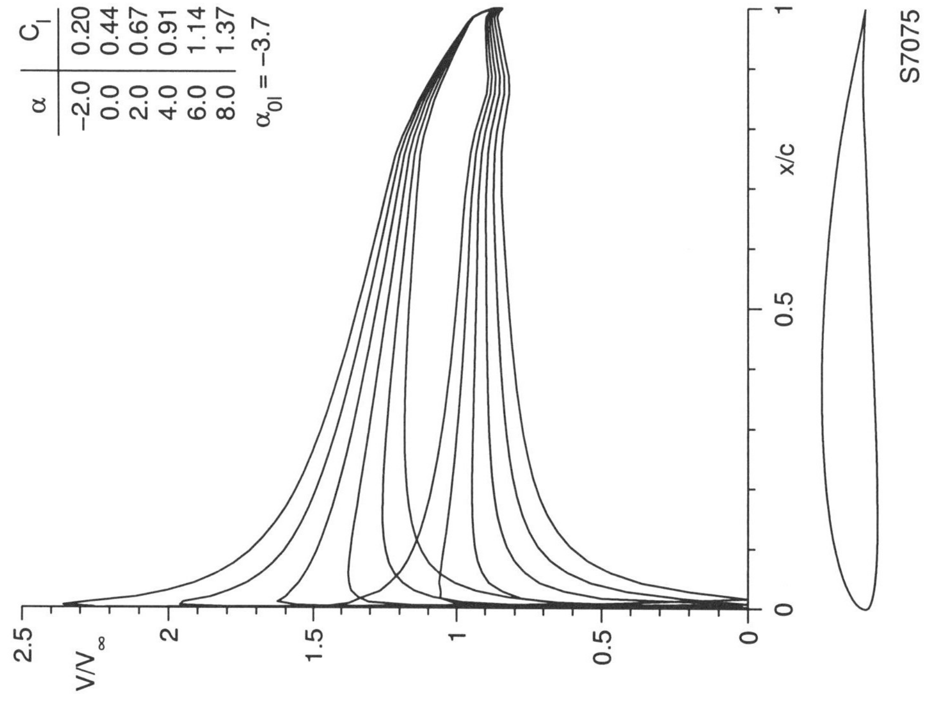
α | C_l
−2.0 | 0.20
0.0 | 0.44
2.0 | 0.67
4.0 | 0.91
6.0 | 1.14
8.0 | 1.37
α_{0l} = −3.7
V/V_∞
2.5
2
1.5
1
0.5
0
x/c
0.5
1
S7075

S7075 (A)

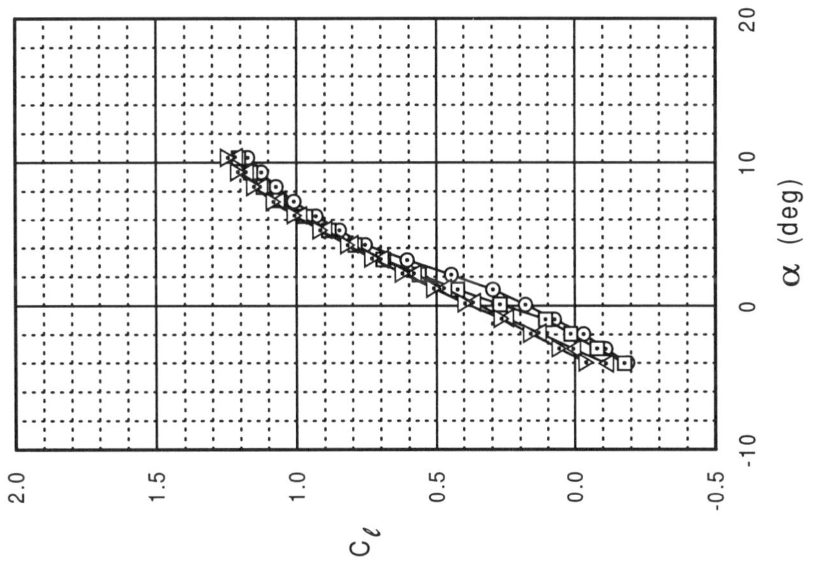
α (deg)
C_ℓ
2.0
1.5
1.0
0.5
0.0
-0.5
-10
0
10
20

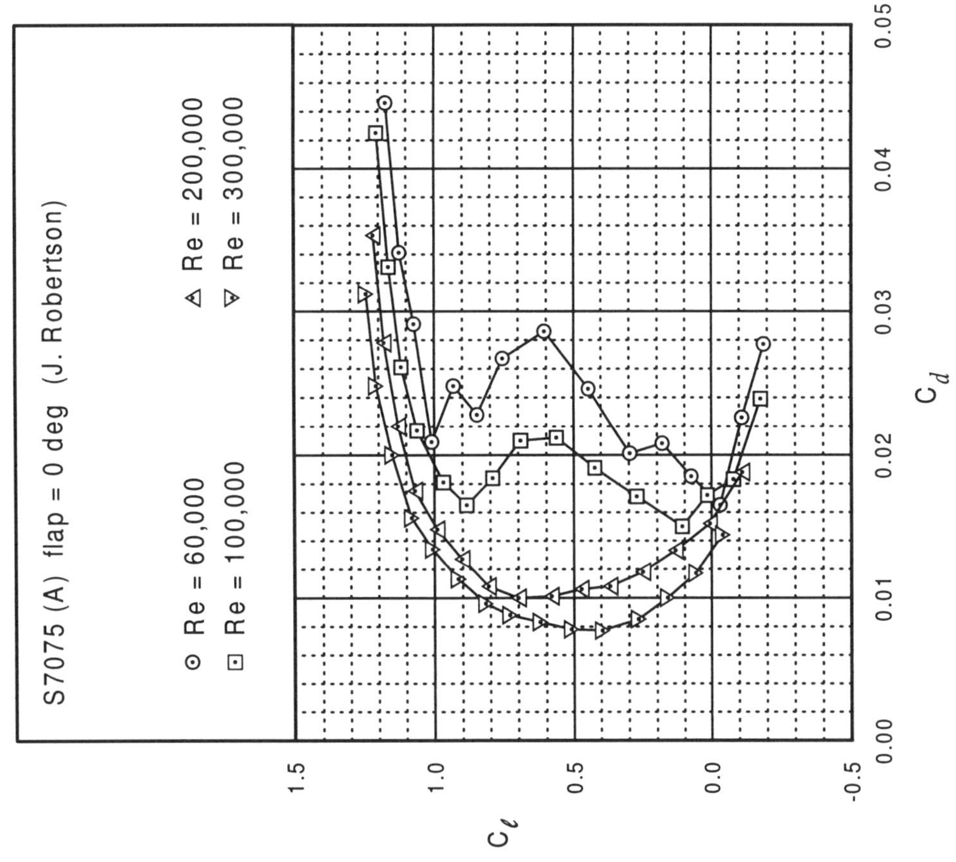
S7075 (A) flap = 0 deg (J. Robertson)
Re = 60,000
Re = 100,000
Re = 200,000
Re = 300,000
C_d
C_ℓ
1.5
1.0
0.5
0.0
-0.5
0.00
0.01
0.02
0.03
0.04
0.05

S7075 (A)

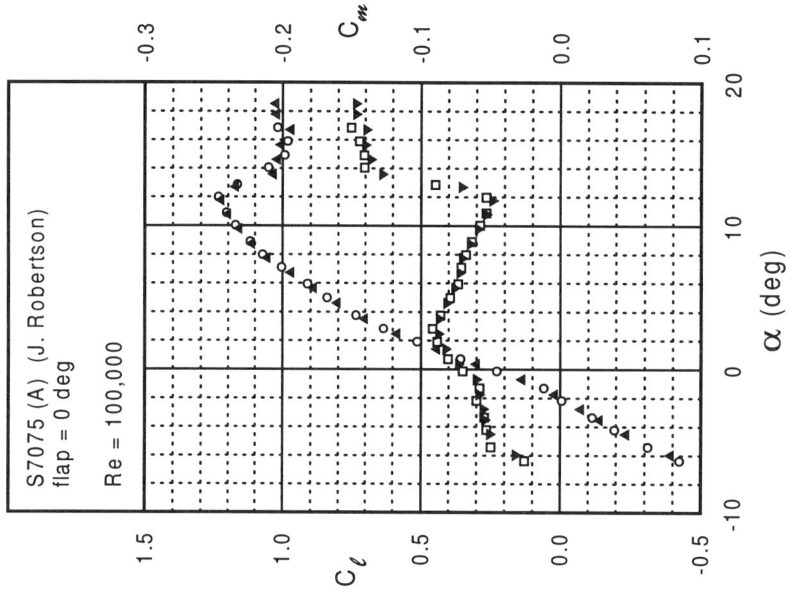
S7075 (A) (J. Robertson)
flap = 0 deg
Re = 100,000
C_ℓ
C_m
α (deg)
1.5
1.0
0.5
0.0
-0.5
-0.3
-0.2
-0.1
0.0
0.1
-10
0
10
20

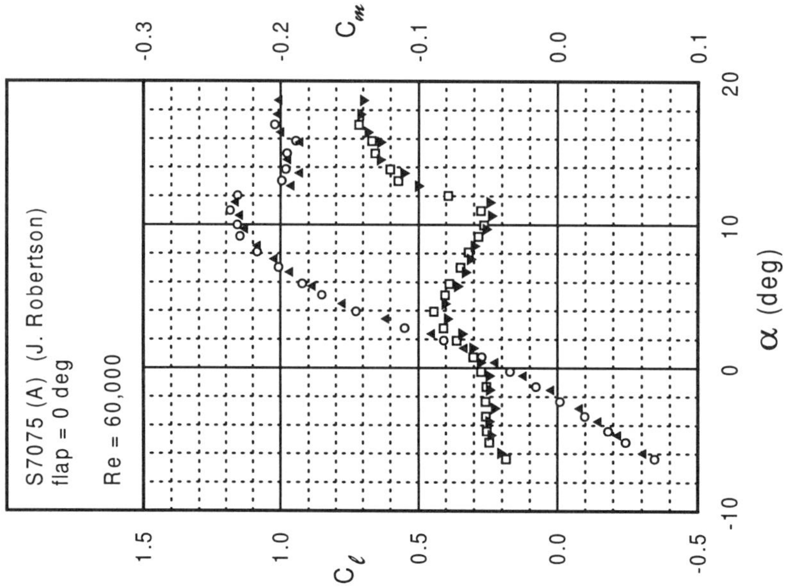
S7075 (A) (J. Robertson)
flap = 0 deg
Re = 60,000
C_ℓ
C_m
α (deg)
1.5
1.0
0.5
0.0
-0.5
-0.3
-0.2
-0.1
0.0
0.1
-10
0
10
20

S7075 (A)

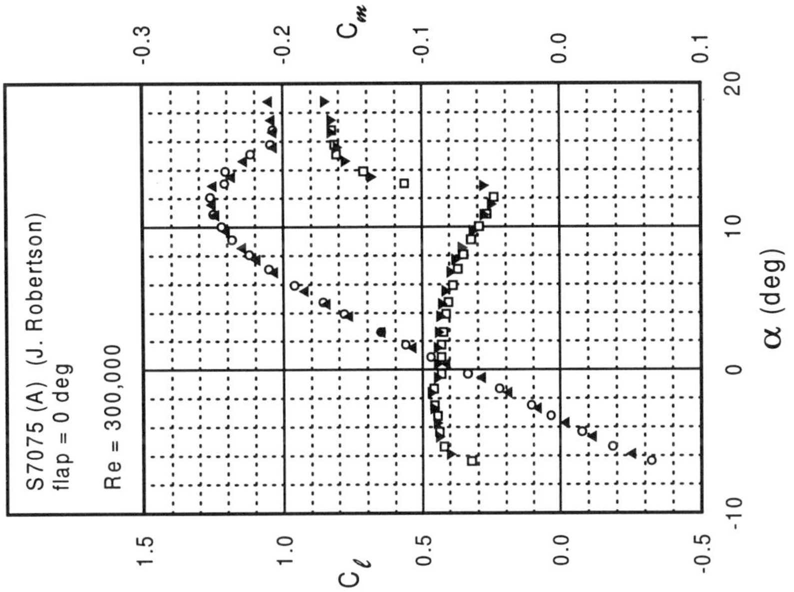
S7075 (A) (J. Robertson)
flap = 0 deg
Re = 300,000
C_ℓ
C_m
α (deg)

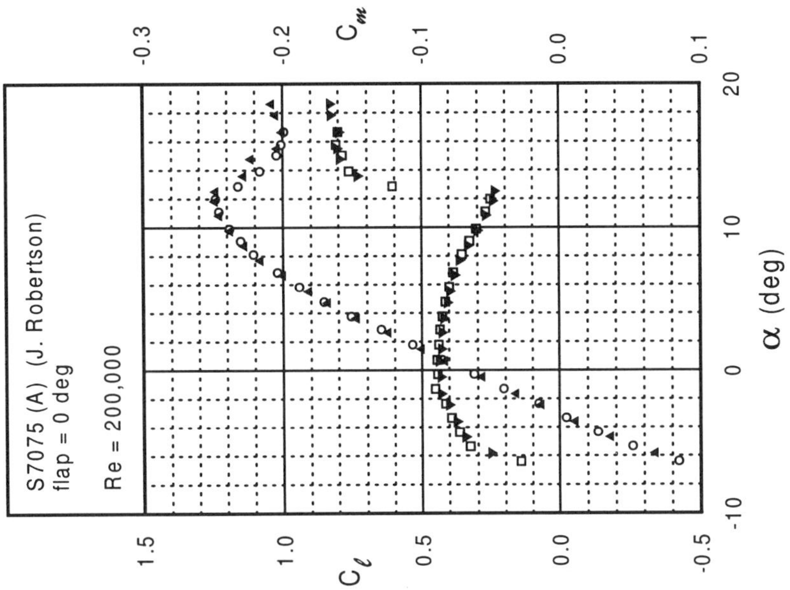
S7075 (A) (J. Robertson)
flap = 0 deg
Re = 200,000
C_ℓ
C_m
α (deg)

S7075 (A)

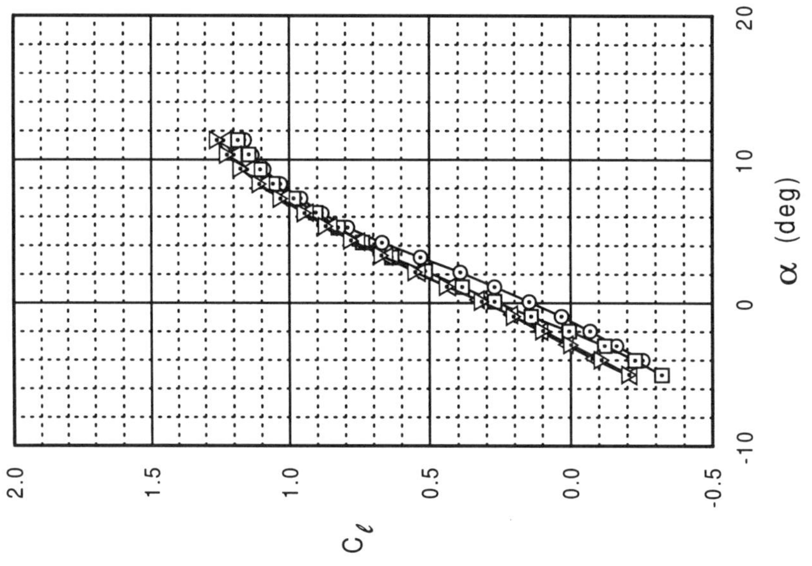
α (deg)
Cℓ

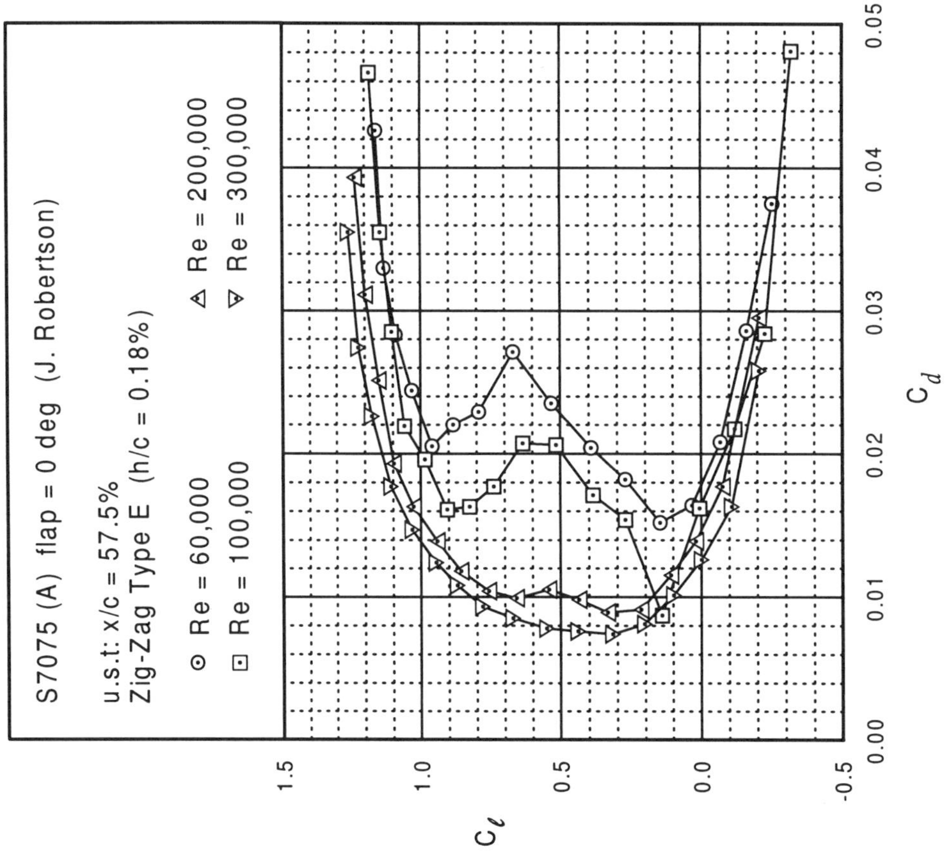
S7075 (A) flap = 0 deg (J. Robertson)
u.s.t: x/c = 57.5%
Zig-Zag Type E (h/c = 0.18%)
Re = 60,000
Re = 100,000
Re = 200,000
Re = 300,000
Cd
Cℓ

S7075 (A)

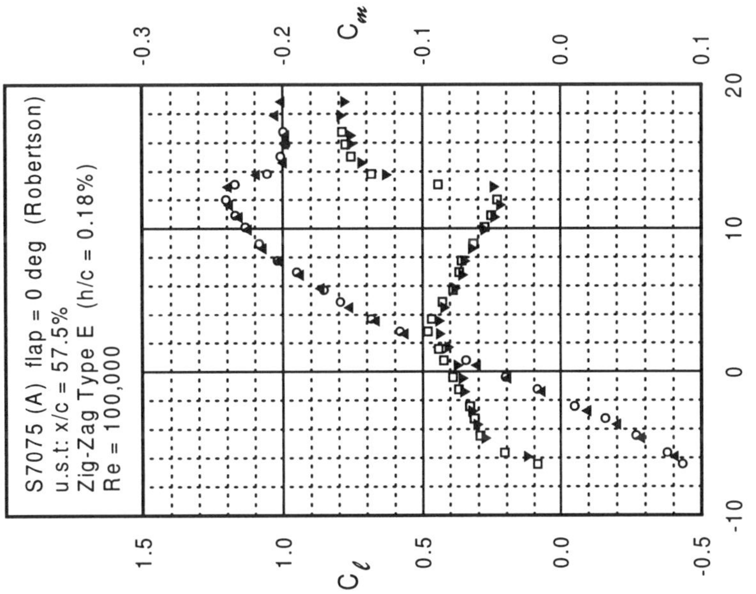
S7075 (A) flap = 0 deg (Robertson)
u.s.t: x/c = 57.5%
Zig-Zag Type E (h/c = 0.18%)
Re = 100,000
C_m
-0.3
-0.2
-0.1
0.0
0.1
C_ℓ
1.5
1.0
0.5
0.0
-0.5
-10
0
10
20
α (deg)

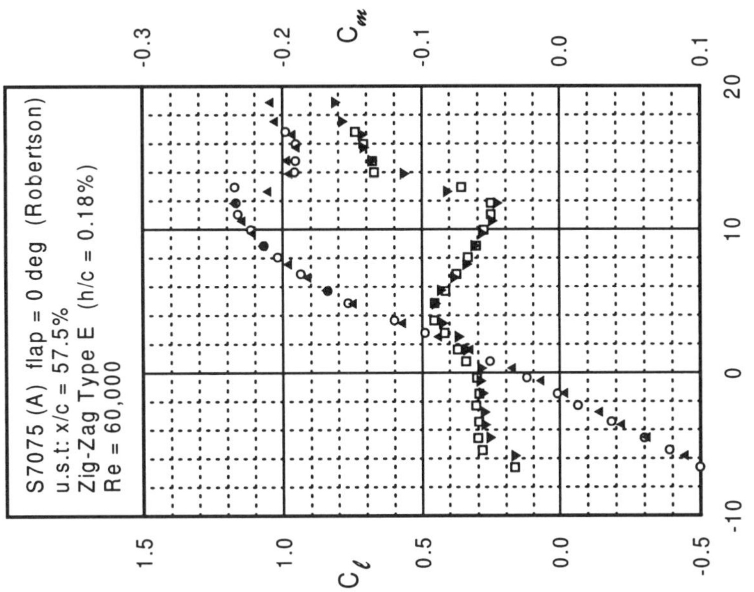
S7075 (A) flap = 0 deg (Robertson)
u.s.t: x/c = 57.5%
Zig-Zag Type E (h/c = 0.18%)
Re = 60,000
C_m
-0.3
-0.2
-0.1
0.0
0.1
C_ℓ
1.5
1.0
0.5
0.0
-0.5
-10
0
10
20
α (deg)

S7075 (A)

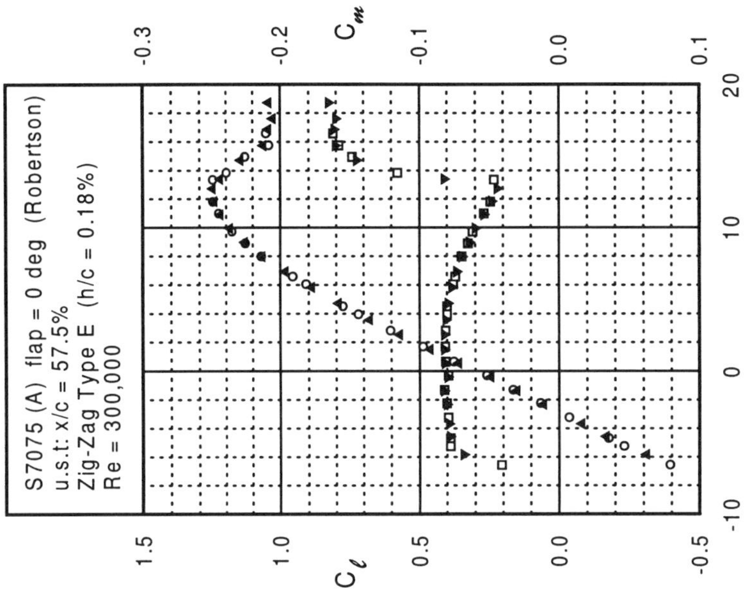
S7075 (A) flap = 0 deg (Robertson)
u.s.t: x/c = 57.5%
Zig-Zag Type E (h/c = 0.18%)
Re = 300,000
C_m
C_l
α (deg)

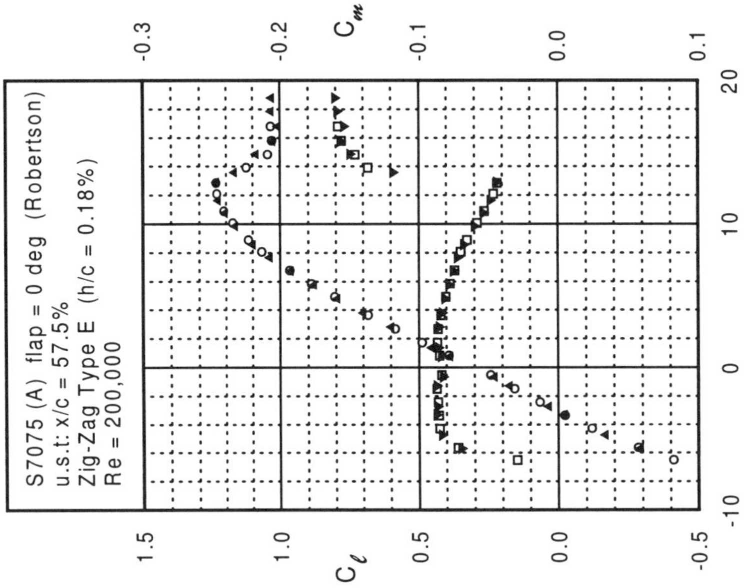
S7075 (A) flap = 0 deg (Robertson)
u.s.t: x/c = 57.5%
Zig-Zag Type E (h/c = 0.18%)
Re = 200,000
C_m
C_l
α (deg)

S7075 (B)

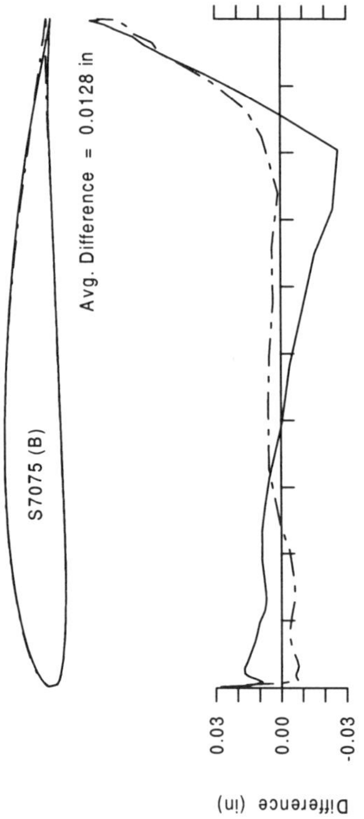
S7075 (B)
Avg. Difference = 0.0128 in
0.03
0.00
-0.03
Difference (in)

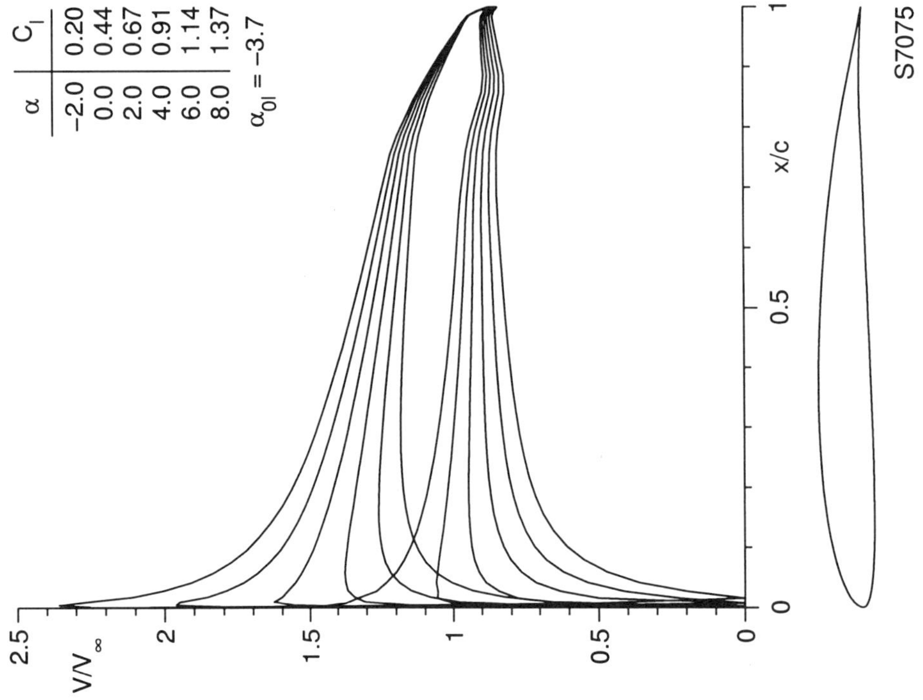
α
C_l
−2.0 0.20
0.0 0.44
2.0 0.67
4.0 0.91
6.0 1.14
8.0 1.37
α_0l = −3.7
2.5
V/V_∞
2
1.5
1
0.5
0
0.5
1
x/c
S7075

S7075 (B)

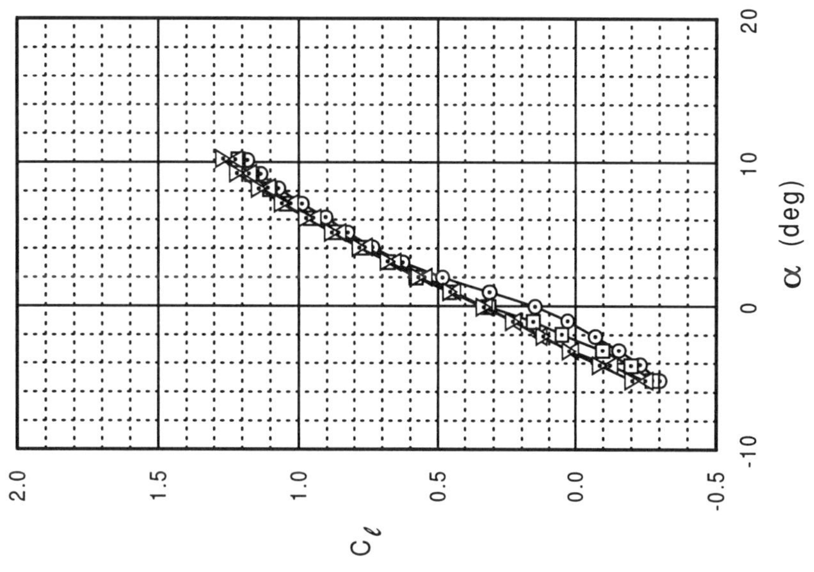
2.0
1.5
1.0
0.5
0.0
-0.5
C_ℓ
-10
0
10
20
α (deg)

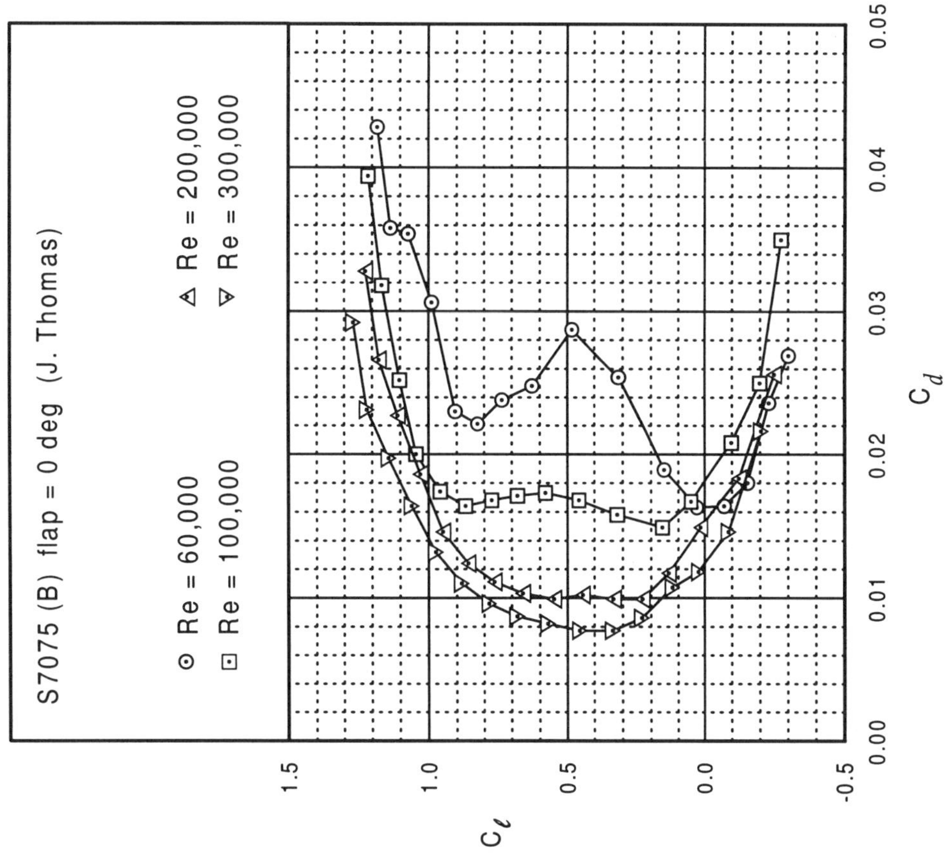
S7075 (B) flap = 0 deg (J. Thomas)
Re = 60,000
Re = 100,000
Re = 200,000
Re = 300,000
1.5
1.0
0.5
0.0
-0.5
C_ℓ
0.00
0.01
0.02
0.03
0.04
0.05
C_d

S7075 (B)

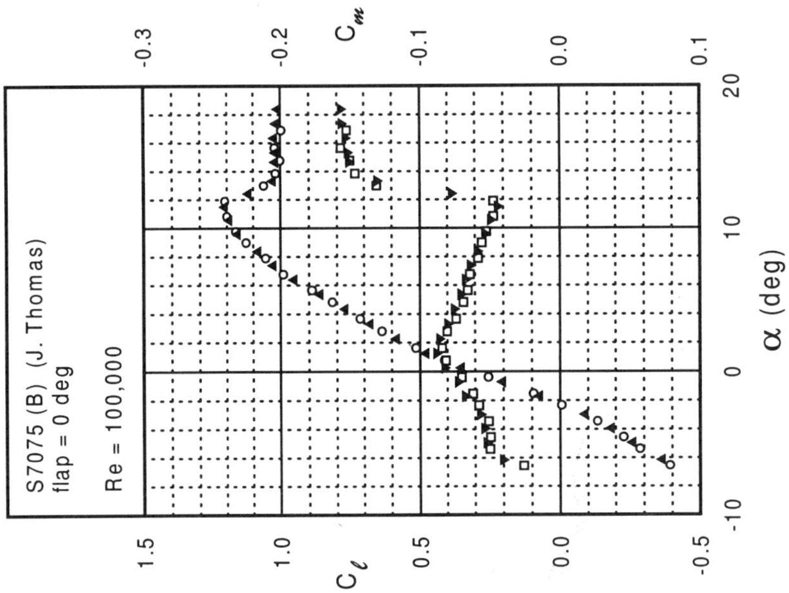
S7075 (B) (J. Thomas)
flap = 0 deg
Re = 100,000
C_m
-0.3
-0.2
-0.1
0.0
0.1
C_ℓ
1.5
1.0
0.5
0.0
-0.5
-10
0
10
20
α (deg)

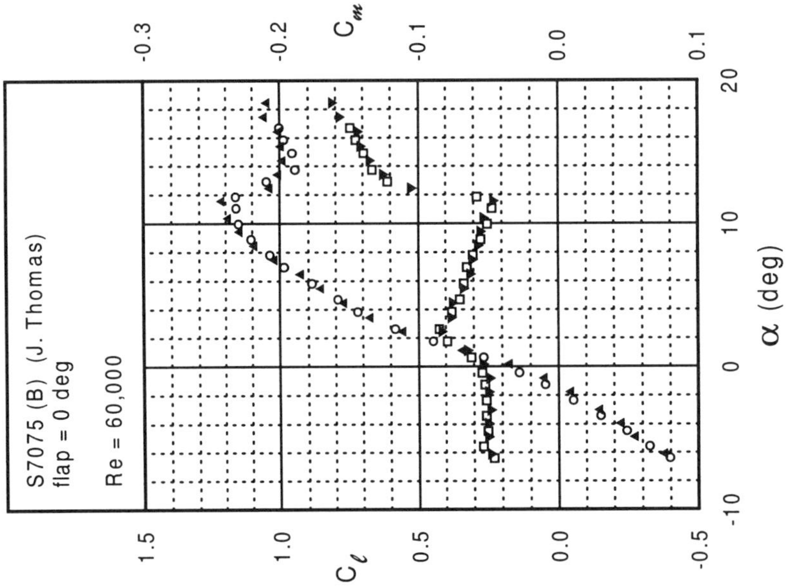
S7075 (B) (J. Thomas)
flap = 0 deg
Re = 60,000
C_m
-0.3
-0.2
-0.1
0.0
0.1
C_ℓ
1.5
1.0
0.5
0.0
-0.5
-10
0
10
20
α (deg)

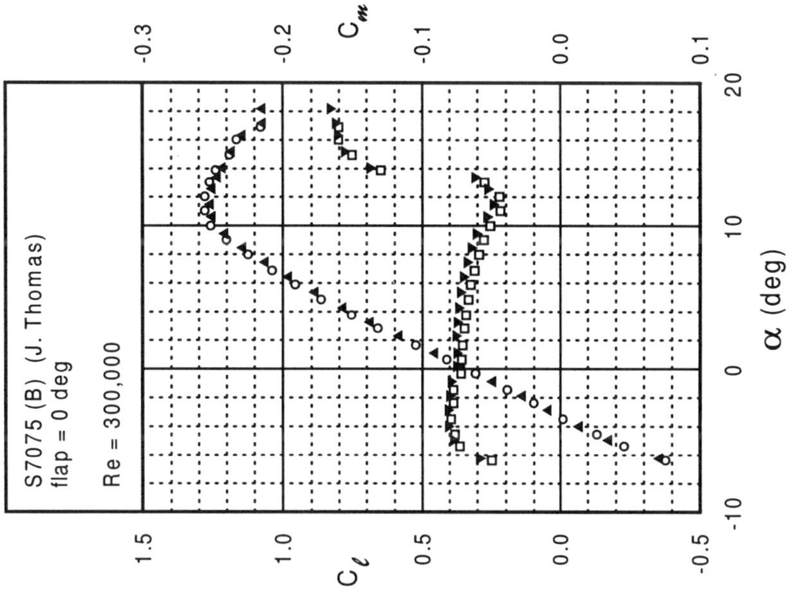
S7075 (B) (J. Thomas)
flap = 0 deg
Re = 300,000
C_ℓ
C_m
α (deg)
1.5
1.0
0.5
0.0
-0.5
-0.3
-0.2
-0.1
0.0
0.1
-10
0
10
20

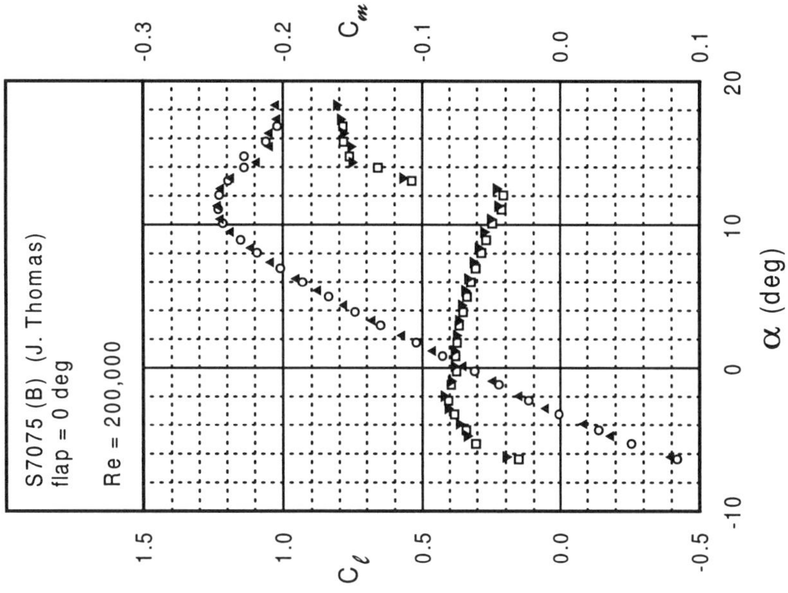
S7075 (B) (J. Thomas)
flap = 0 deg
Re = 200,000
C_ℓ
C_m
α (deg)
1.5
1.0
0.5
0.0
-0.5
-0.3
-0.2
-0.1
0.0
0.1
-10
0
10
20

S8036

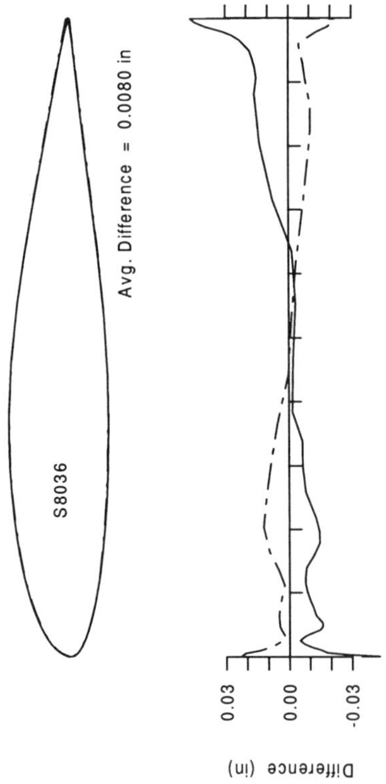
S8036
Avg. Difference = 0.0080 in
0.03
0.00
-0.03
Difference (in)

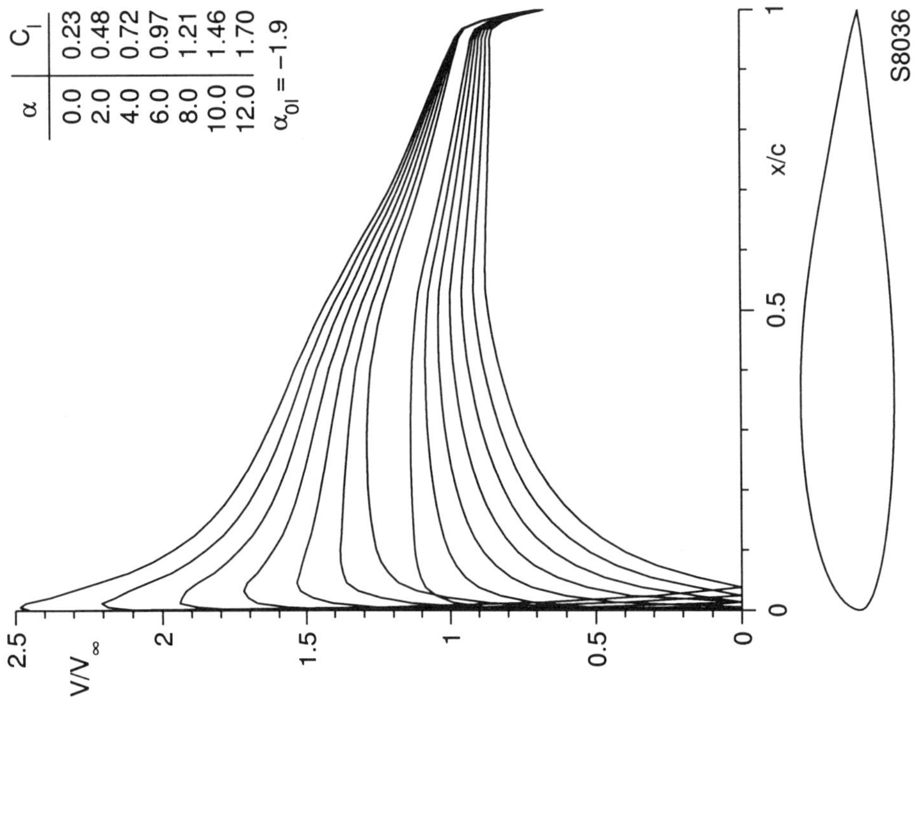
α | C_l
0.0 | 0.23
2.0 | 0.48
4.0 | 0.72
6.0 | 0.97
8.0 | 1.21
10.0 | 1.46
12.0 | 1.70
α_{0l} = −1.9
V/V_∞
2.5
2
1.5
1
0.5
0
x/c
0
0.5
1
S8036

S8036

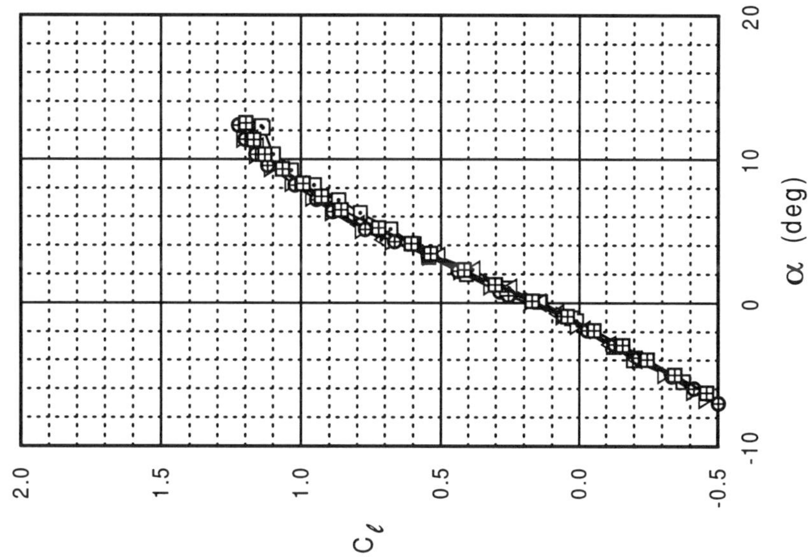
2.0
1.5
1.0
0.5
0.0
-0.5
C_ℓ
-10
0
10
20
α (deg)

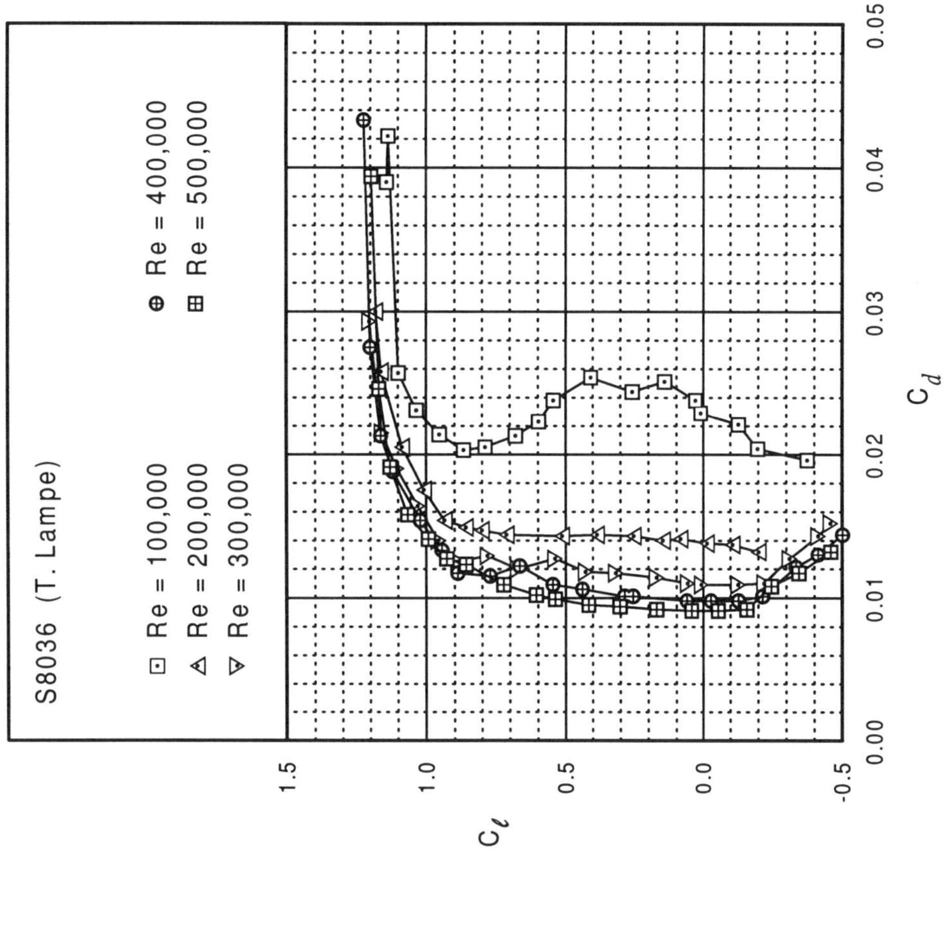
S8036 (T. Lampe)
Re = 100,000
Re = 200,000
Re = 300,000
Re = 400,000
Re = 500,000
1.5
1.0
0.5
0.0
-0.5
C_ℓ
0.00
0.01
0.02
0.03
0.04
0.05
C_d

S8036

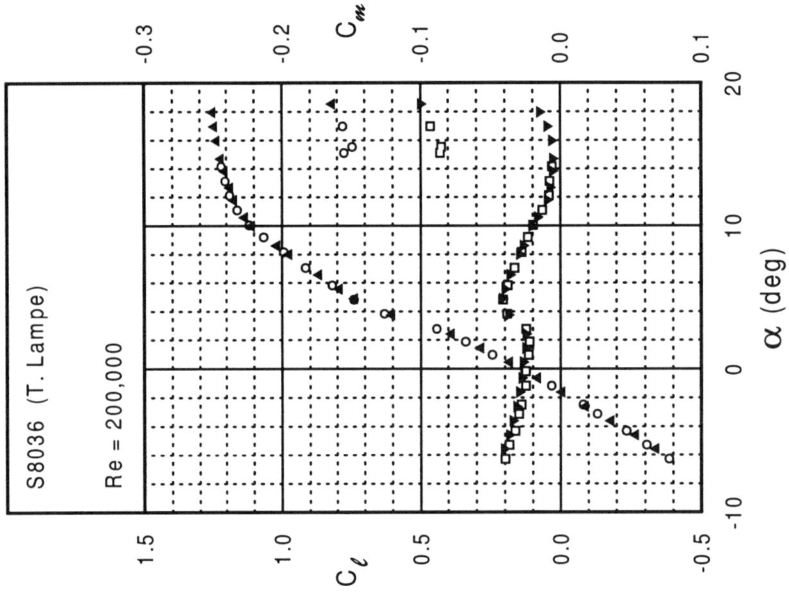
S8036 (T. Lampe)
Re = 200,000
C_ℓ
C_m
α (deg)

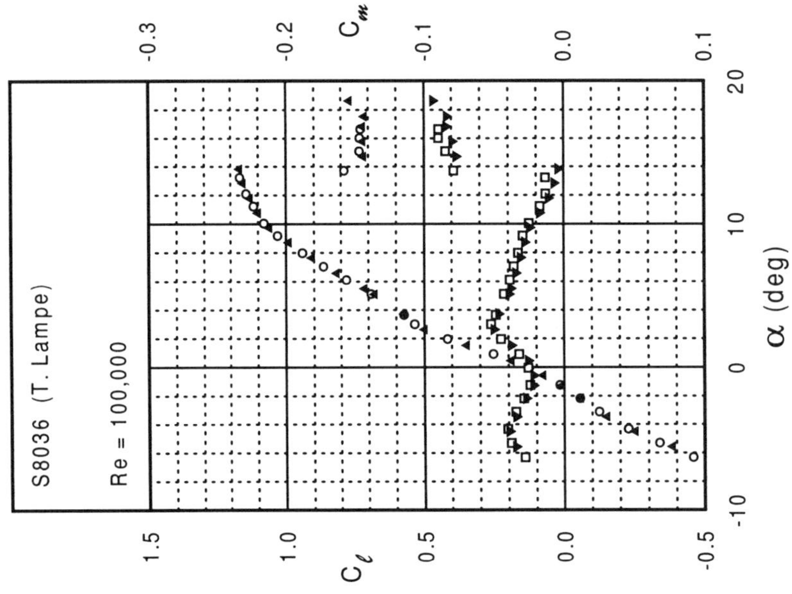
S8036 (T. Lampe)
Re = 100,000
C_ℓ
C_m
α (deg)

S8036

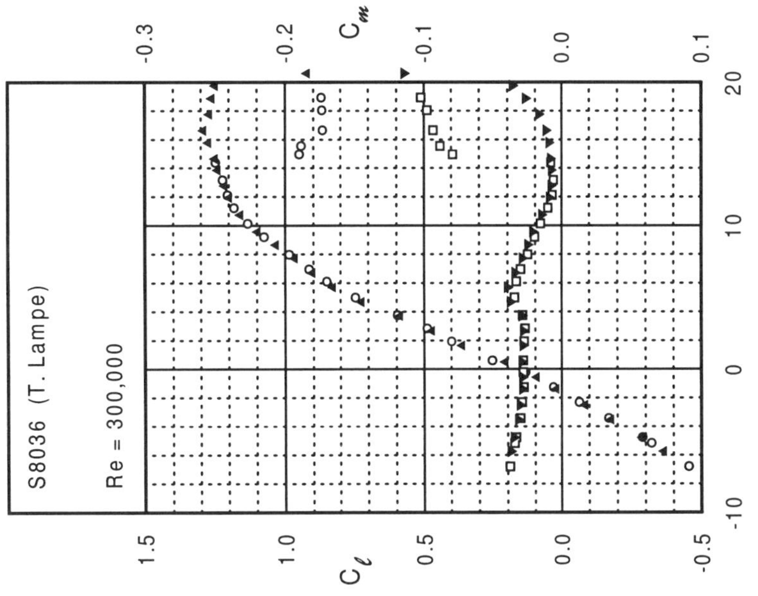
S8036 (T. Lampe)
Re = 300,000
C_ℓ
C_m
α (deg)

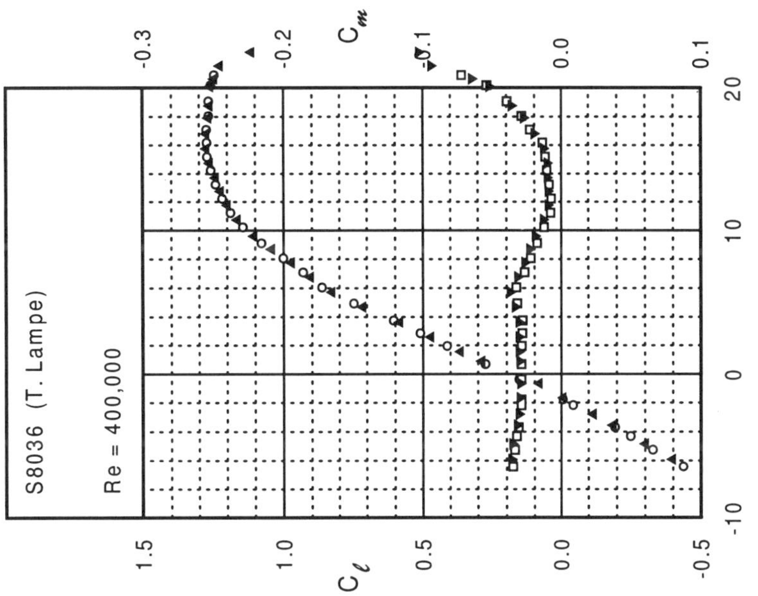
S8036 (T. Lampe)
Re = 400,000
C_ℓ
C_m
α (deg)

S8036

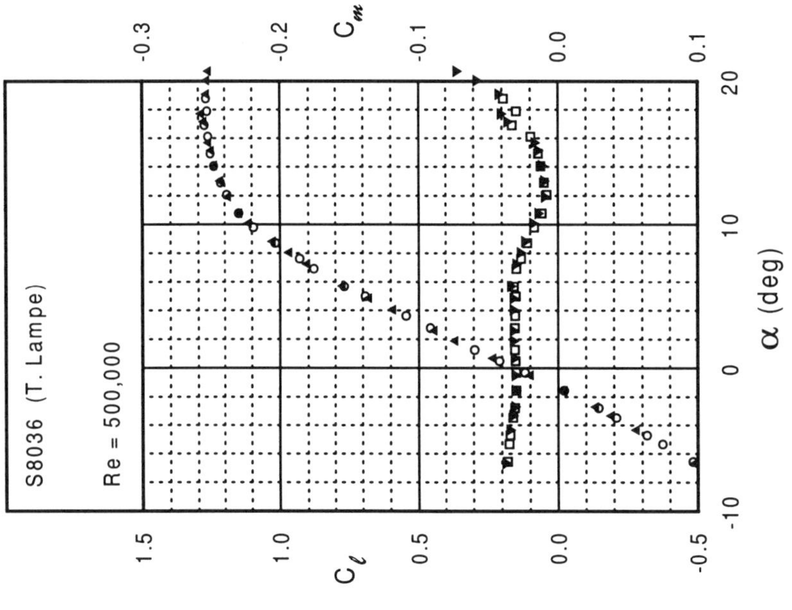
S8036 (T. Lampe)
Re = 500,000
C_ℓ
1.5
1.0
0.5
0.0
-0.5
C_m
-0.3
-0.2
-0.1
0.0
0.1
-10
0
10
20
α (deg)

S8037

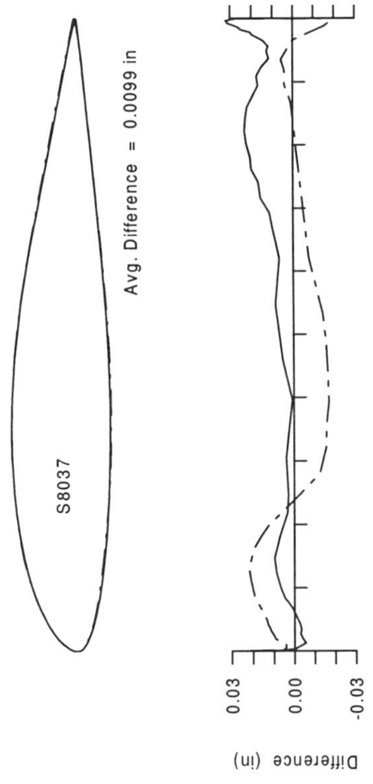
S8037
Avg. Difference = 0.0099 in
0.03
0.00
-0.03
Difference (in)

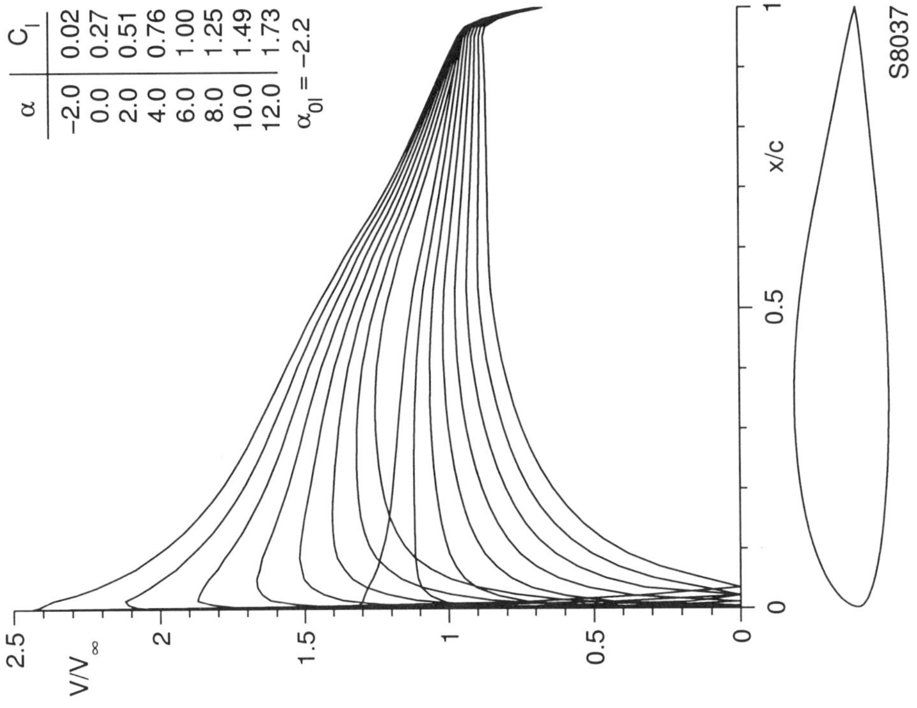
α
C_l
-2.0 0.02
0.0 0.27
2.0 0.51
4.0 0.76
6.0 1.00
8.0 1.25
10.0 1.49
12.0 1.73
α_0l = -2.2
V/V_∞
2.5
2
1.5
1
0.5
0
x/c
0
0.5
1
S8037

S8037

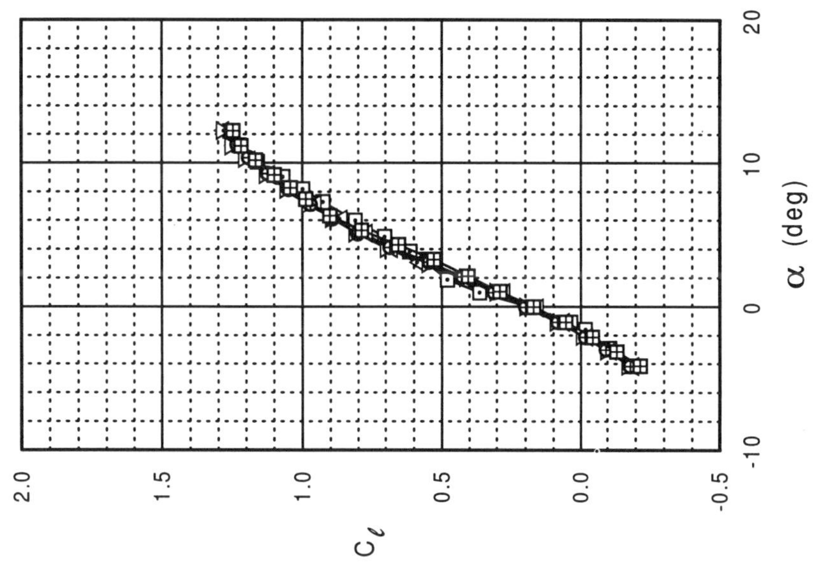

α (deg)
C_ℓ

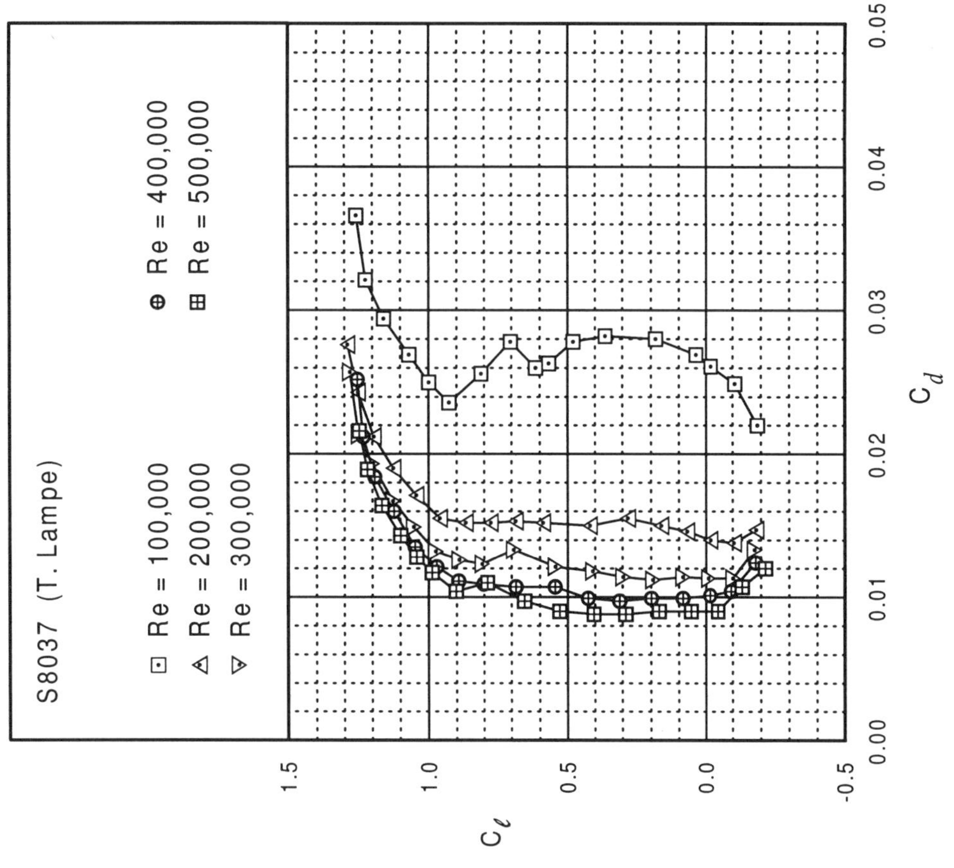

S8037 (T. Lampe)
Re = 100,000
Re = 200,000
Re = 300,000
Re = 400,000
Re = 500,000
C_d
C_ℓ

S8037

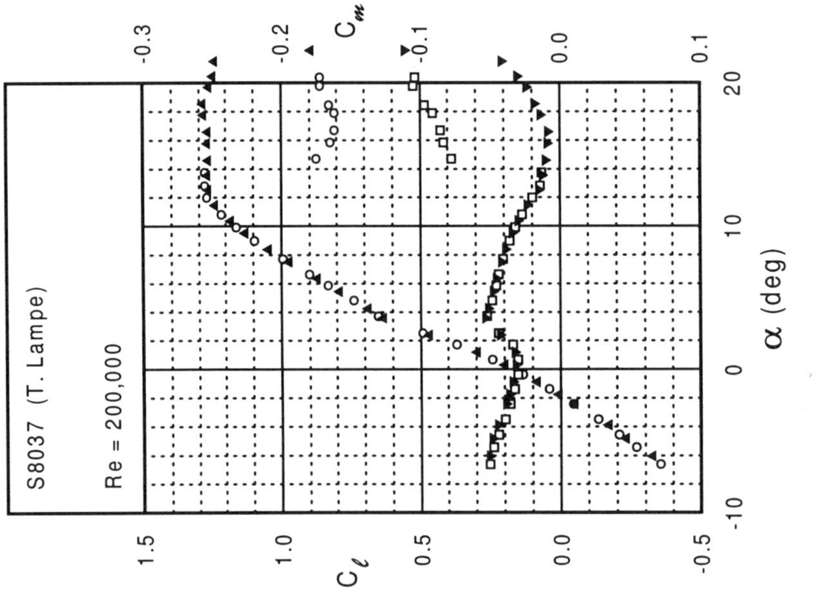
S8037 (T. Lampe)
Re = 200,000
C_ℓ
C_m
α (deg)
-0.3
-0.2
-0.1
0.0
0.1
1.5
1.0
0.5
0.0
-0.5
-10
0
10
20

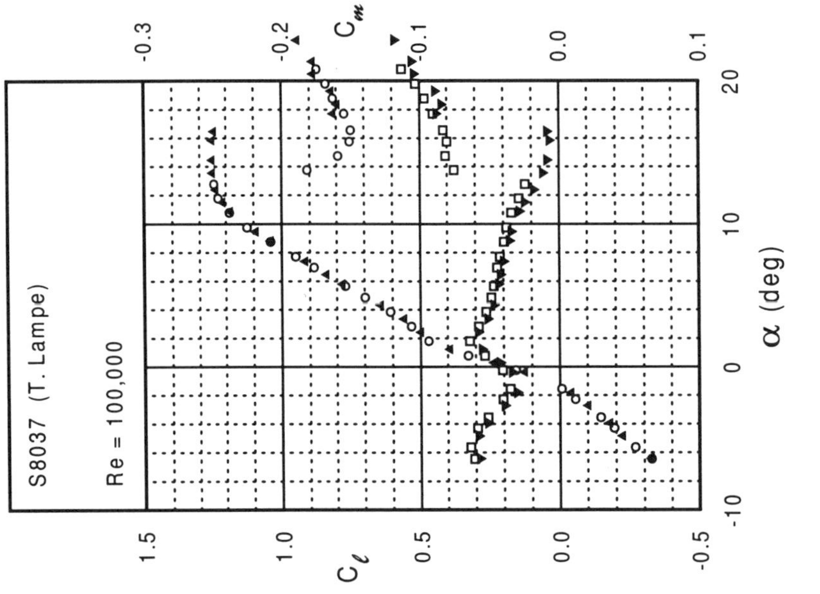
S8037 (T. Lampe)
Re = 100,000
C_ℓ
C_m
α (deg)
-0.3
-0.2
-0.1
0.0
0.1
1.5
1.0
0.5
0.0
-0.5
-10
0
10
20

S8037

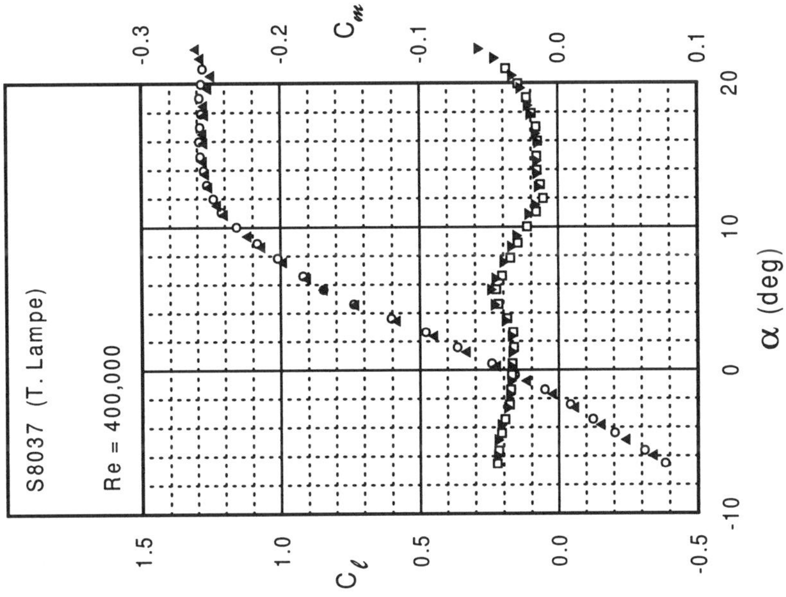
S8037 (T. Lampe)
Re = 400,000
C_ℓ
C_m
α (deg)

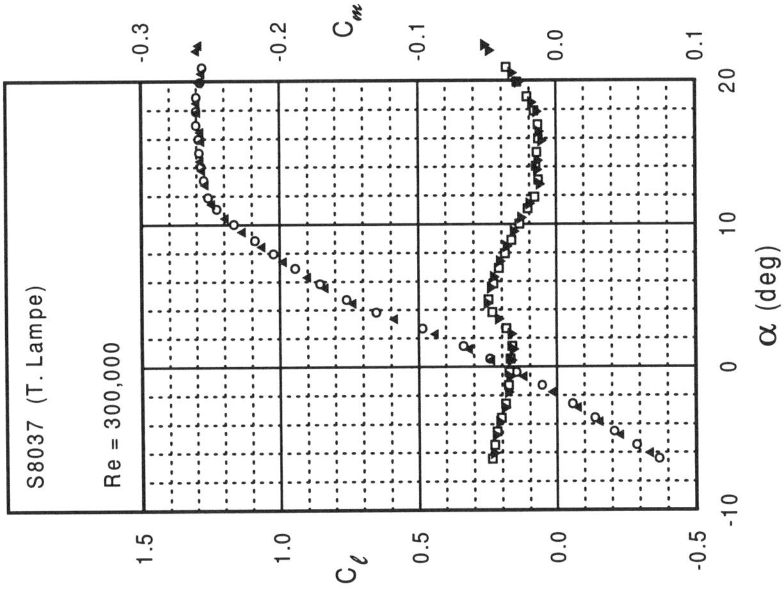
S8037 (T. Lampe)
Re = 300,000
C_ℓ
C_m
α (deg)

S8037

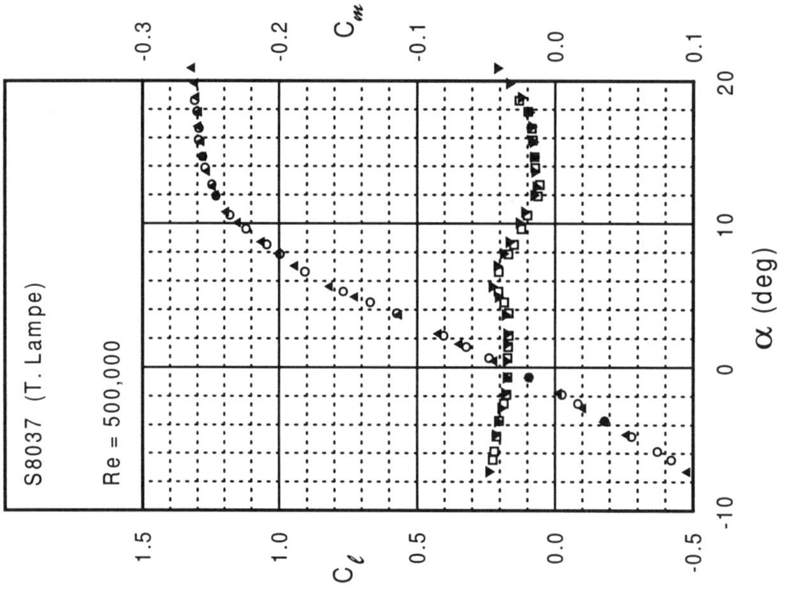
S8037 (T. Lampe)
Re = 500,000
C_ℓ
C_m
α (deg)
1.5
1.0
0.5
0.0
-0.5
-0.3
-0.2
-0.1
0.0
0.1
-10
0
10
20

S8052

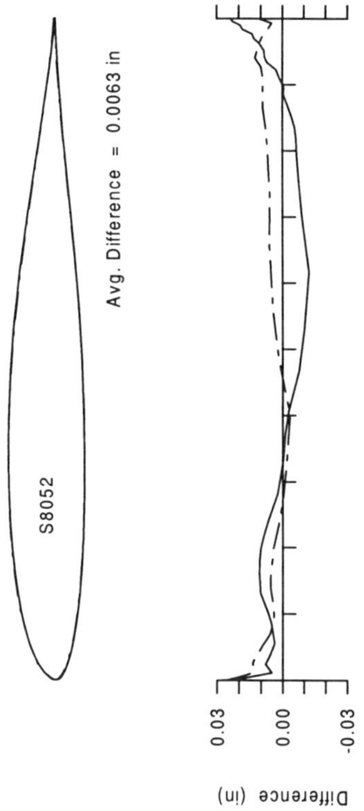
S8052
Avg. Difference = 0.0063 in
Difference (in)
0.03
0.00
-0.03

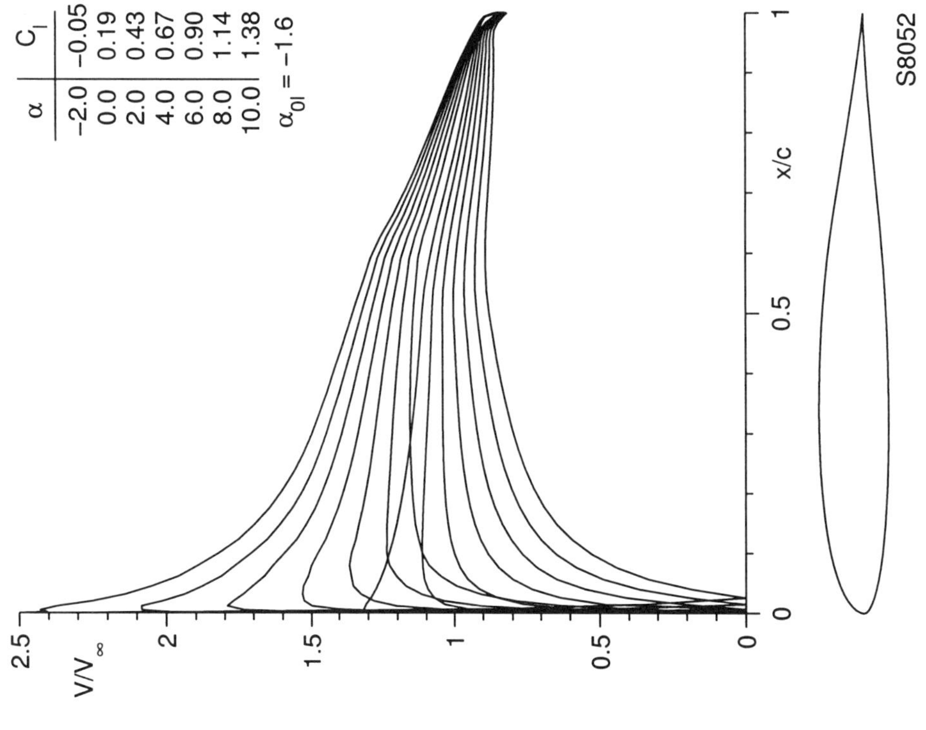
α | C_l
−2.0 | −0.05
0.0 | 0.19
2.0 | 0.43
4.0 | 0.67
6.0 | 0.90
8.0 | 1.14
10.0 | 1.38
$\alpha_{0l} = -1.6$
V/V_∞
2.5
2
1.5
1
0.5
0
0
0.5
1
x/c
S8052

S8052

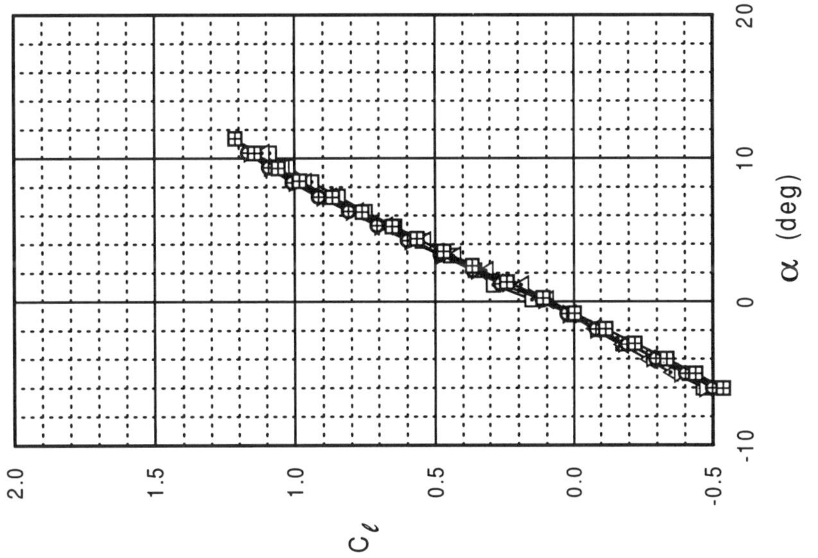

2.0
1.5
1.0
0.5
0.0
-0.5
C_ℓ
-10
0
10
20
α (deg)

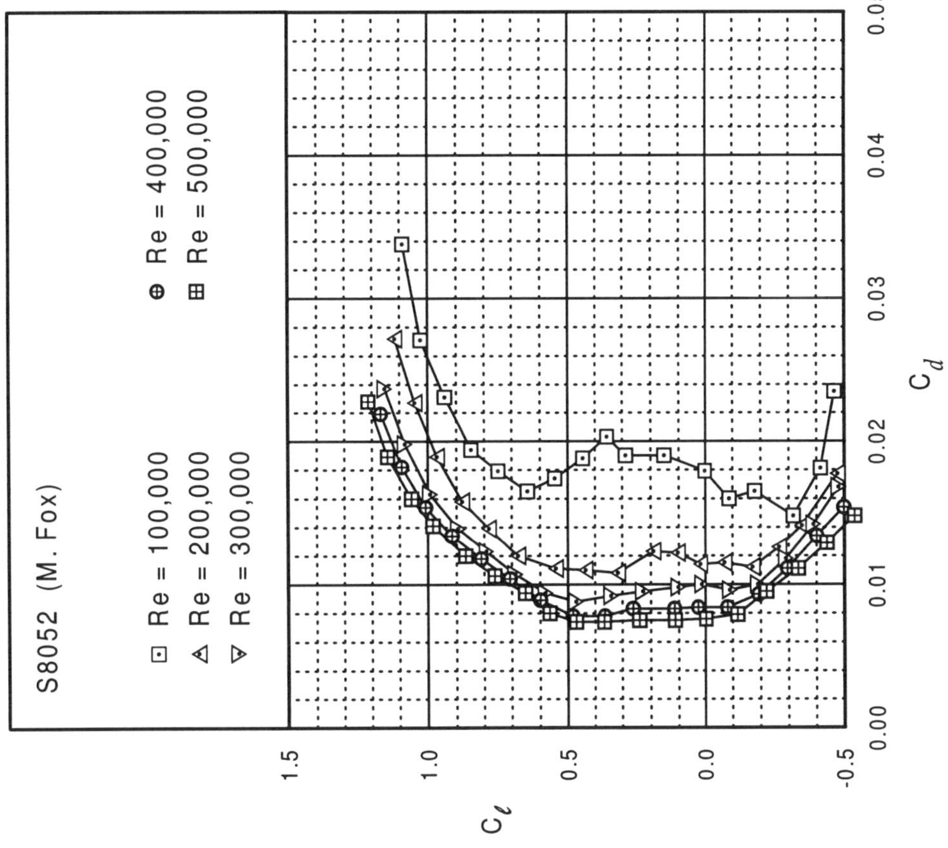

S8052 (M. Fox)
Re = 100,000
Re = 200,000
Re = 300,000
Re = 400,000
Re = 500,000
1.5
1.0
0.5
0.0
-0.5
C_ℓ
0.00
0.01
0.02
0.03
0.04
0.05
C_d

S8052

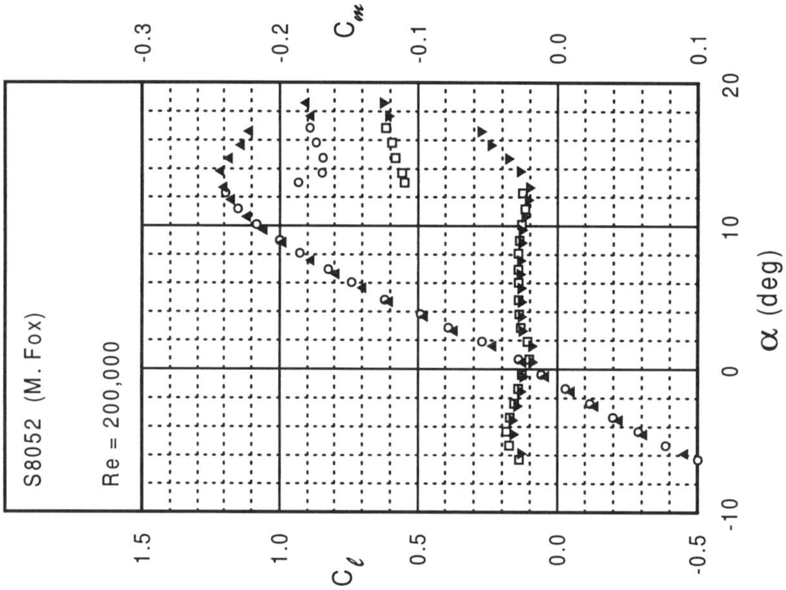

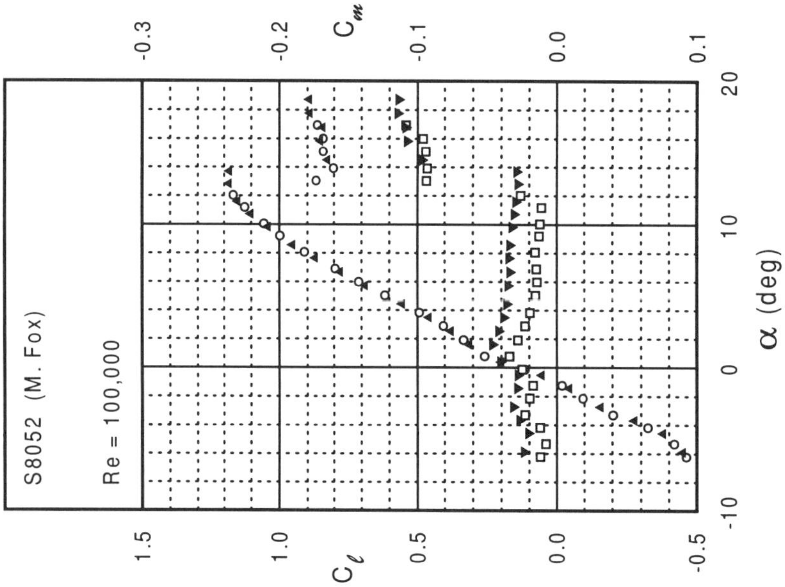

S8052

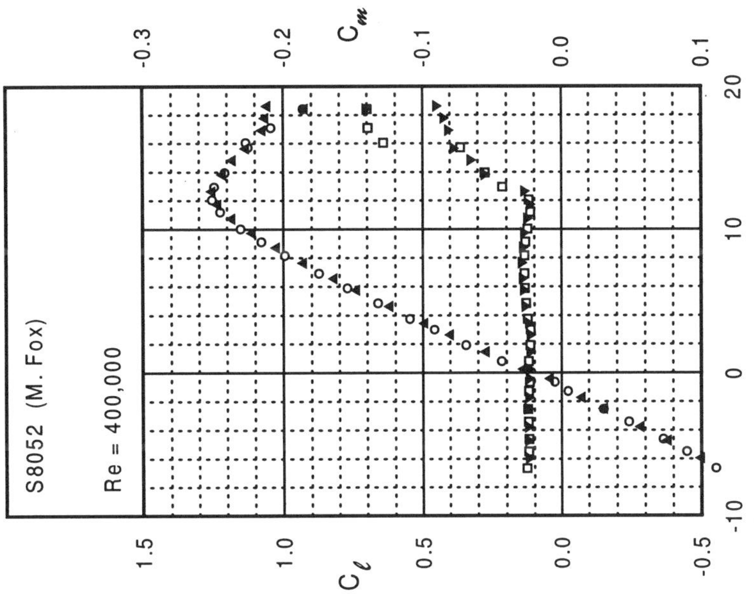
S8052 (M. Fox)
Re = 400,000
C_ℓ
C_m
α (deg)

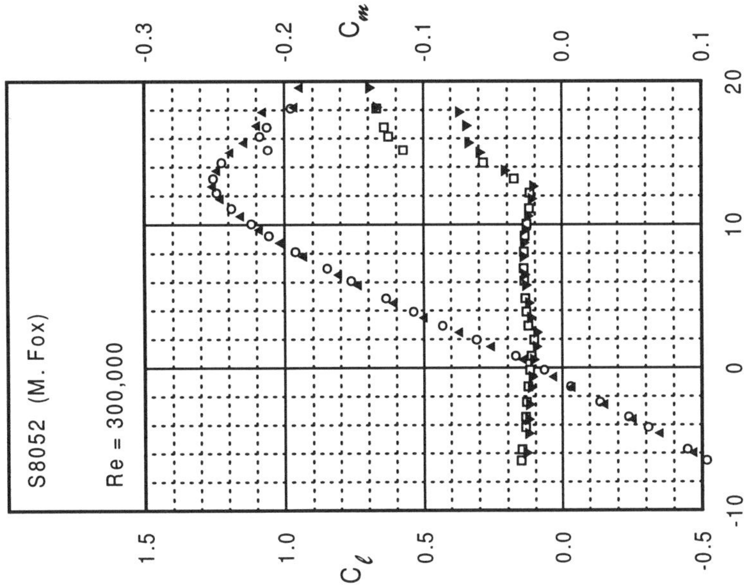
S8052 (M. Fox)
Re = 300,000
C_ℓ
C_m
α (deg)

S8052

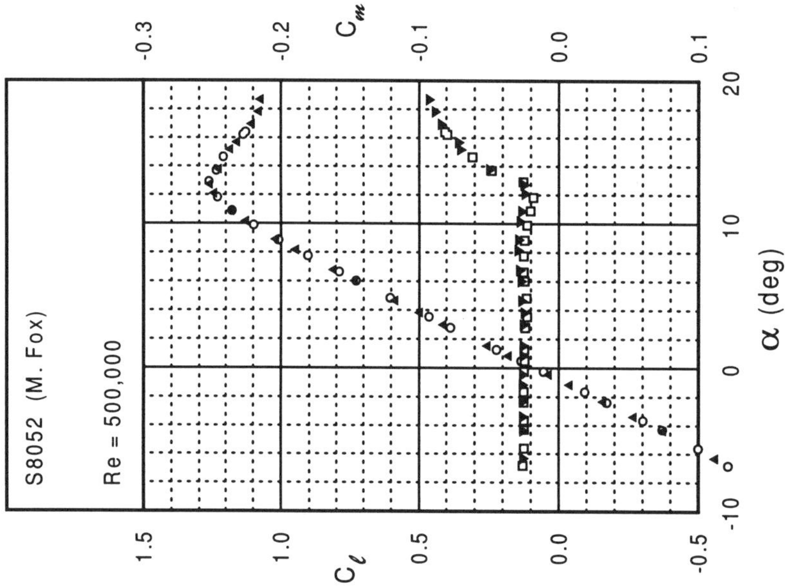
S8052 (M. Fox)
Re = 500,000
C_ℓ
C_m
α (deg)
1.5
1.0
0.5
0.0
-0.5
-0.3
-0.2
-0.1
0.0
0.1
-10
0
10
20

SA7035

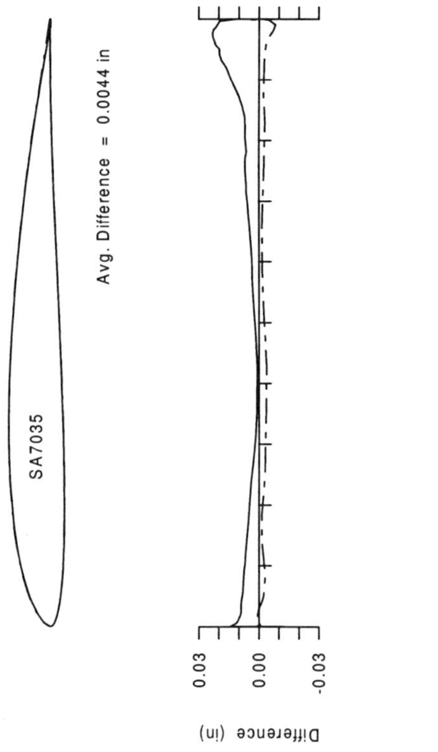
SA7035
Avg. Difference = 0.0044 in
0.03
0.00
-0.03
Difference (in)

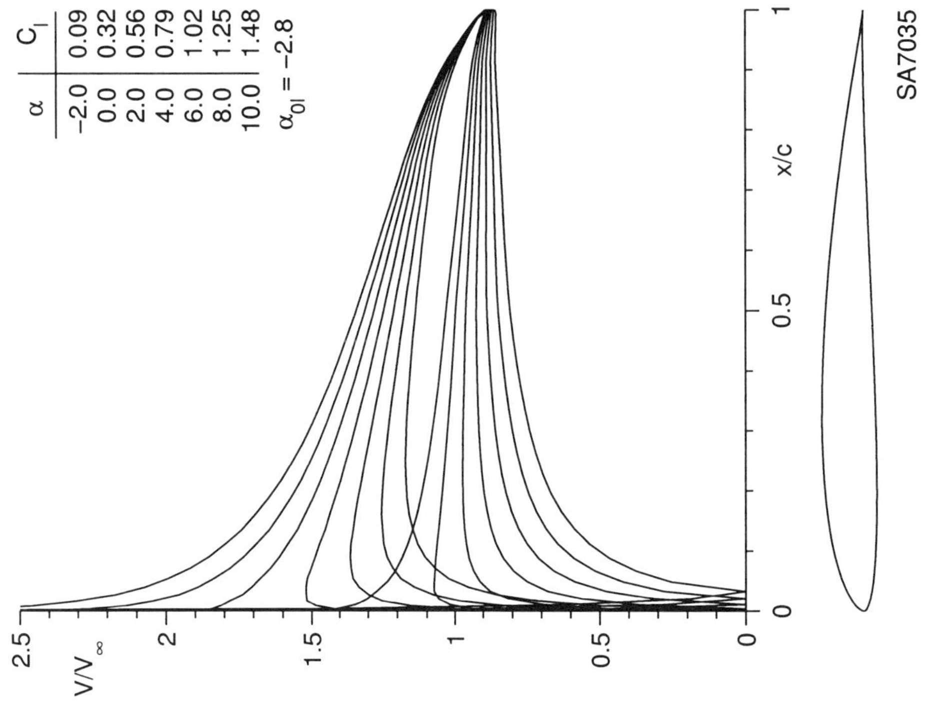
α | C_l
−2.0 | 0.09
0.0 | 0.32
2.0 | 0.56
4.0 | 0.79
6.0 | 1.02
8.0 | 1.25
10.0 | 1.48
α_0l = −2.8
V/V_∞
2.5
2
1.5
1
0.5
0
0.5
1
x/c
SA7035

SA7035

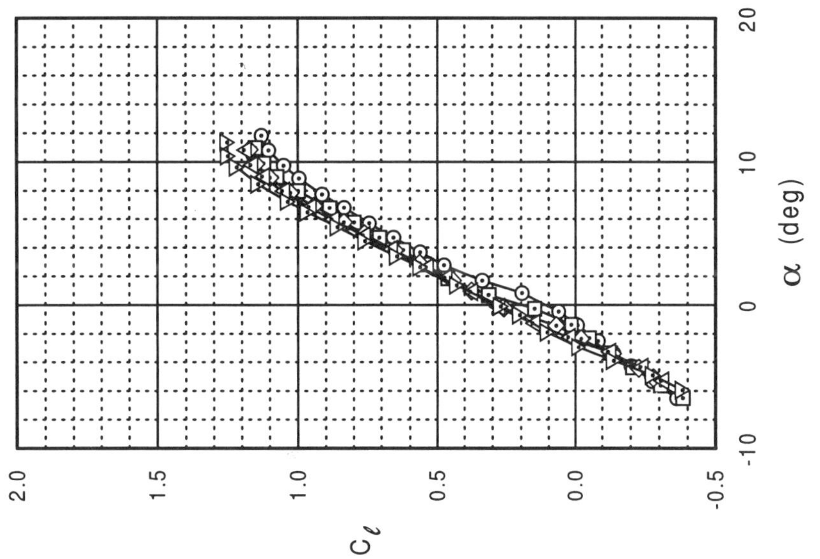

2.0
1.5
1.0
0.5
0.0
-0.5
C_ℓ
-10
0
10
20
α (deg)

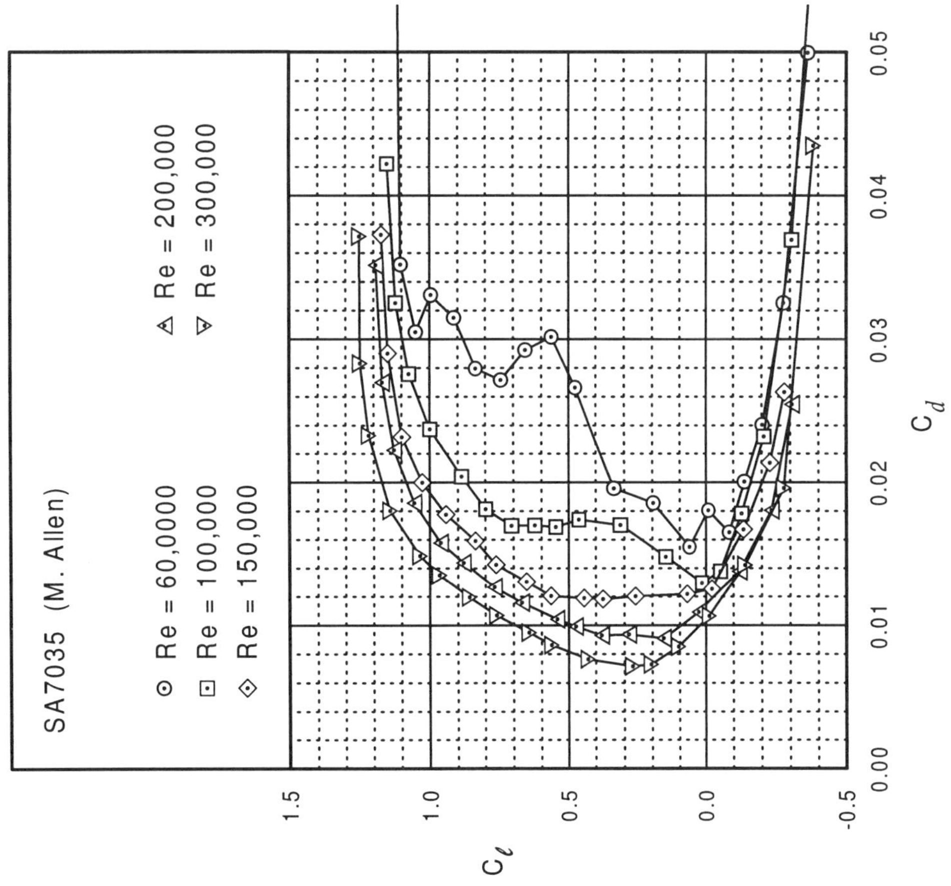

SA7035 (M. Allen)
Re = 60,0000
Re = 100,000
Re = 150,000
Re = 200,000
Re = 300,000
1.5
1.0
0.5
0.0
-0.5
C_ℓ
0.00
0.01
0.02
0.03
0.04
0.05
C_d

SA7035

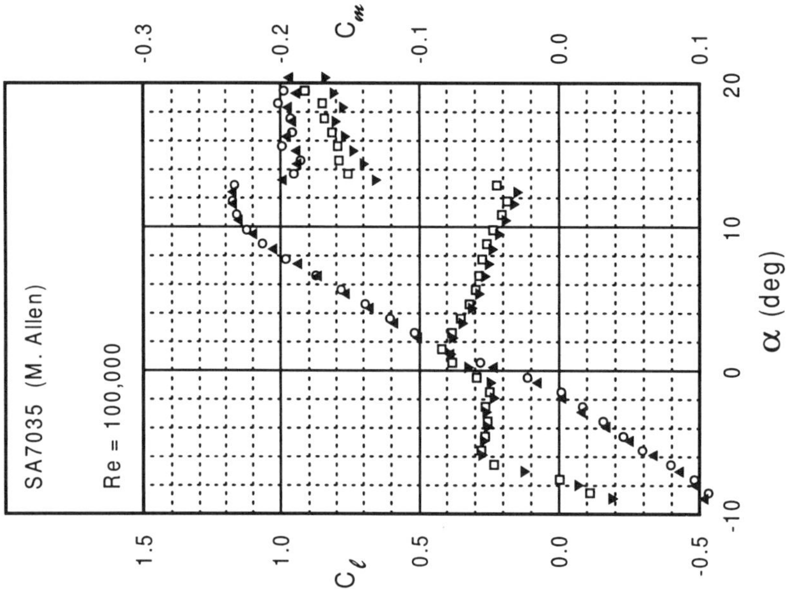
SA7035 (M. Allen)
Re = 100,000
C_m
-0.3
-0.2
-0.1
0.0
0.1
C_ℓ
1.5
1.0
0.5
0.0
-0.5
-10
0
10
20
α (deg)

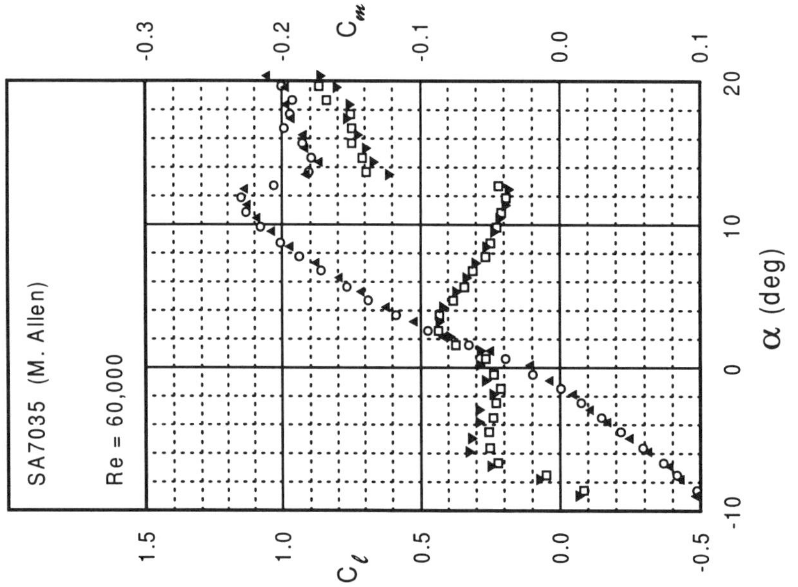
SA7035 (M. Allen)
Re = 60,000
C_m
-0.3
-0.2
-0.1
0.0
0.1
C_ℓ
1.5
1.0
0.5
0.0
-0.5
-10
0
10
20
α (deg)

SA7035

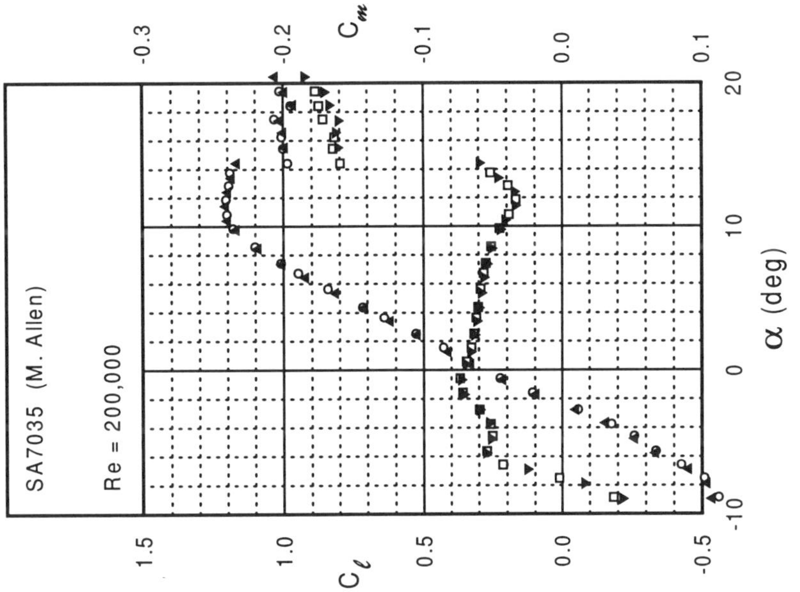
SA7035 (M. Allen)
Re = 200,000
C_ℓ
C_m
α (deg)

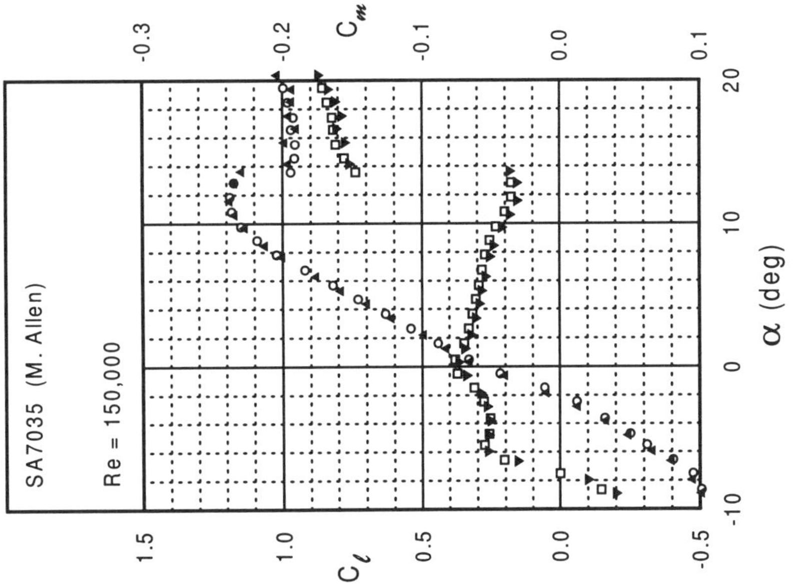
SA7035 (M. Allen)
Re = 150,000
C_ℓ
C_m
α (deg)

SA7035

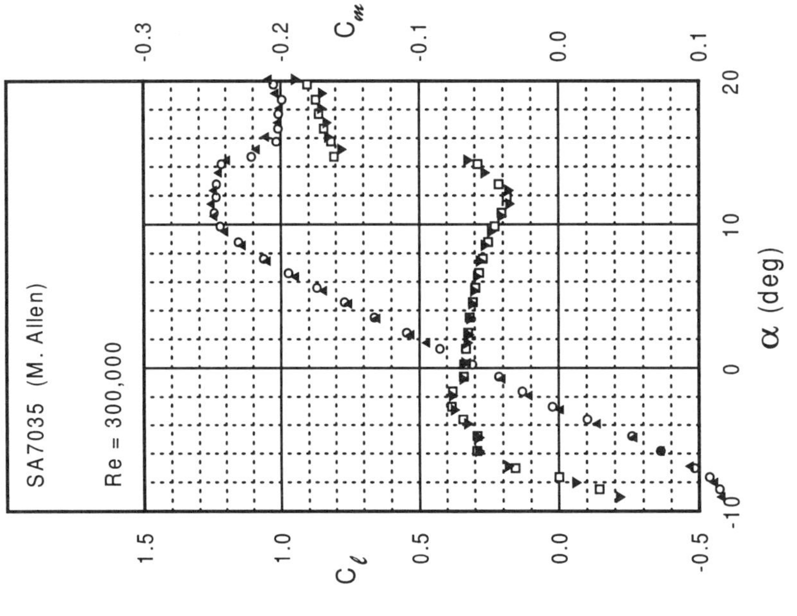
SA7035 (M. Allen)
Re = 300,000
C_m
-0.3
-0.2
-0.1
0.0
0.1
C_ℓ
1.5
1.0
0.5
0.0
-0.5
-10
0
10
20
α (deg)

SA7036 (A)

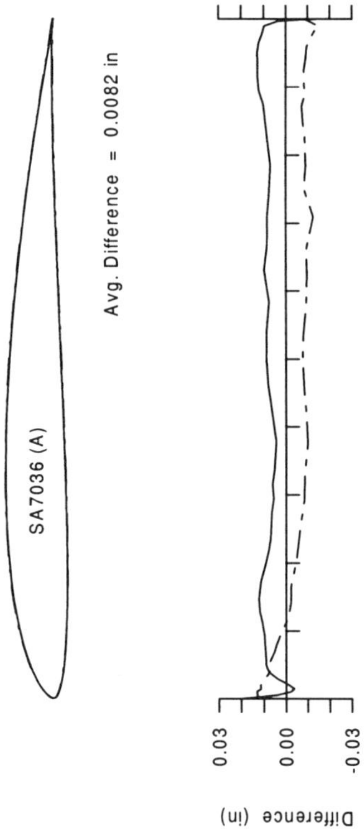
SA7036 (A)
Avg. Difference = 0.0082 in
0.03
0.00
-0.03
Difference (in)

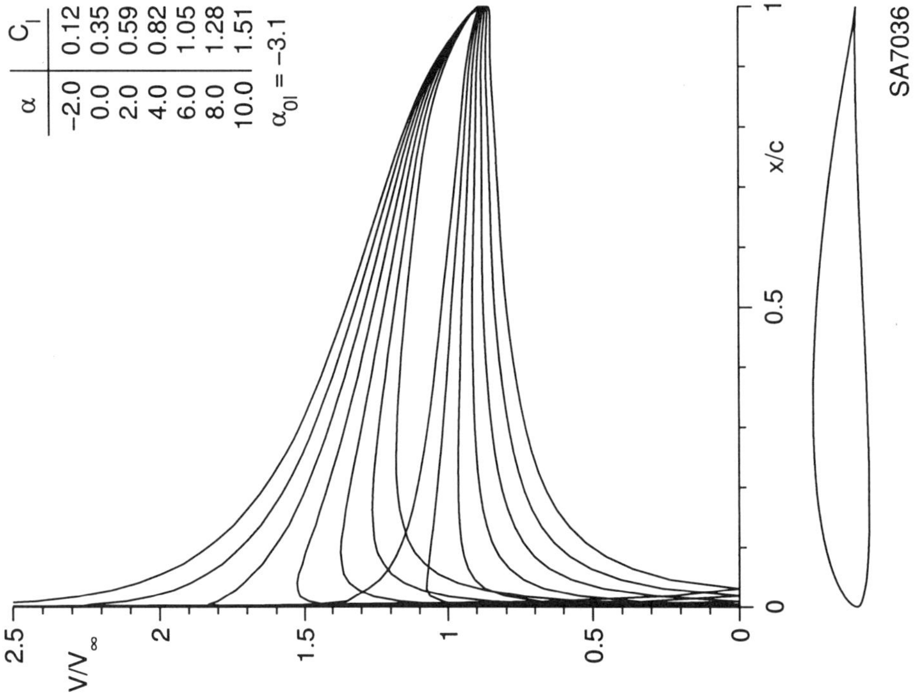
α | C_l
−2.0 | 0.12
0.0 | 0.35
2.0 | 0.59
4.0 | 0.82
6.0 | 1.05
8.0 | 1.28
10.0 | 1.51
α_{0l} = −3.1
2.5
V/V_∞
2
1.5
1
0.5
0
0
0.5
1
x/c
SA7036

SA7036 (A)

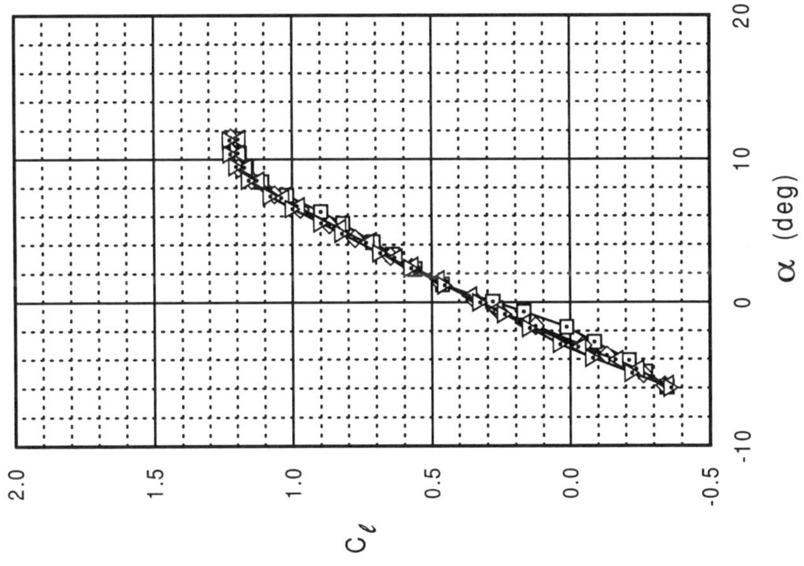
α (deg)
C_ℓ

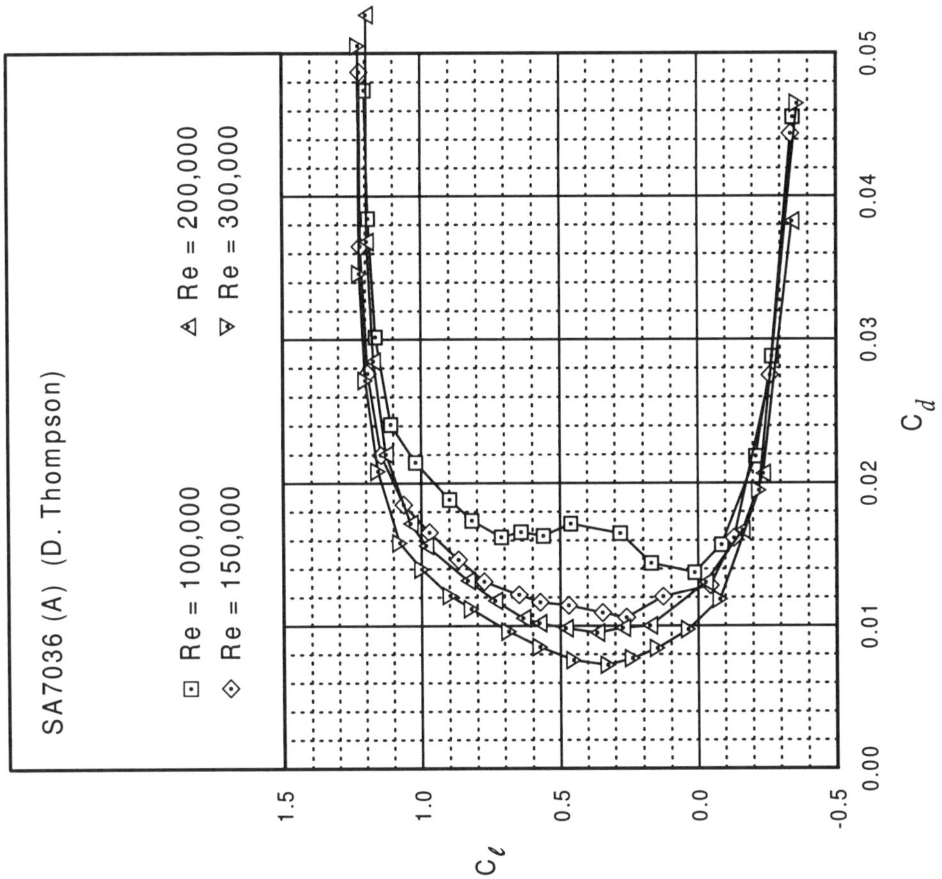
SA7036 (A) (D. Thompson)
Re = 100,000
Re = 150,000
Re = 200,000
Re = 300,000
C_d
C_ℓ

SA7036 (A)

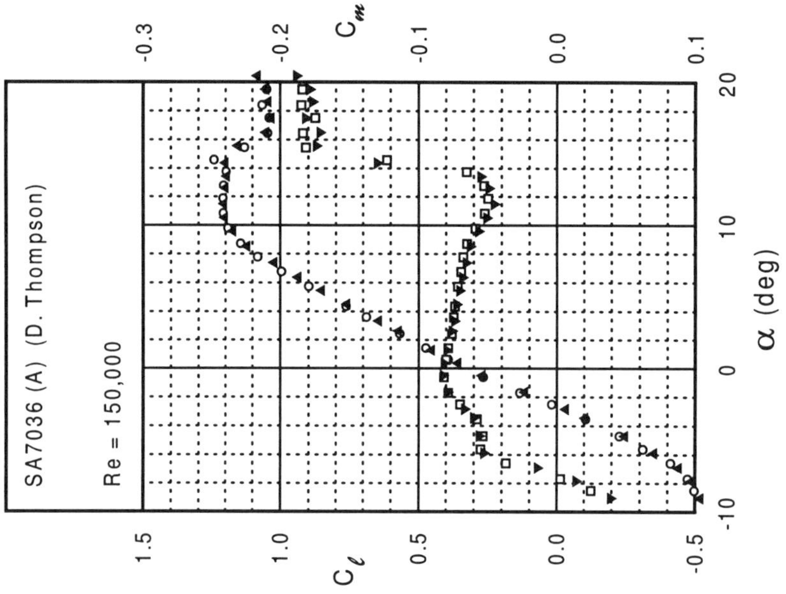
SA7036 (A) (D. Thompson)
Re = 150,000
C_m
C_ℓ
α (deg)

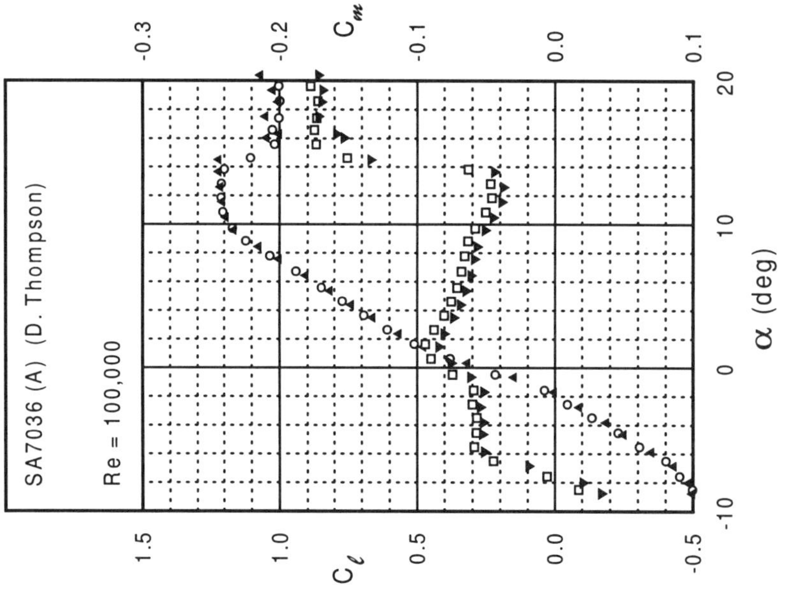
SA7036 (A) (D. Thompson)
Re = 100,000
C_m
C_ℓ
α (deg)

SA7036 (A)

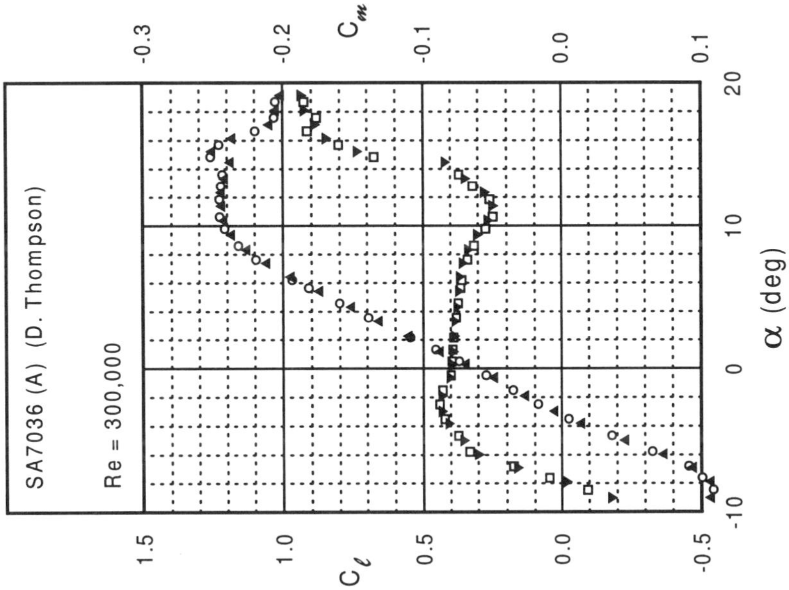
SA7036 (A) (D. Thompson)
Re = 300,000
C_ℓ
C_m
α (deg)

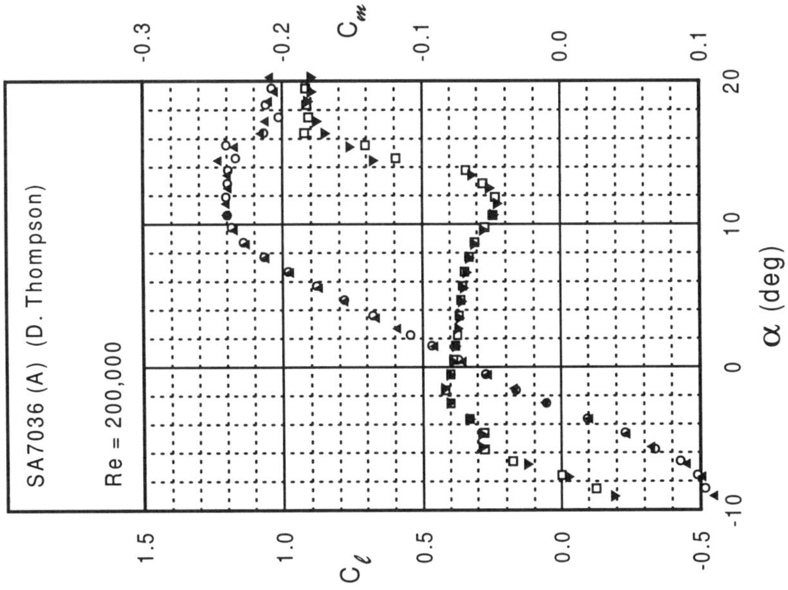
SA7036 (A) (D. Thompson)
Re = 200,000
C_ℓ
C_m
α (deg)

SA7036 (B)

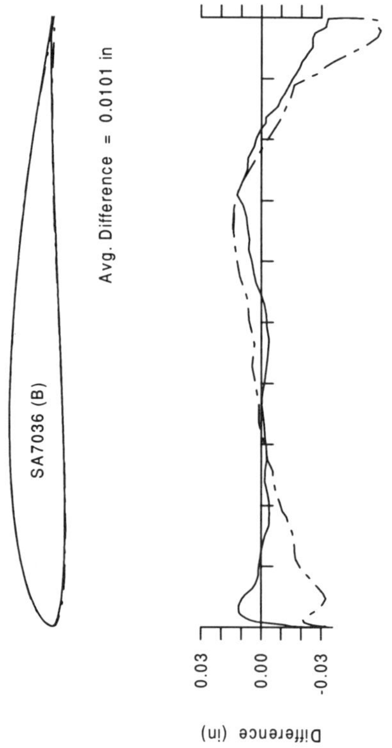
SA7036 (B)
Avg. Difference = 0.0101 in
0.03
0.00
-0.03
Difference (in)

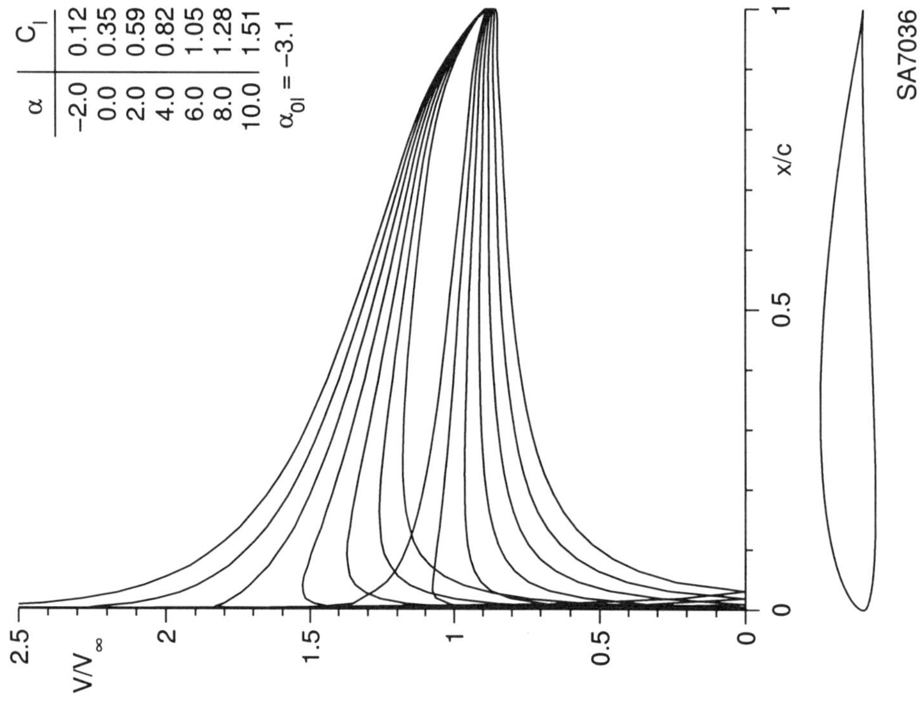
α
C$_l$
−2.0 0.12
0.0 0.35
2.0 0.59
4.0 0.82
6.0 1.05
8.0 1.28
10.0 1.51
$\alpha_{0l} = -3.1$
V/V$_\infty$
2.5
2
1.5
1
0.5
0
0.5
1
x/c
SA7036

SA7036 (B)

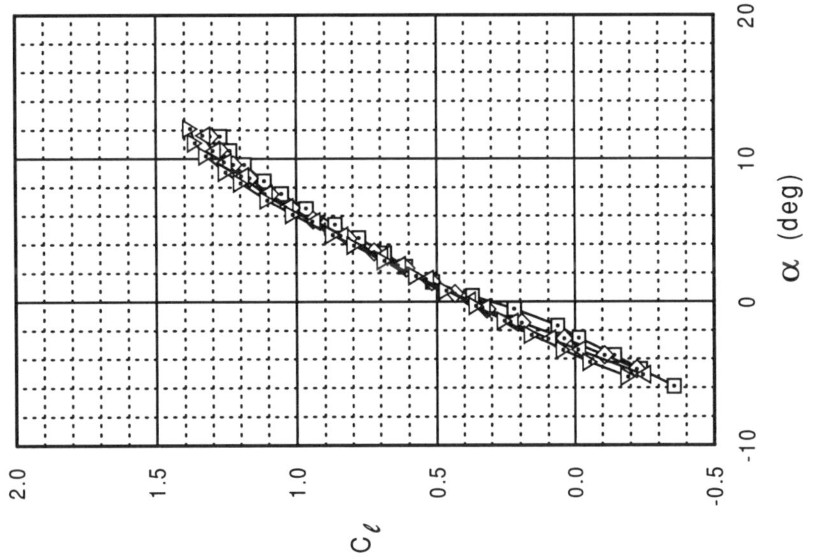
2.0
1.5
1.0
0.5
0.0
-0.5
C_ℓ
-10
0
10
20
α (deg)

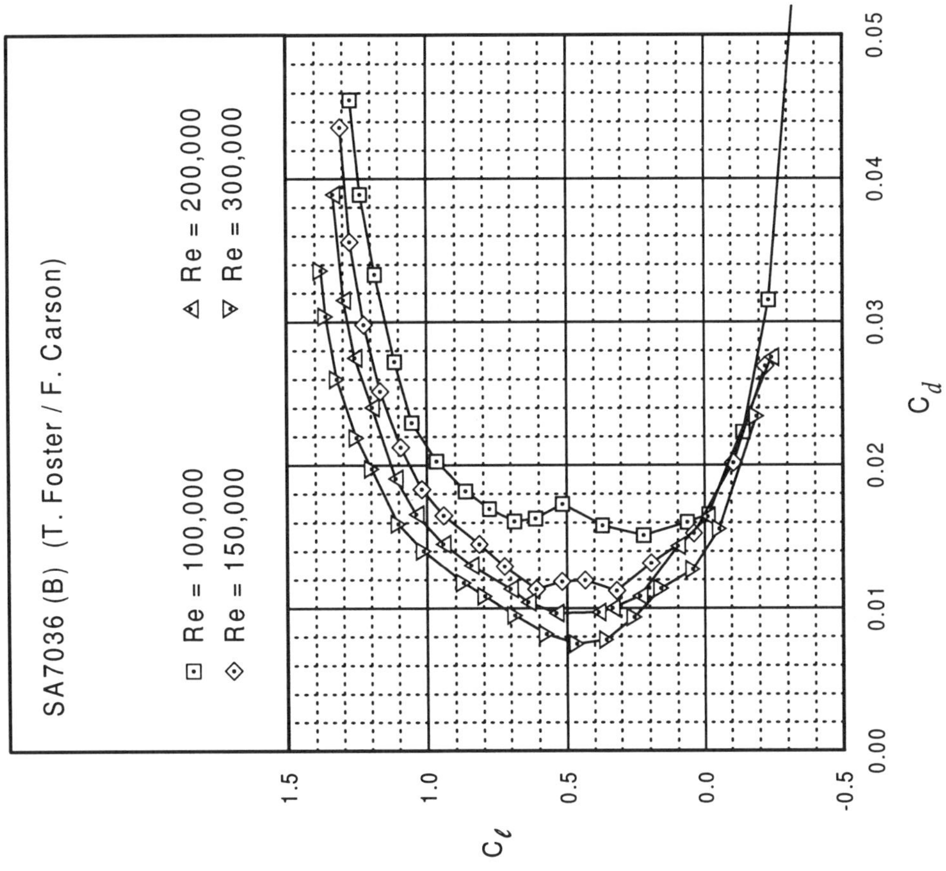
SA7036 (B) (T. Foster / F. Carson)
Re = 100,000
Re = 150,000
Re = 200,000
Re = 300,000
1.5
1.0
0.5
0.0
-0.5
C_ℓ
0.00
0.01
0.02
0.03
0.04
0.05
C_d

SA7036 (B)

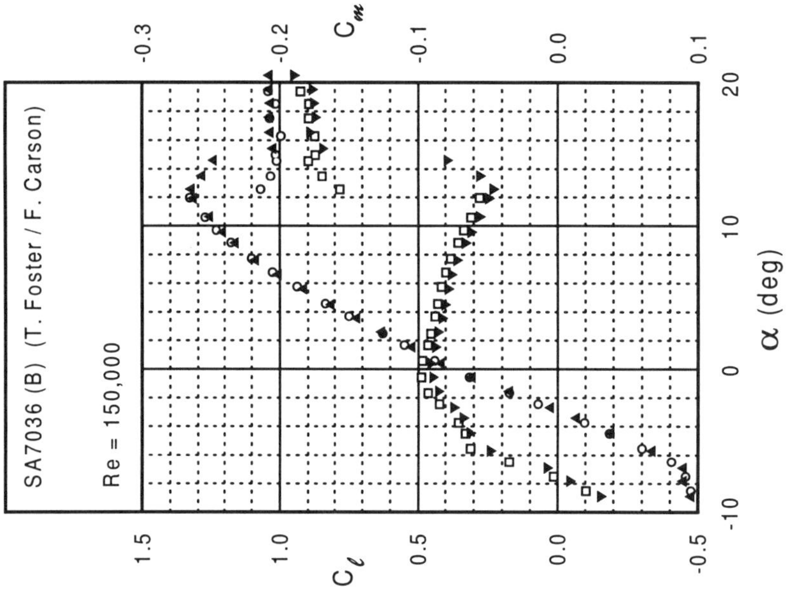
SA7036 (B) (T. Foster / F. Carson)
Re = 150,000
C_m
-0.3
-0.2
-0.1
0.0
0.1
C_ℓ
1.5
1.0
0.5
0.0
-0.5
-10
0
10
20
α (deg)

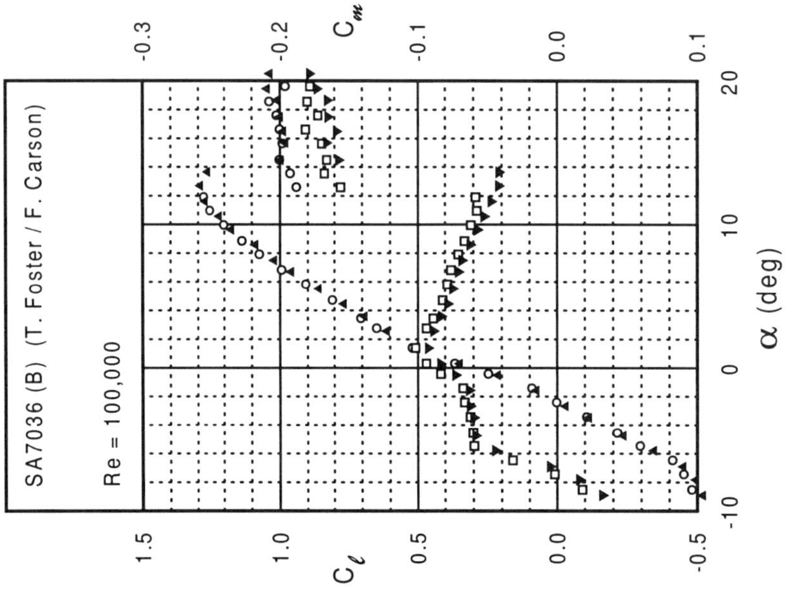
SA7036 (B) (T. Foster / F. Carson)
Re = 100,000
C_m
-0.3
-0.2
-0.1
0.0
0.1
C_ℓ
1.5
1.0
0.5
0.0
-0.5
-10
0
10
20
α (deg)

SA7036 (B)

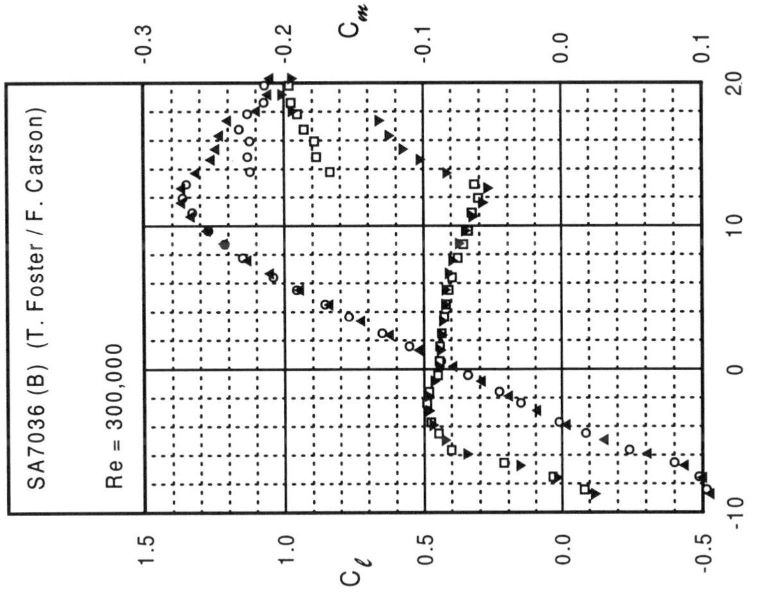
SA7036 (B) (T. Foster / F. Carson)
Re = 300,000
C_ℓ
1.5
1.0
0.5
0.0
-0.5
C_m
-0.3
-0.2
-0.1
0.0
0.1
-10
0
10
20
α (deg)

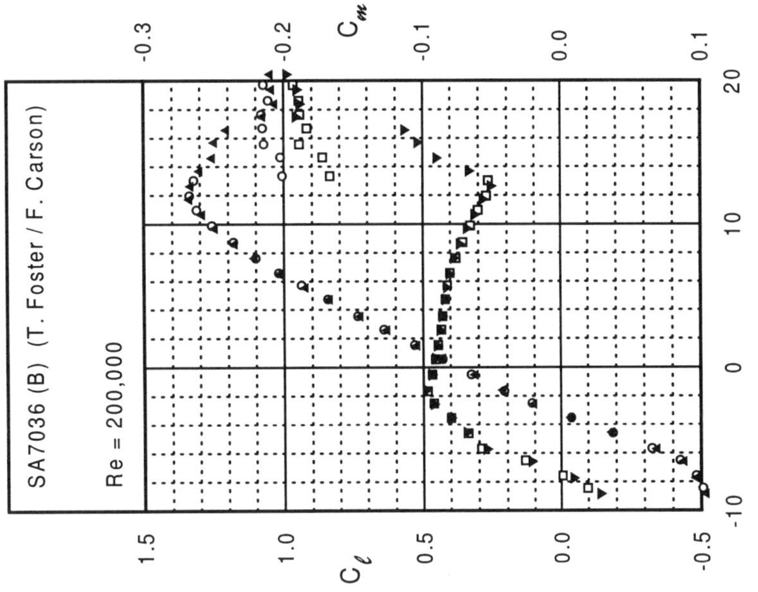
SA7036 (B) (T. Foster / F. Carson)
Re = 200,000
C_ℓ
1.5
1.0
0.5
0.0
-0.5
C_m
-0.3
-0.2
-0.1
0.0
0.1
-10
0
10
20
α (deg)

SA7038

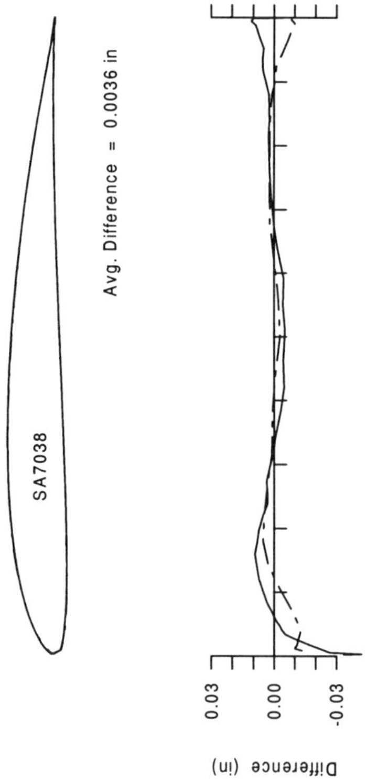
SA7038
Avg. Difference = 0.0036 in
0.03
0.00
-0.03
Difference (in)

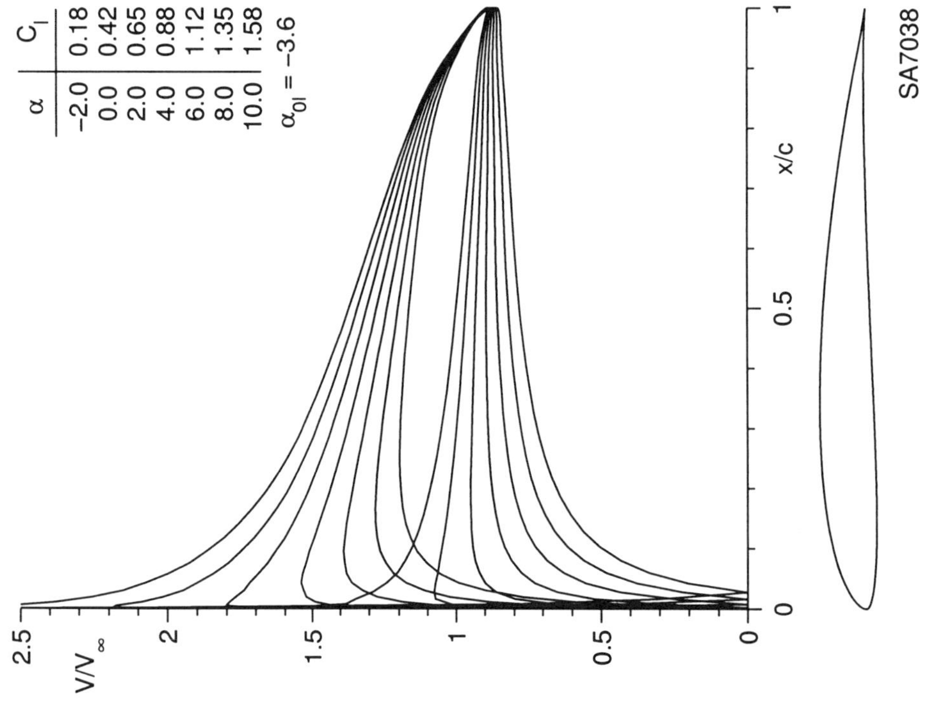
α | C_l
−2.0 | 0.18
0.0 | 0.42
2.0 | 0.65
4.0 | 0.88
6.0 | 1.12
8.0 | 1.35
10.0 | 1.58
α_{0l} = −3.6
V/V_∞
2.5
2
1.5
1
0.5
0
x/c
0.5
1
SA7038

SA7038

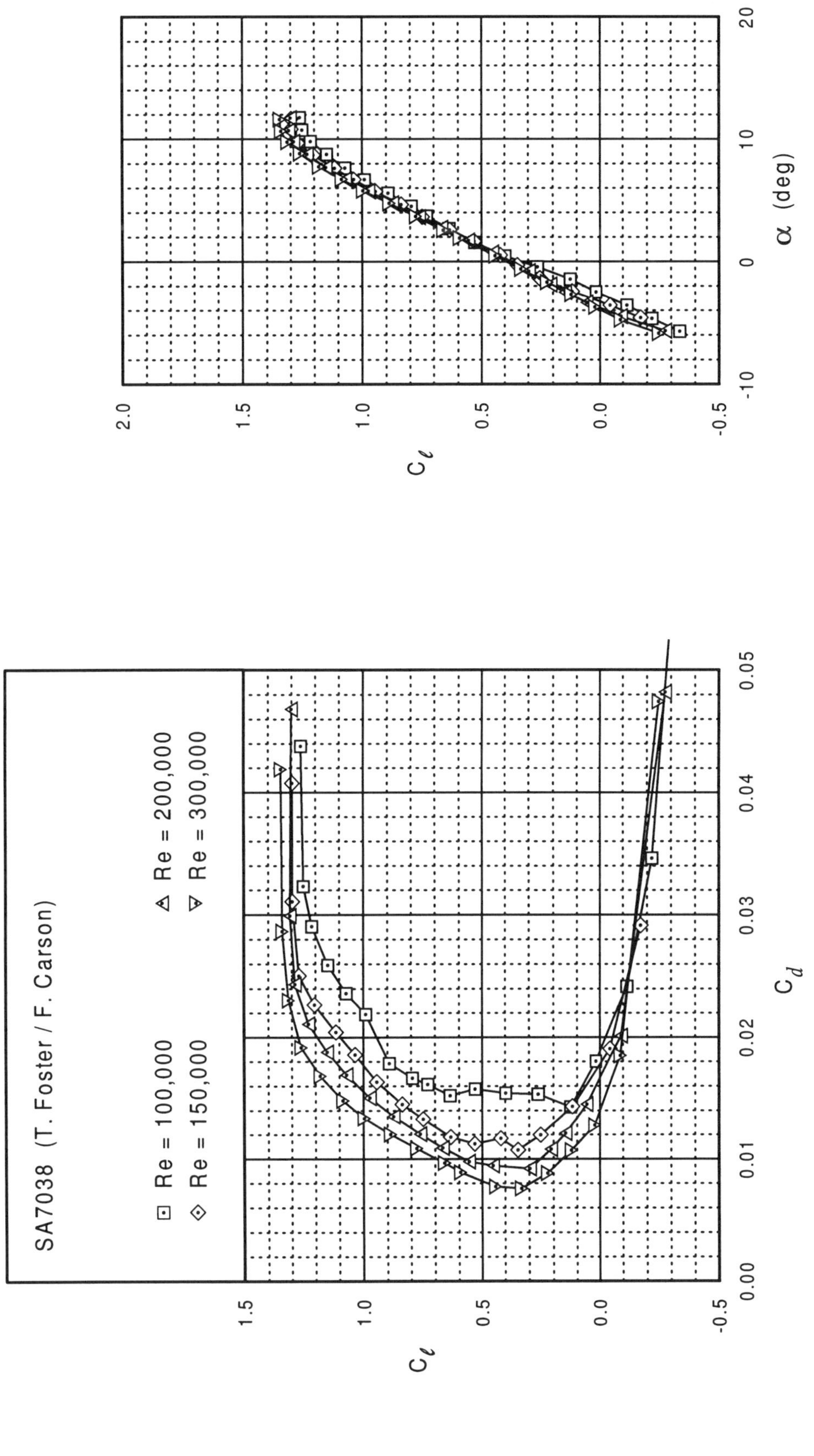
SA7038 (T. Foster / F. Carson)
Re = 100,000
Re = 150,000
Re = 200,000
Re = 300,000
C_ℓ
C_d
α (deg)
2.0
1.5
1.0
0.5
0.0
-0.5
-10
0
10
20
0.00
0.01
0.02
0.03
0.04
0.05

SA7038

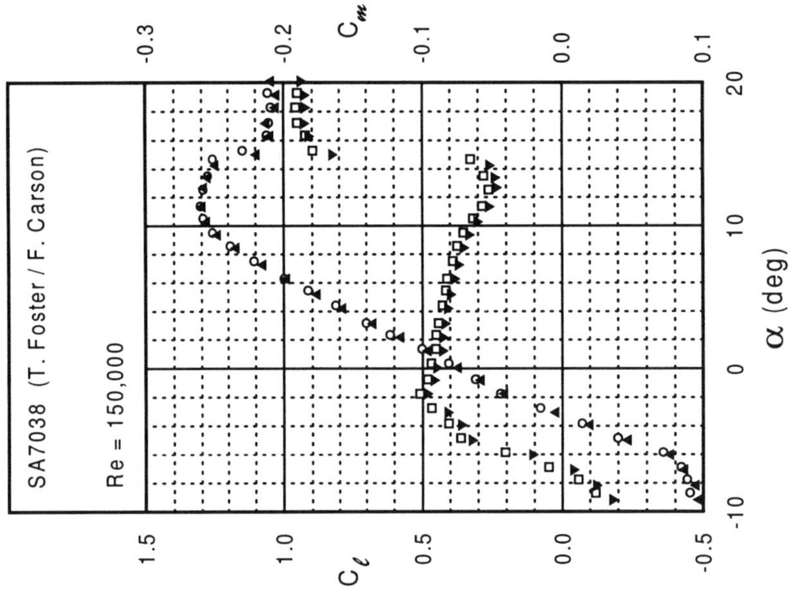
SA7038 (T. Foster / F. Carson)
Re = 150,000
C_ℓ
1.5
1.0
0.5
0.0
-0.5
C_m
-0.3
-0.2
-0.1
0.0
0.1
α (deg)
-10
0
10
20

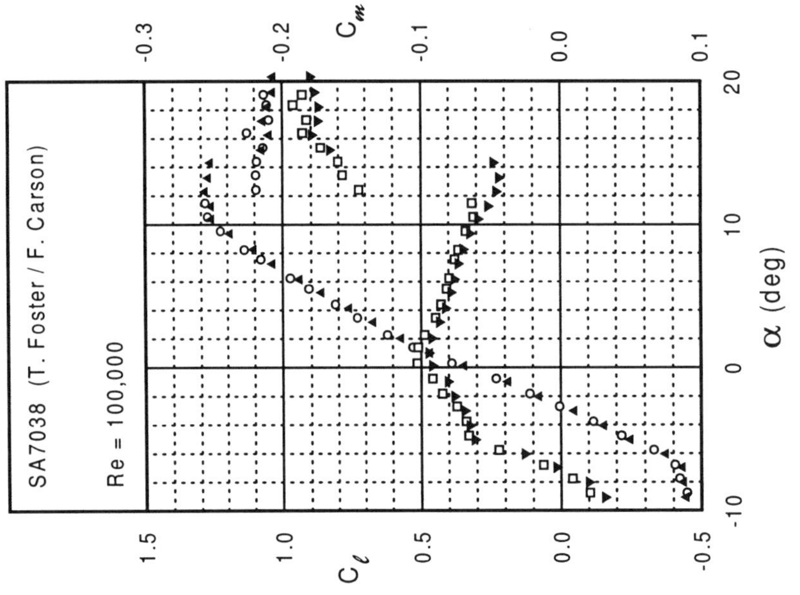
SA7038 (T. Foster / F. Carson)
Re = 100,000
C_ℓ
1.5
1.0
0.5
0.0
-0.5
C_m
-0.3
-0.2
-0.1
0.0
0.1
α (deg)
-10
0
10
20

SA7038

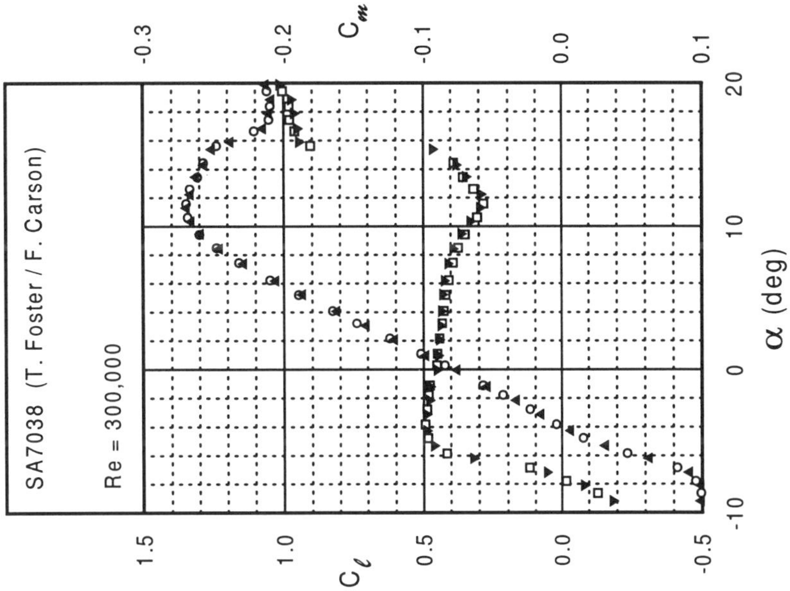
SA7038 (T. Foster / F. Carson)
Re = 300,000
C_ℓ
C_m
α (deg)

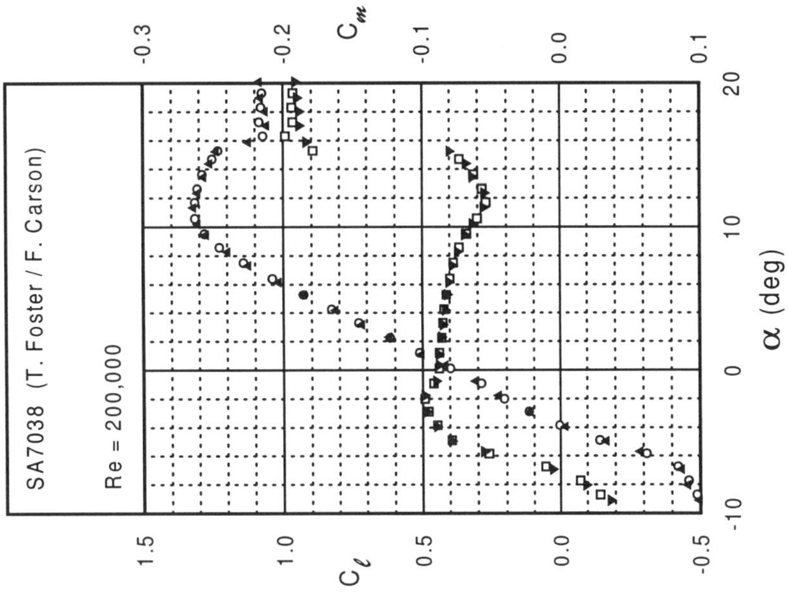
SA7038 (T. Foster / F. Carson)
Re = 200,000
C_ℓ
C_m
α (deg)

SD7032 (D)

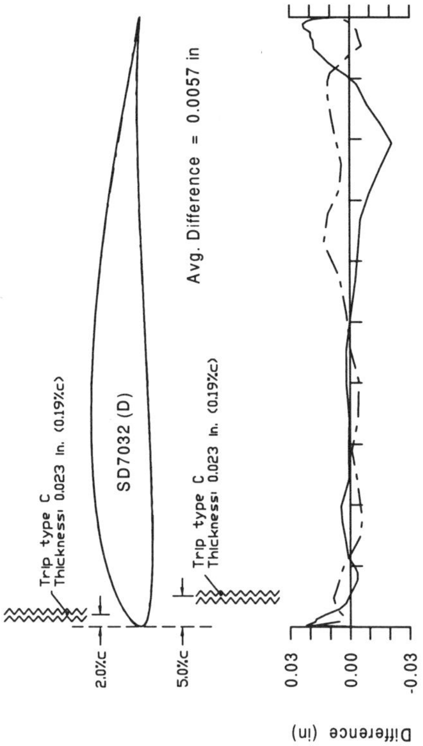
Trip type C
Thickness: 0.023 in. (0.19%c)
SD7032 (D)
Trip type C
Thickness: 0.023 in. (0.19%c)
Avg. Difference = 0.0057 in
2.0%c
5.0%c
0.03
0.00
-0.03
Difference (in)

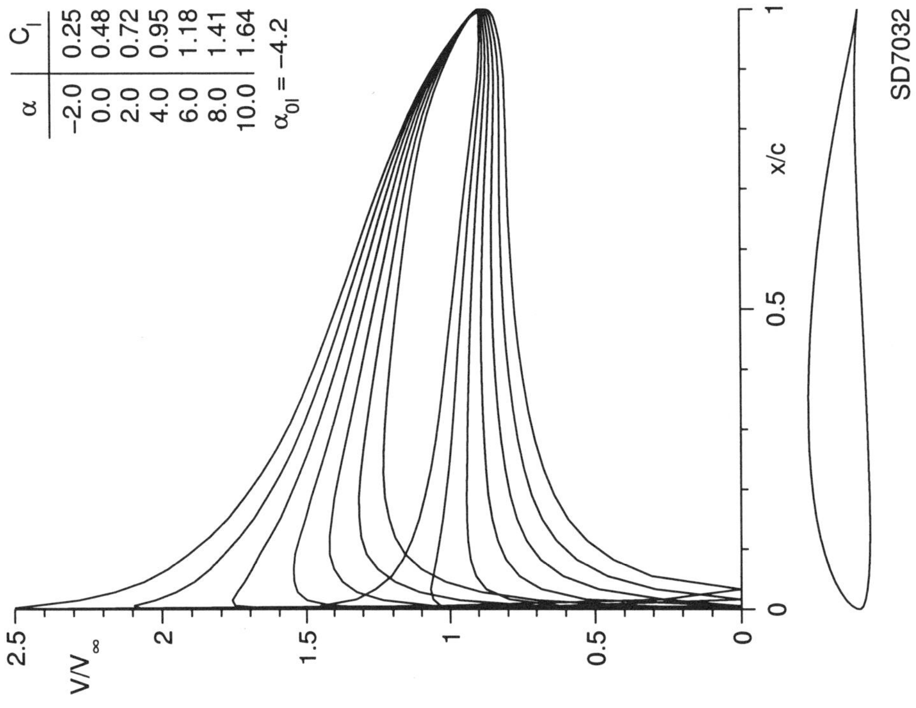
α
C_l
−2.0 0.25
0.0 0.48
2.0 0.72
4.0 0.95
6.0 1.18
8.0 1.41
10.0 1.64
α_0l = −4.2
V/V_∞
2.5
2
1.5
1
0.5
0
0
0.5
1
x/c
SD7032

SD7032 (D)

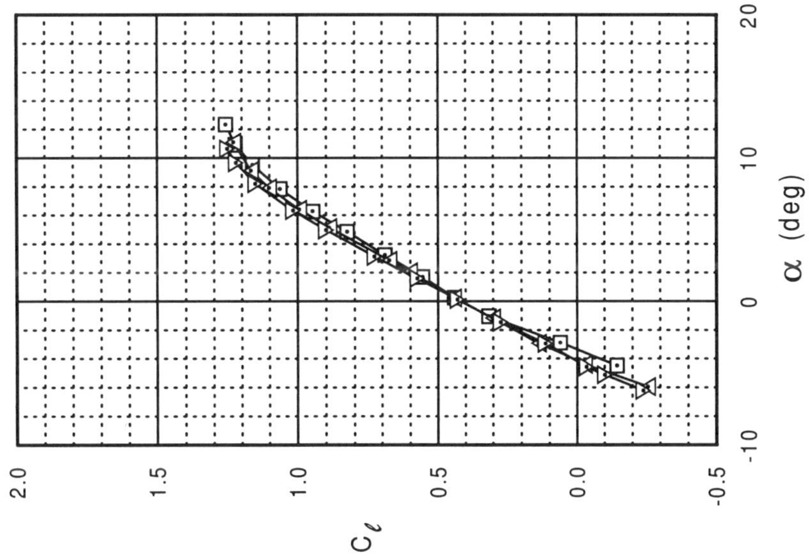
C_ℓ
α (deg)
2.0
1.5
1.0
0.5
0.0
-0.5
-10
0
10
20

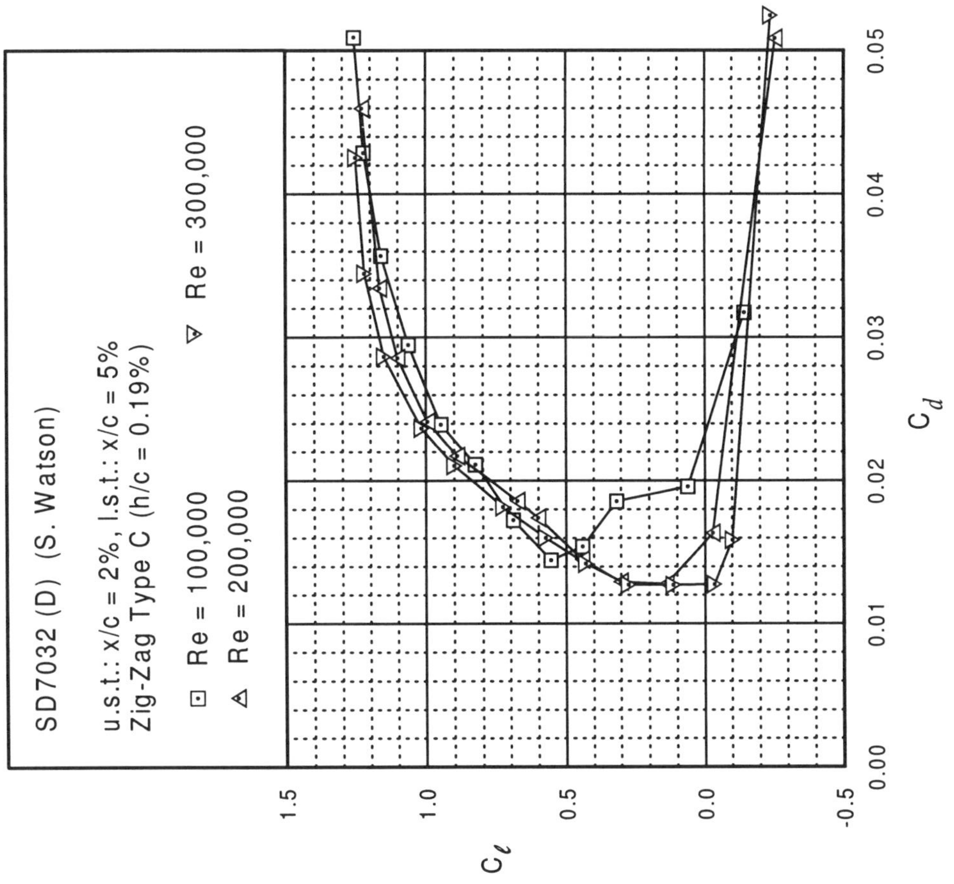
SD7032 (D) (S. Watson)
u.s.t.: x/c = 2%, l.s.t.: x/c = 5%
Zig-Zag Type C (h/c = 0.19%)
Re = 100,000
Re = 200,000
Re = 300,000
C_ℓ
C_d
1.5
1.0
0.5
0.0
-0.5
0.00
0.01
0.02
0.03
0.04
0.05

SD7032 (D)

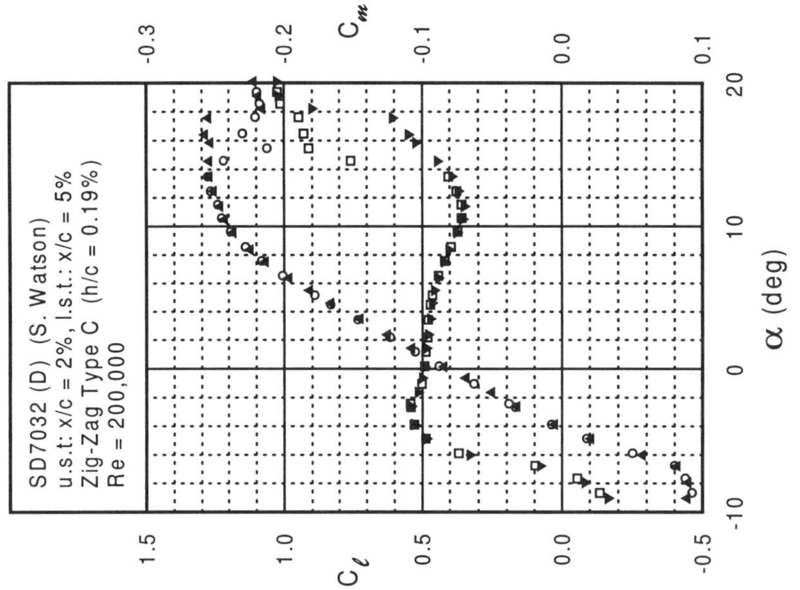
SD7032 (D) (S. Watson)
u.s.t.: x/c = 2%, l.s.t.: x/c = 5%
Zig-Zag Type C (h/c = 0.19%)
Re = 200,000
C_m
-0.3
-0.2
-0.1
0.0
0.1
C_ℓ
1.5
1.0
0.5
0.0
-0.5
-10
0
10
20
α (deg)

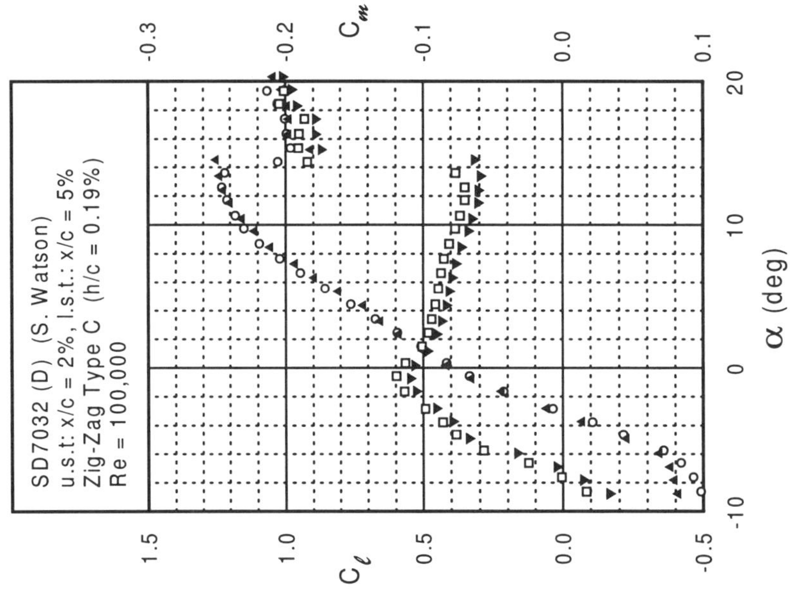
SD7032 (D) (S. Watson)
u.s.t.: x/c = 2%, l.s.t.: x/c = 5%
Zig-Zag Type C (h/c = 0.19%)
Re = 100,000
C_m
-0.3
-0.2
-0.1
0.0
0.1
C_ℓ
1.5
1.0
0.5
0.0
-0.5
-10
0
10
20
α (deg)

SD7032 (D)

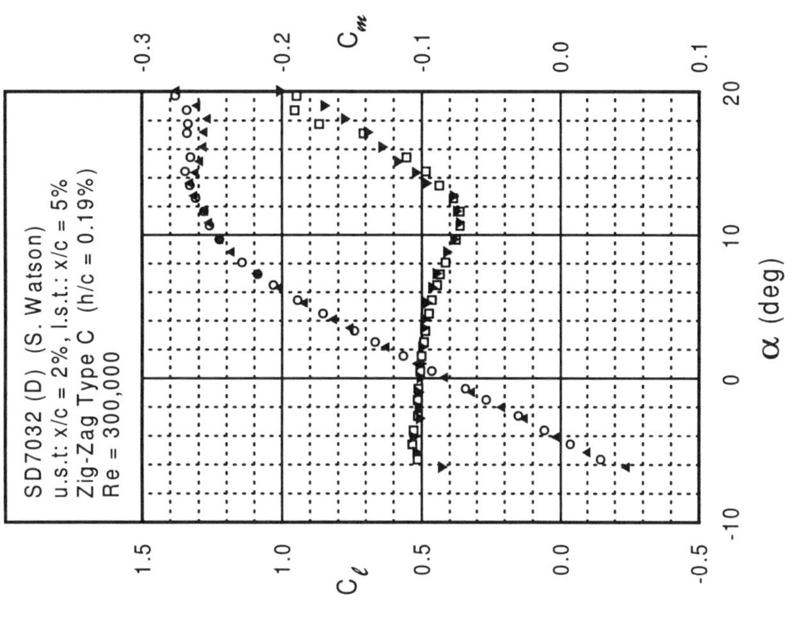
SD7032 (D) (S. Watson)
u.s.t: x/c = 2%, l.s.t.: x/c = 5%
Zig-Zag Type C (h/c = 0.19%)
Re = 300,000
C_m
-0.3
-0.2
-0.1
0.0
0.1
C_ℓ
1.5
1.0
0.5
0.0
-0.5
-10
0
10
20
α (deg)

SD7037 (B)

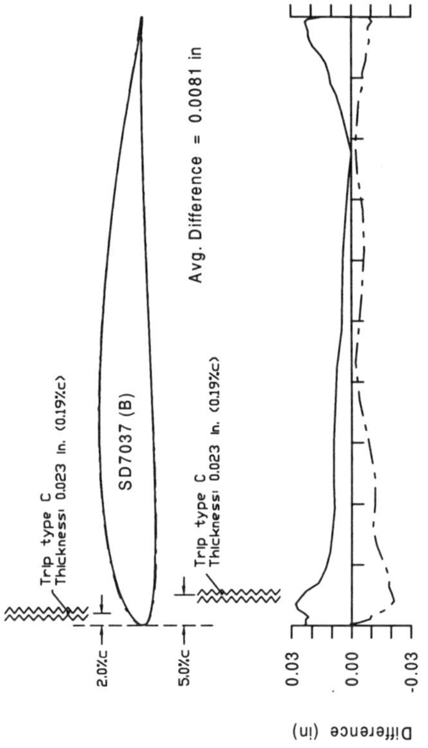
SD7037 (B)
Trip type C
Thickness: 0.023 in. (0.19%c)
Trip type C
Thickness: 0.023 in. (0.19%c)
2.0%c
5.0%c
Avg. Difference = 0.0081 in
0.03
0.00
-0.03
Difference (in)

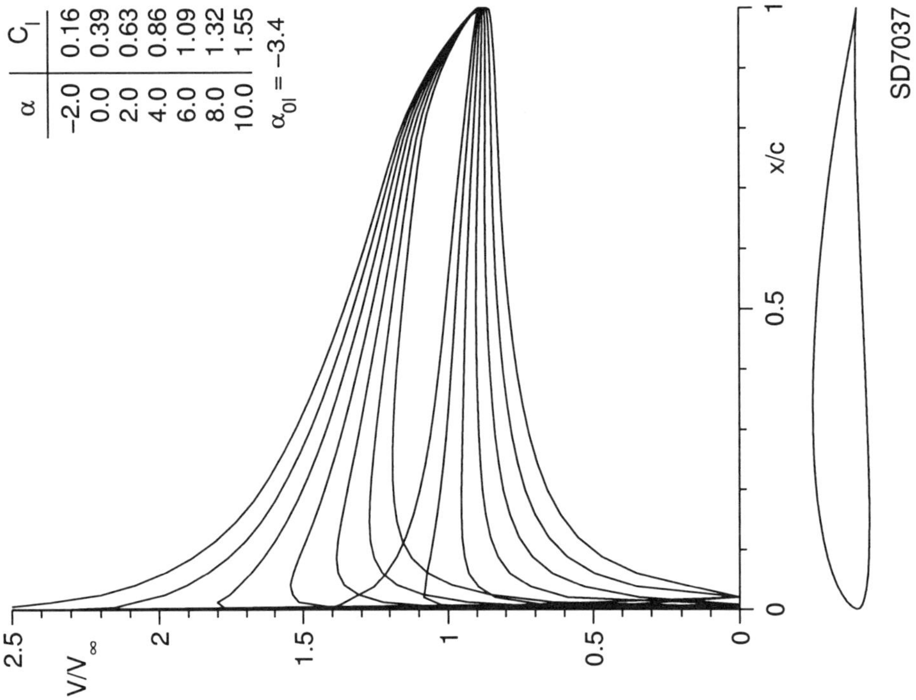
α
C_l
−2.0 0.16
0.0 0.39
2.0 0.63
4.0 0.86
6.0 1.09
8.0 1.32
10.0 1.55
α_0l = −3.4
V/V_∞
2.5
2
1.5
1
0.5
0
x/c
0
0.5
1
SD7037

SD7037 (B)

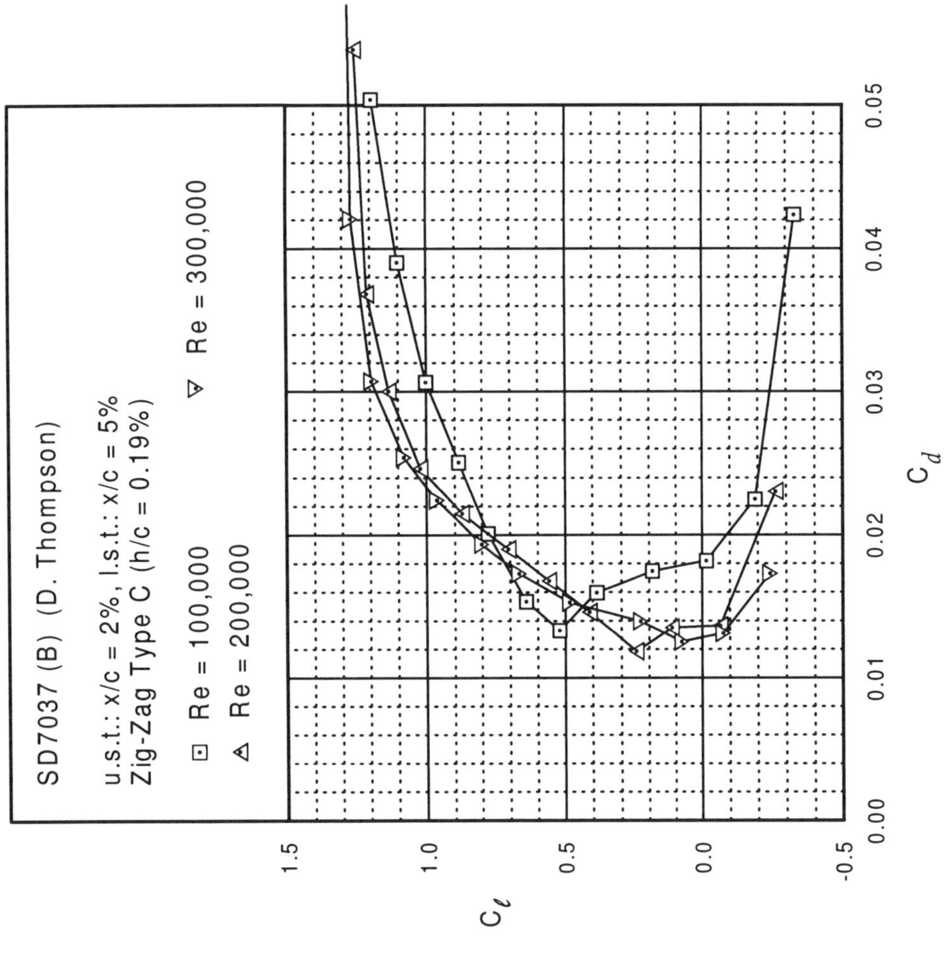
SD7037 (B) (D. Thompson)
u.s.t.: x/c = 2%, l.s.t.: x/c = 5%
Zig-Zag Type C (h/c = 0.19%)
Re = 100,000
Re = 200,000
Re = 300,000
C_ℓ
C_d
1.5
1.0
0.5
0.0
-0.5
0.00
0.01
0.02
0.03
0.04
0.05

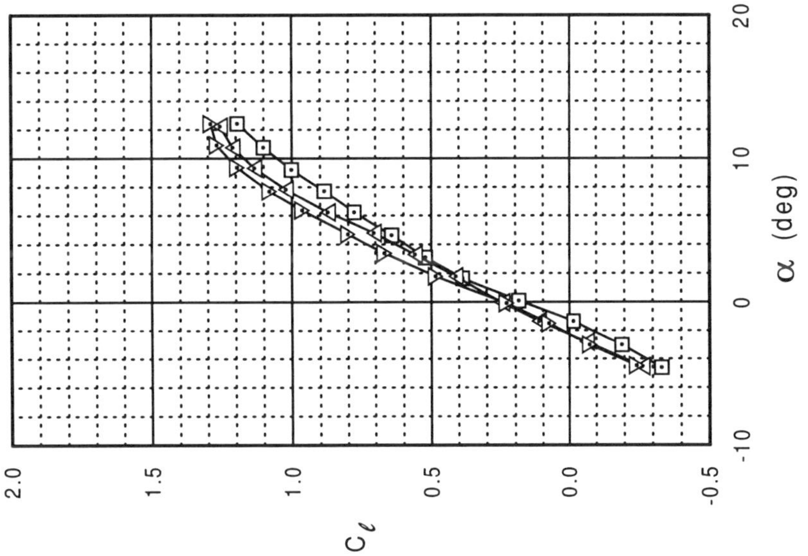
C_ℓ
α (deg)
2.0
1.5
1.0
0.5
0.0
-0.5
-10
0
10
20

SD7037 (B)

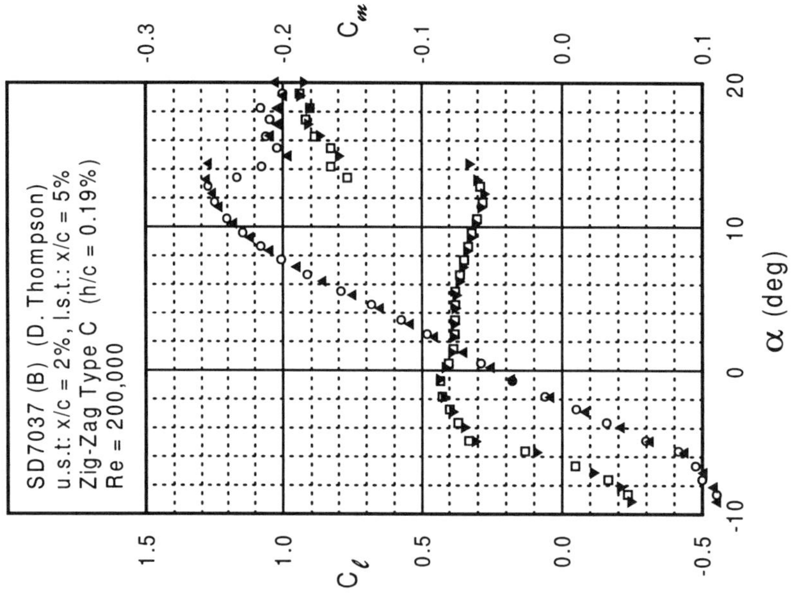
SD7037 (B) (D. Thompson)
u.s.t: x/c = 2%, l.s.t.: x/c = 5%
Zig-Zag Type C (h/c = 0.19%)
Re = 200,000
C_m
-0.3
-0.2
-0.1
0.0
0.1
C_ℓ
1.5
1.0
0.5
0.0
-0.5
-10
0
10
20
α (deg)

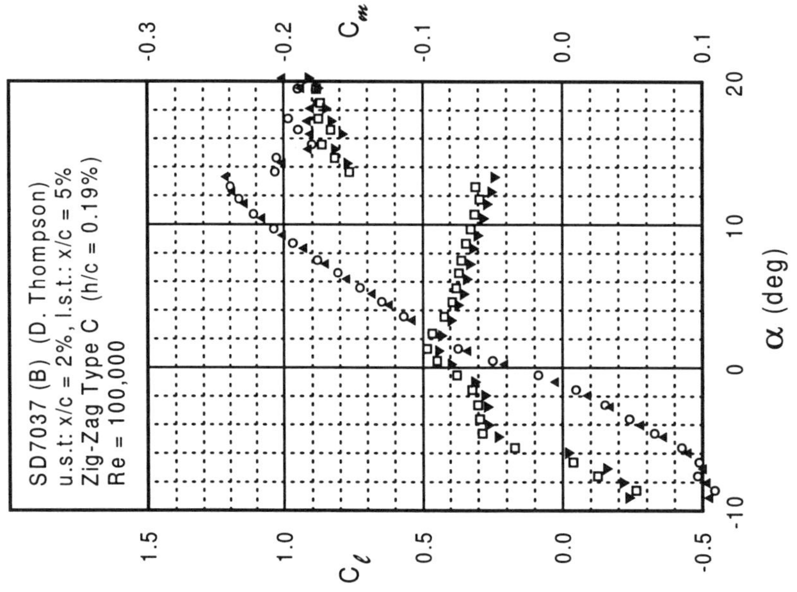
SD7037 (B) (D. Thompson)
u.s.t: x/c = 2%, l.s.t.: x/c = 5%
Zig-Zag Type C (h/c = 0.19%)
Re = 100,000
C_m
-0.3
-0.2
-0.1
0.0
0.1
C_ℓ
1.5
1.0
0.5
0.0
-0.5
-10
0
10
20
α (deg)

SD7037 (B)

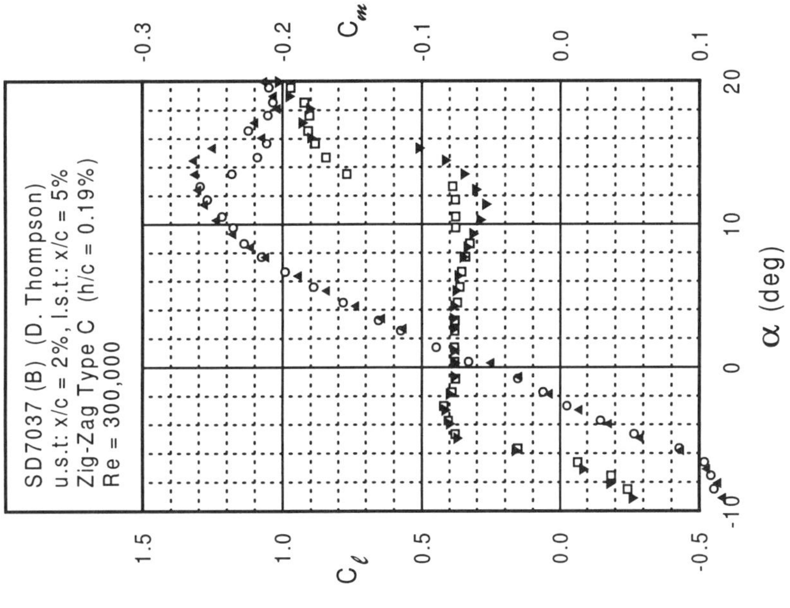
SD7037 (B) (D. Thompson)
u.s.t: x/c = 2%, l.s.t.: x/c = 5%
Zig-Zag Type C (h/c = 0.19%)
Re = 300,000
C
-0.3
-0.2
-0.1
0.0
0.1
C
1.5
1.0
0.5
0.0
-0.5
α (deg)
-10
0
10
20

SD7037 (D)

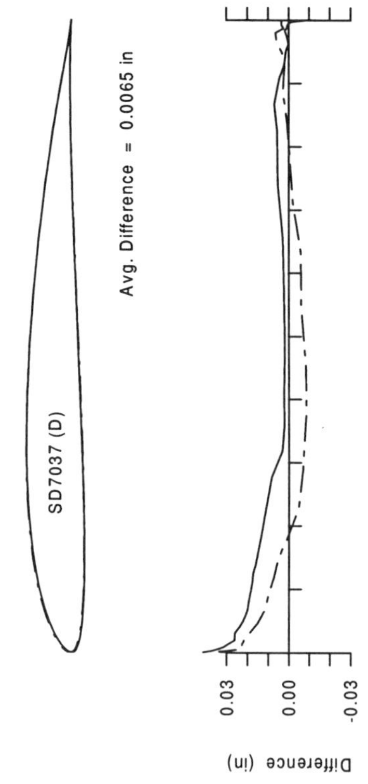
SD7037 (D)
Avg. Difference = 0.0065 in
0.03
0.00
-0.03
Difference (in)

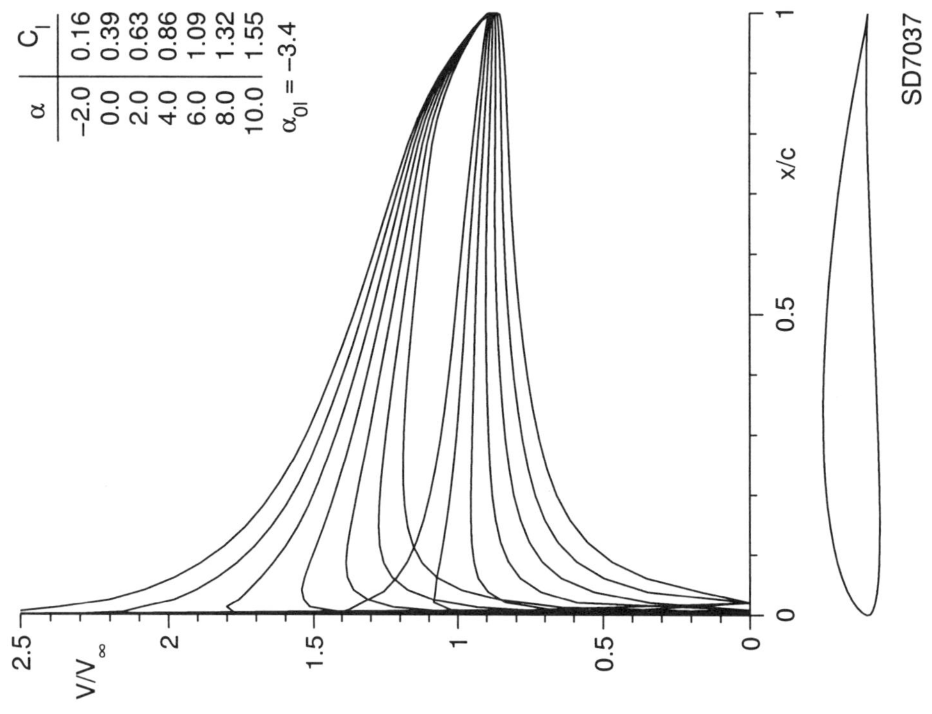
α
C_l
−2.0 0.16
0.0 0.39
2.0 0.63
4.0 0.86
6.0 1.09
8.0 1.32
10.0 1.55
α_0l = −3.4
V/V_∞
2.5
2
1.5
1
0.5
0
0.5
1
x/c
SD7037

SD7037 (D)

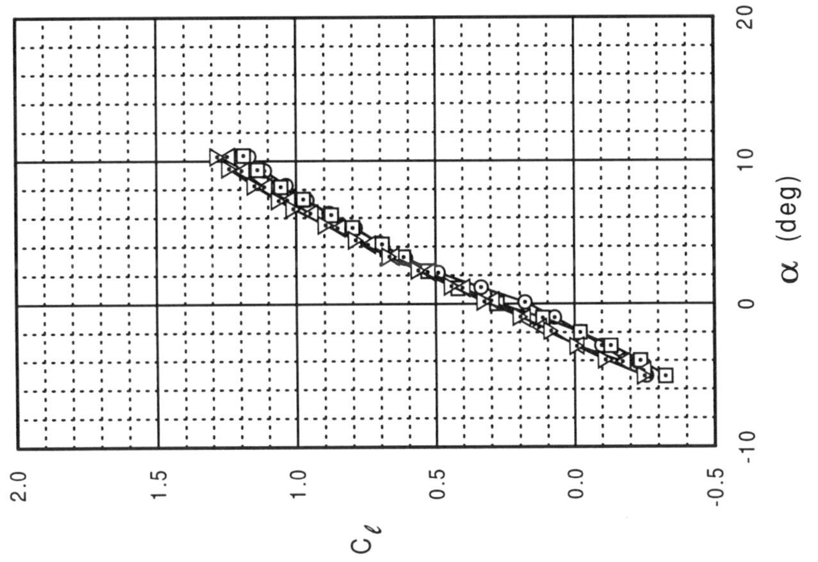
2.0
1.5
1.0
0.5
0.0
-0.5
-10
0
10
20
α (deg)

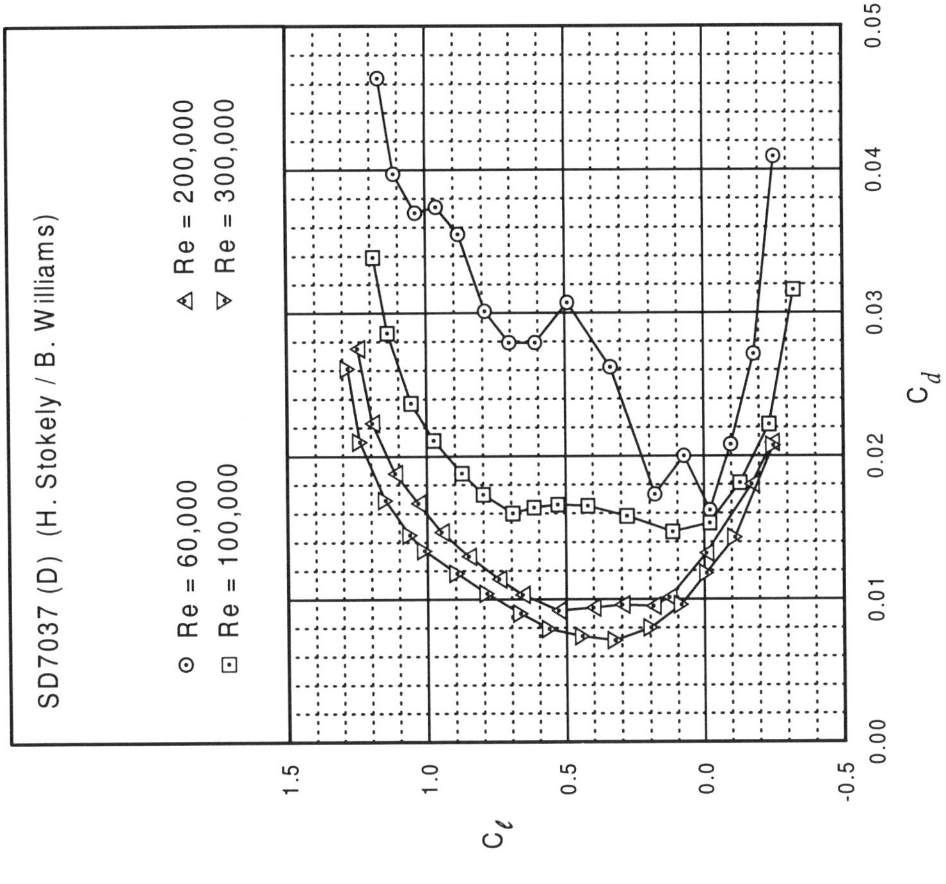
SD7037 (D) (H. Stokely / B. Williams)
Re = 60,000
Re = 100,000
Re = 200,000
Re = 300,000
1.5
1.0
0.5
0.0
-0.5
0.00
0.01
0.02
0.03
0.04
0.05

SD7037 (D)

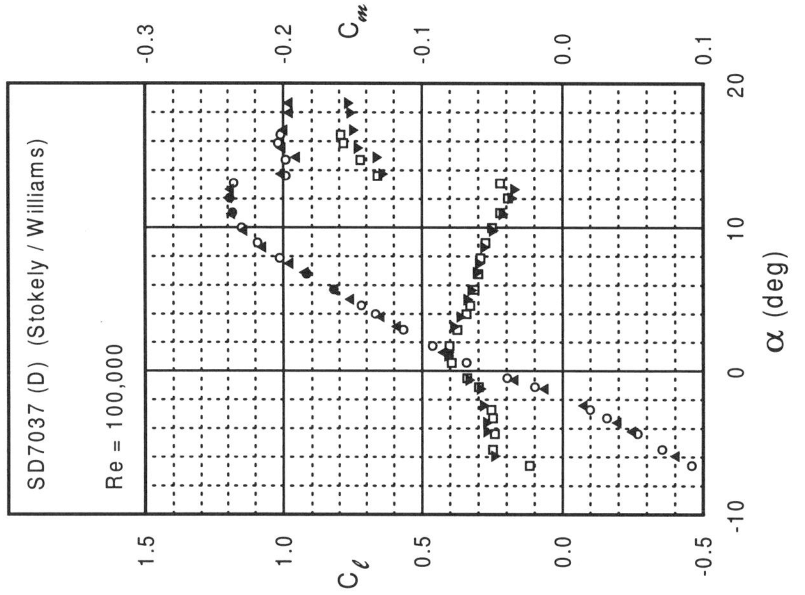
SD7037 (D) (Stokely / Williams)
Re = 100,000
C_ℓ
C_m
α (deg)
-0.3
-0.2
-0.1
0.0
0.1
1.5
1.0
0.5
0.0
-0.5
-10
0
10
20

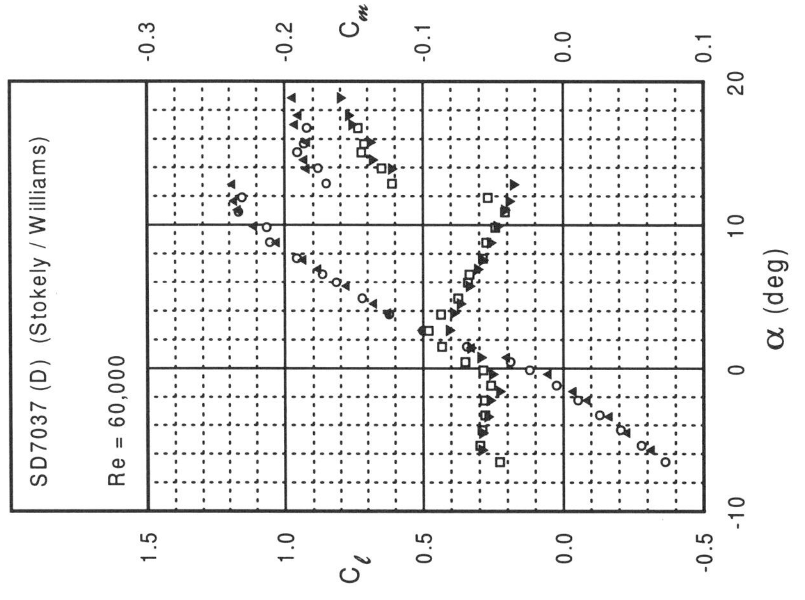
SD7037 (D) (Stokely / Williams)
Re = 60,000
C_ℓ
C_m
α (deg)
-0.3
-0.2
-0.1
0.0
0.1
1.5
1.0
0.5
0.0
-0.5
-10
0
10
20

SD7037 (D)

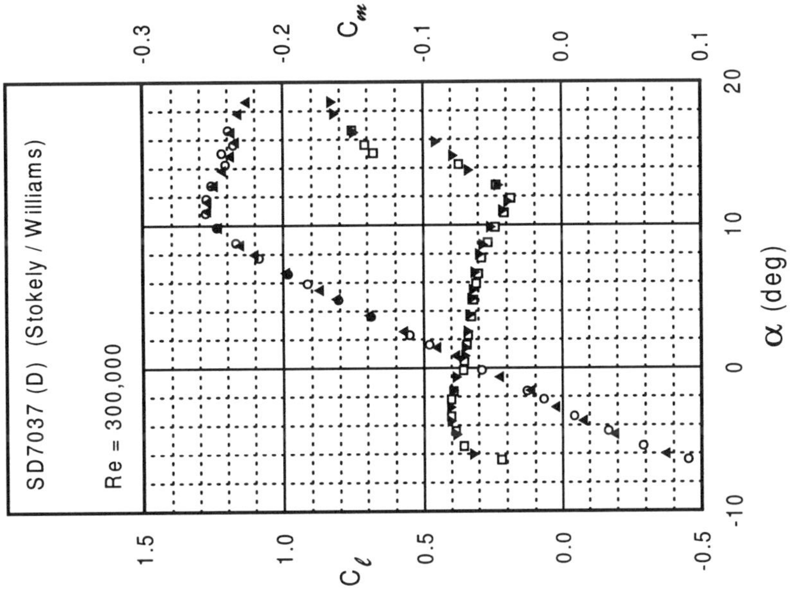
SD7037 (D) (Stokely / Williams)
Re = 300,000
1.5
1.0
0.5
0.0
-0.5
C_ℓ
-0.3
-0.2
-0.1
0.0
0.1
C_m
-10
0
10
20
α (deg)

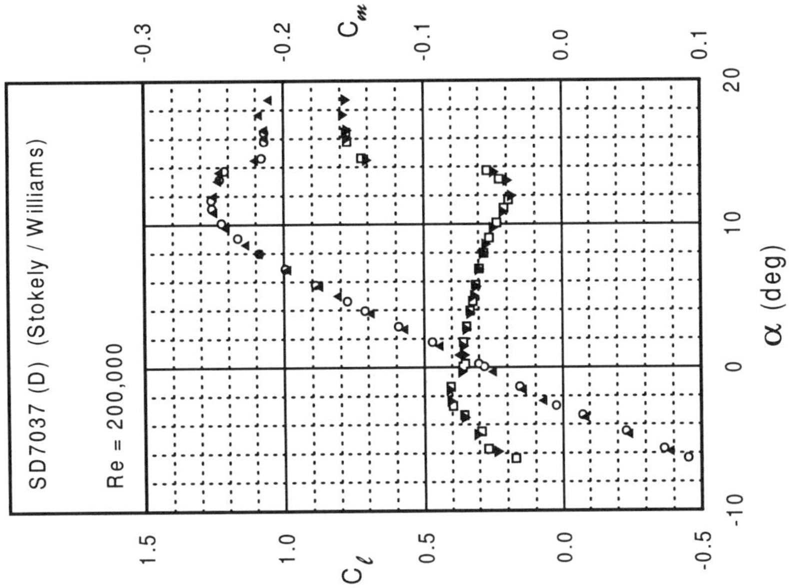
SD7037 (D) (Stokely / Williams)
Re = 200,000
1.5
1.0
0.5
0.0
-0.5
C_ℓ
-0.3
-0.2
-0.1
0.0
0.1
C_m
-10
0
10
20
α (deg)

SD7037 (D)

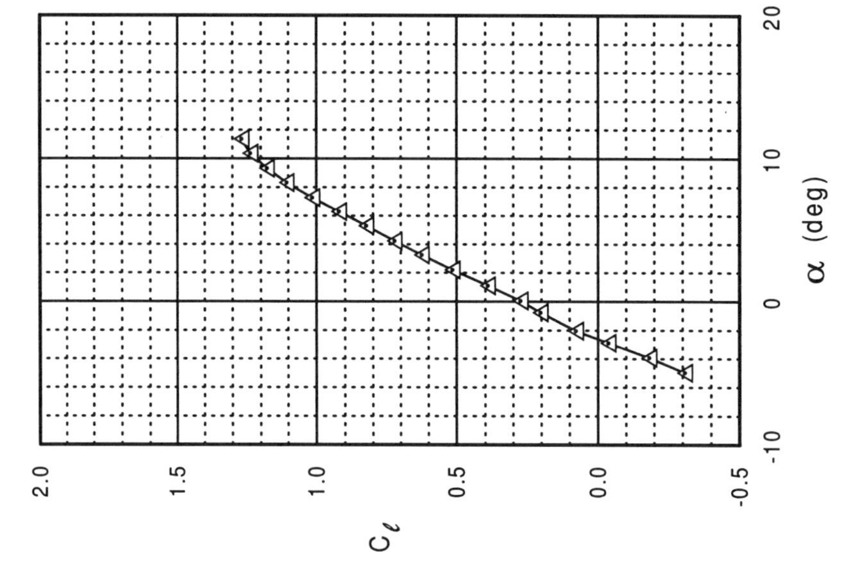
2.0
1.5
1.0
0.5
0.0
-0.5
C_ℓ
-10
0
10
20
α (deg)

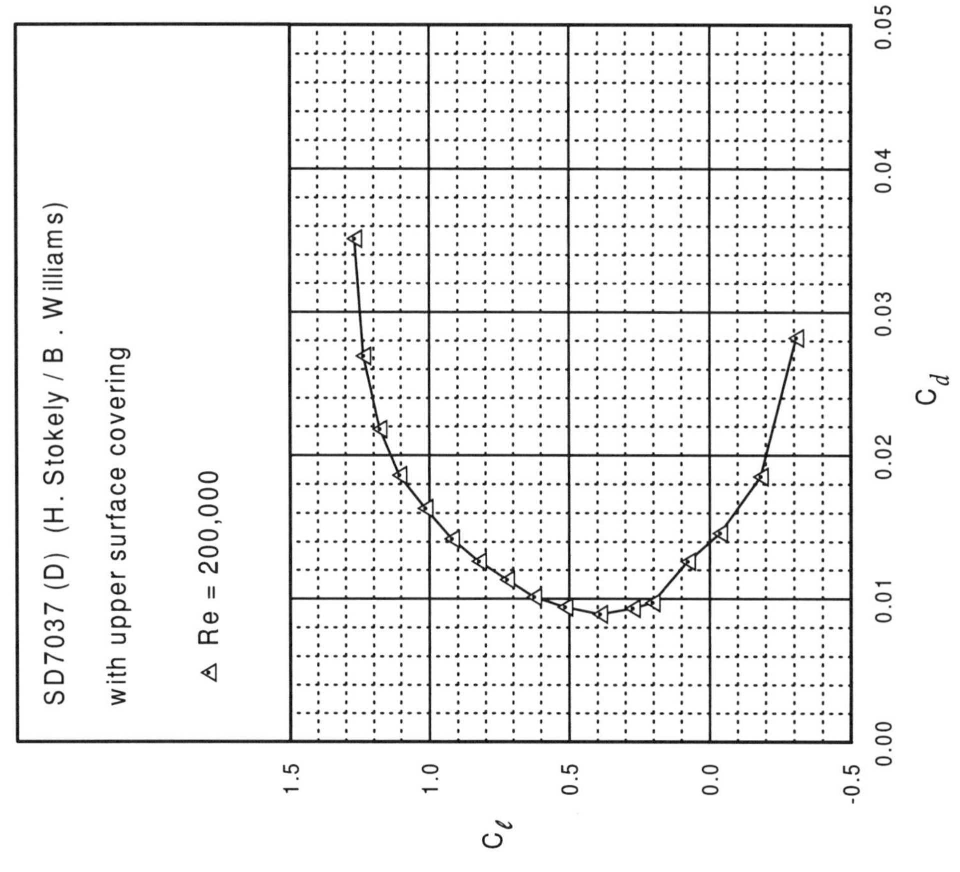
SD7037 (D) (H. Stokely / B. Williams)
with upper surface covering
Re = 200,000
1.5
1.0
0.5
0.0
-0.5
C_ℓ
0.00
0.01
0.02
0.03
0.04
0.05
C_d

SD7037 (D)

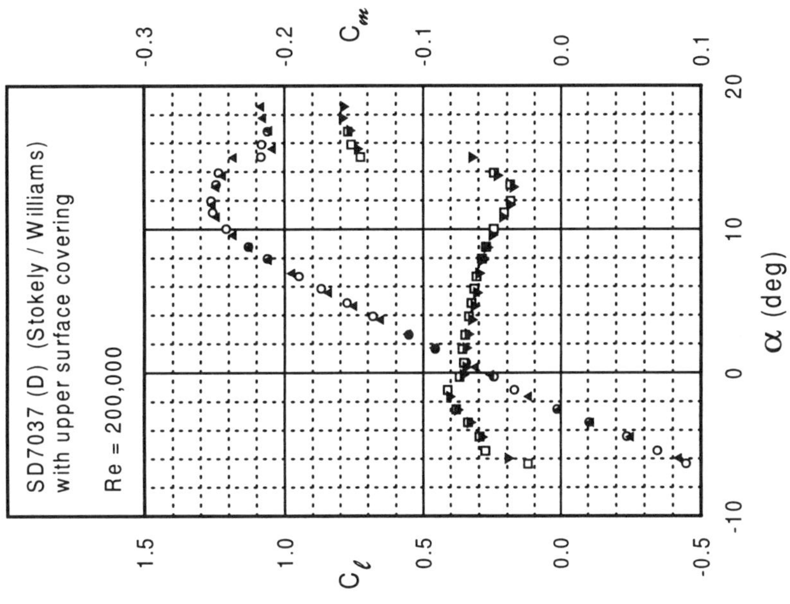
SD7037 (D) (Stokely / Williams)
with upper surface covering
Re = 200,000
C_m
-0.3
-0.2
-0.1
0.0
0.1
C_ℓ
1.5
1.0
0.5
0.0
-0.5
-10
0
10
20
α (deg)

SD7037 (E)

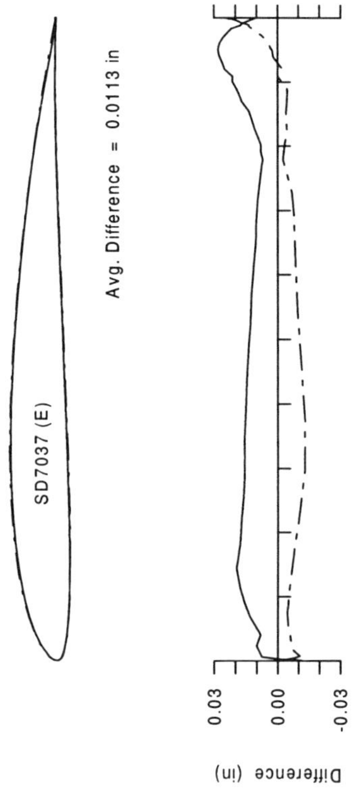
SD7037 (E)
Avg. Difference = 0.0113 in
0.03
0.00
-0.03
Difference (in)

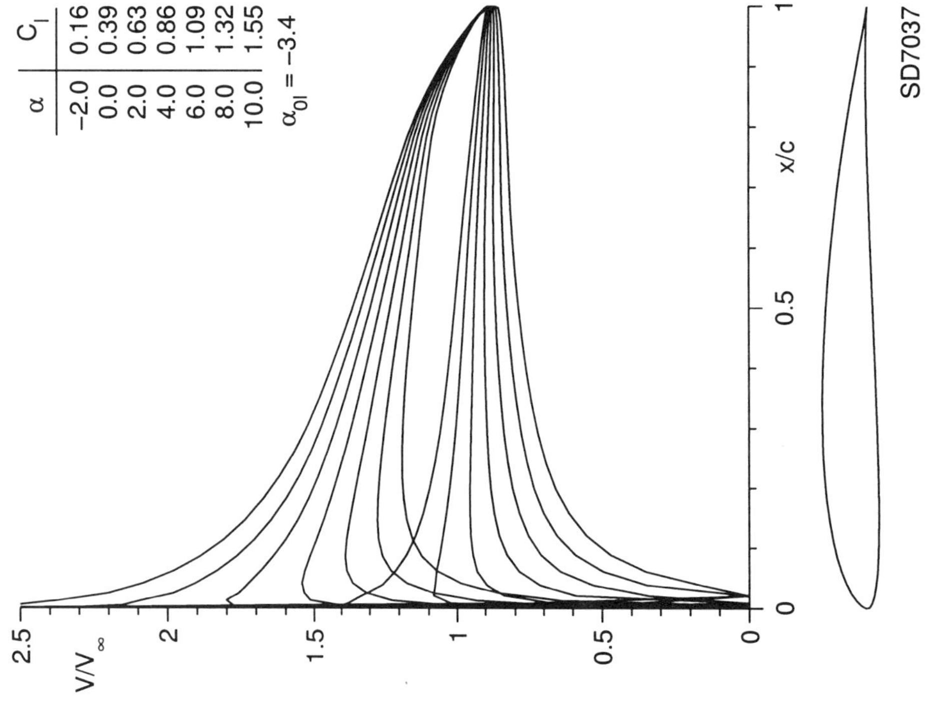
α | C_l
-2.0 | 0.16
0.0 | 0.39
2.0 | 0.63
4.0 | 0.86
6.0 | 1.09
8.0 | 1.32
10.0 | 1.55
α_{0l} = -3.4
V/V_∞
2.5
2
1.5
1
0.5
0
x/c
0
0.5
1
SD7037

SD7037 (E)

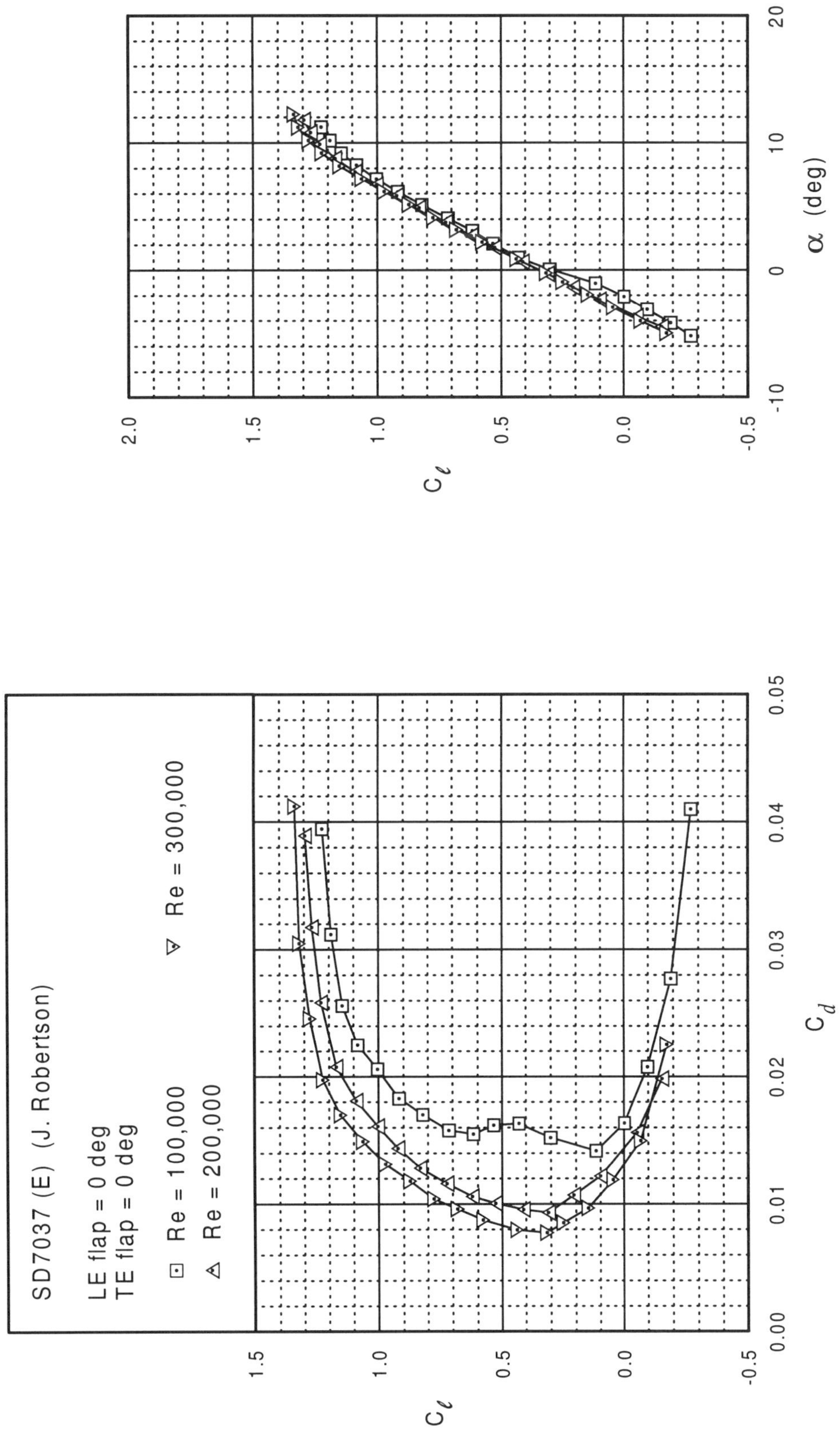
2.0
1.5
1.0
0.5
0.0
-0.5
C_ℓ
-10
0
10
20
α (deg)
SD7037 (E) (J. Robertson)
LE flap = 0 deg
TE flap = 0 deg
Re = 100,000
Re = 200,000
Re = 300,000
1.5
1.0
0.5
0.0
-0.5
C_ℓ
0.00
0.01
0.02
0.03
0.04
0.05
C_d

SD7037 (E)

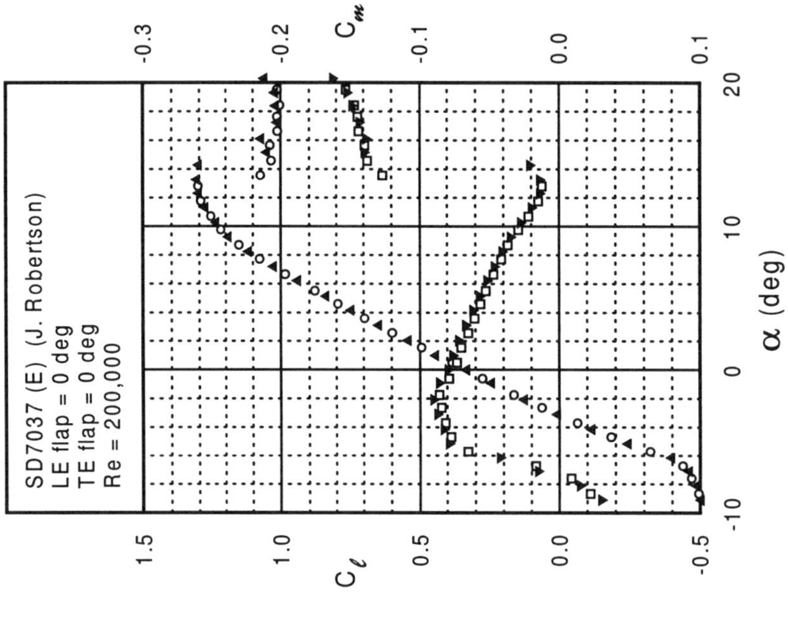
SD7037 (E) (J. Robertson)
LE flap = 0 deg
TE flap = 0 deg
Re = 200,000
C_ℓ
1.5
1.0
0.5
0.0
-0.5
C_m
-0.3
-0.2
-0.1
0.0
0.1
-10
0
10
20
α (deg)

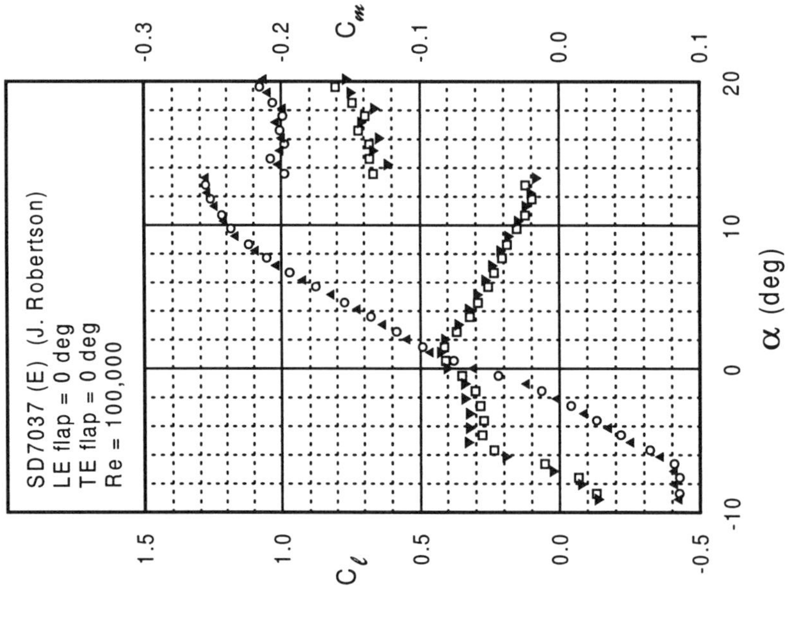
SD7037 (E) (J. Robertson)
LE flap = 0 deg
TE flap = 0 deg
Re = 100,000
C_ℓ
1.5
1.0
0.5
0.0
-0.5
C_m
-0.3
-0.2
-0.1
0.0
0.1
-10
0
10
20
α (deg)

SD7037 (E)

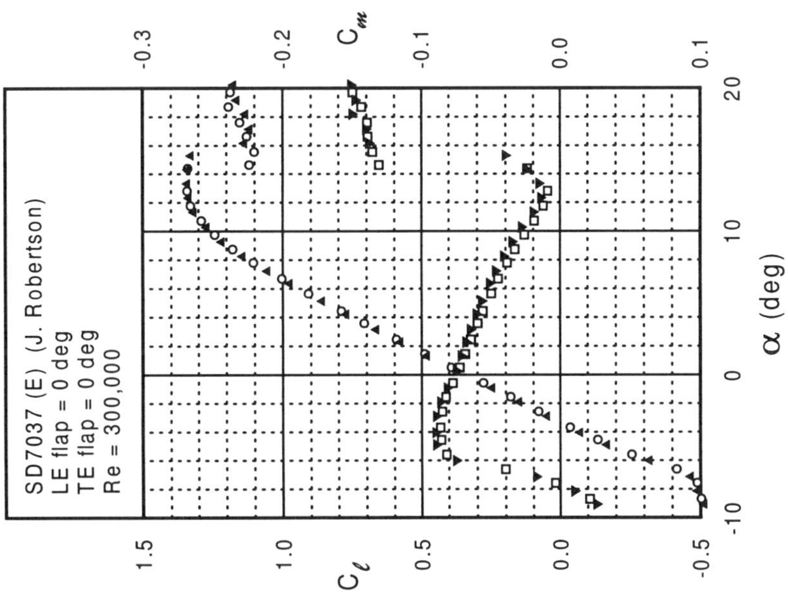
SD7037 (E) (J. Robertson)
LE flap = 0 deg
TE flap = 0 deg
Re = 300,000
C_m
-0.3
-0.2
-0.1
0.0
0.1
C_ℓ
1.5
1.0
0.5
0.0
-0.5
-10
0
10
20
α (deg)

SD7037 (E)

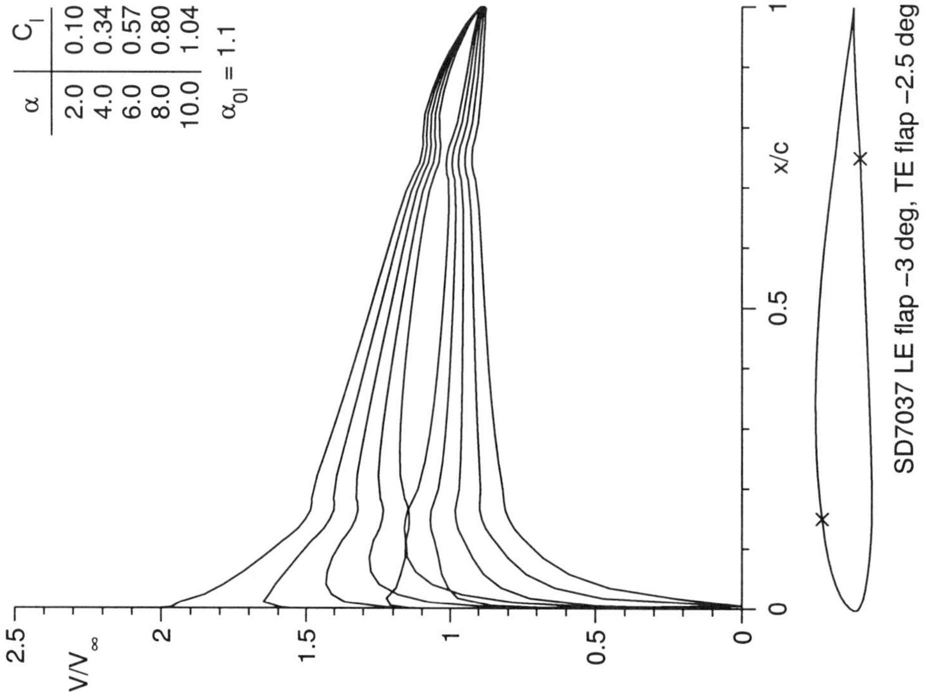
α | C_l
2.0 | 0.10
4.0 | 0.34
6.0 | 0.57
8.0 | 0.80
10.0 | 1.04
α_{0l} = 1.1
V/V∞
2.5
2
1.5
1
0.5
0
x/c
0.5
SD7037 LE flap −3 deg, TE flap −2.5 deg

SD7037 (E)

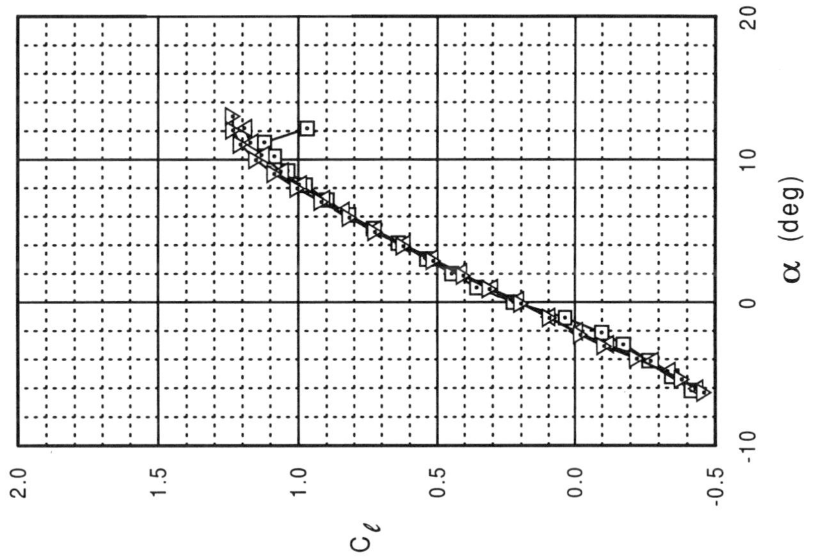
2.0
1.5
1.0
0.5
0.0
-0.5
-10
0
10
20
α (deg)
C_ℓ

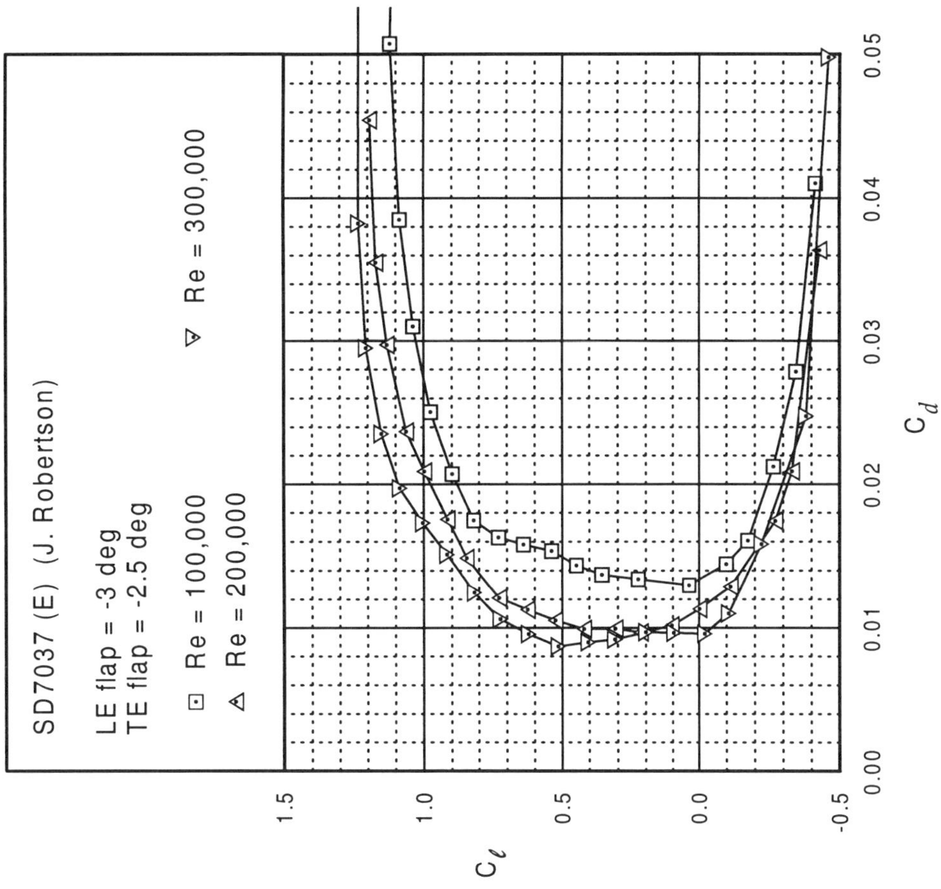
SD7037 (E) (J. Robertson)
LE flap = -3 deg
TE flap = -2.5 deg
Re = 100,000
Re = 200,000
Re = 300,000
1.5
1.0
0.5
0.0
-0.5
0.00
0.01
0.02
0.03
0.04
0.05
C_d
C_ℓ

SD7037 (E)

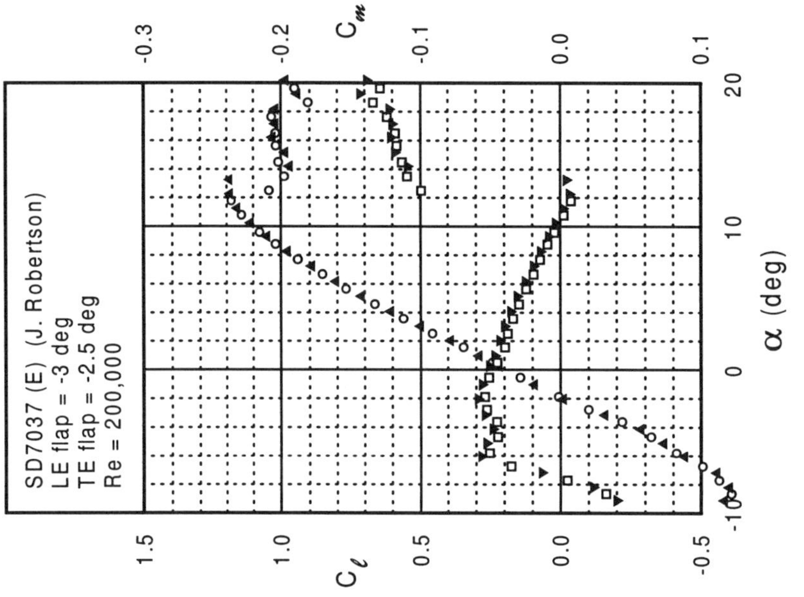
SD7037 (E) (J. Robertson)
LE flap = -3 deg
TE flap = -2.5 deg
Re = 200,000
C_ℓ
C_m
α (deg)

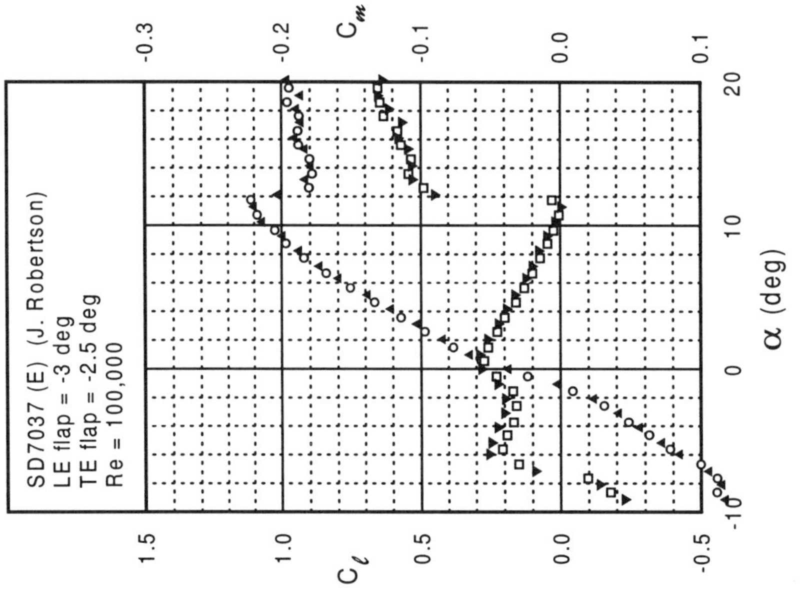
SD7037 (E) (J. Robertson)
LE flap = -3 deg
TE flap = -2.5 deg
Re = 100,000
C_ℓ
C_m
α (deg)

SD7037 (E)

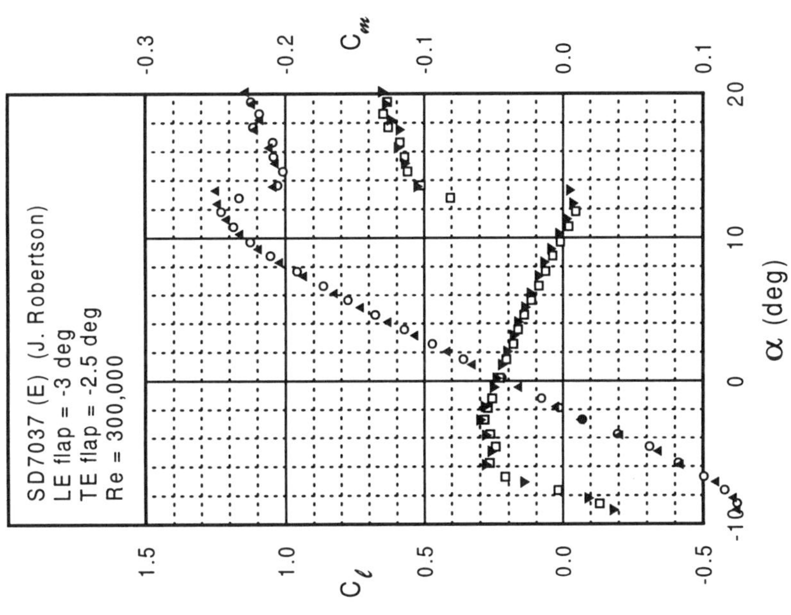
SD7037 (E) (J. Robertson)
LE flap = -3 deg
TE flap = -2.5 deg
Re = 300,000
C_ℓ
C_m
α (deg)
1.5
1.0
0.5
0.0
-0.5
-0.3
-0.2
-0.1
0.0
0.1
-10
0
10
20

SD7037 (E)

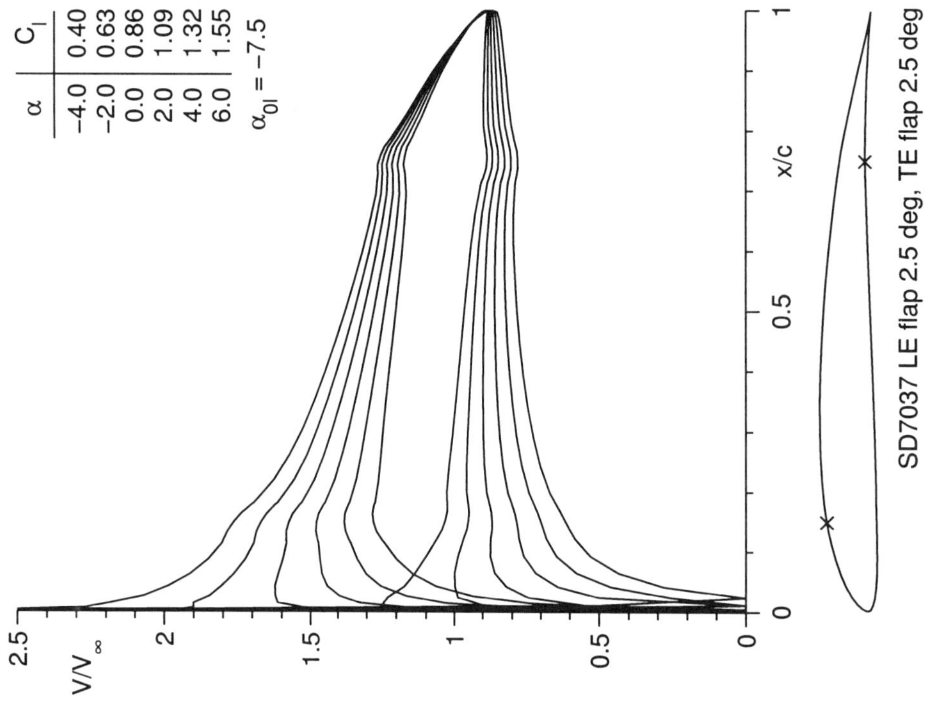
α | C_l
−4.0 | 0.40
−2.0 | 0.63
0.0 | 0.86
2.0 | 1.09
4.0 | 1.32
6.0 | 1.55
α_{0l} = −7.5
V/V_∞
2.5
2
1.5
1
0.5
0
x/c
0.5
1
SD7037 LE flap 2.5 deg, TE flap 2.5 deg

SD7037 (E)

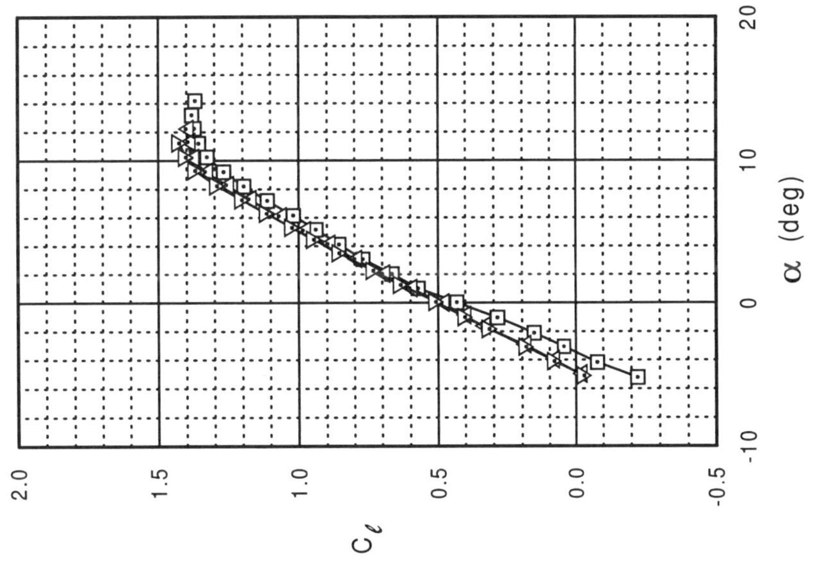
α (deg)
C_ℓ

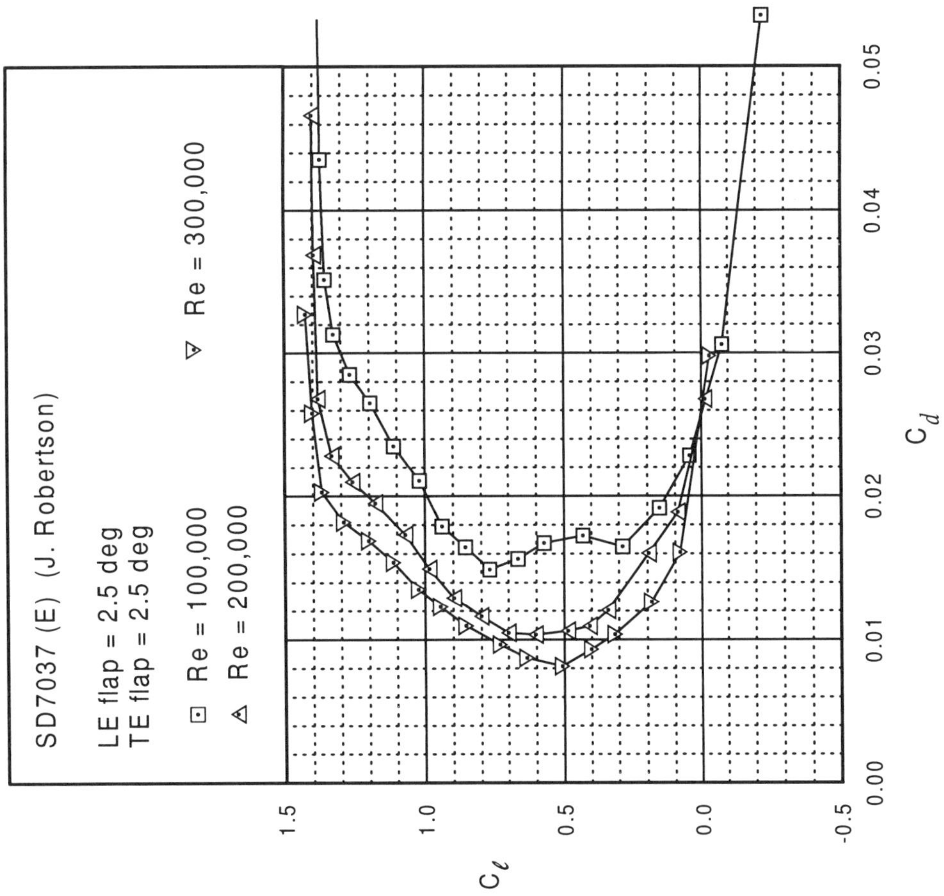
SD7037 (E) (J. Robertson)
LE flap = 2.5 deg
TE flap = 2.5 deg
Re = 100,000
Re = 200,000
Re = 300,000
C_d
C_ℓ

SD7037 (E)

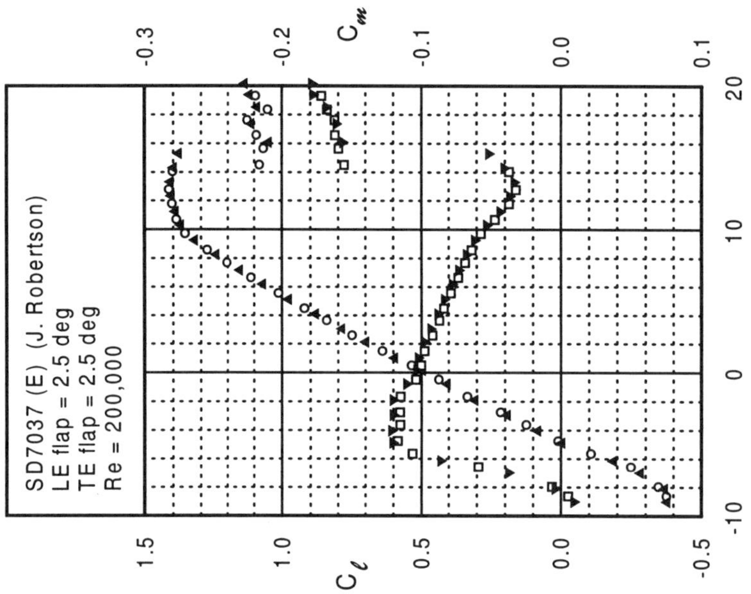
SD7037 (E) (J. Robertson)
LE flap = 2.5 deg
TE flap = 2.5 deg
Re = 200,000
α (deg)
C_ℓ
C_m

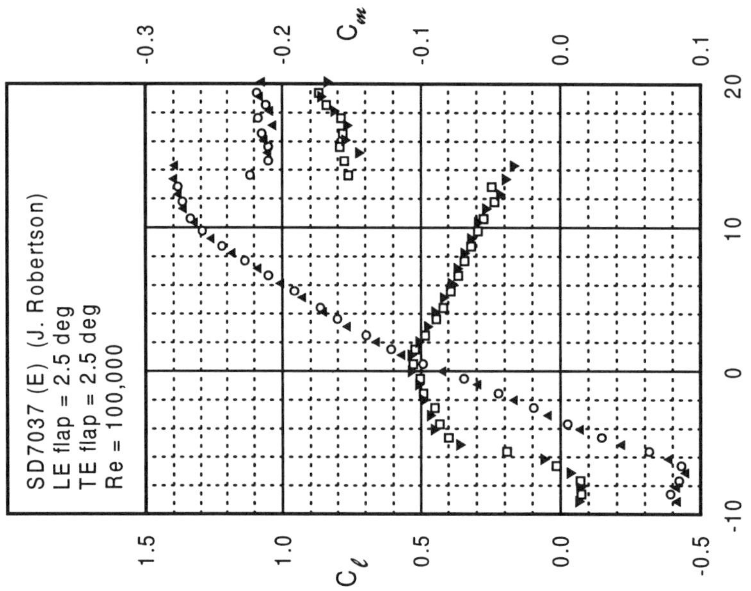
SD7037 (E) (J. Robertson)
LE flap = 2.5 deg
TE flap = 2.5 deg
Re = 100,000
α (deg)
C_ℓ
C_m

SD7037 (E)

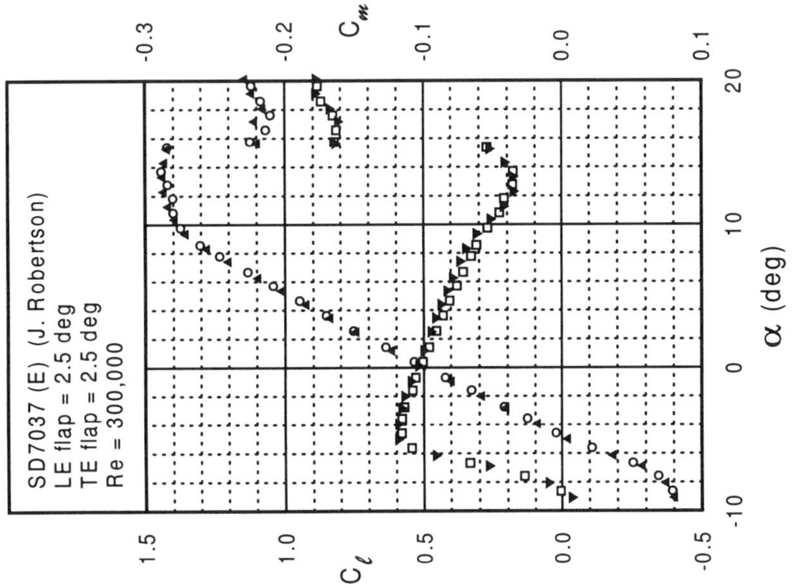
SD7037 (E) (J. Robertson)
LE flap = 2.5 deg
TE flap = 2.5 deg
Re = 300,000
C_ℓ
1.5
1.0
0.5
0.0
-0.5
C_m
-0.3
-0.2
-0.1
0.0
0.1
-10
0
10
20
α (deg)

SD7037 (E)

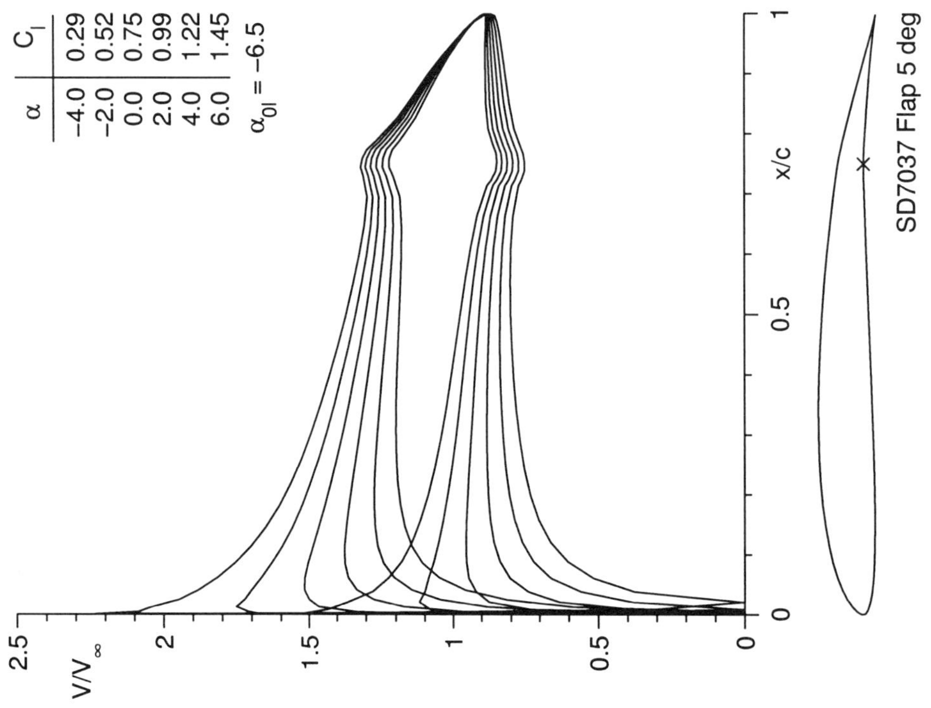
α
C_l
−4.0 0.29
−2.0 0.52
0.0 0.75
2.0 0.99
4.0 1.22
6.0 1.45
α_{0l} = −6.5
V/V∞
2.5
2
1.5
1
0.5
0
0
0.5
1
x/c
SD7037 Flap 5 deg

SD7037 (E)

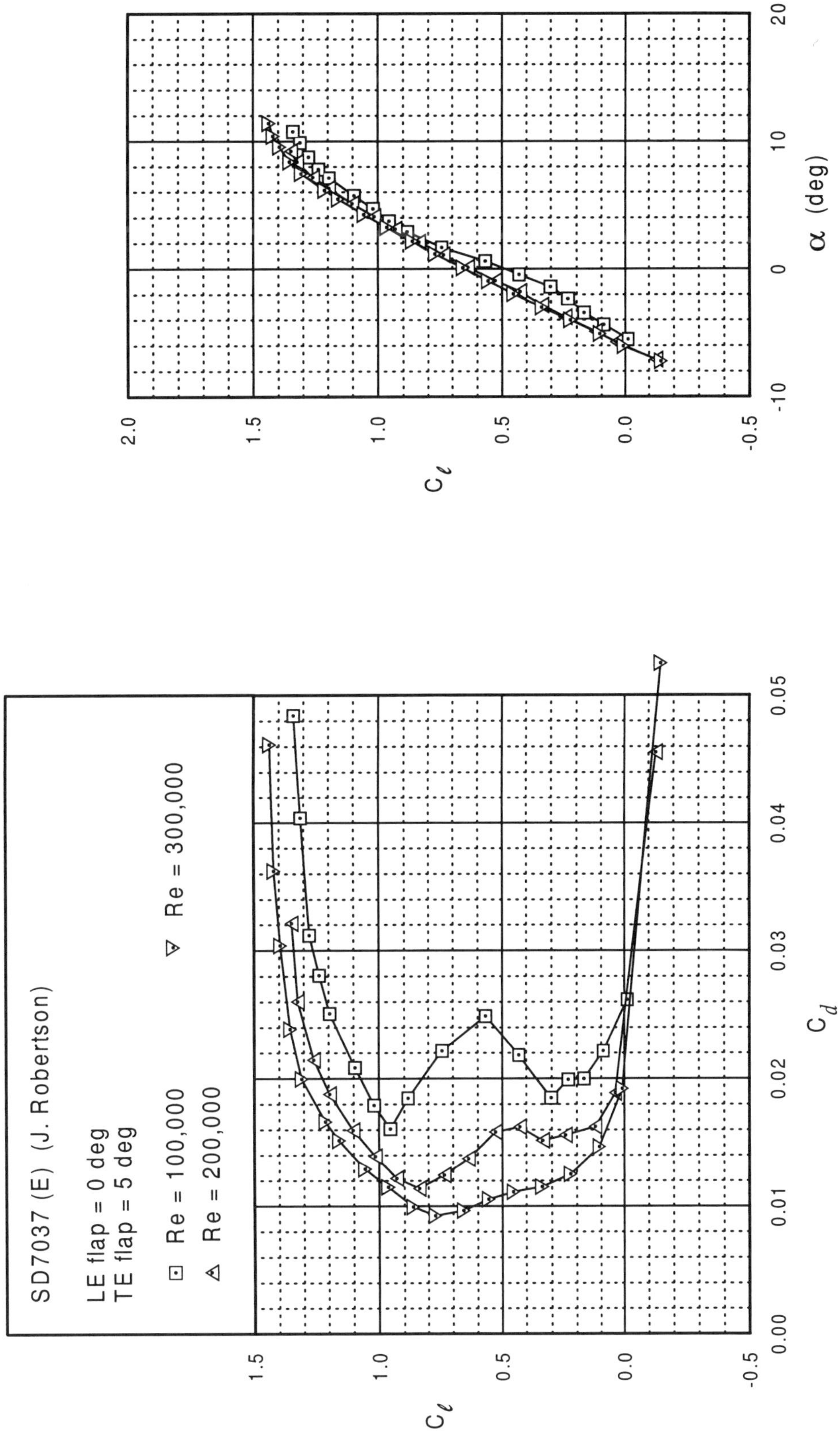
SD7037 (E) (J. Robertson)
LE flap = 0 deg
TE flap = 5 deg
Re = 100,000
Re = 200,000
Re = 300,000
C_ℓ
C_d
α (deg)
1.5
1.0
0.5
0.0
-0.5
2.0
0.00
0.01
0.02
0.03
0.04
0.05
-10
0
10
20

SD7037 (E)

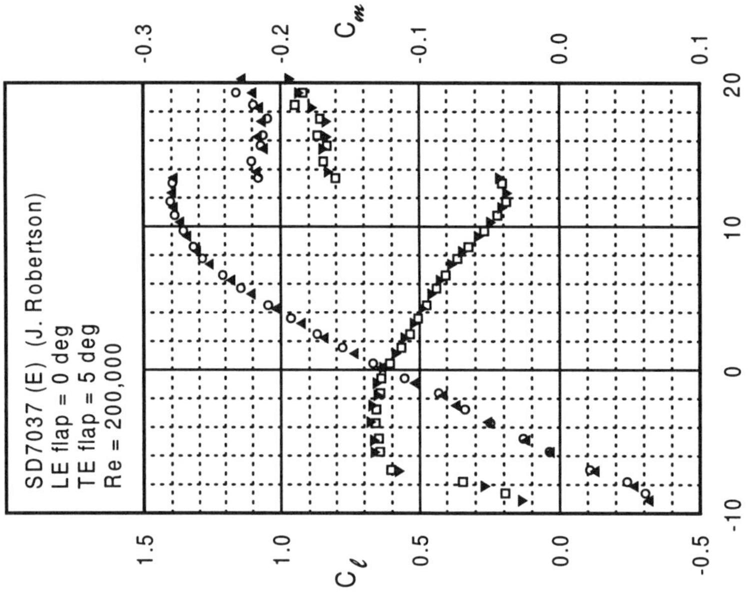
SD7037 (E) (J. Robertson)
LE flap = 0 deg
TE flap = 5 deg
Re = 200,000
α (deg)
C_m
C_l

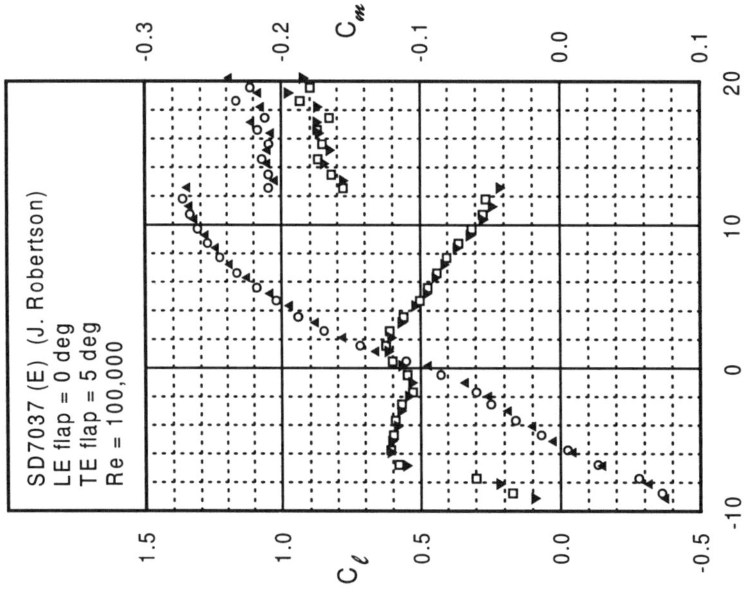
SD7037 (E) (J. Robertson)
LE flap = 0 deg
TE flap = 5 deg
Re = 100,000
α (deg)
C_m
C_l

SD7037 (E)

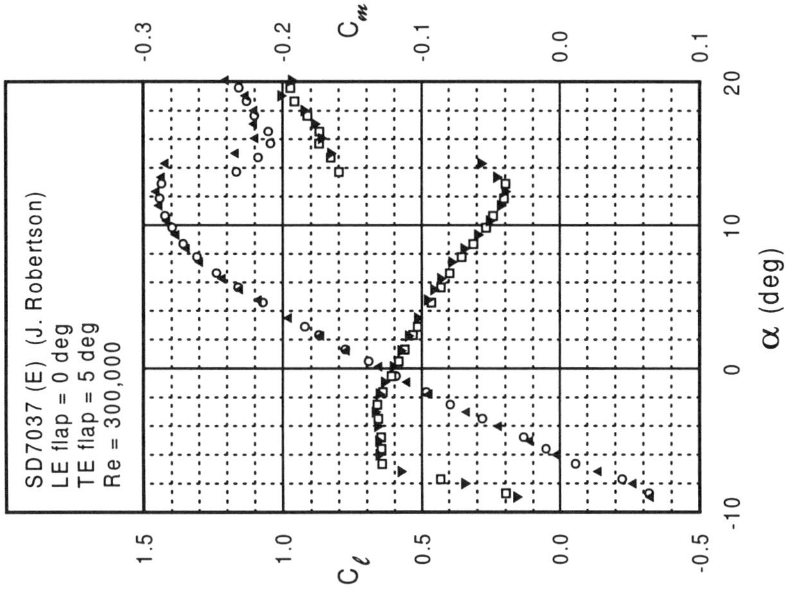
SD7037 (E) (J. Robertson)
LE flap = 0 deg
TE flap = 5 deg
Re = 300,000
C_ℓ
1.5
1.0
0.5
0.0
-0.5
C_m
-0.3
-0.2
-0.1
0.0
0.1
-10
0
10
20
α (deg)

SD7062 (B)

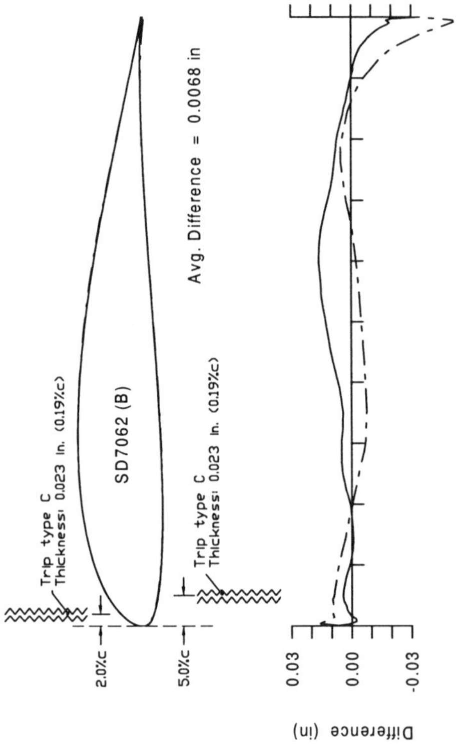
Trip type C
Thickness: 0.023 in. (0.19%c)
SD7062 (B)
Trip type C
Thickness: 0.023 in. (0.19%c)
2.0%c
5.0%c
Avg. Difference = 0.0068 in
0.03
0.00
-0.03
Difference (in)

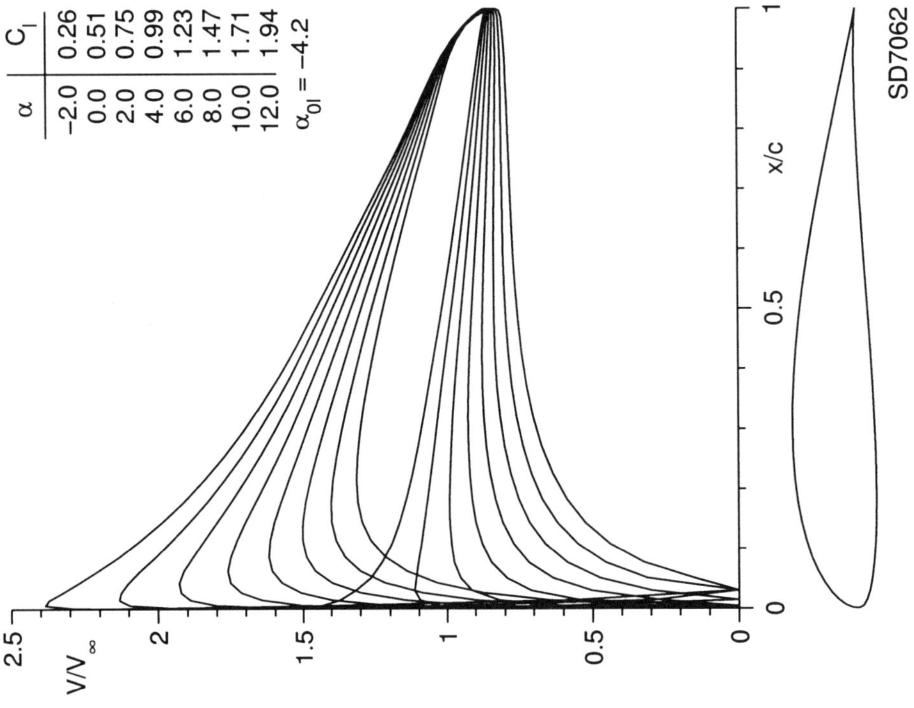
α
C_l
−2.0 0.26
0.0 0.51
2.0 0.75
4.0 0.99
6.0 1.23
8.0 1.47
10.0 1.71
12.0 1.94
α_0l = −4.2
V/V_∞
2.5
2
1.5
1
0.5
0
0
0.5
1
x/c
SD7062

SD7062 (B)

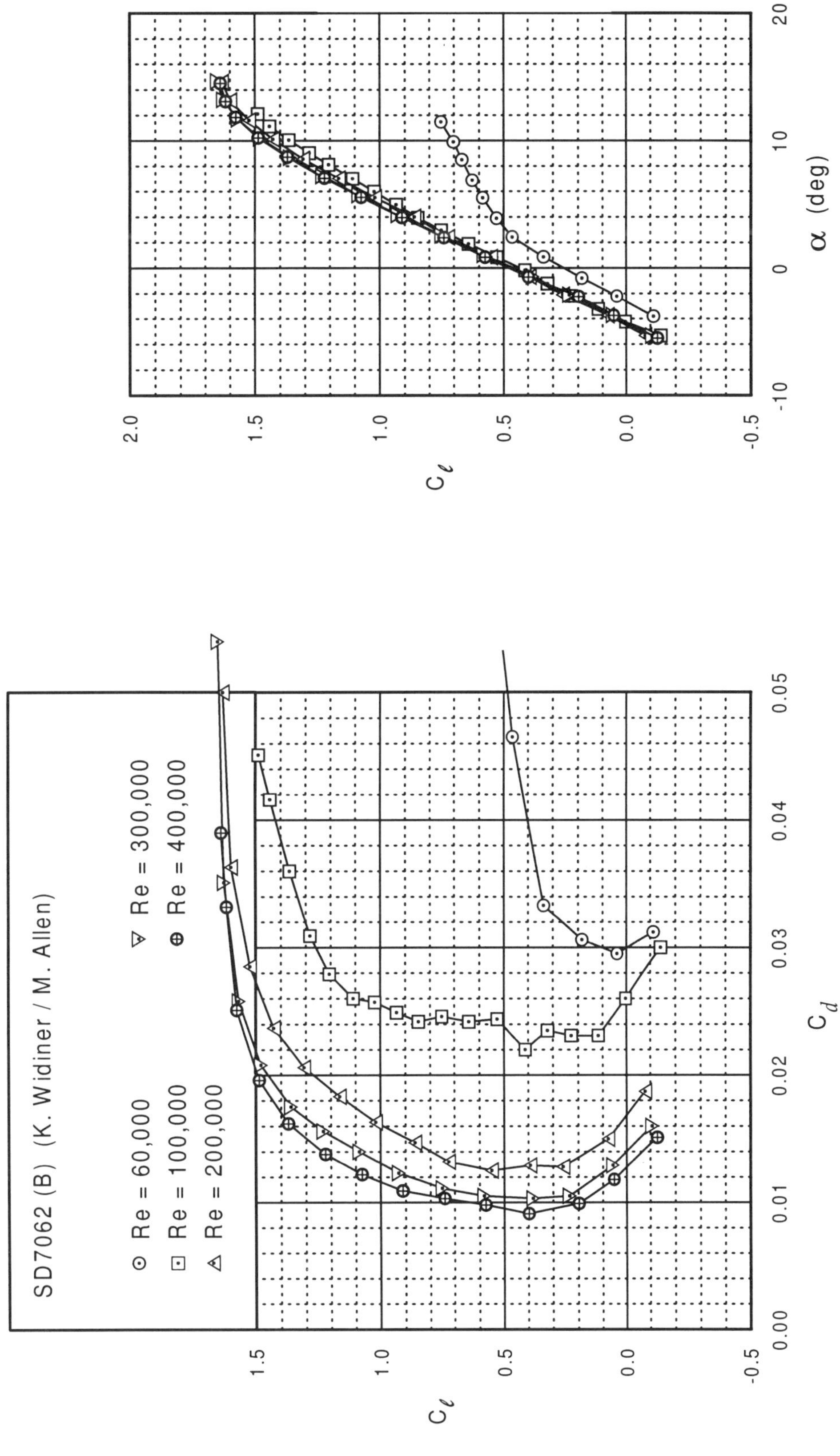
SD7062 (B) (K. Widiner / M. Allen)
Re = 60,000
Re = 100,000
Re = 200,000
Re = 300,000
Re = 400,000
C_ℓ
C_d
α (deg)
-0.5
0.0
0.5
1.0
1.5
2.0
0.00
0.01
0.02
0.03
0.04
0.05
-10
0
10
20

SD7062 (B)

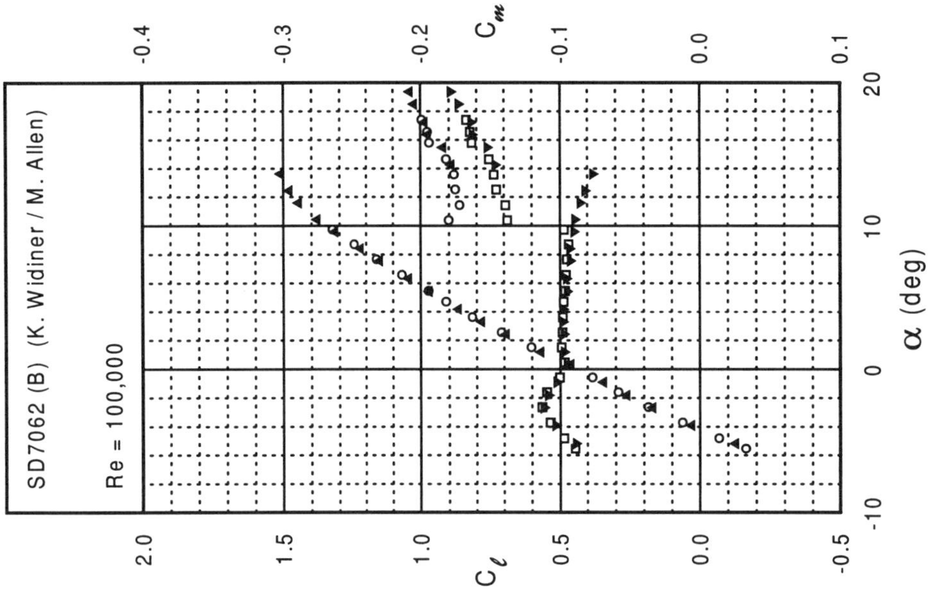
SD7062 (B) (K. Widiner / M. Allen)
Re = 100,000
α (deg)
C_ℓ
C_m

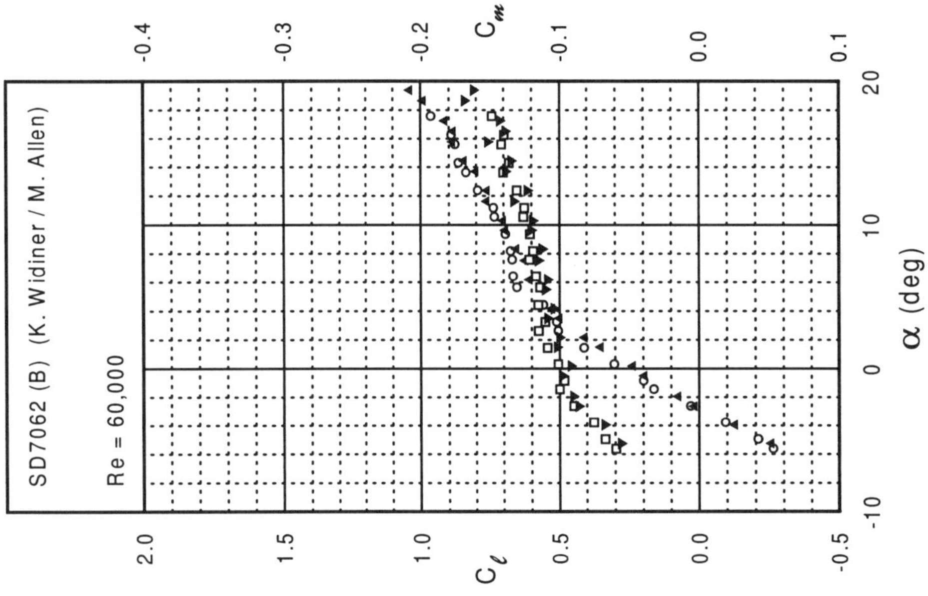
SD7062 (B) (K. Widiner / M. Allen)
Re = 60,000
α (deg)
C_ℓ
C_m

SD7062 (B)

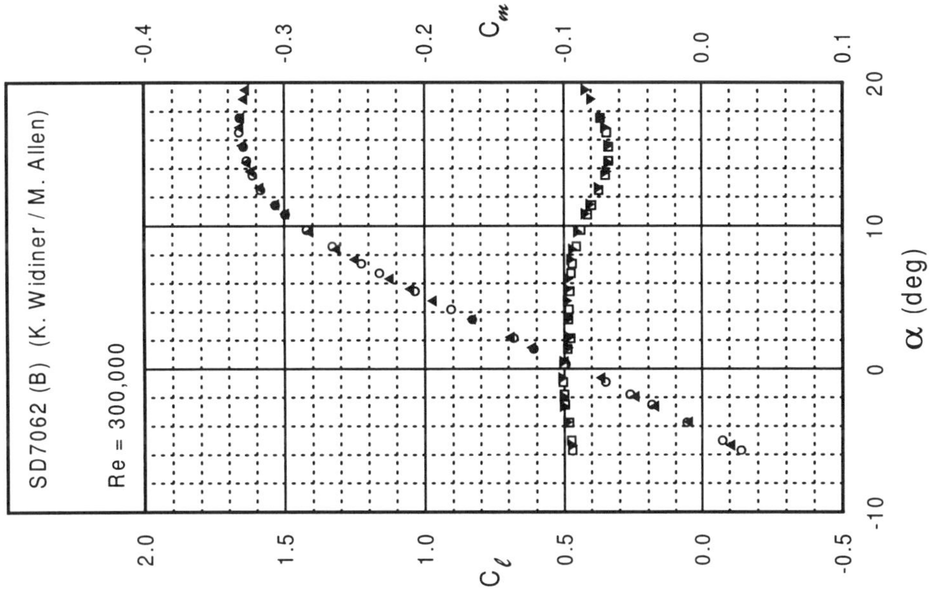
SD7062 (B) (K. Widiner / M. Allen)
Re = 300,000
C_l
C_m
α (deg)
2.0
1.5
1.0
0.5
0.0
-0.5
-0.4
-0.3
-0.2
-0.1
0.0
0.1
-10
0
10
20

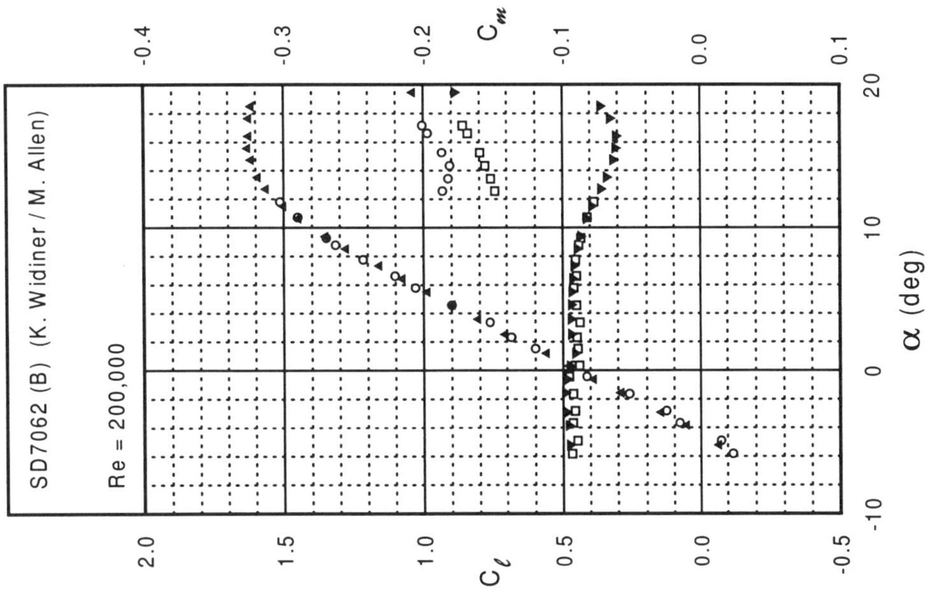
SD7062 (B) (K. Widiner / M. Allen)
Re = 200,000
C_l
C_m
α (deg)
2.0
1.5
1.0
0.5
0.0
-0.5
-0.4
-0.3
-0.2
-0.1
0.0
0.1
-10
0
10
20

SD7062 (B)

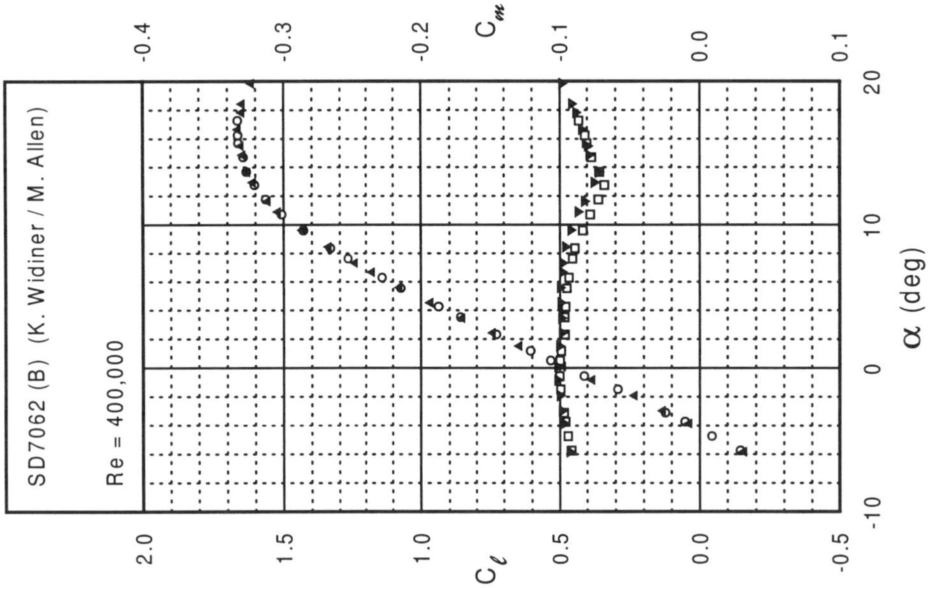
SD7062 (B) (K. Widiner / M. Allen)
Re = 400,000
C_ℓ
2.0
1.5
1.0
0.5
0.0
-0.5
C_m
-0.4
-0.3
-0.2
-0.1
0.0
0.1
α (deg)
-10
0
10
20

SD7062 (B)

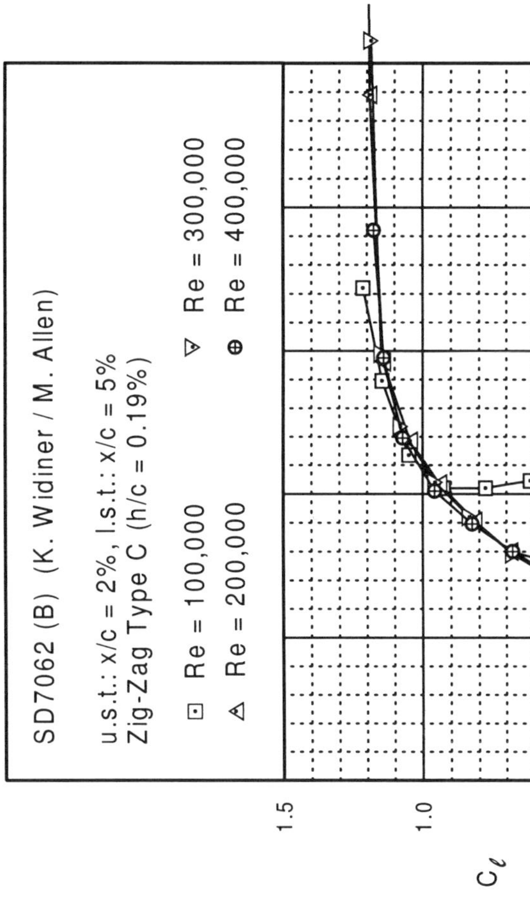
SD7062 (B) (K. Widiner / M. Allen)
u.s.t.: x/c = 2%, l.s.t.: x/c = 5%
Zig-Zag Type C (h/c = 0.19%)
Re = 100,000
Re = 200,000
Re = 300,000
Re = 400,000
1.5
1.0
0.5
0.0
-0.5
C_ℓ
0.00
0.01
0.02
0.03
0.04
0.05
C_d

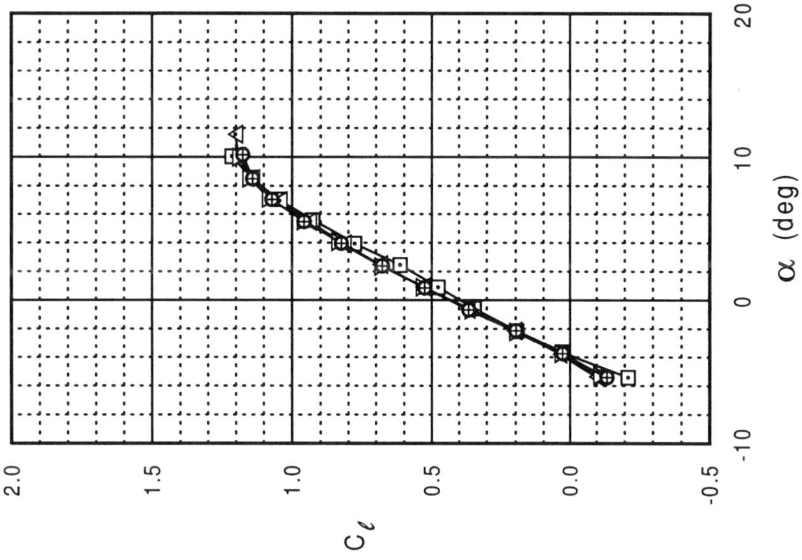
2.0
1.5
1.0
0.5
0.0
-0.5
C_ℓ
-10
0
10
20
α (deg)

SD7062 (B)

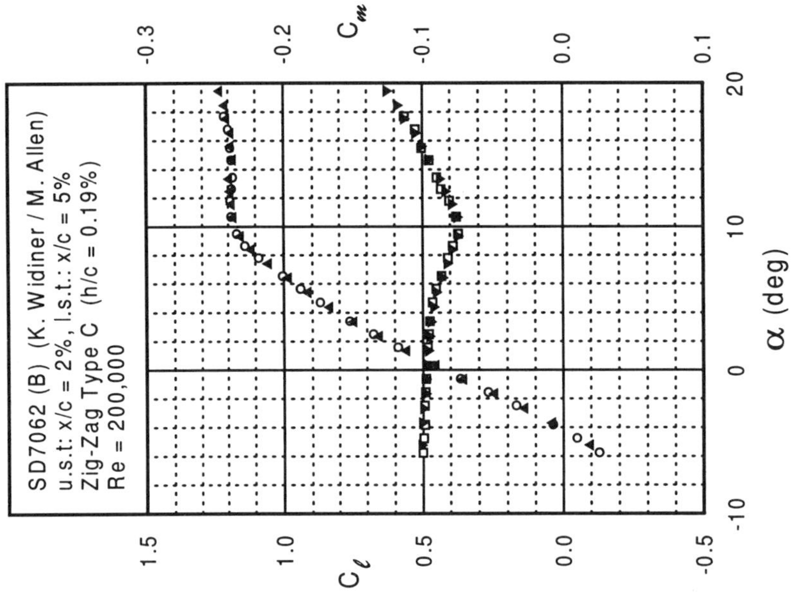
SD7062 (B) (K. Widiner / M. Allen)
u.s.t: x/c = 2%, l.s.t.: x/c = 5%
Zig-Zag Type C (h/c = 0.19%)
Re = 200,000
C_m
C_l
α (deg)

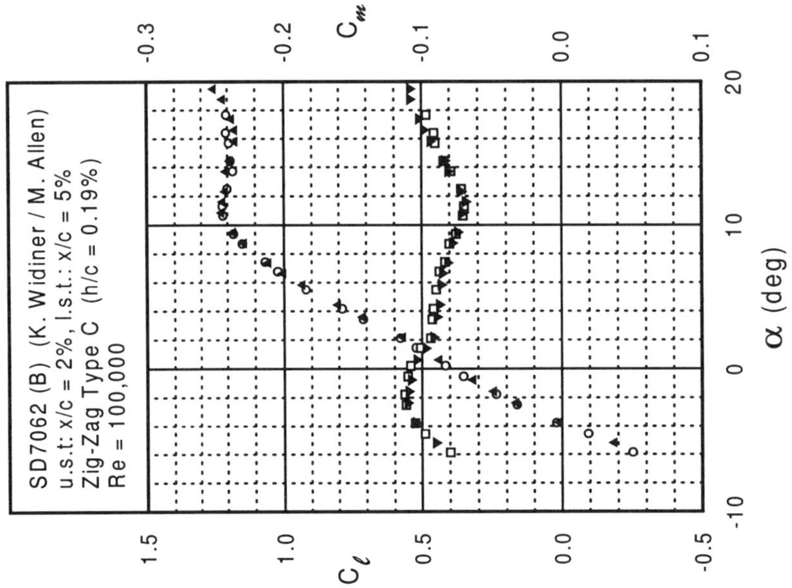
SD7062 (B) (K. Widiner / M. Allen)
u.s.t: x/c = 2%, l.s.t.: x/c = 5%
Zig-Zag Type C (h/c = 0.19%)
Re = 100,000
C_m
C_l
α (deg)

SD7062 (B)

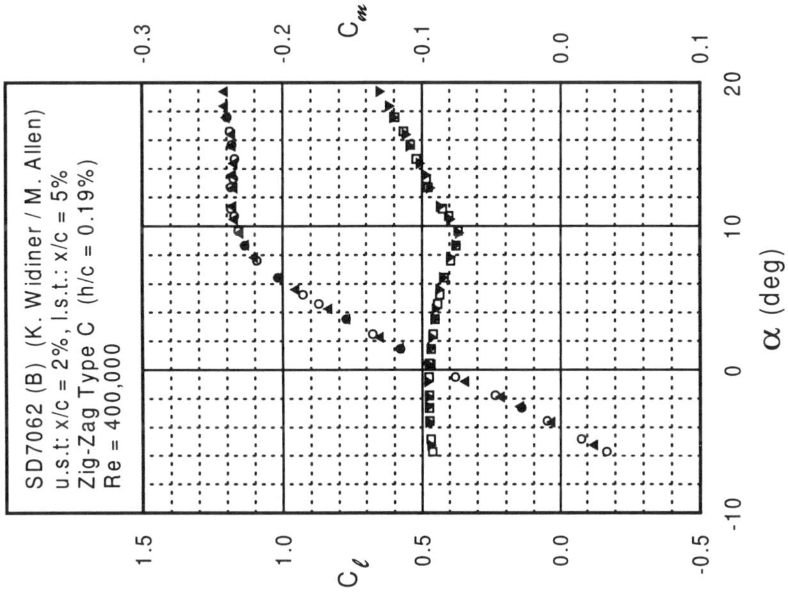
SD7062 (B) (K. Widiner / M. Allen)
u.s.t: x/c = 2%, l.s.t.: x/c = 5%
Zig-Zag Type C (h/c = 0.19%)
Re = 400,000
α (deg)
-0.3
-0.2
-0.1
0.0
0.1
1.5
1.0
0.5
0.0
-0.5
-10
0
10
20

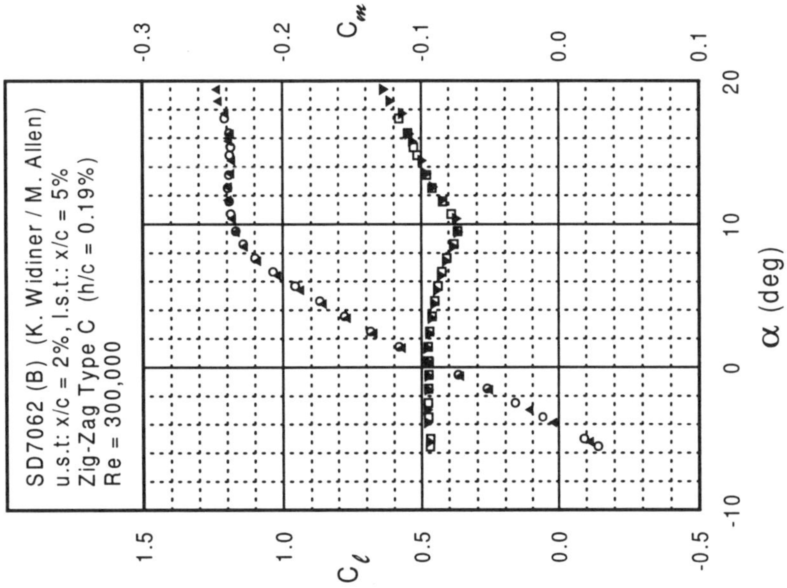
SD7062 (B) (K. Widiner / M. Allen)
u.s.t: x/c = 2%, l.s.t.: x/c = 5%
Zig-Zag Type C (h/c = 0.19%)
Re = 300,000
α (deg)
-0.3
-0.2
-0.1
0.0
0.1
1.5
1.0
0.5
0.0
-0.5
-10
0
10
20

SD7080

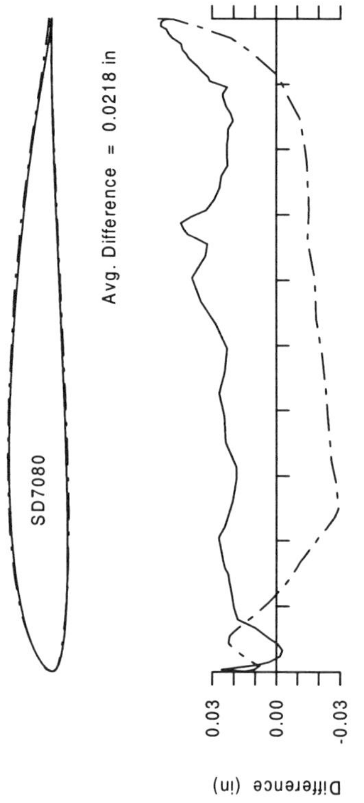
SD7080
Avg. Difference = 0.0218 in
0.03
0.00
-0.03
Difference (in)

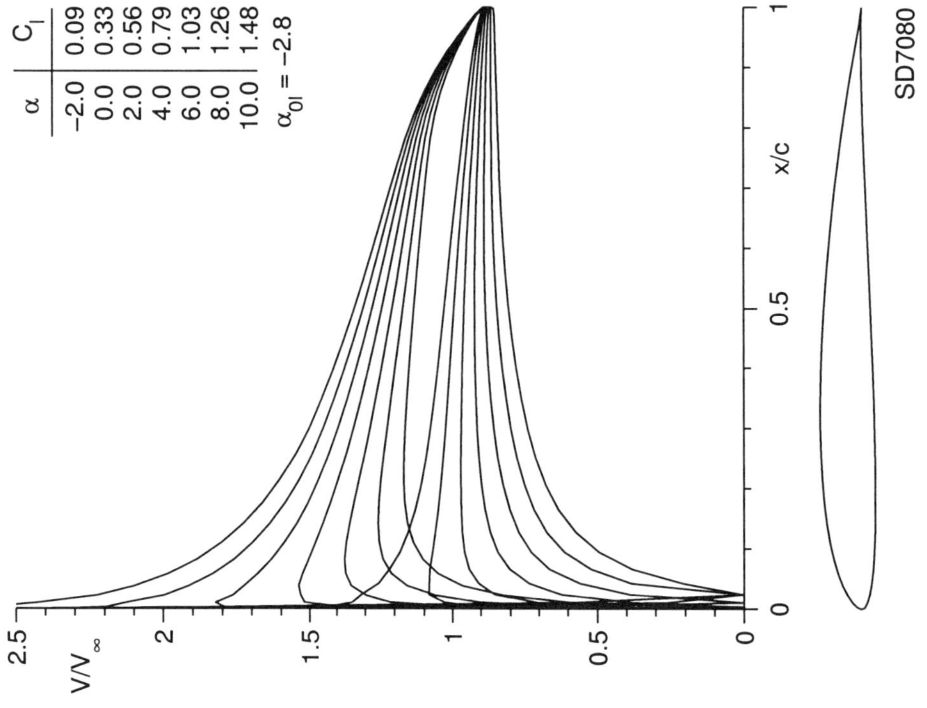
α | C_l
−2.0 | 0.09
0.0 | 0.33
2.0 | 0.56
4.0 | 0.79
6.0 | 1.03
8.0 | 1.26
10.0 | 1.48
$\alpha_{0l} = -2.8$
V/V_∞
2.5
2
1.5
1
0.5
0
x/c
0.5
1
SD7080

SD7080

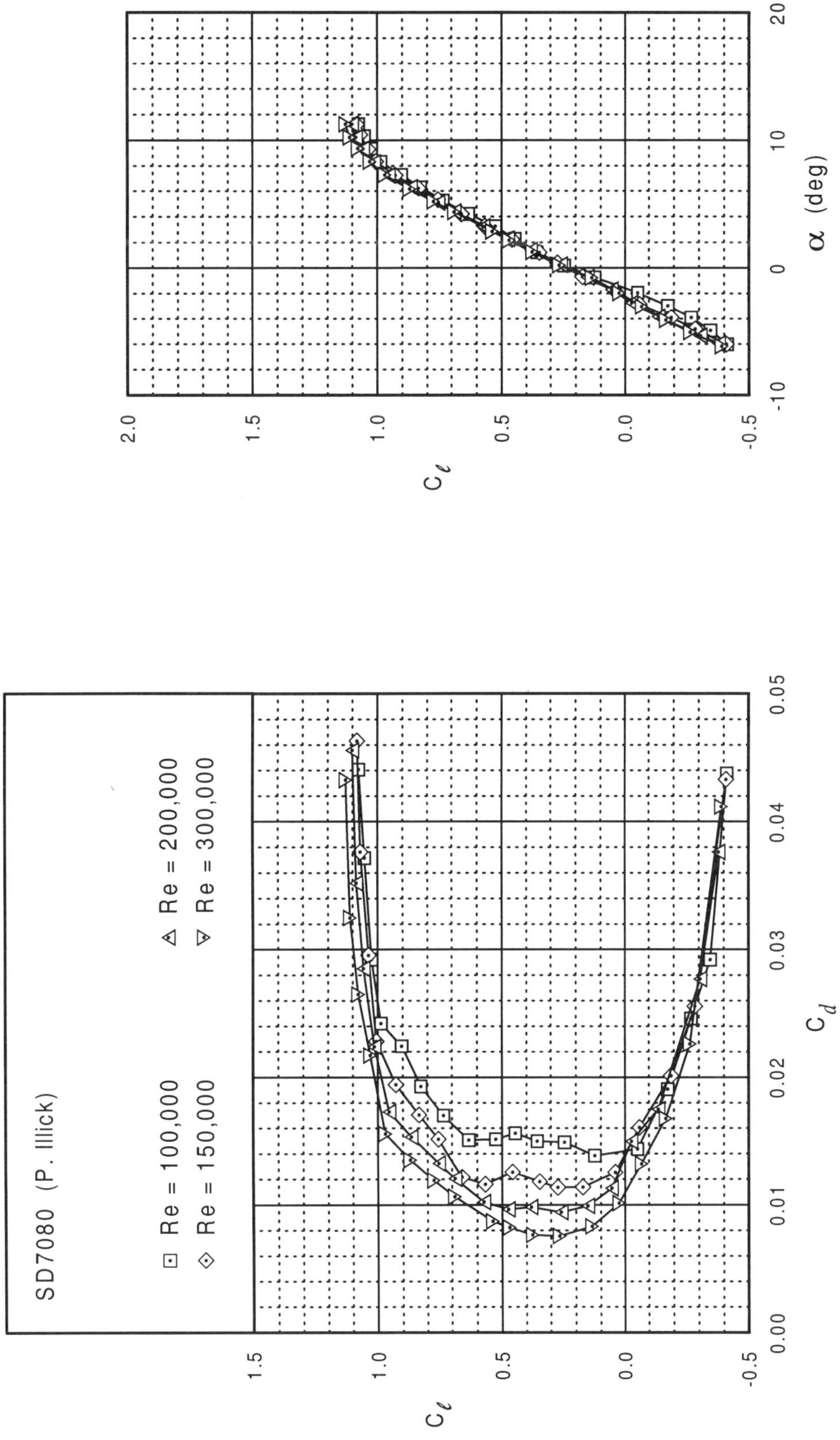
SD7080 (P. Illick)
Re = 100,000
Re = 150,000
Re = 200,000
Re = 300,000
C_ℓ
C_d
α (deg)
1.5
1.0
0.5
0.0
-0.5
2.0
0.00
0.01
0.02
0.03
0.04
0.05
-10
0
10
20

SD7080

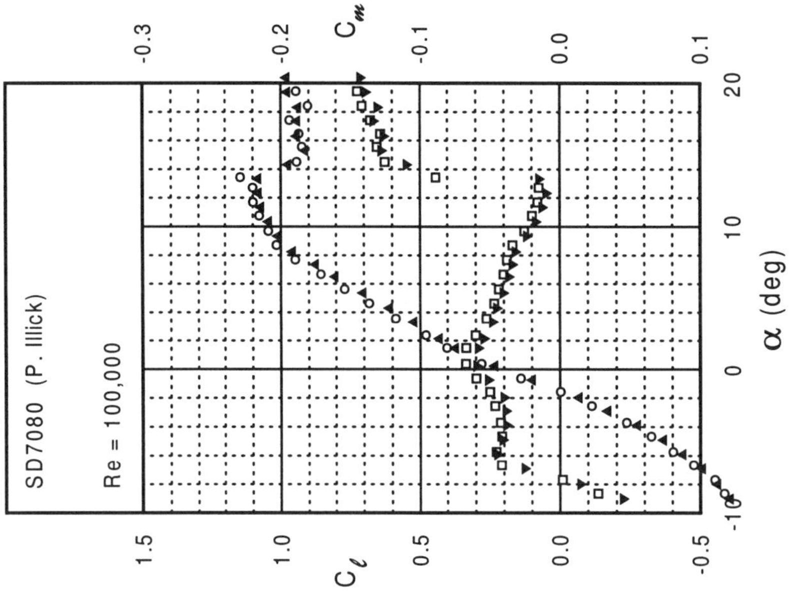
SD7080 (P. Illick)
Re = 100,000
α (deg)
C_l
C_m

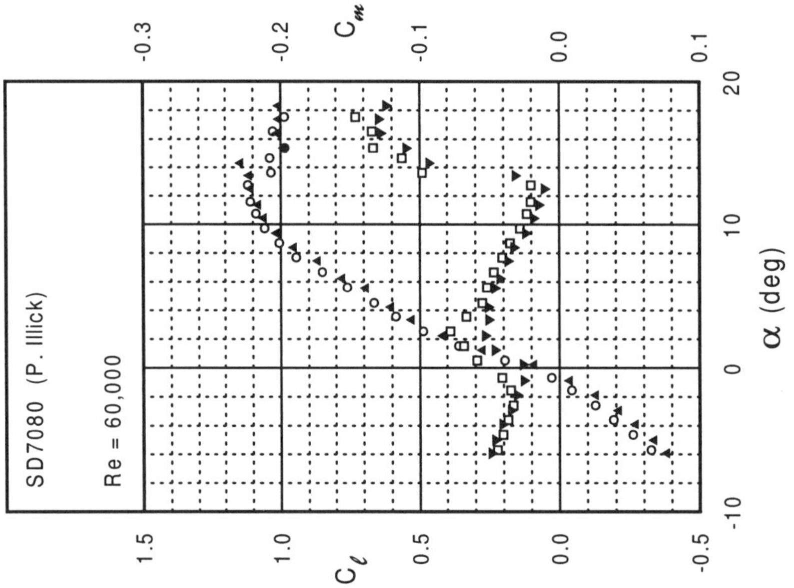
SD7080 (P. Illick)
Re = 60,000
α (deg)
C_l
C_m

SD7080

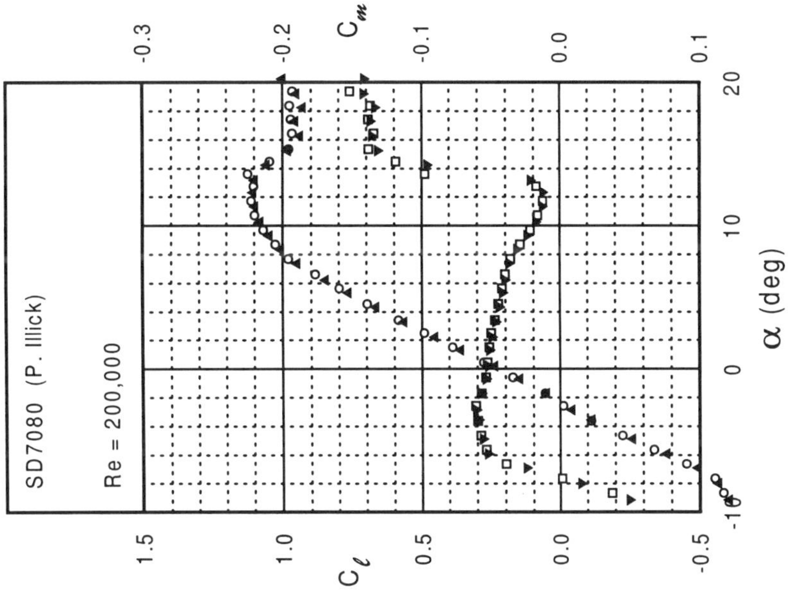
SD7080 (P. Illick)
Re = 200,000
C_ℓ
C_m
α (deg)

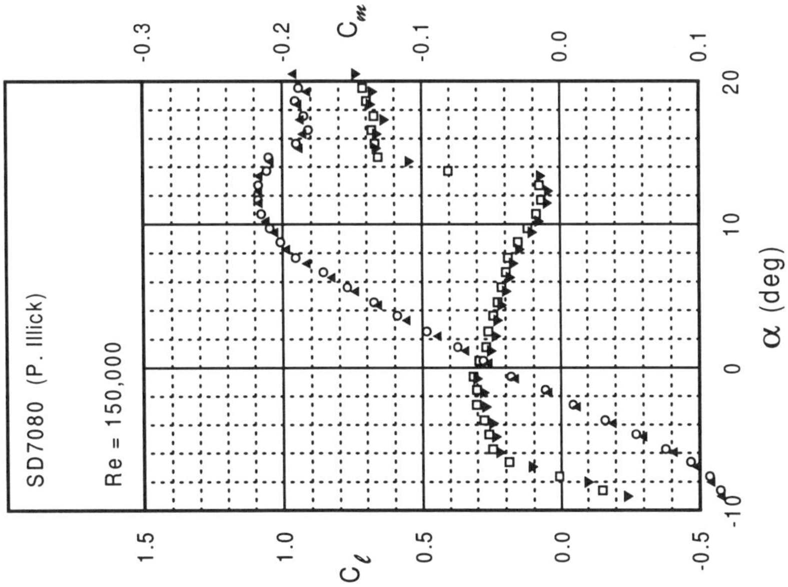
SD7080 (P. Illick)
Re = 150,000
C_ℓ
C_m
α (deg)

SD7080

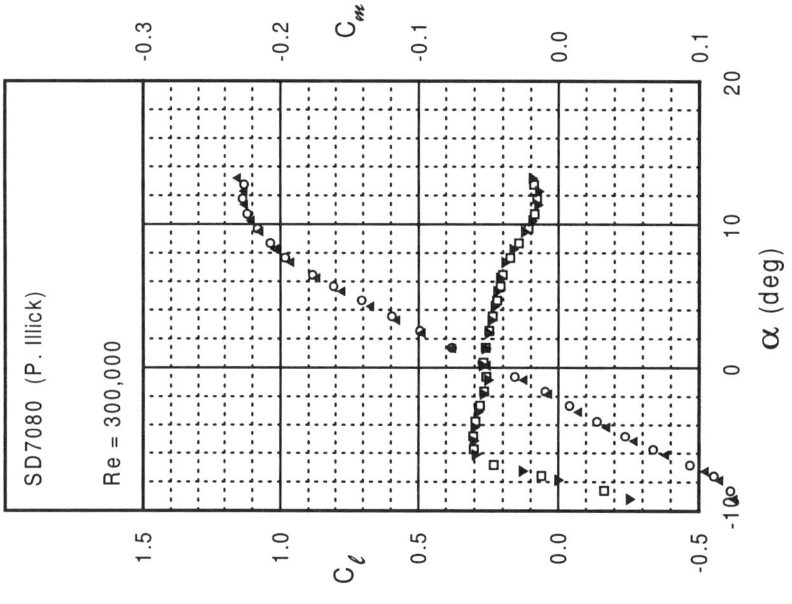
SD7080 (P. Illick)
Re = 300,000
C_ℓ
C_m
α (deg)
1.5
1.0
0.5
0.0
-0.5
-0.3
-0.2
-0.1
0.0
0.1
-10
0
10
20

SG6040

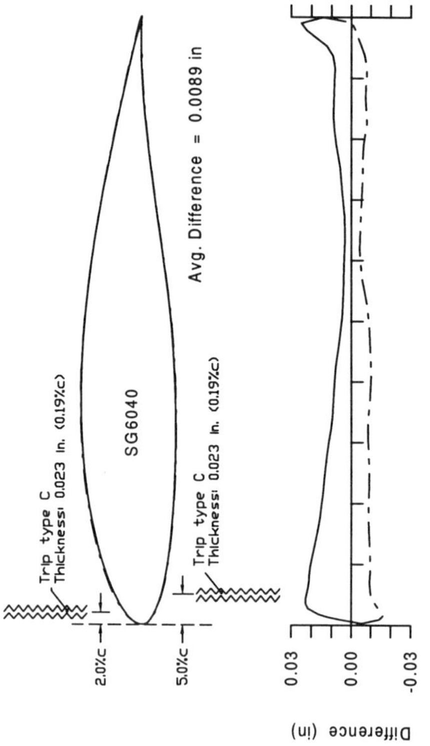
Trip type C
Thickness: 0.023 in. (0.19%c)
SG6040
Trip type C
Thickness: 0.023 in. (0.19%c)
Avg. Difference = 0.0089 in
2.0%c
5.0%c
0.03
0.00
-0.03
Difference (in)

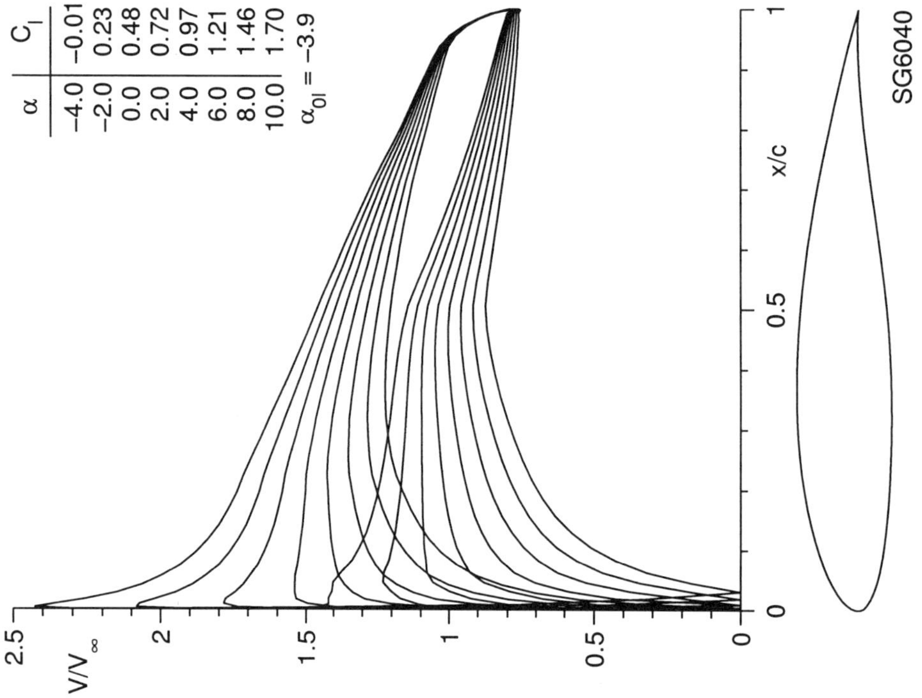
α
C_l
−4.0 −0.01
−2.0 0.23
0.0 0.48
2.0 0.72
4.0 0.97
6.0 1.21
8.0 1.46
10.0 1.70
α_{0l} = −3.9
2.5
2
1.5
1
0.5
0
V/V_∞
0
0.5
1
x/c
SG6040

SG6040

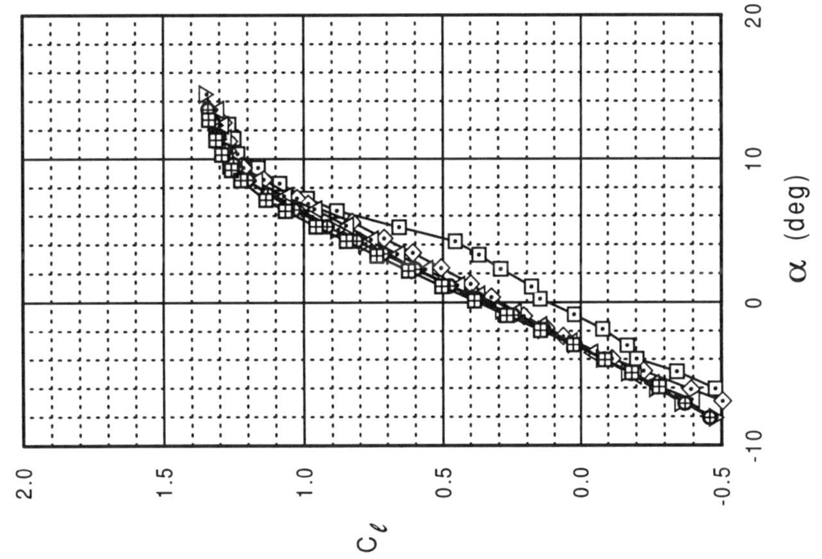

C_ℓ
α (deg)

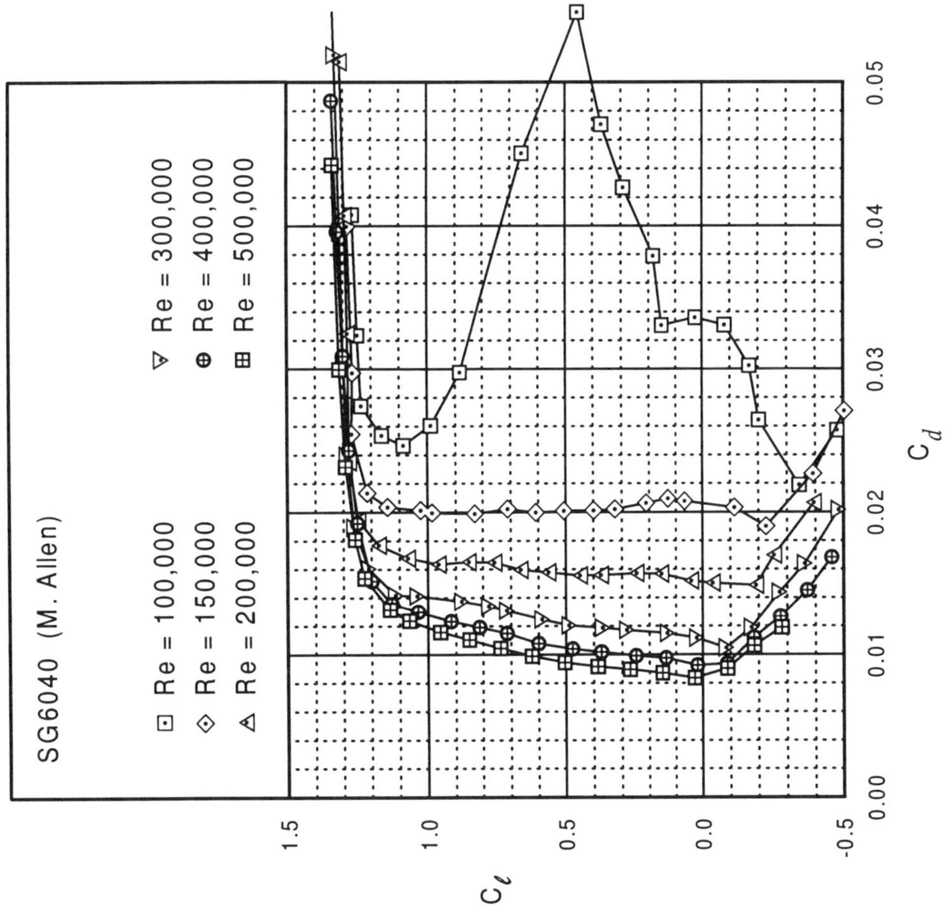

SG6040 (M. Allen)
Re = 100,000
Re = 150,000
Re = 200,000
Re = 300,000
Re = 400,000
Re = 500,000
C_ℓ
C_d

SG6040

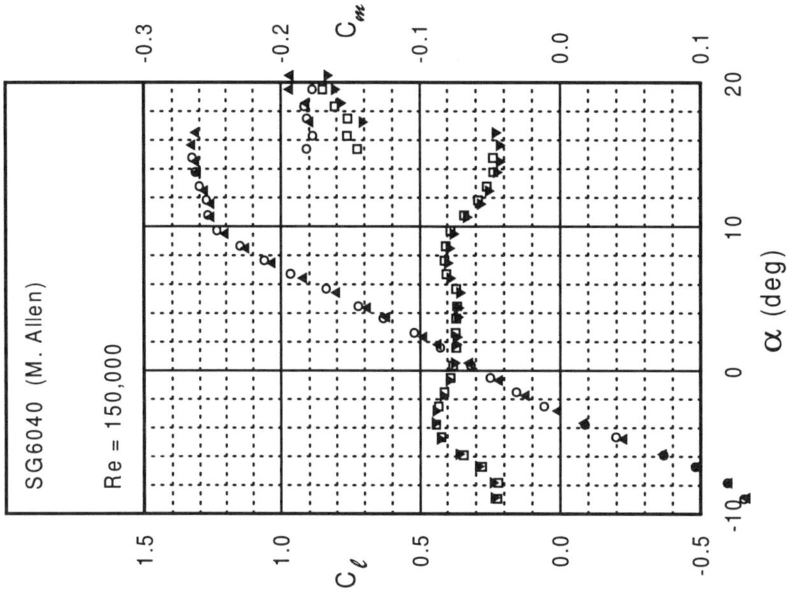
SG6040 (M. Allen)
Re = 150,000
C_ℓ
C_m
α (deg)
-10
0
10
20
1.5
1.0
0.5
0.0
-0.5
-0.3
-0.2
-0.1
0.0
0.1

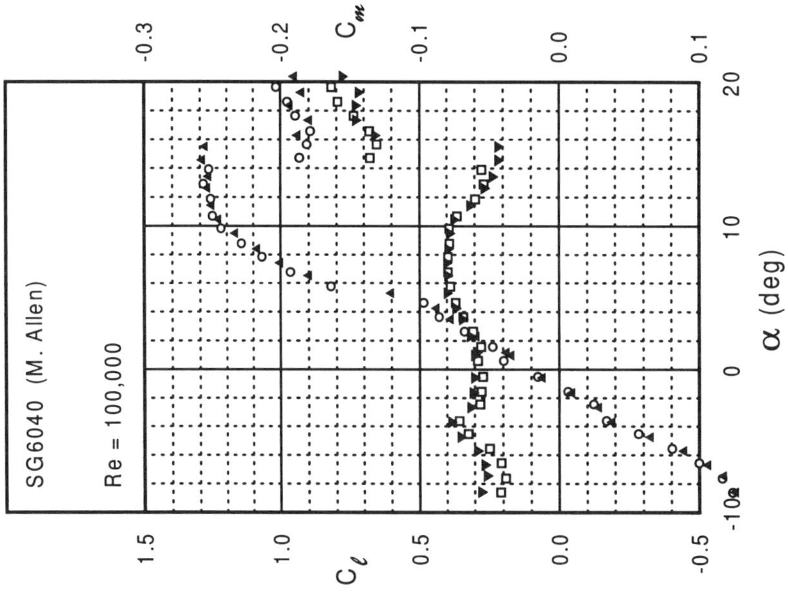
SG6040 (M. Allen)
Re = 100,000
C_ℓ
C_m
α (deg)
-10
0
10
20
1.5
1.0
0.5
0.0
-0.5
-0.3
-0.2
-0.1
0.0
0.1

SG6040

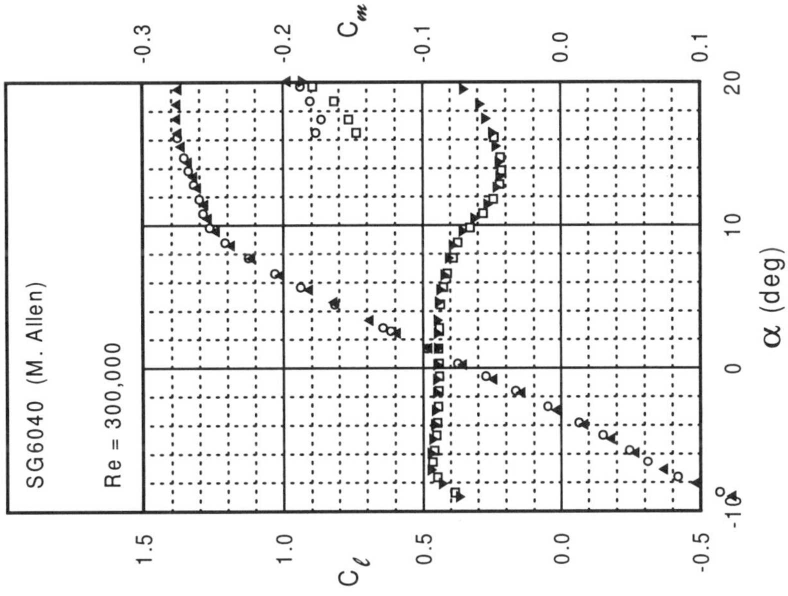

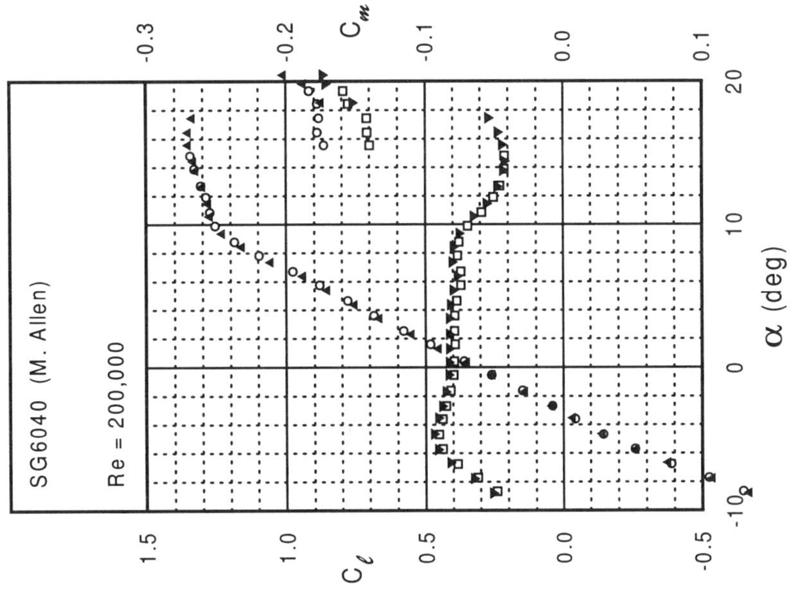

SG6040

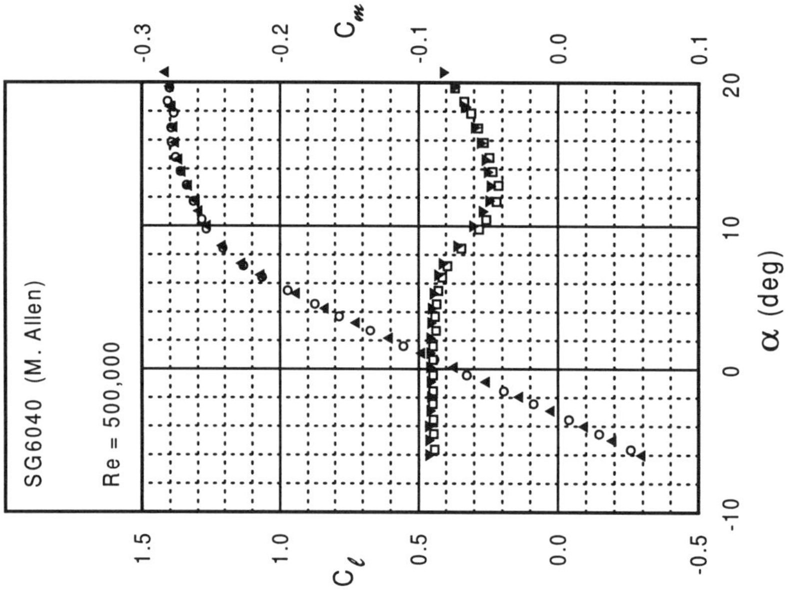
SG6040 (M. Allen)
Re = 500,000
C_m
-0.3
-0.2
-0.1
0.0
0.1
C_ℓ
1.5
1.0
0.5
0.0
-0.5
-10
0
10
20
α (deg)

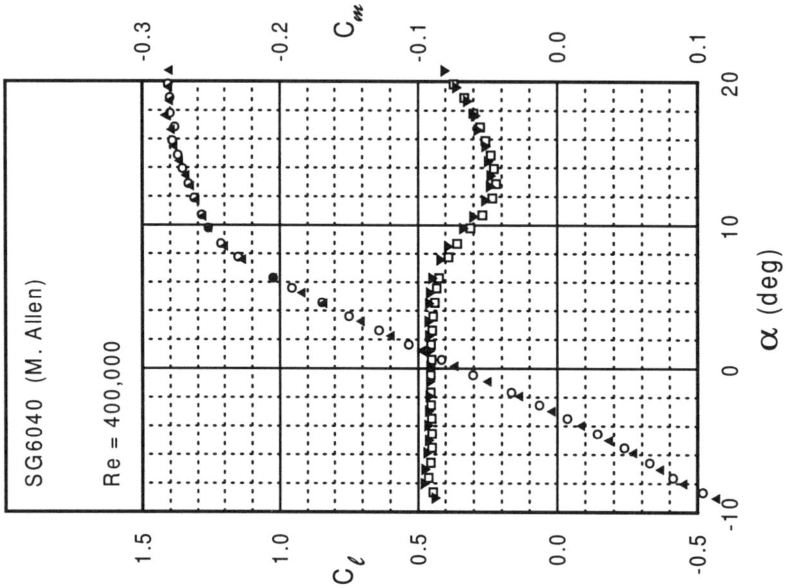
SG6040 (M. Allen)
Re = 400,000
C_m
-0.3
-0.2
-0.1
0.0
0.1
C_ℓ
1.5
1.0
0.5
0.0
-0.5
-10
0
10
20
α (deg)

SG6040

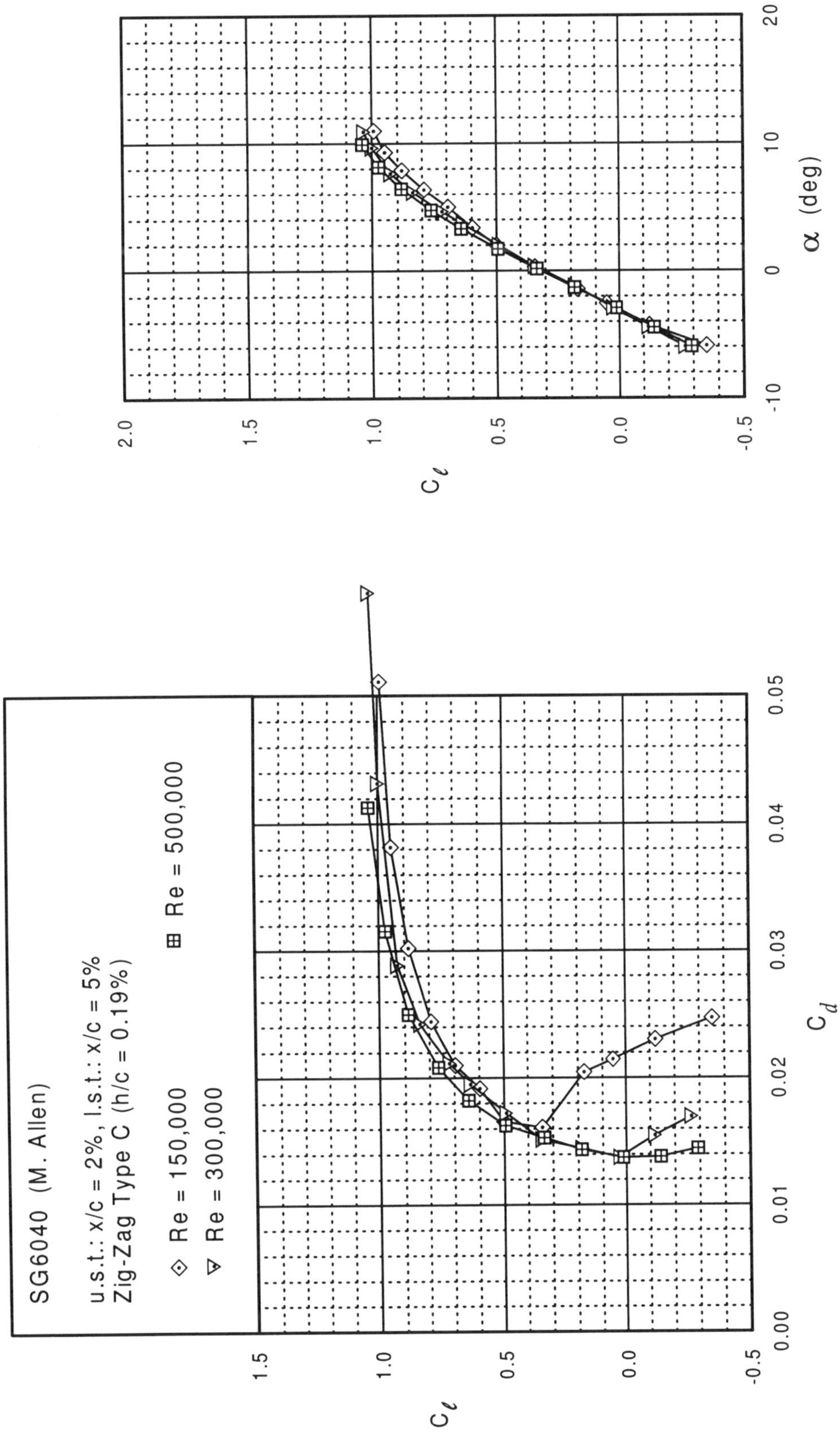
SG6040 (M. Allen)
u.s.t.: x/c = 2%, l.s.t.: x/c = 5%
Zig-Zag Type C (h/c = 0.19%)
Re = 150,000
Re = 300,000
Re = 500,000
C_l
C_d
α (deg)

SG6040

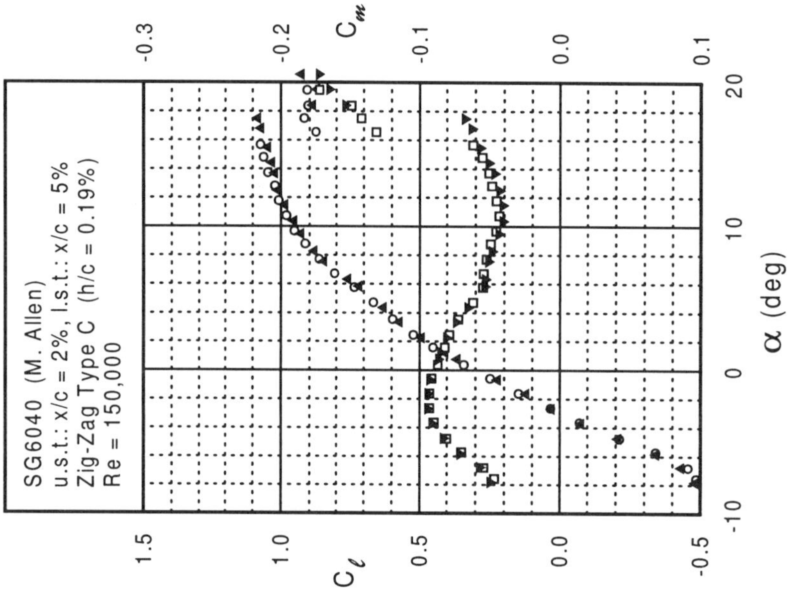
SG6040 (M. Allen)
u.s.t.: x/c = 2%, l.s.t.: x/c = 5%
Zig-Zag Type C (h/c = 0.19%)
Re = 150,000
C_m
-0.3
-0.2
-0.1
0.0
0.1
C_ℓ
1.5
1.0
0.5
0.0
-0.5
-10
0
10
20
α (deg)

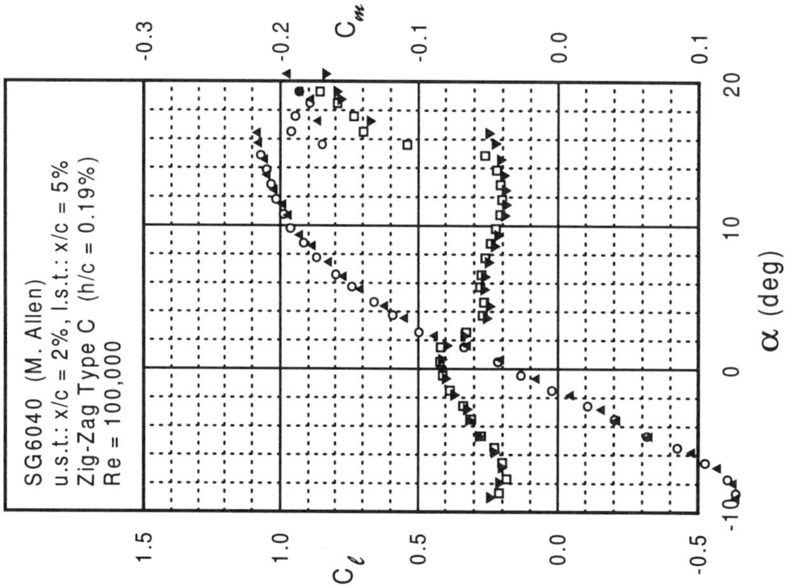
SG6040 (M. Allen)
u.s.t.: x/c = 2%, l.s.t.: x/c = 5%
Zig-Zag Type C (h/c = 0.19%)
Re = 100,000
C_m
-0.3
-0.2
-0.1
0.0
0.1
C_ℓ
1.5
1.0
0.5
0.0
-0.5
-10
0
10
20
α (deg)

SG6040

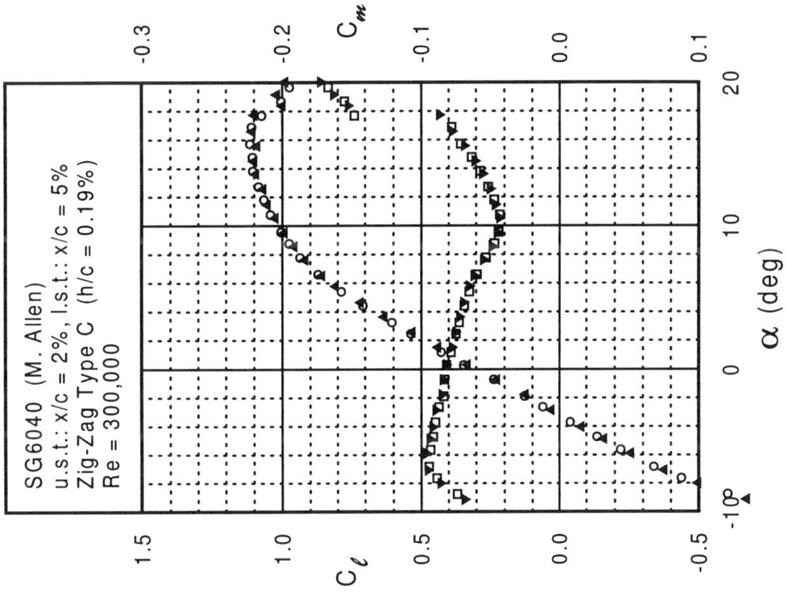
SG6040 (M. Allen)
u.s.t.: x/c = 2%, l.s.t.: x/c = 5%
Zig-Zag Type C (h/c = 0.19%)
Re = 300,000
C_ℓ
1.5
1.0
0.5
0.0
-0.5
C_m
-0.3
-0.2
-0.1
0.0
0.1
-10
0
10
20
α (deg)

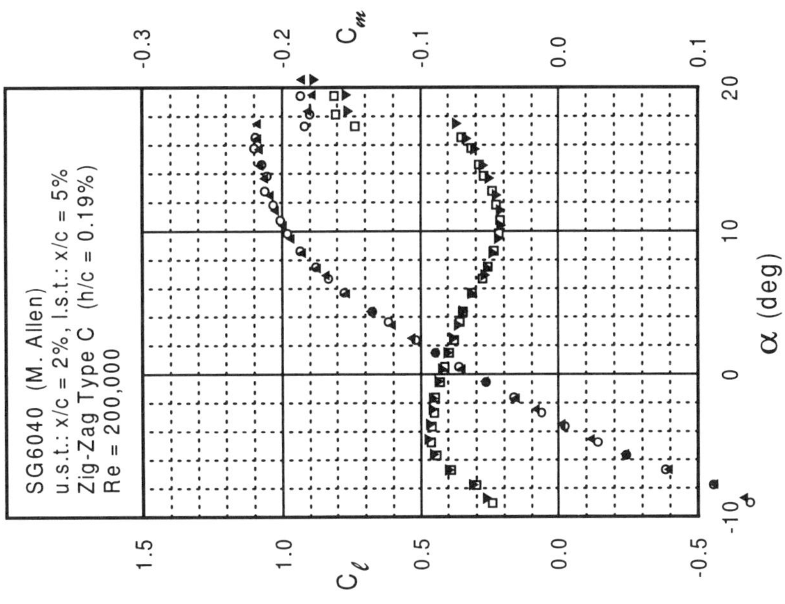
SG6040 (M. Allen)
u.s.t.: x/c = 2%, l.s.t.: x/c = 5%
Zig-Zag Type C (h/c = 0.19%)
Re = 200,000
C_ℓ
1.5
1.0
0.5
0.0
-0.5
C_m
-0.3
-0.2
-0.1
0.0
0.1
-10
0
10
20
α (deg)

SG6040

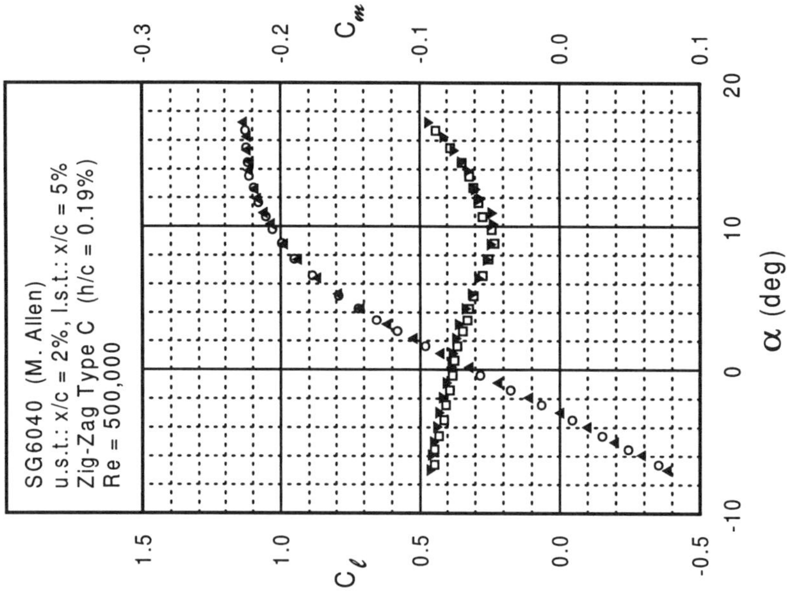
SG6040 (M. Allen)
u.s.t.: x/c = 2%, l.s.t.: x/c = 5%
Zig-Zag Type C (h/c = 0.19%)
Re = 500,000
C_m
C_ℓ
α (deg)

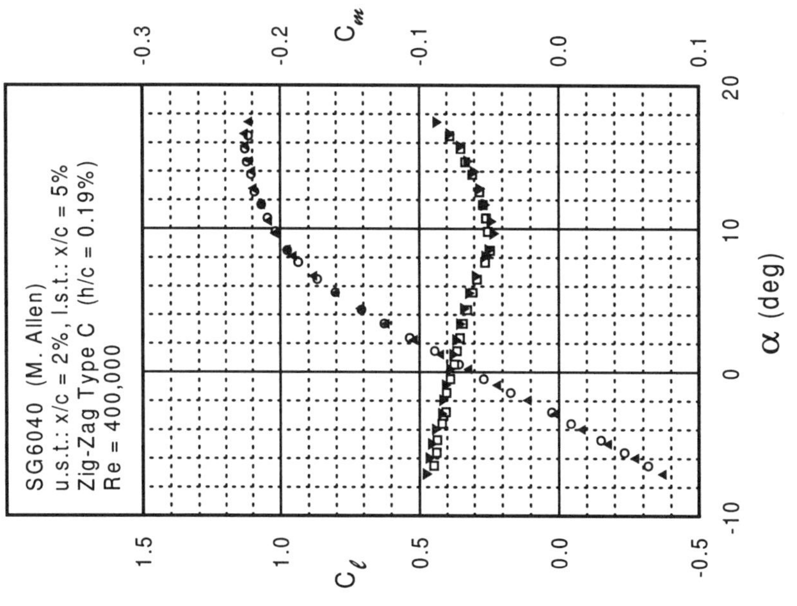
SG6040 (M. Allen)
u.s.t.: x/c = 2%, l.s.t.: x/c = 5%
Zig-Zag Type C (h/c = 0.19%)
Re = 400,000
C_m
C_ℓ
α (deg)

SG6041

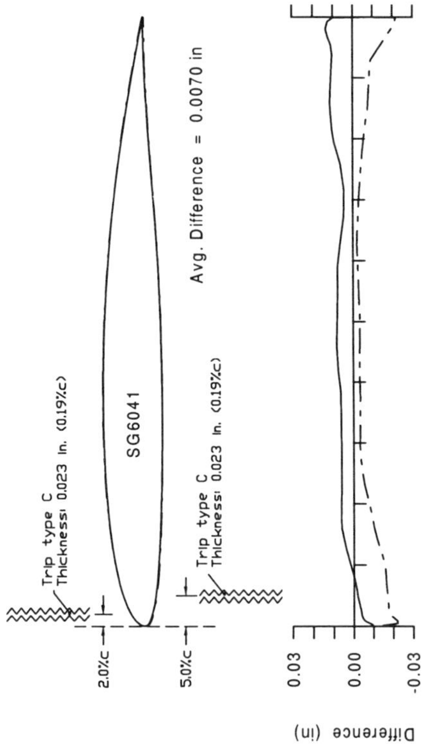
SG6041
Trip type C
Thickness: 0.023 in. (0.19%c)
2.0%c
5.0%c
Avg. Difference = 0.0070 in
0.03
0.00
-0.03
Difference (in)

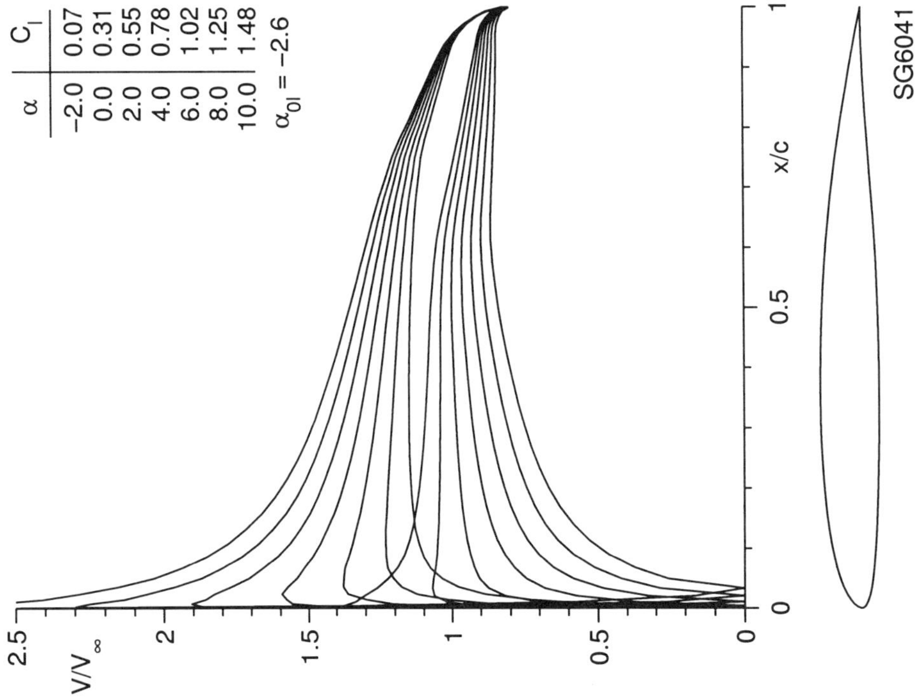
α
C_l
−2.0 0.07
0.0 0.31
2.0 0.55
4.0 0.78
6.0 1.02
8.0 1.25
10.0 1.48
α_0l = −2.6
V/V_∞
2.5
2
1.5
1
0.5
0
x/c
SG6041

SG6041

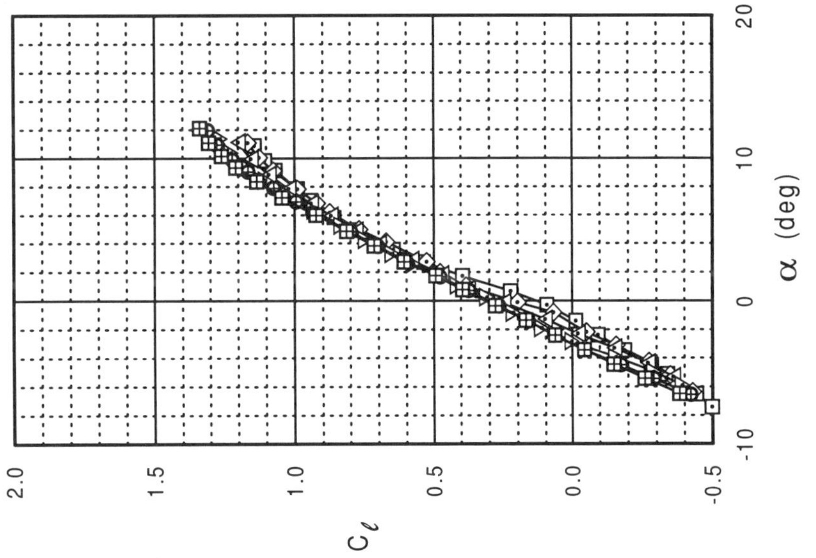

20
10
0
-10
α (deg)
2.0
1.5
1.0
0.5
0.0
-0.5
C_ℓ

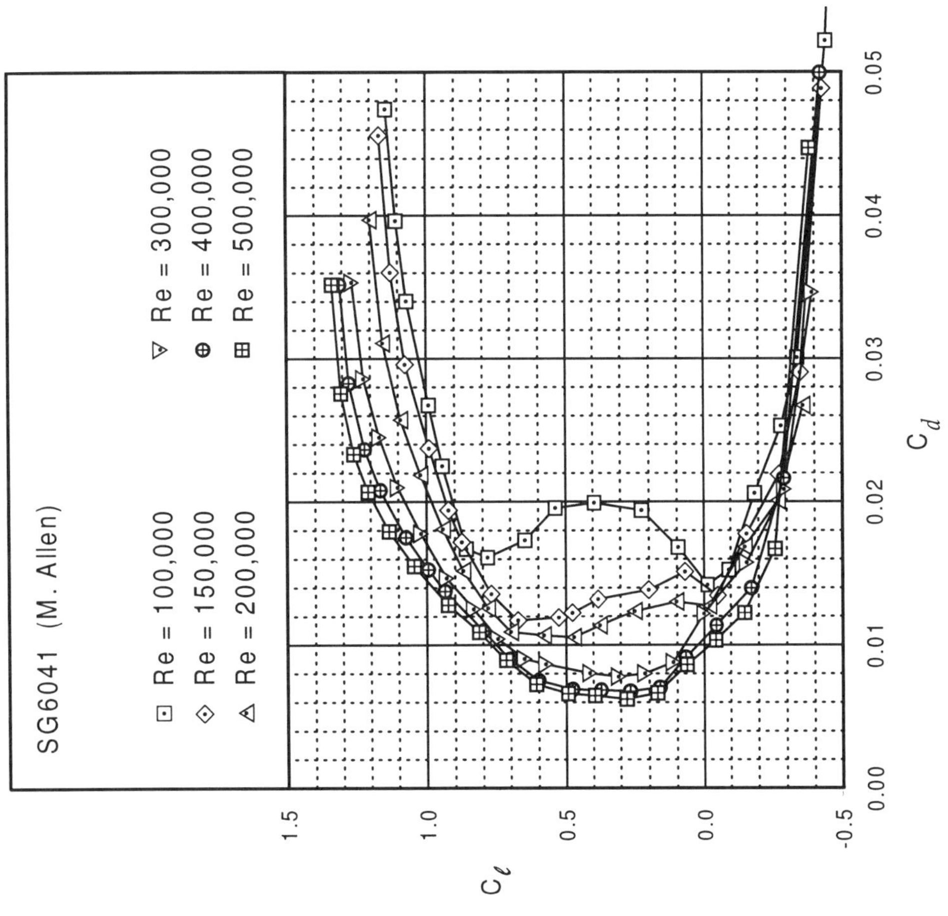

SG6041 (M. Allen)
Re = 100,000
Re = 150,000
Re = 200,000
Re = 300,000
Re = 400,000
Re = 500,000
0.00
0.01
0.02
0.03
0.04
0.05
C_d
1.5
1.0
0.5
0.0
-0.5
C_ℓ

SG6041

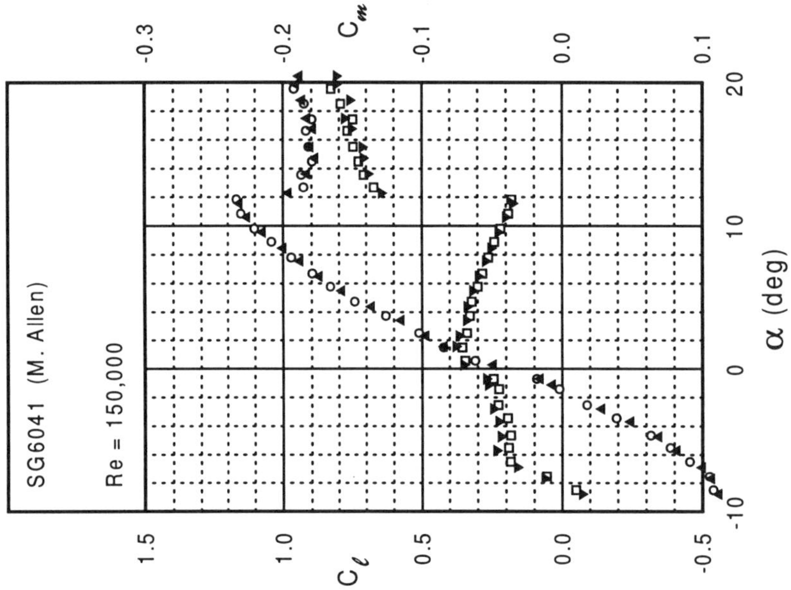
SG6041 (M. Allen)
Re = 150,000
C_ℓ
C_m
α (deg)

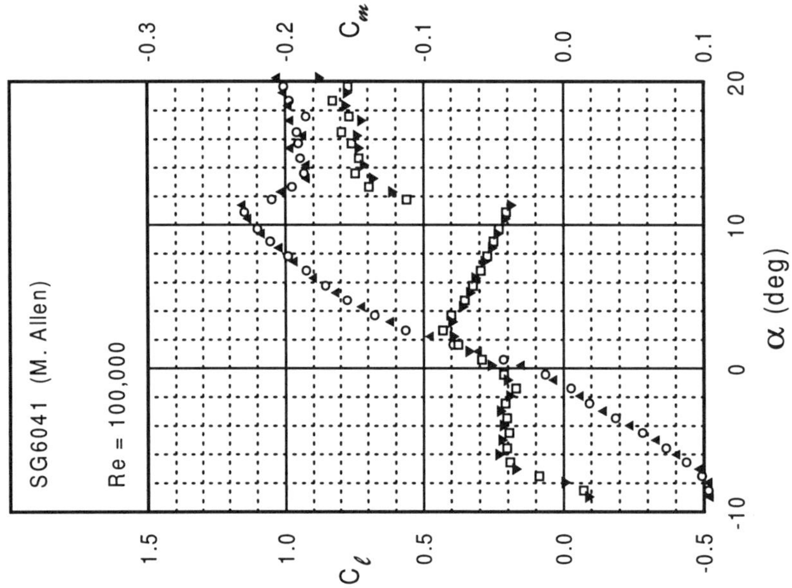
SG6041 (M. Allen)
Re = 100,000
C_ℓ
C_m
α (deg)

SG6041

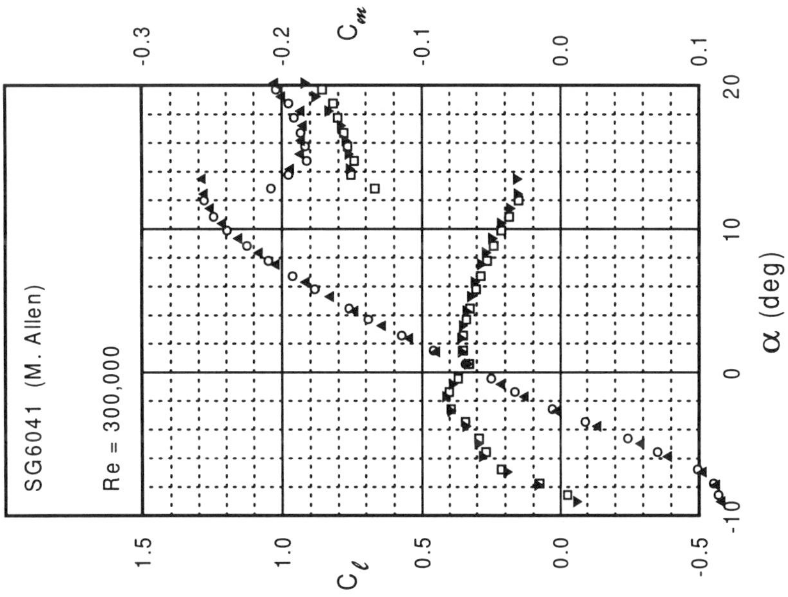
SG6041 (M. Allen)
Re = 300,000
C_l
C_m
α (deg)
1.5
1.0
0.5
0.0
-0.5
-0.3
-0.2
-0.1
0.0
0.1
-10
0
10
20

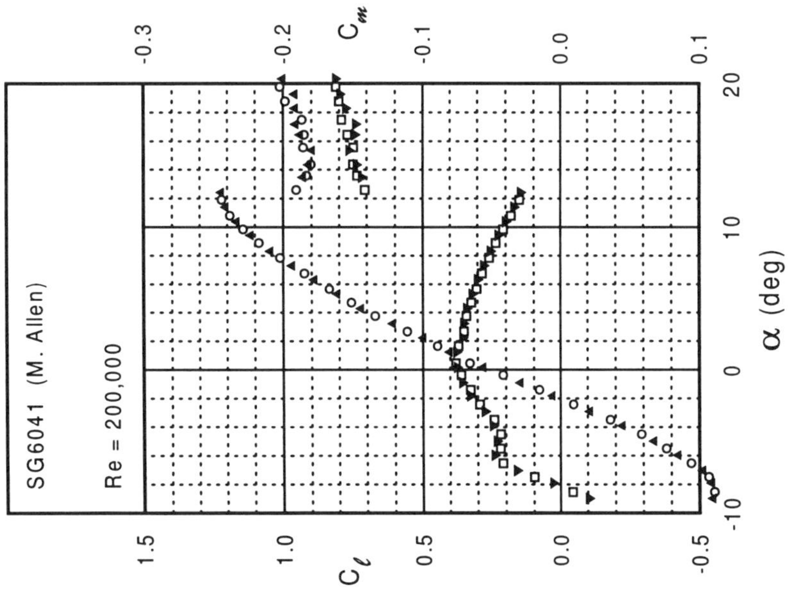
SG6041 (M. Allen)
Re = 200,000
C_l
C_m
α (deg)
1.5
1.0
0.5
0.0
-0.5
-0.3
-0.2
-0.1
0.0
0.1
-10
0
10
20

SG6041

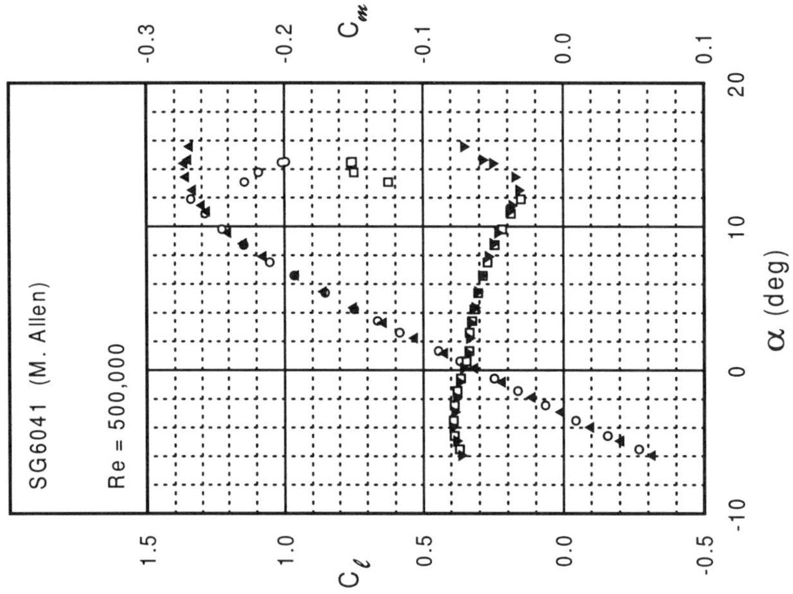
SG6041 (M. Allen)
Re = 500,000
C_ℓ
C_m
α (deg)

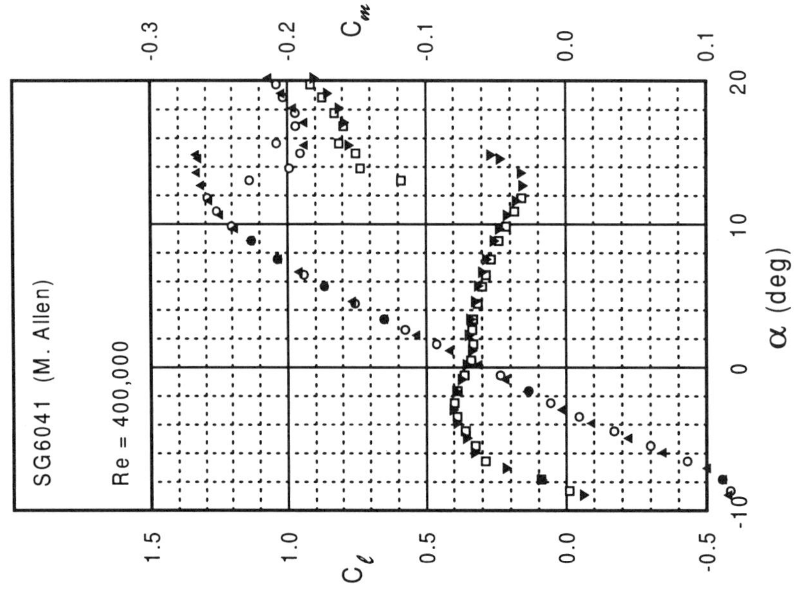
SG6041 (M. Allen)
Re = 400,000
C_ℓ
C_m
α (deg)

SG6041

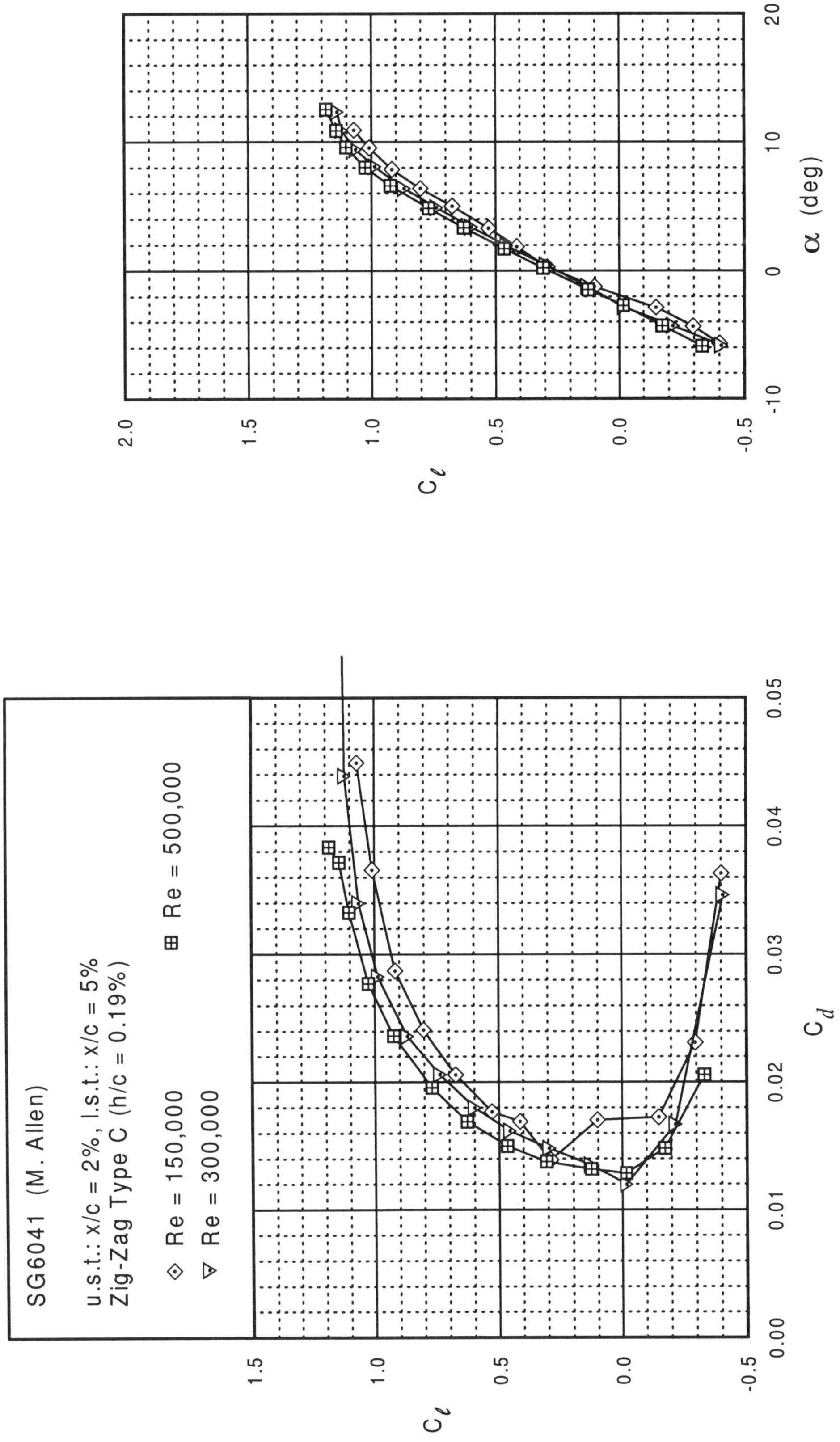
SG6041 (M. Allen)
u.s.t.: x/c = 2%, l.s.t.: x/c = 5%
Zig-Zag Type C (h/c = 0.19%)
Re = 150,000
Re = 300,000
Re = 500,000
C_ℓ
C_d
α (deg)
1.5
1.0
0.5
0.0
-0.5
2.0
0.00
0.01
0.02
0.03
0.04
0.05
-10
0
10
20

SG6041

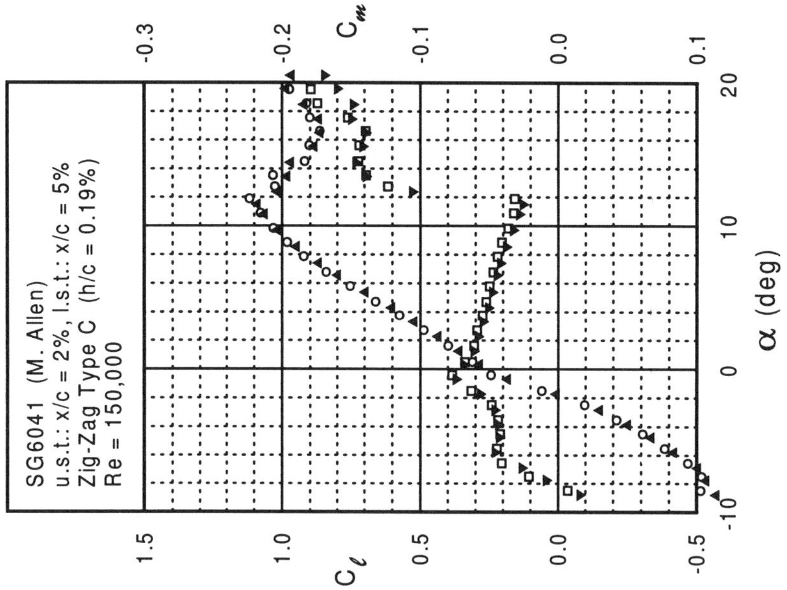
SG6041 (M. Allen)
u.s.t.: x/c = 2%, l.s.t.: x/c = 5%
Zig-Zag Type C (h/c = 0.19%)
Re = 150,000
-0.3
-0.2
-0.1
0.0
0.1
C_m
1.5
1.0
0.5
0.0
-0.5
C_ℓ
-10
0
10
20
α (deg)

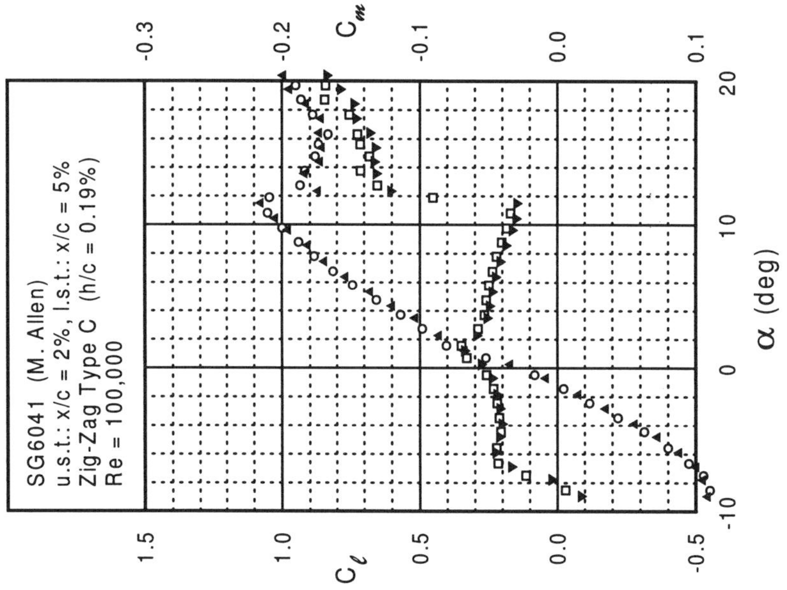
SG6041 (M. Allen)
u.s.t.: x/c = 2%, l.s.t.: x/c = 5%
Zig-Zag Type C (h/c = 0.19%)
Re = 100,000
-0.3
-0.2
-0.1
0.0
0.1
C_m
1.5
1.0
0.5
0.0
-0.5
C_ℓ
-10
0
10
20
α (deg)

SG6041

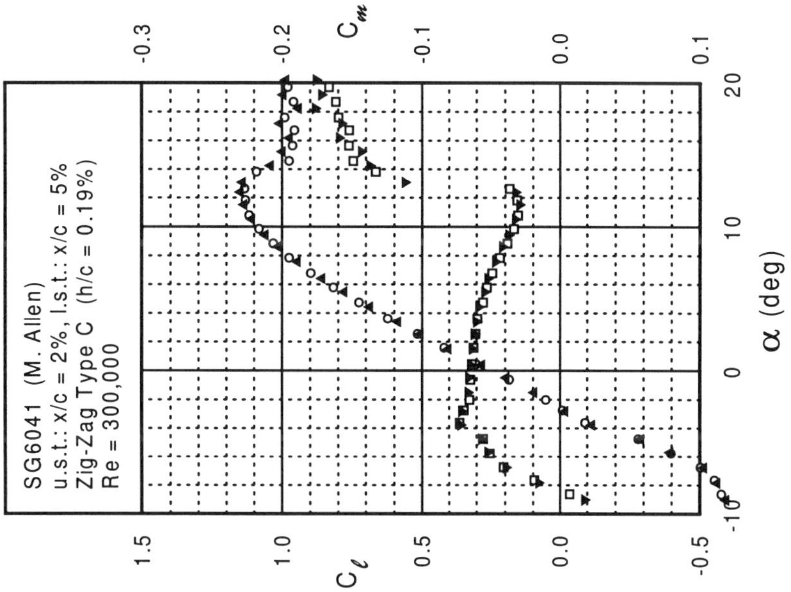
SG6041 (M. Allen)
u.s.t.: x/c = 2%, l.s.t.: x/c = 5%
Zig-Zag Type C (h/c = 0.19%)
Re = 300,000
-0.3
-0.2
-0.1
0.0
0.1
C_m
1.5
1.0
0.5
0.0
-0.5
C_ℓ
-10
0
10
20
α (deg)

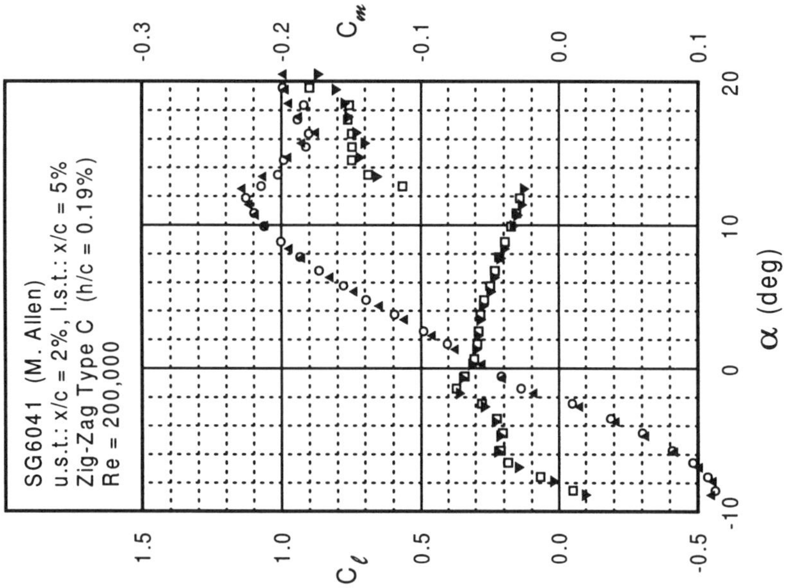
SG6041 (M. Allen)
u.s.t.: x/c = 2%, l.s.t.: x/c = 5%
Zig-Zag Type C (h/c = 0.19%)
Re = 200,000
-0.3
-0.2
-0.1
0.0
0.1
C_m
1.5
1.0
0.5
0.0
-0.5
C_ℓ
-10
0
10
20
α (deg)

SG6041

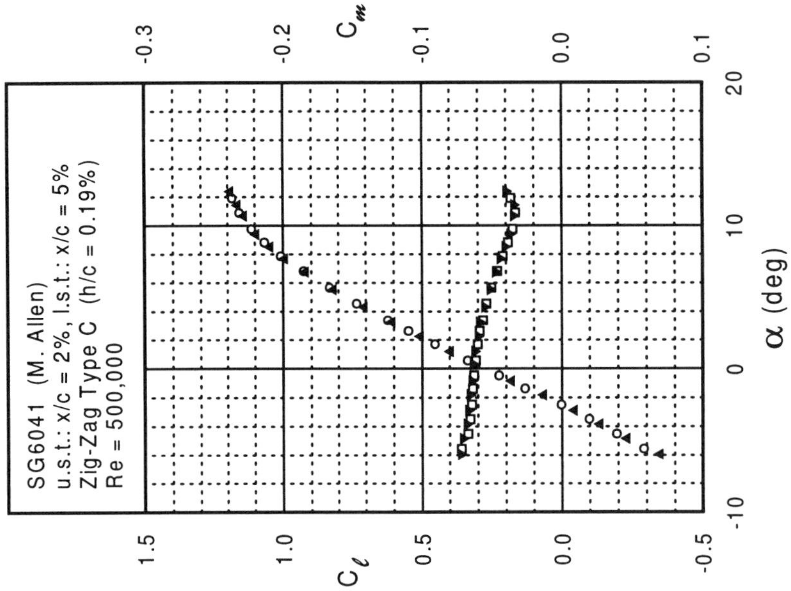
SG6041 (M. Allen)
u.s.t.: x/c = 2%, l.s.t.: x/c = 5%
Zig-Zag Type C (h/c = 0.19%)
Re = 500,000
C_m
-0.3
-0.2
-0.1
0.0
0.1
C_ℓ
1.5
1.0
0.5
0.0
-0.5
-10
0
10
20
α (deg)

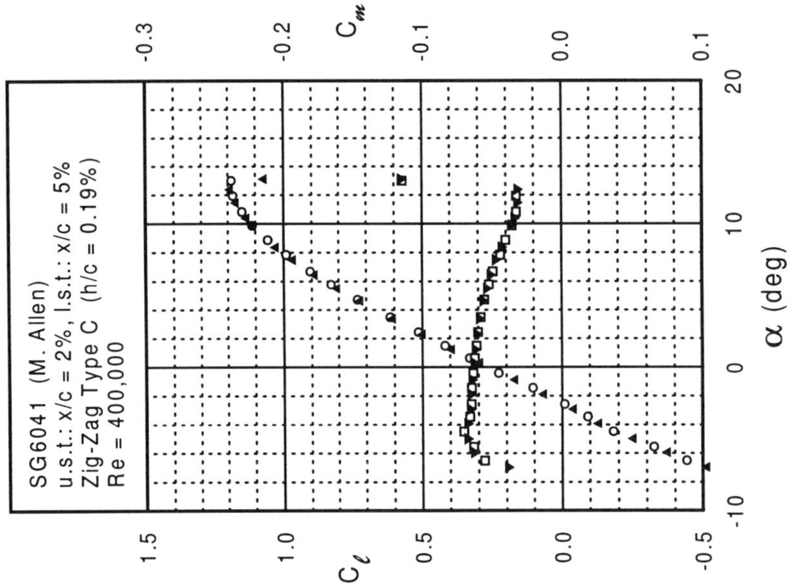
SG6041 (M. Allen)
u.s.t.: x/c = 2%, l.s.t.: x/c = 5%
Zig-Zag Type C (h/c = 0.19%)
Re = 400,000
C_m
-0.3
-0.2
-0.1
0.0
0.1
C_ℓ
1.5
1.0
0.5
0.0
-0.5
-10
0
10
20
α (deg)

SG6042

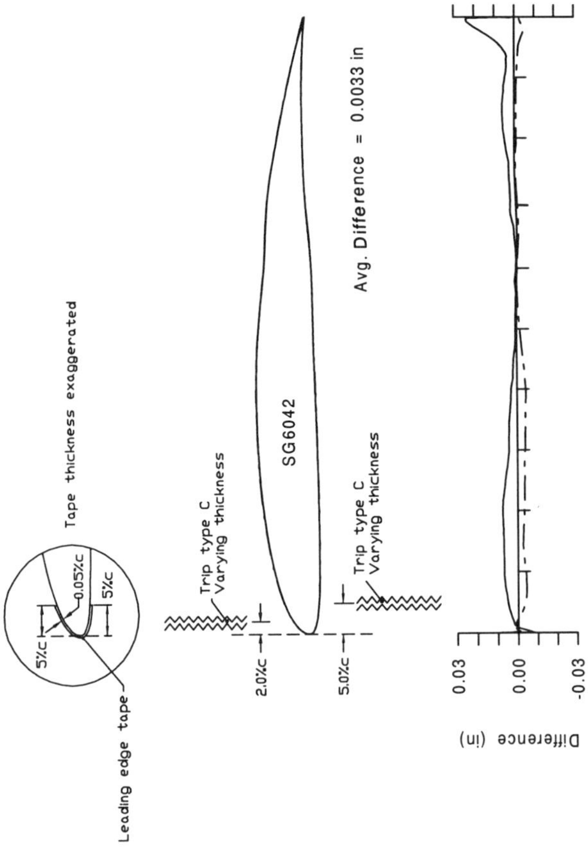
Tape thickness exaggerated
0.05%c
5%c
5%c
Leading edge tape
Trip type C
Varying thickness
2.0%c
SG6042
Trip type C
Varying thickness
5.0%c
Avg. Difference = 0.0033 in
0.03
0.00
-0.03
Difference (in)

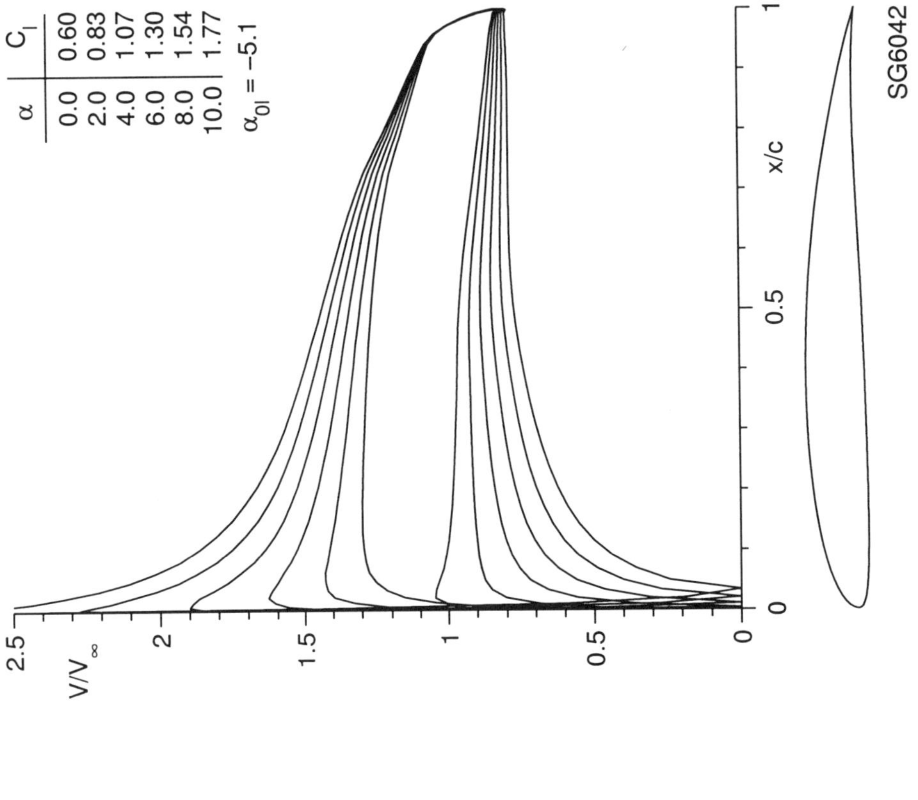
α
0.0
2.0
4.0
6.0
8.0
10.0
C_l
0.60
0.83
1.07
1.30
1.54
1.77
α_{0l} = −5.1
2.5
V/V_∞
2
1.5
1
0.5
0
0
0.5
x/c
1
SG6042

SG6042

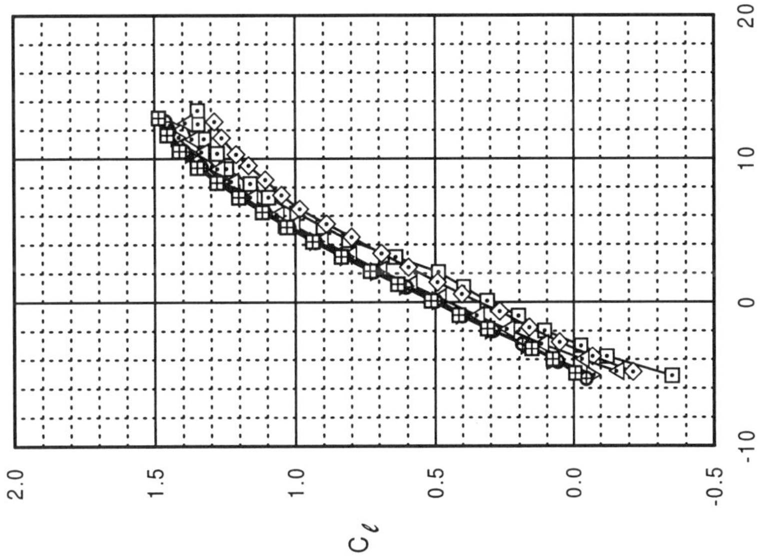

2.0
1.5
1.0
0.5
0.0
-0.5
c_ℓ
-10
0
10
20
α (deg)

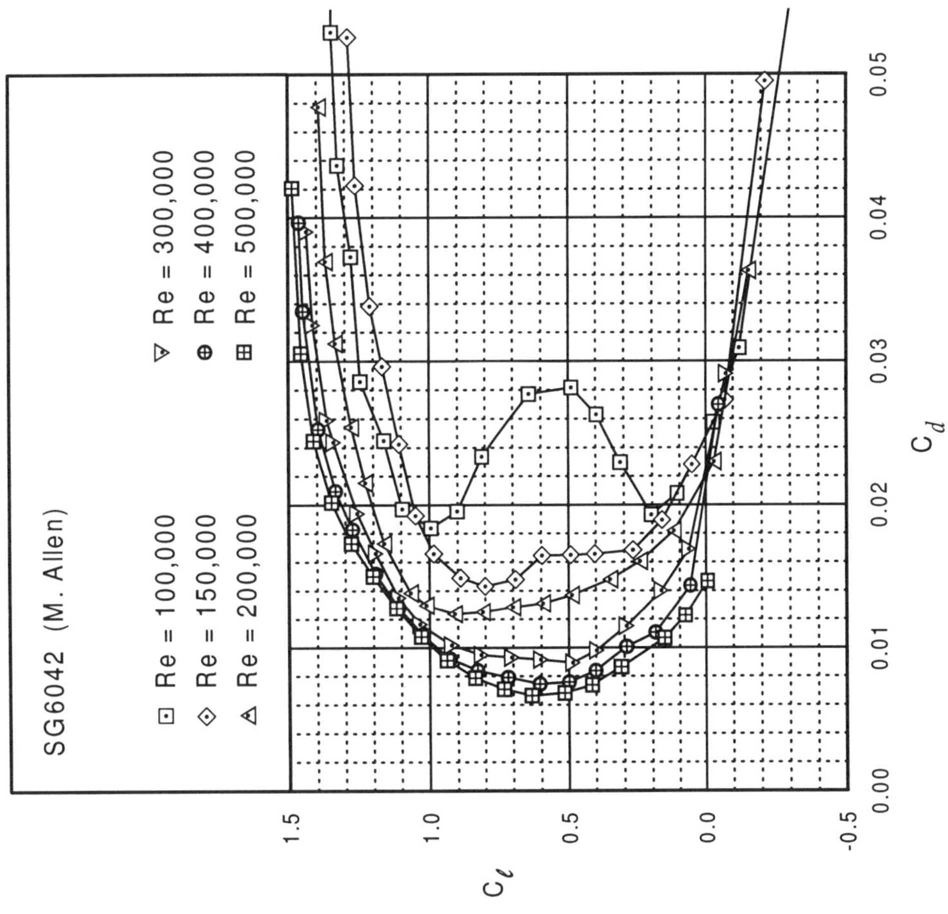

SG6042 (M. Allen)
Re = 100,000
Re = 150,000
Re = 200,000
Re = 300,000
Re = 400,000
Re = 500,000
1.5
1.0
0.5
0.0
-0.5
c_ℓ
0.00
0.01
0.02
0.03
0.04
0.05
c_d

SG6042

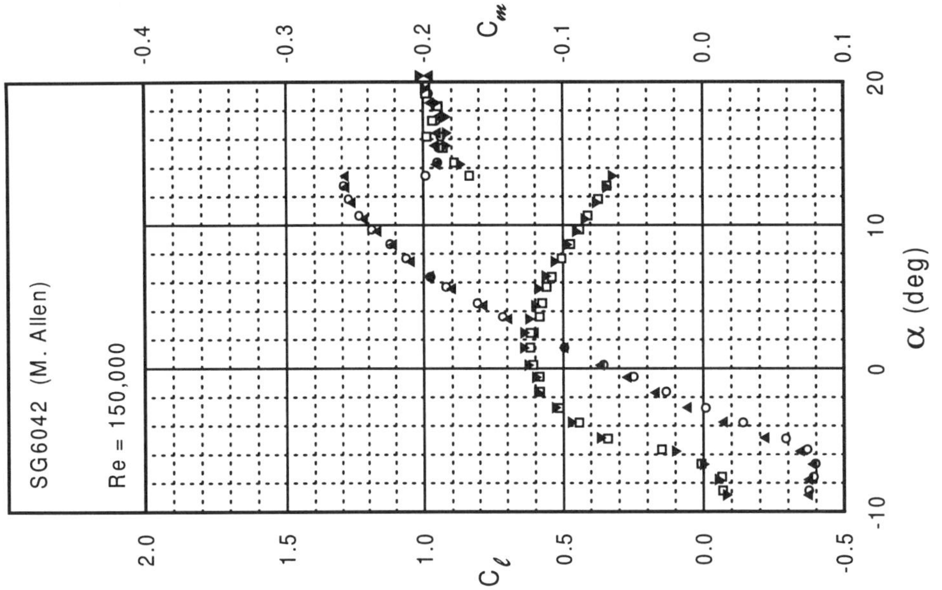
SG6042 (M. Allen)
Re = 150,000
C_ℓ
C_m
α (deg)

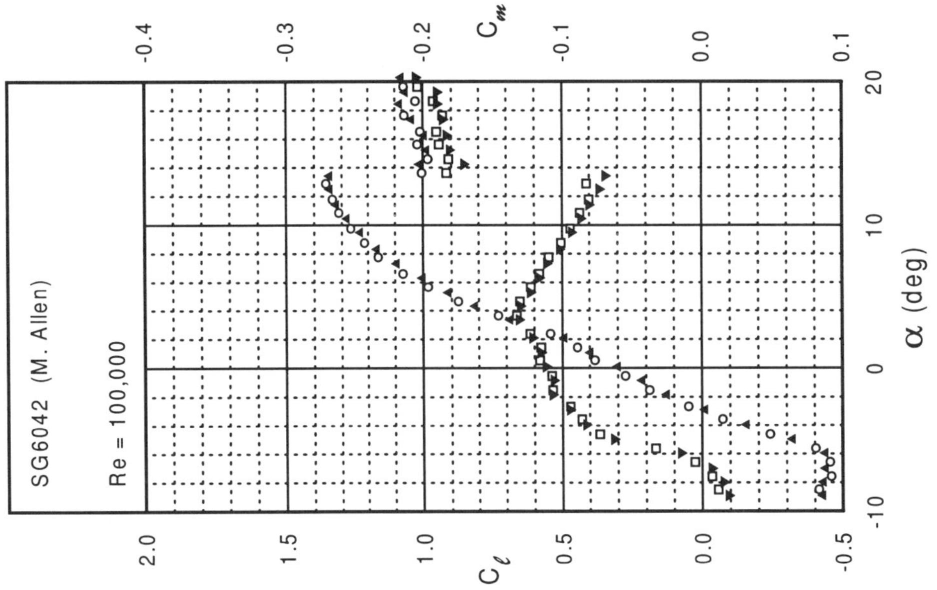
SG6042 (M. Allen)
Re = 100,000
C_ℓ
C_m
α (deg)

SG6042

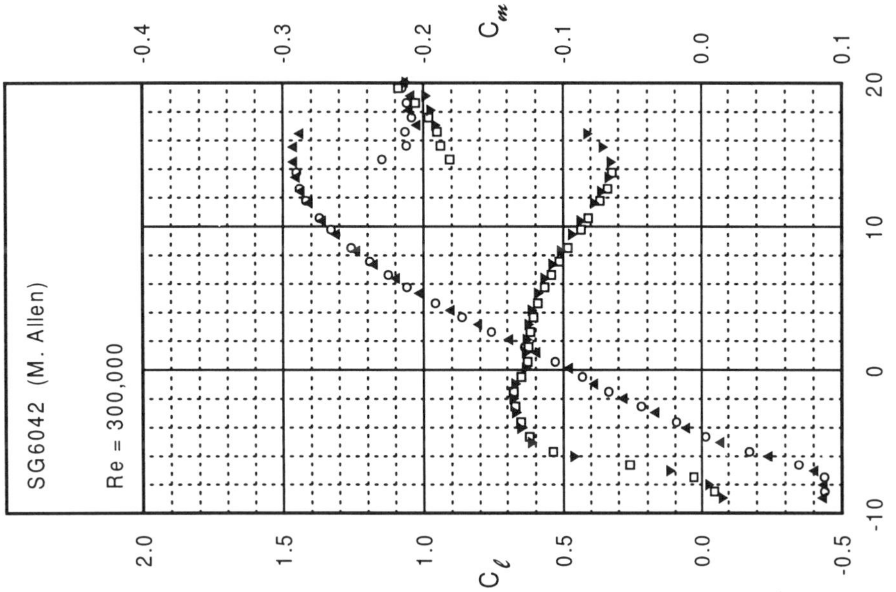
SG6042 (M. Allen)
Re = 300,000
C_ℓ
C_m
α (deg)
2.0
1.5
1.0
0.5
0.0
-0.5
-0.4
-0.3
-0.2
-0.1
0.0
0.1
-10
0
10
20

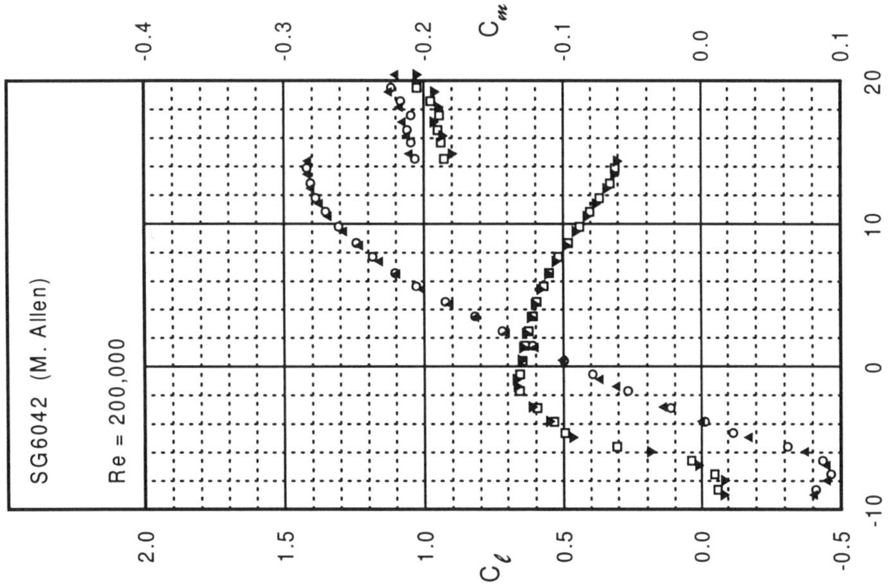
SG6042 (M. Allen)
Re = 200,000
C_ℓ
C_m
α (deg)
2.0
1.5
1.0
0.5
0.0
-0.5
-0.4
-0.3
-0.2
-0.1
0.0
0.1
-10
0
10
20

SG6042

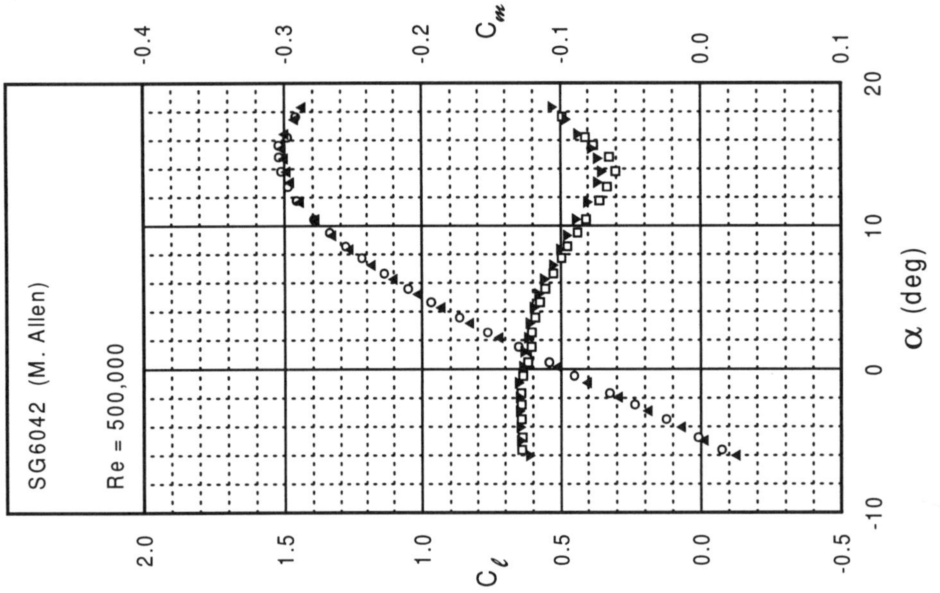
SG6042 (M. Allen)
Re = 500,000
C_m
C_ℓ
α (deg)

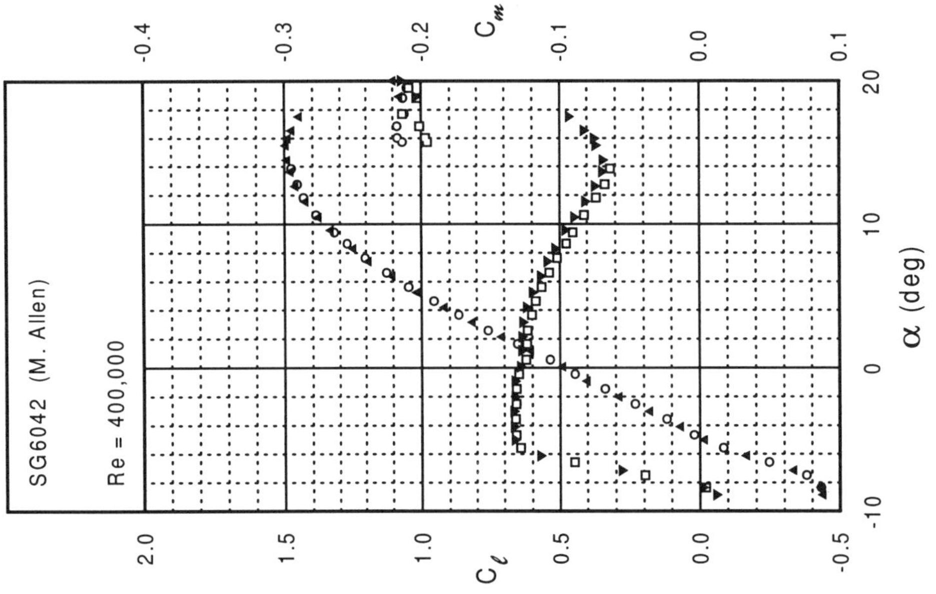
SG6042 (M. Allen)
Re = 400,000
C_m
C_ℓ
α (deg)

SG6042

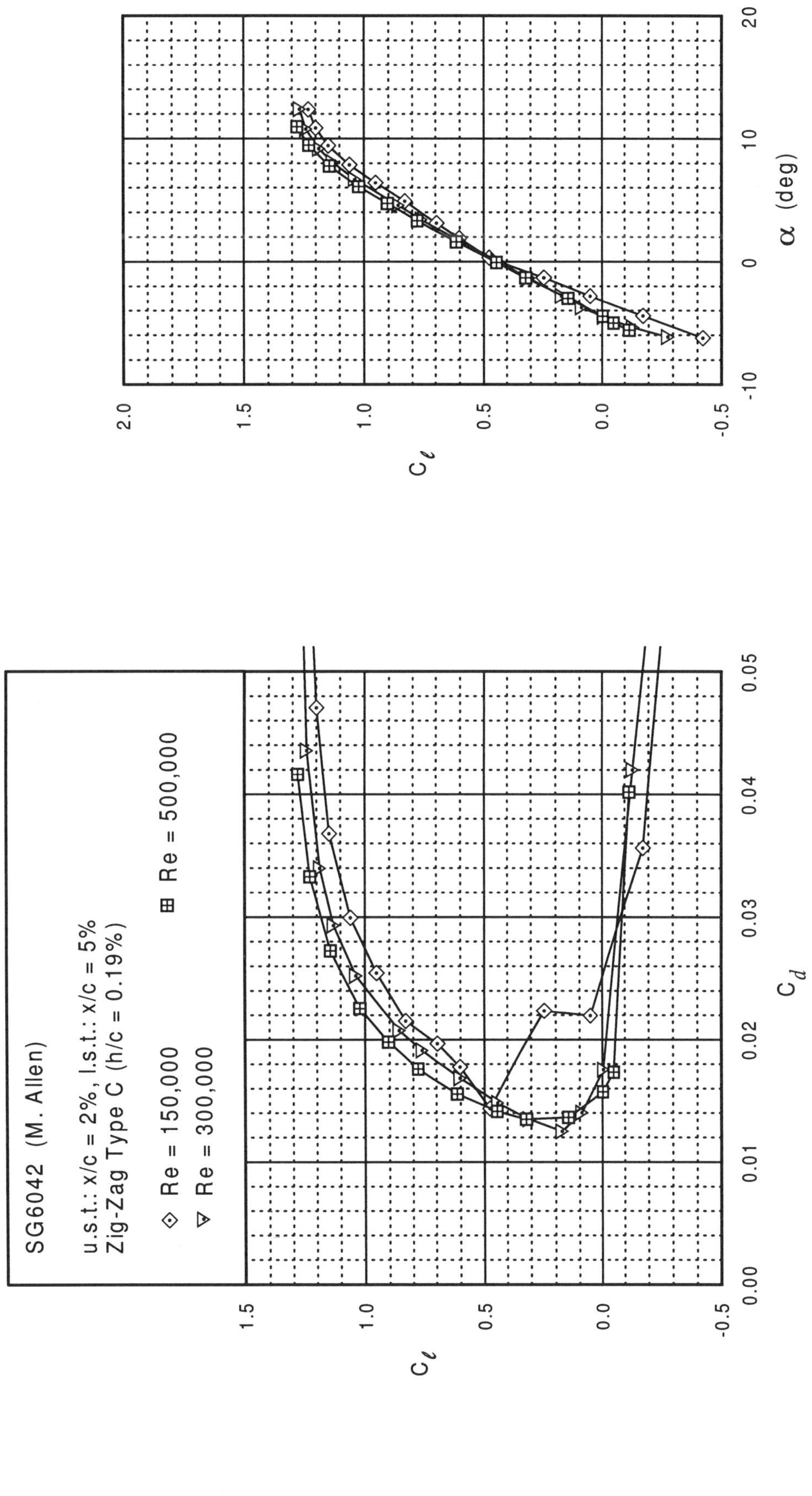
SG6042 (M. Allen)
u.s.t.: x/c = 2%, l.s.t.: x/c = 5%
Zig-Zag Type C (h/c = 0.19%)
Re = 150,000
Re = 300,000
Re = 500,000
C_d
C_ℓ
α (deg)

SG6042

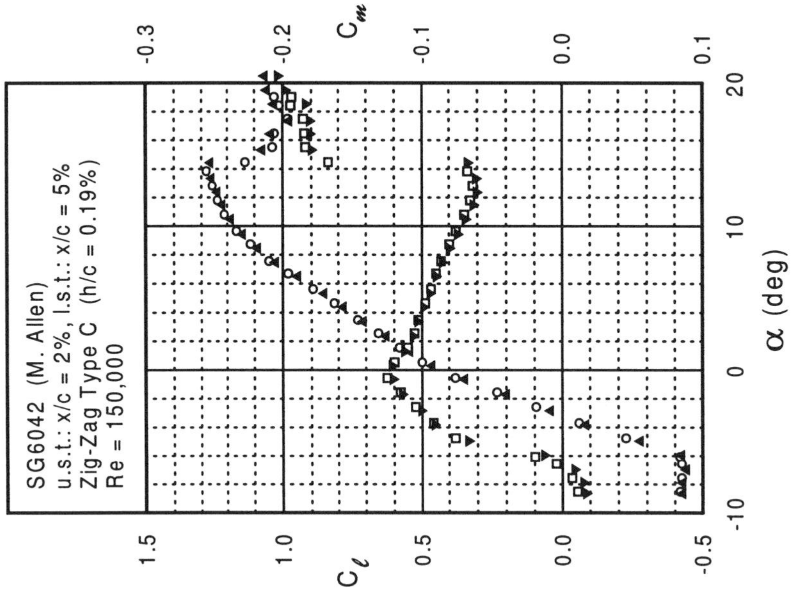
SG6042 (M. Allen)
u.s.t.: x/c = 2%, l.s.t.: x/c = 5%
Zig-Zag Type C (h/c = 0.19%)
Re = 150,000
C_m
C_ℓ
α (deg)

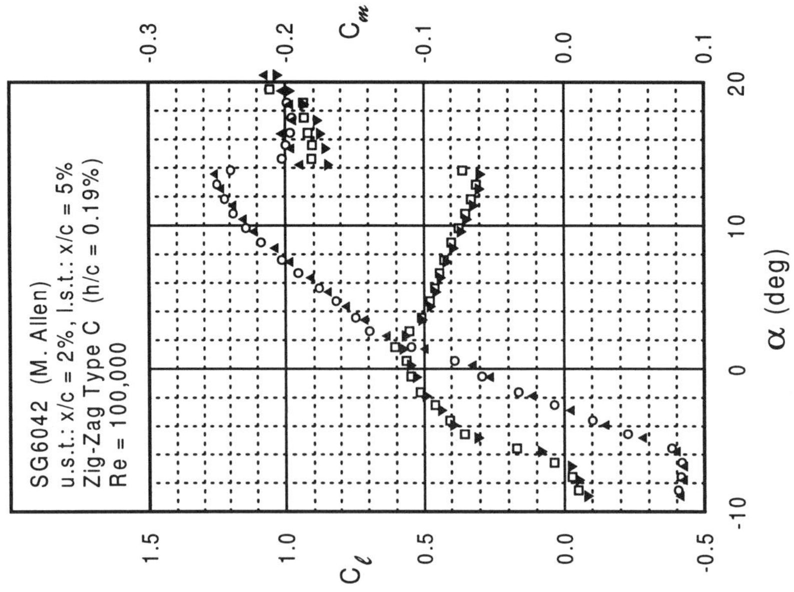
SG6042 (M. Allen)
u.s.t.: x/c = 2%, l.s.t.: x/c = 5%
Zig-Zag Type C (h/c = 0.19%)
Re = 100,000
C_m
C_ℓ
α (deg)

SG6042

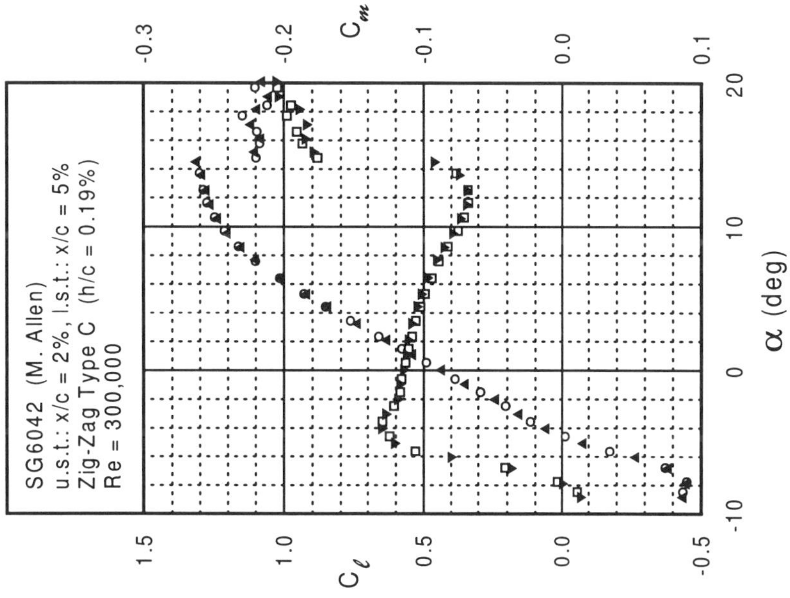
SG6042 (M. Allen)
u.s.t.: x/c = 2%, l.s.t.: x/c = 5%
Zig-Zag Type C (h/c = 0.19%)
Re = 300,000
C_ℓ
C_m
α (deg)

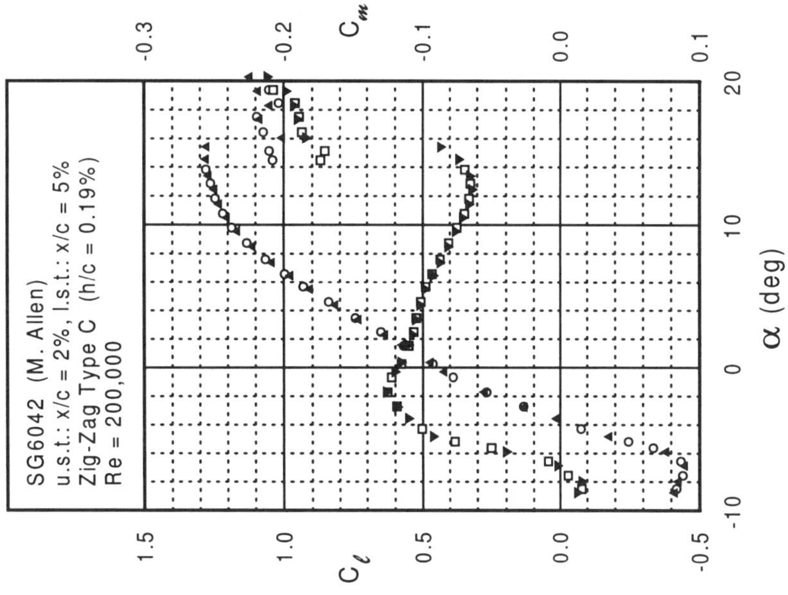
SG6042 (M. Allen)
u.s.t.: x/c = 2%, l.s.t.: x/c = 5%
Zig-Zag Type C (h/c = 0.19%)
Re = 200,000
C_ℓ
C_m
α (deg)

SG6042

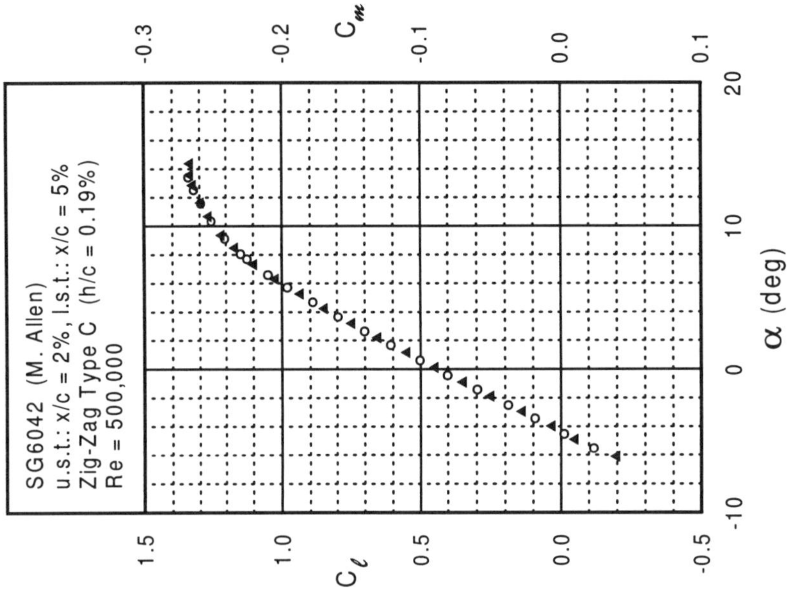
SG6042 (M. Allen)
u.s.t.: x/c = 2%, l.s.t.: x/c = 5%
Zig-Zag Type C (h/c = 0.19%)
Re = 500,000
C_ℓ
C_m
α (deg)

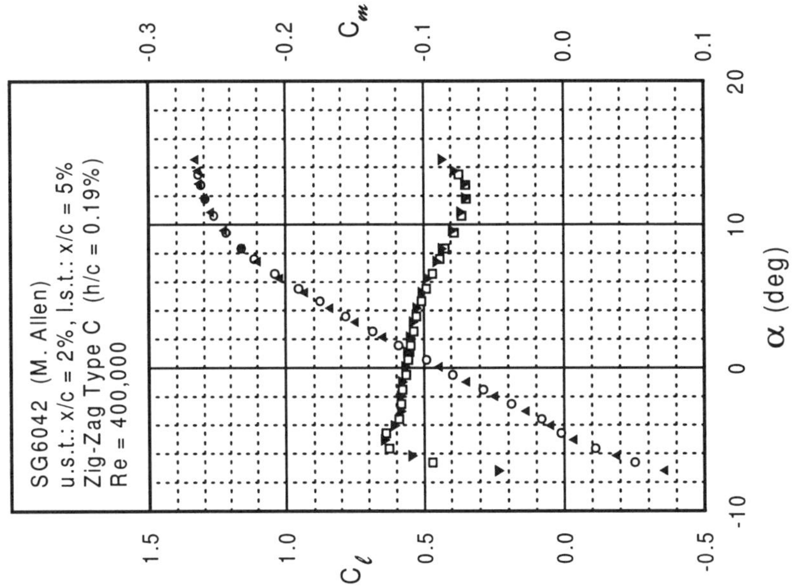
SG6042 (M. Allen)
u.s.t.: x/c = 2%, l.s.t.: x/c = 5%
Zig-Zag Type C (h/c = 0.19%)
Re = 400,000
C_ℓ
C_m
α (deg)

SG6042

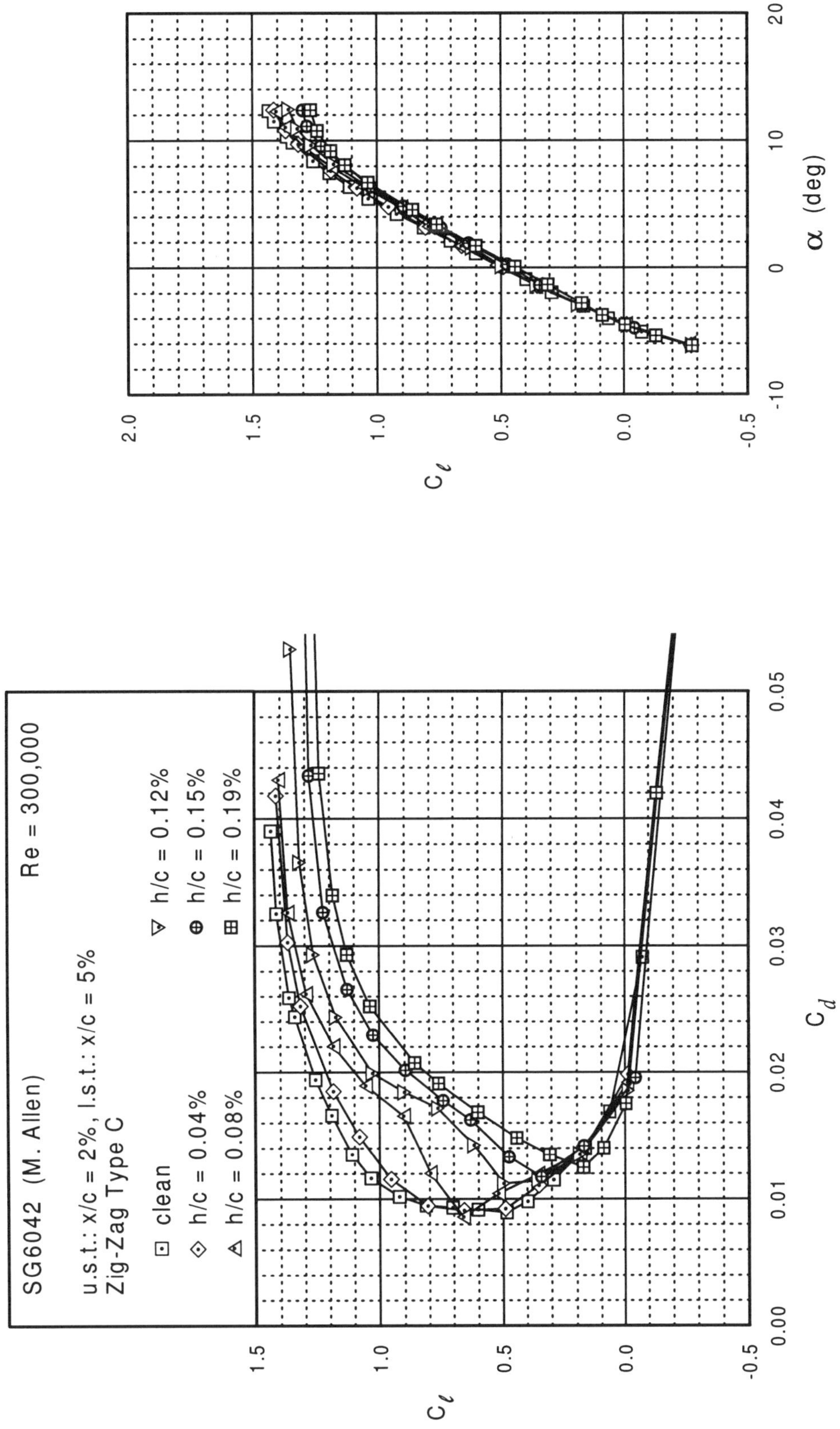
SG6042 (M. Allen)
Re = 300,000
u.s.t.: x/c = 2%, l.s.t.: x/c = 5%
Zig-Zag Type C
clean
h/c = 0.04%
h/c = 0.08%
h/c = 0.12%
h/c = 0.15%
h/c = 0.19%
C_ℓ
C_d
α (deg)

SG6042

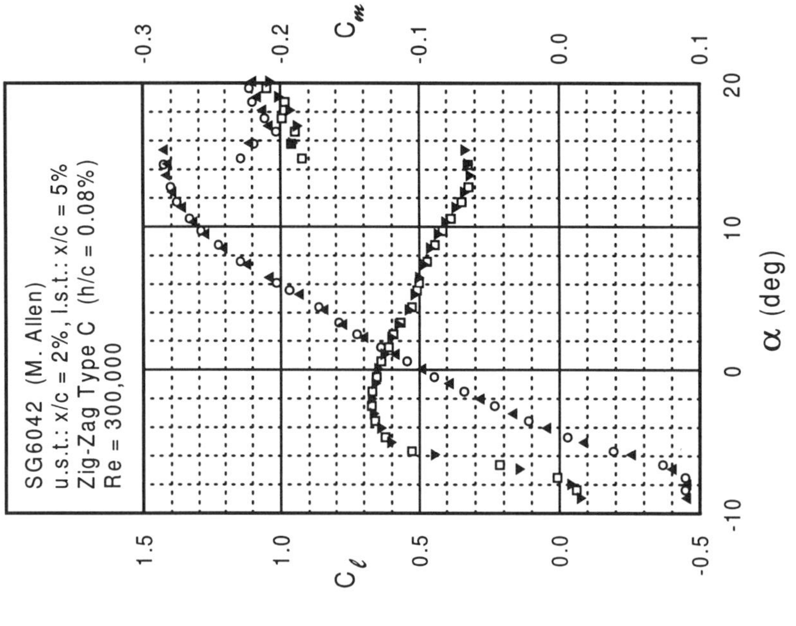
SG6042 (M. Allen)
u.s.t.: x/c = 2%, l.s.t.: x/c = 5%
Zig-Zag Type C (h/c = 0.08%)
Re = 300,000
C_ℓ
C_m
α (deg)

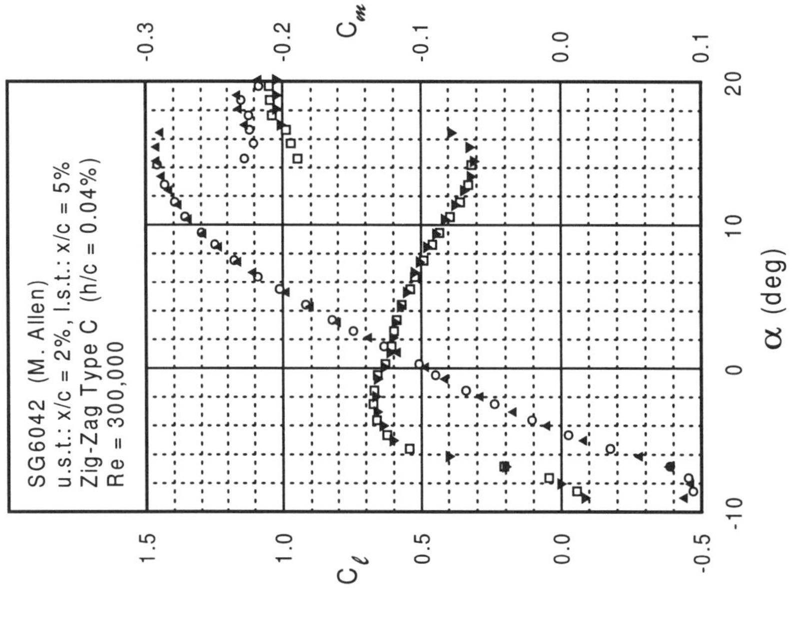
SG6042 (M. Allen)
u.s.t.: x/c = 2%, l.s.t.: x/c = 5%
Zig-Zag Type C (h/c = 0.04%)
Re = 300,000
C_ℓ
C_m
α (deg)

SG6042

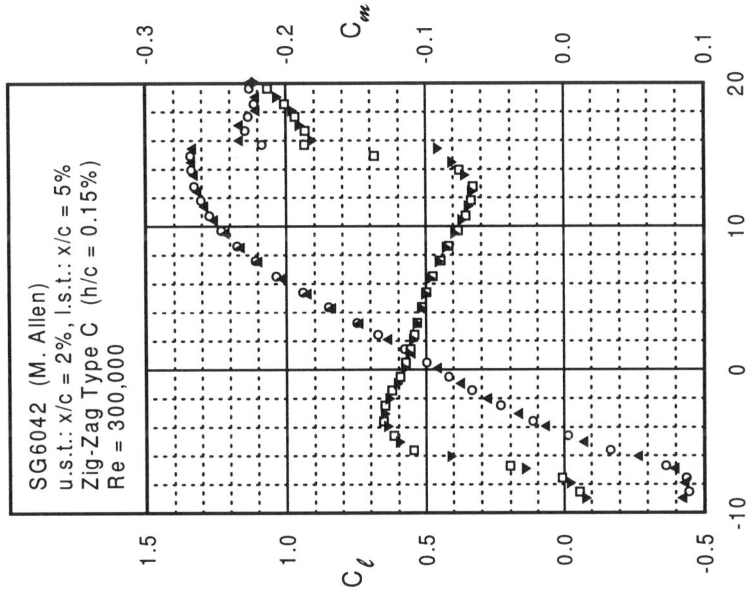
SG6042 (M. Allen)
u.s.t.: x/c = 2%, l.s.t.: x/c = 5%
Zig-Zag Type C (h/c = 0.15%)
Re = 300,000
C_ℓ
C_m
α (deg)

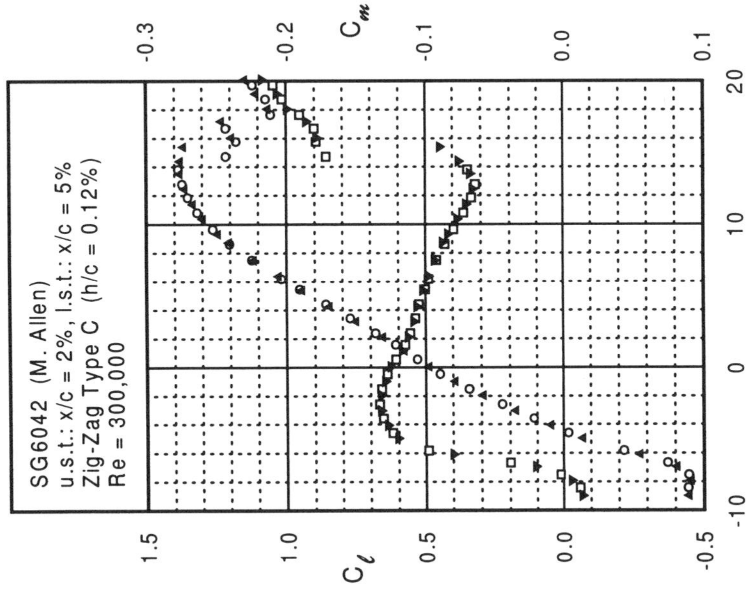
SG6042 (M. Allen)
u.s.t.: x/c = 2%, l.s.t.: x/c = 5%
Zig-Zag Type C (h/c = 0.12%)
Re = 300,000
C_ℓ
C_m
α (deg)

SG6042

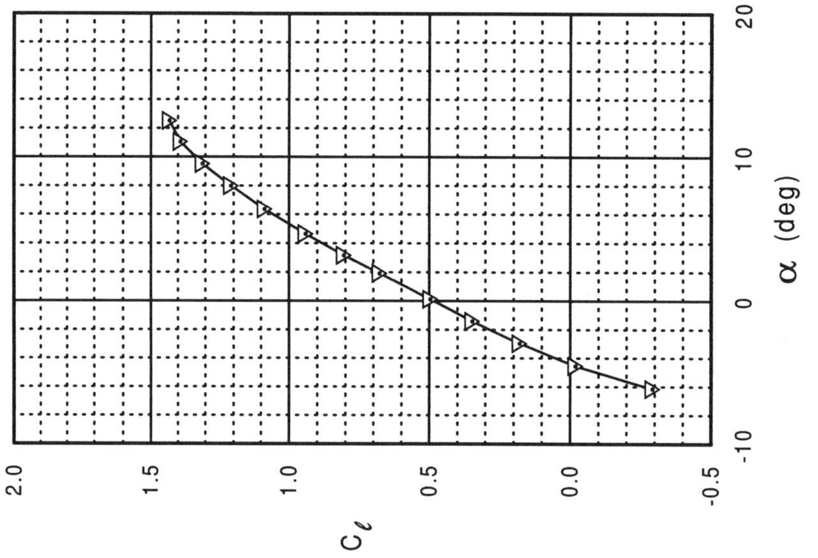
2.0
1.5
1.0
0.5
0.0
-0.5
-10
0
10
20
α (deg)

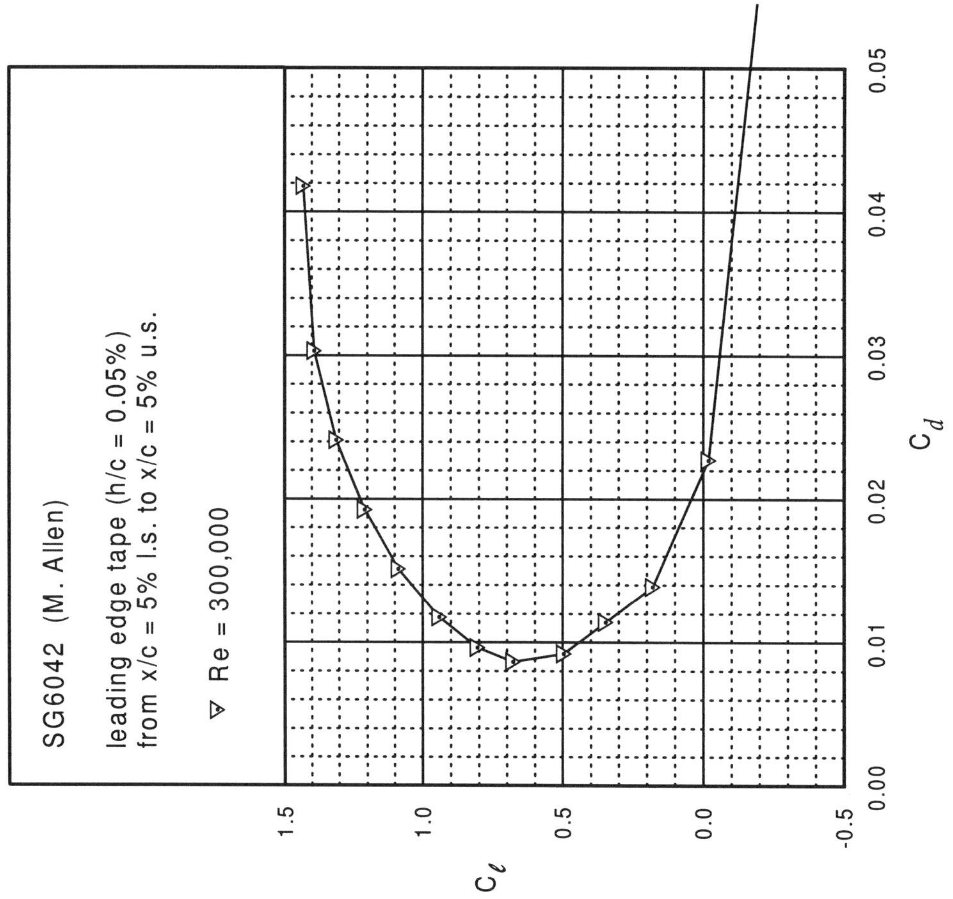
SG6042 (M. Allen)
leading edge tape (h/c = 0.05%)
from x/c = 5% l.s. to x/c = 5% u.s.
Re = 300,000
1.5
1.0
0.5
0.0
-0.5
0.00
0.01
0.02
0.03
0.04
0.05

SG6042

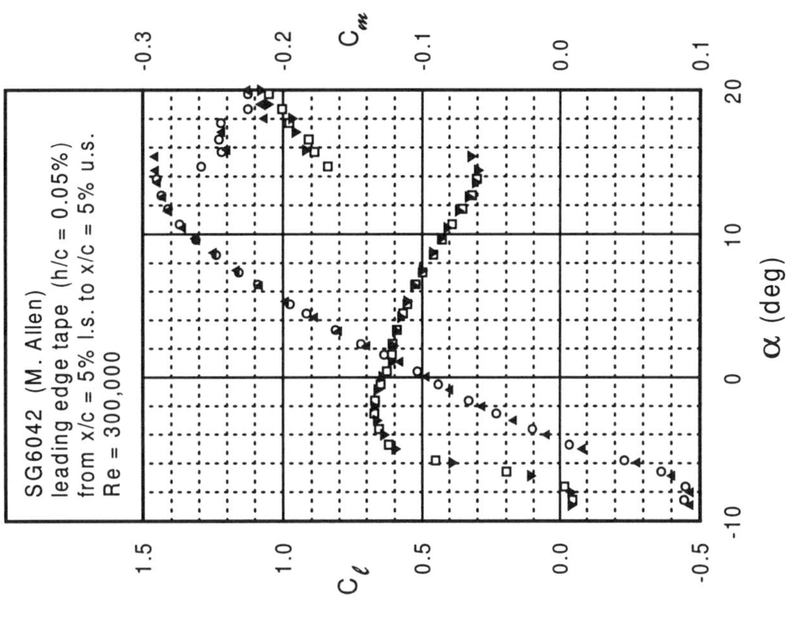
SG6042 (M. Allen)
leading edge tape (h/c = 0.05%)
from x/c = 5% l.s. to x/c = 5% u.s.
Re = 300,000
-0.3
-0.2
-0.1
0.0
0.1
C_m
1.5
1.0
0.5
0.0
-0.5
C_l
-10
0
10
20
α (deg)

SG6043

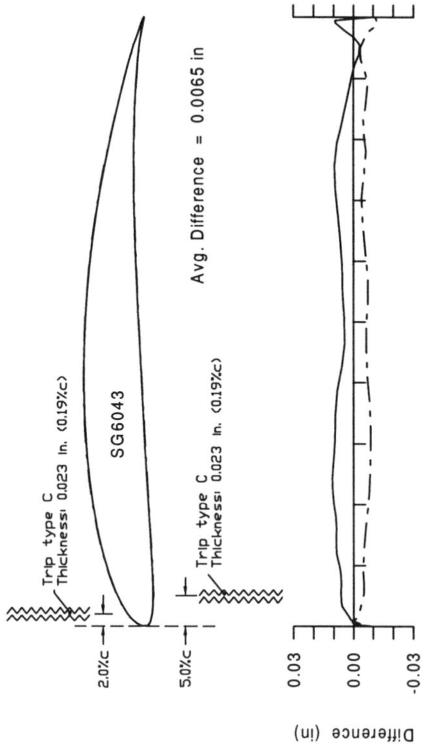
Trip type C
Thickness: 0.023 in. (0.19%c)
SG6043
Trip type C
Thickness: 0.023 in. (0.19%c)
Avg. Difference = 0.0065 in
2.0%c
5.0%c
0.03
0.00
-0.03
Difference (in)

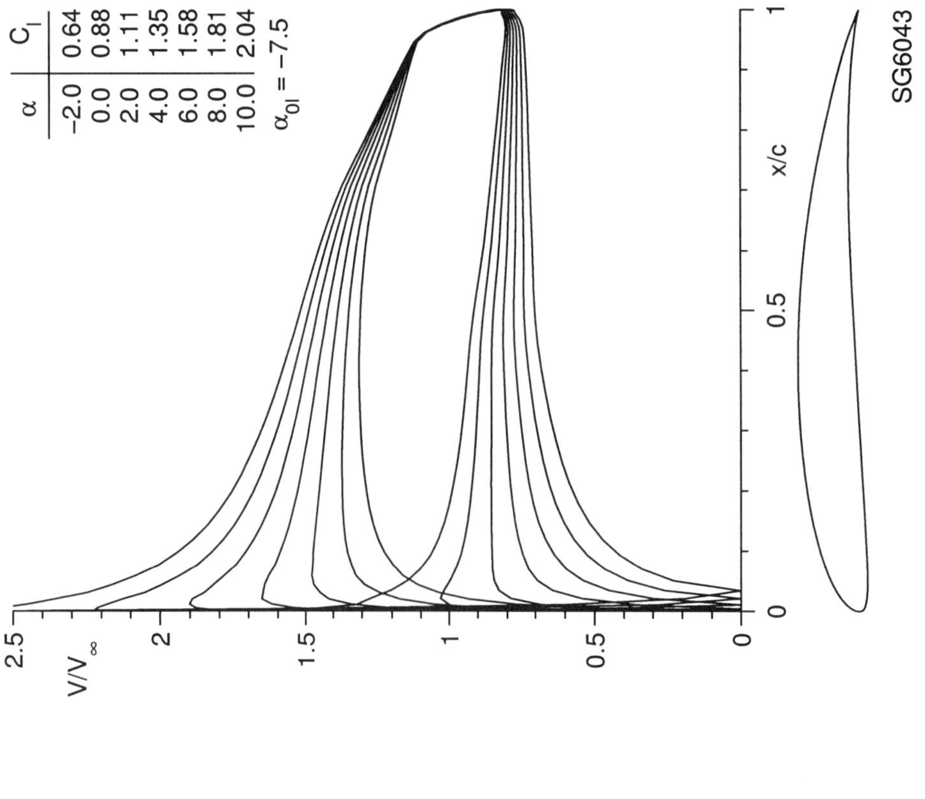
α | C_l
−2.0 | 0.64
0.0 | 0.88
2.0 | 1.11
4.0 | 1.35
6.0 | 1.58
8.0 | 1.81
10.0 | 2.04
α_{0l} = −7.5
V/V_∞
2.5
2
1.5
1
0.5
0
0
0.5
1
x/c
SG6043

SG6043

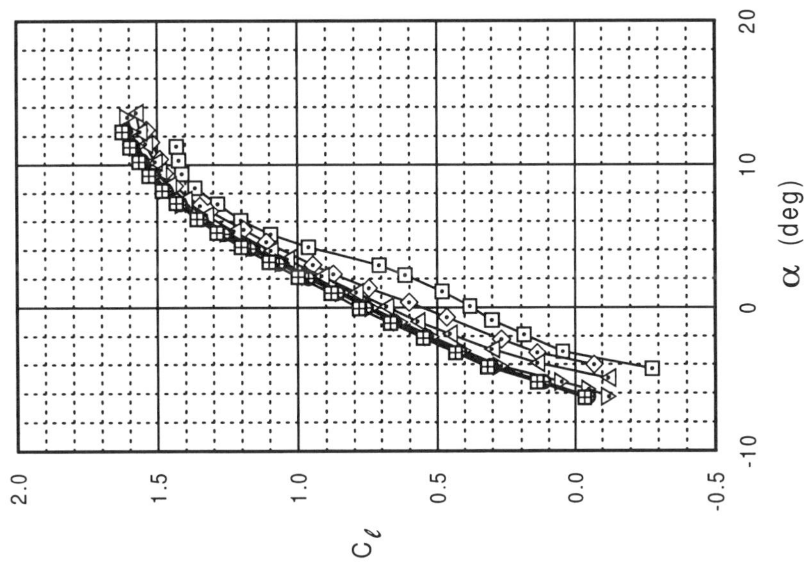
2.0
1.5
1.0
0.5
0.0
-0.5
C_ℓ
-10
0
10
20
α (deg)

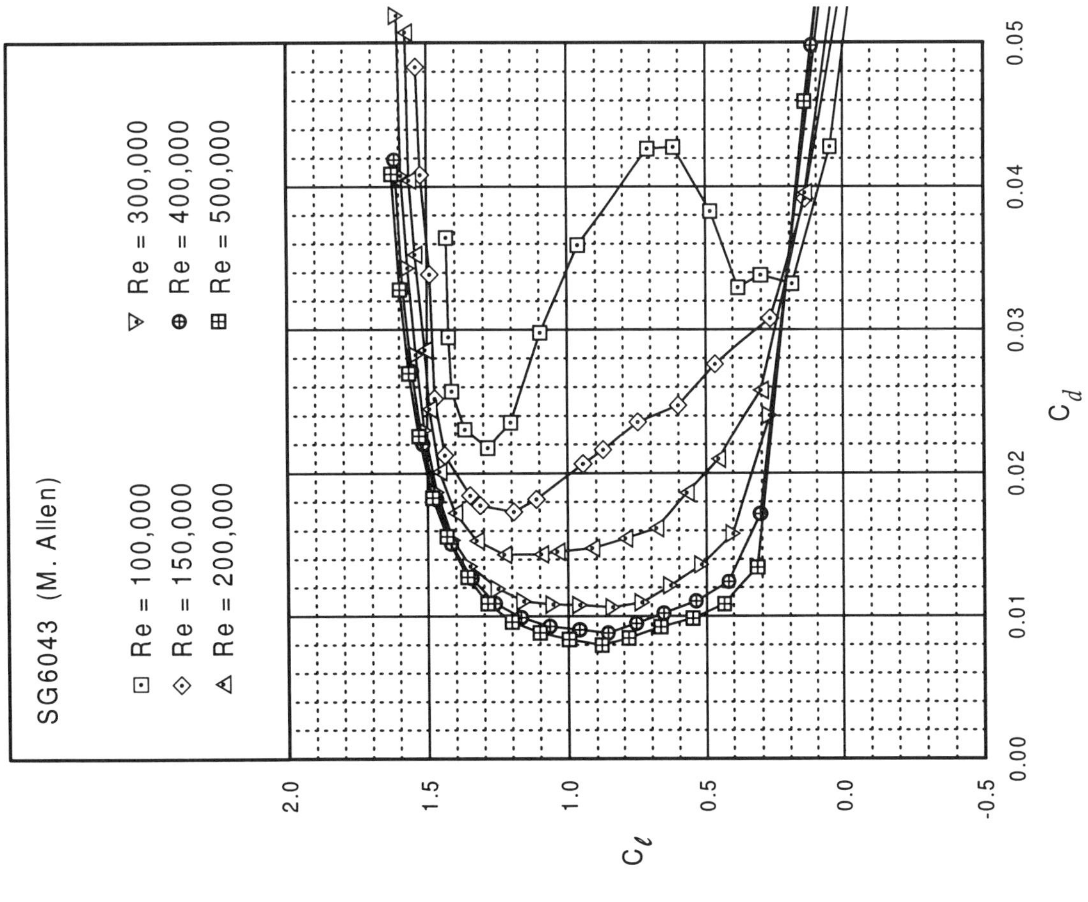
SG6043 (M. Allen)
Re = 100,000
Re = 150,000
Re = 200,000
Re = 300,000
Re = 400,000
Re = 500,000
2.0
1.5
1.0
0.5
0.0
-0.5
C_ℓ
0.00
0.01
0.02
0.03
0.04
0.05
C_d

SG6043

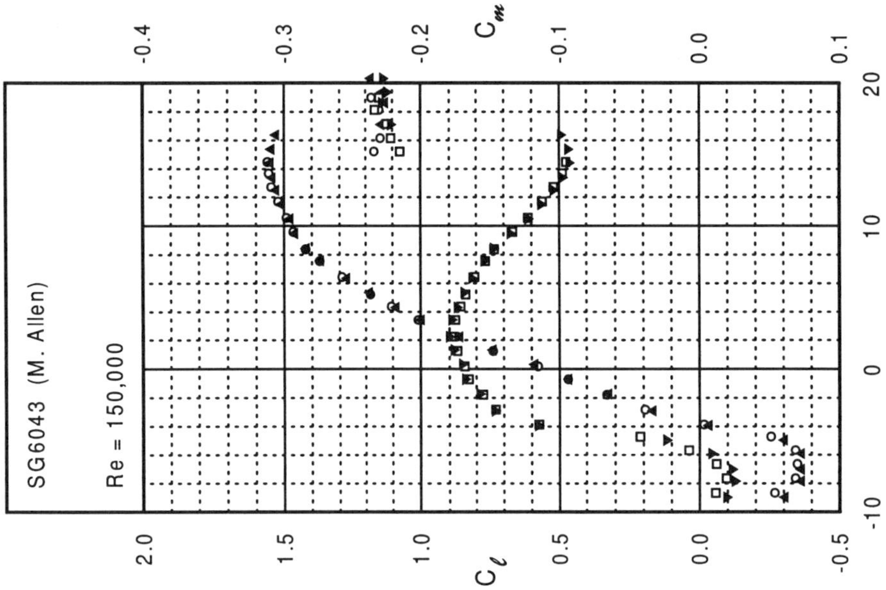
SG6043 (M. Allen)
Re = 150,000
C_l
C_m
α (deg)

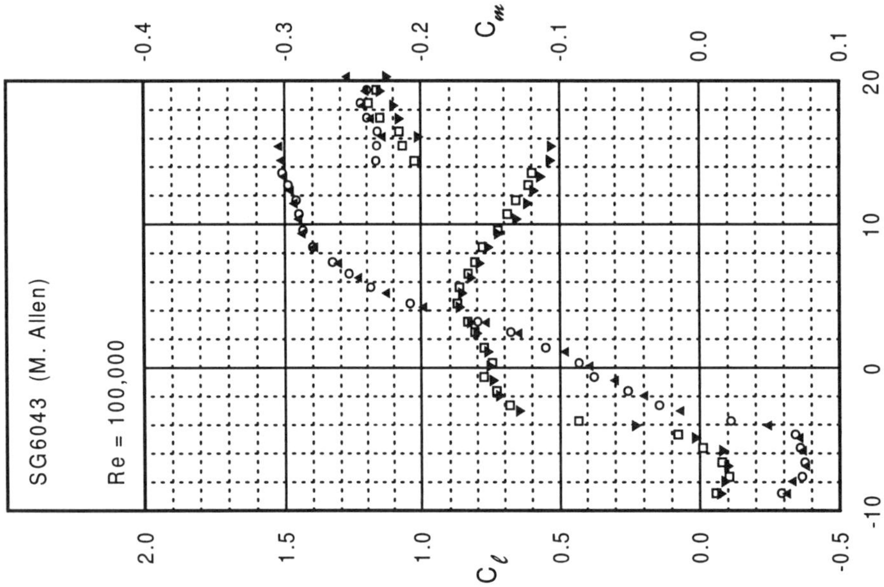
SG6043 (M. Allen)
Re = 100,000
C_l
C_m
α (deg)

SG6043

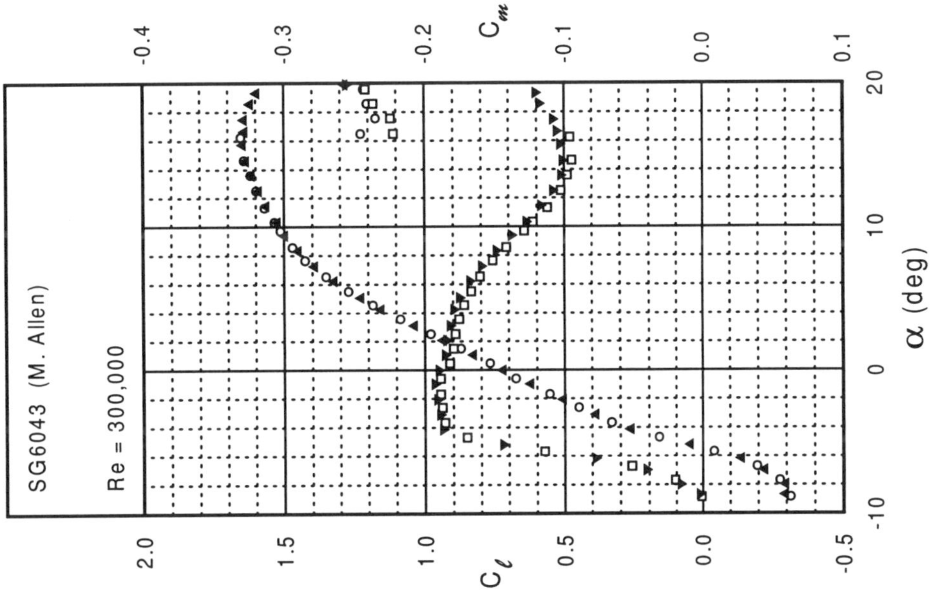
SG6043 (M. Allen)
Re = 300,000
C_ℓ
C_m
α (deg)

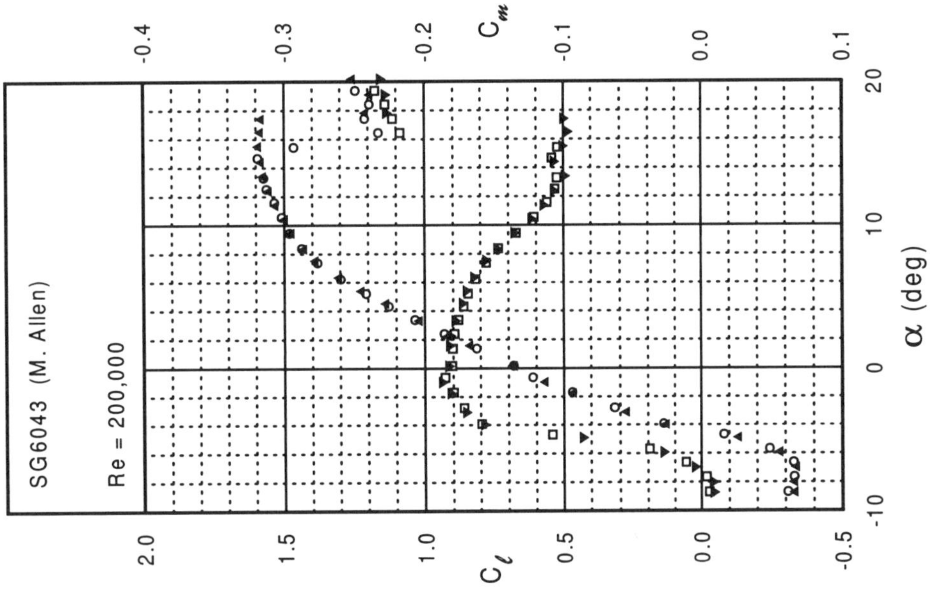
SG6043 (M. Allen)
Re = 200,000
C_ℓ
C_m
α (deg)

SG6043

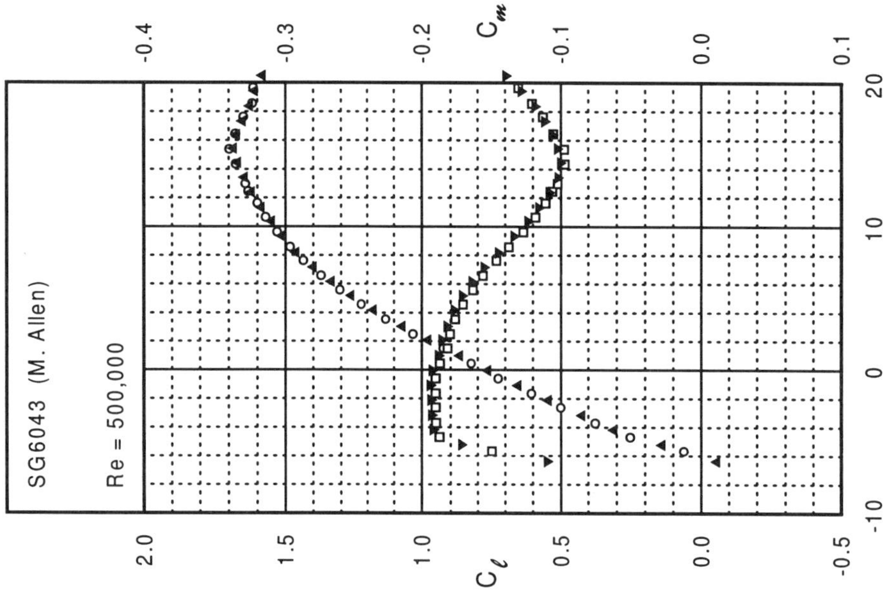
SG6043 (M. Allen)
Re = 500,000
C_m
-0.4
-0.3
-0.2
-0.1
0.0
0.1
C_ℓ
2.0
1.5
1.0
0.5
0.0
-0.5
-10
0
10
20
α (deg)

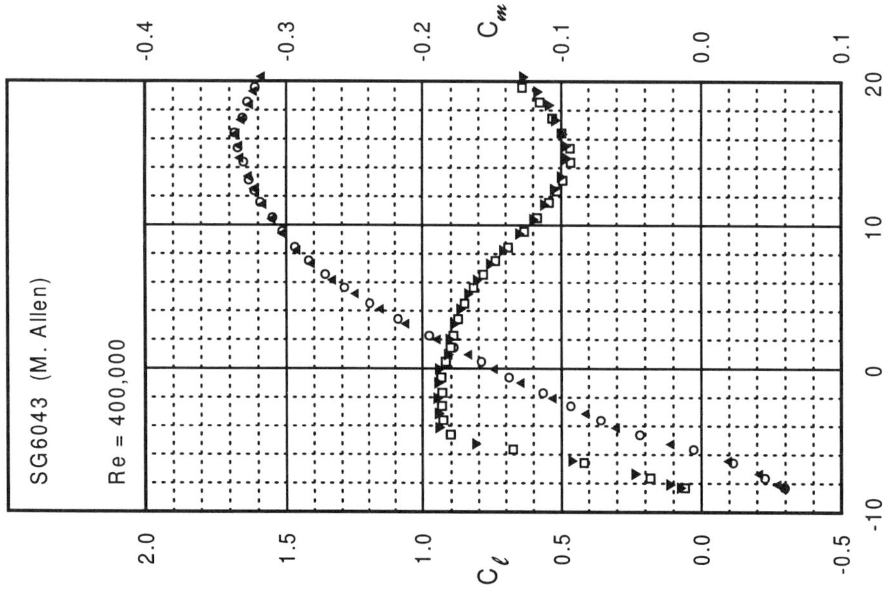
SG6043 (M. Allen)
Re = 400,000
C_m
-0.4
-0.3
-0.2
-0.1
0.0
0.1
C_ℓ
2.0
1.5
1.0
0.5
0.0
-0.5
-10
0
10
20
α (deg)

SG6043

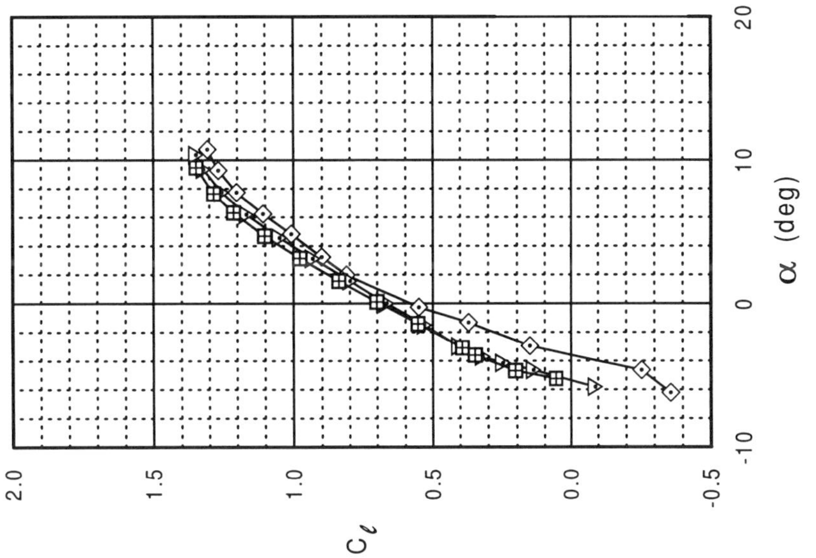
2.0
1.5
1.0
0.5
0.0
-0.5
C_ℓ
-10
0
10
20
α (deg)

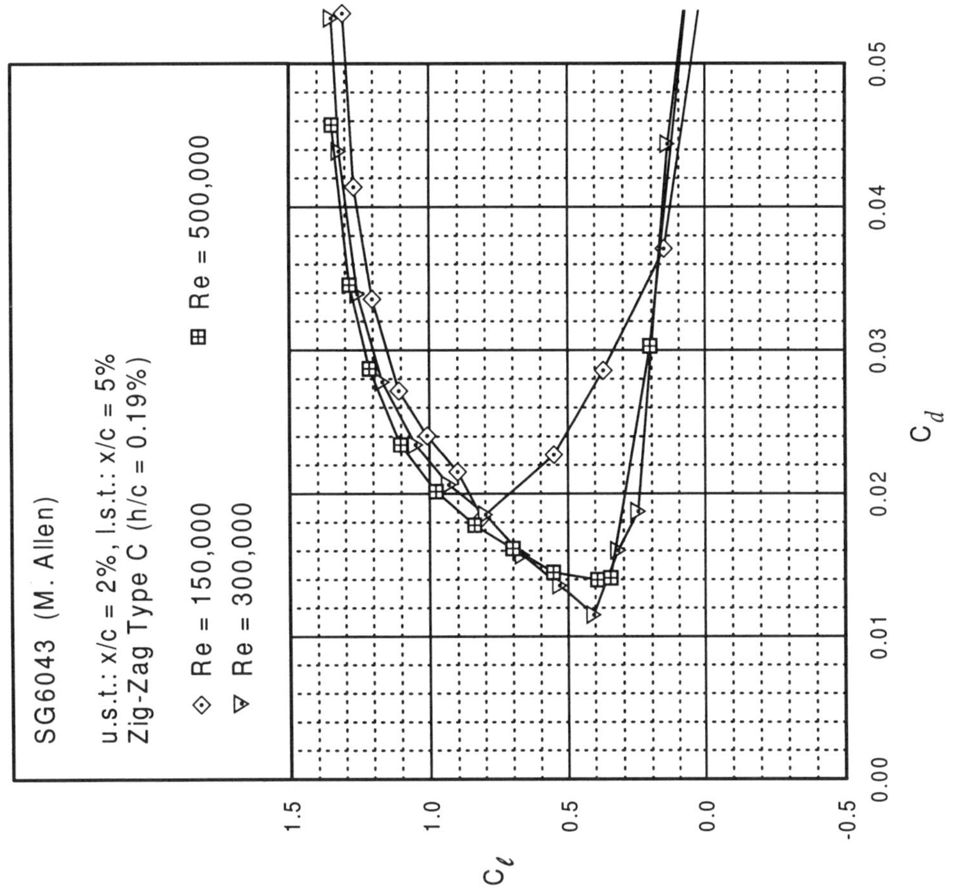
SG6043 (M. Allen)
u.s.t.: x/c = 2%, l.s.t.: x/c = 5%
Zig-Zag Type C (h/c = 0.19%)
Re = 150,000
Re = 300,000
Re = 500,000
1.5
1.0
0.5
0.0
-0.5
C_ℓ
0.00
0.01
0.02
0.03
0.04
0.05
C_d

SG6043

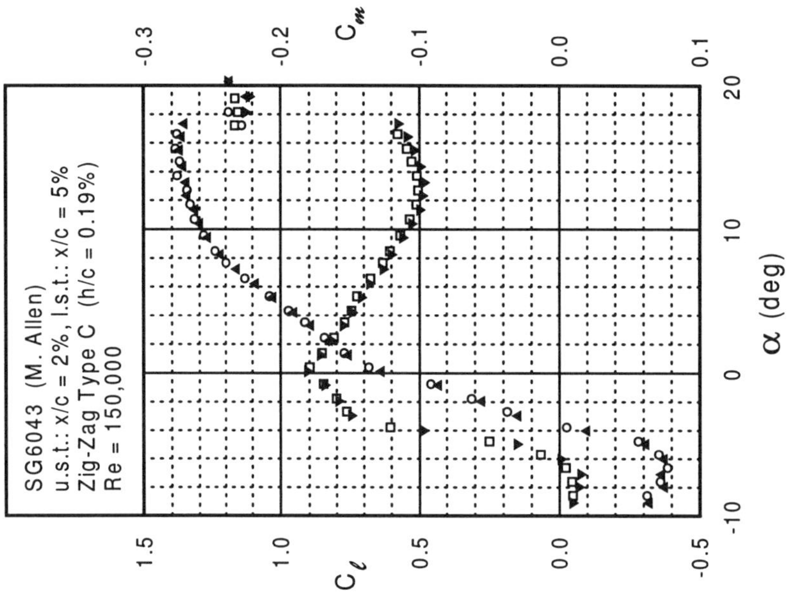
SG6043 (M. Allen)
u.s.t.: x/c = 2%, l.s.t.: x/c = 5%
Zig-Zag Type C (h/c = 0.19%)
Re = 150,000
C_m
C_ℓ
α (deg)

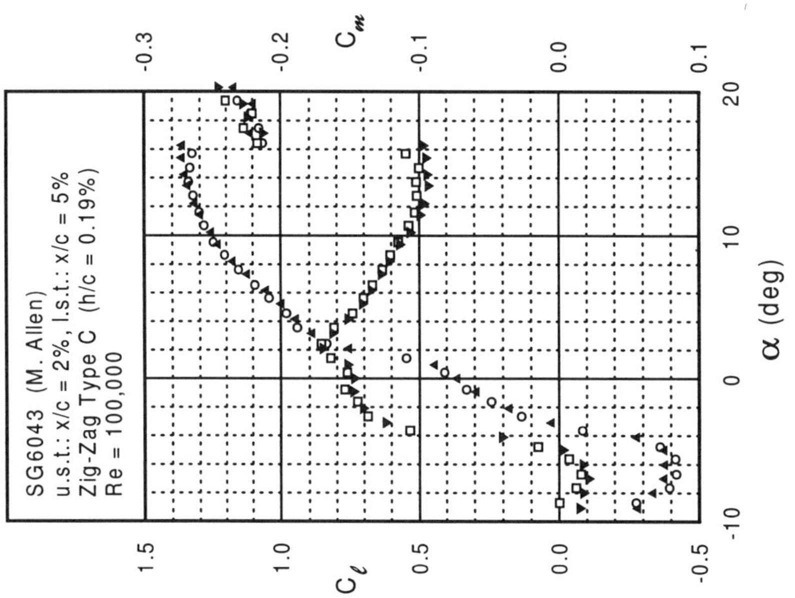
SG6043 (M. Allen)
u.s.t.: x/c = 2%, l.s.t.: x/c = 5%
Zig-Zag Type C (h/c = 0.19%)
Re = 100,000
C_m
C_ℓ
α (deg)

SG6043

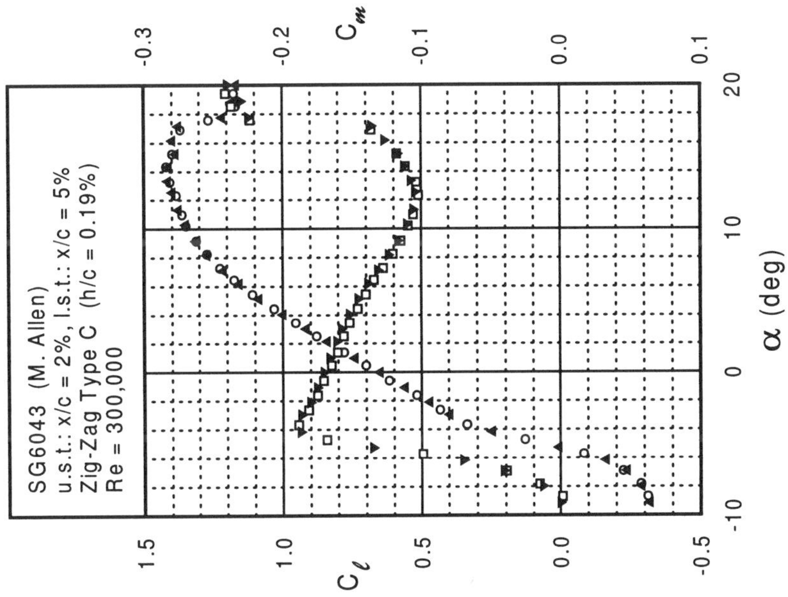
SG6043 (M. Allen)
u.s.t.: x/c = 2%, l.s.t.: x/c = 5%
Zig-Zag Type C (h/c = 0.19%)
Re = 300,000
C_m
-0.3
-0.2
-0.1
0.0
0.1
C_ℓ
1.5
1.0
0.5
0.0
-0.5
-10
0
10
20
α (deg)

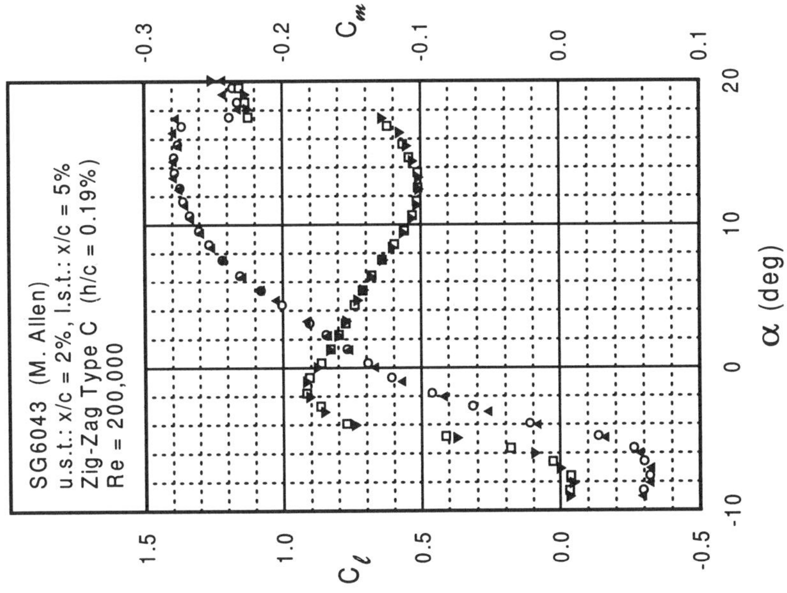
SG6043 (M. Allen)
u.s.t.: x/c = 2%, l.s.t.: x/c = 5%
Zig-Zag Type C (h/c = 0.19%)
Re = 200,000
C_m
-0.3
-0.2
-0.1
0.0
0.1
C_ℓ
1.5
1.0
0.5
0.0
-0.5
-10
0
10
20
α (deg)

SG6043

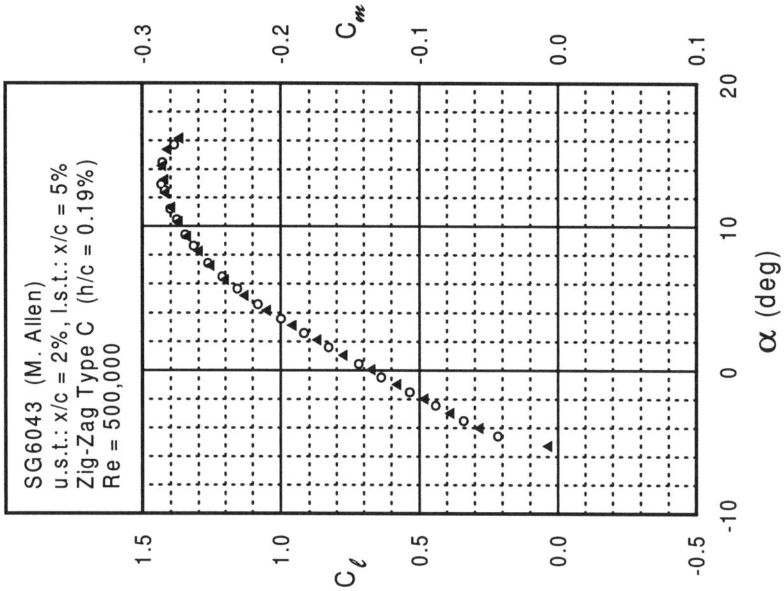
SG6043 (M. Allen)
u.s.t.: x/c = 2%, l.s.t.: x/c = 5%
Zig-Zag Type C (h/c = 0.19%)
Re = 500,000
C_l
C_m
α (deg)

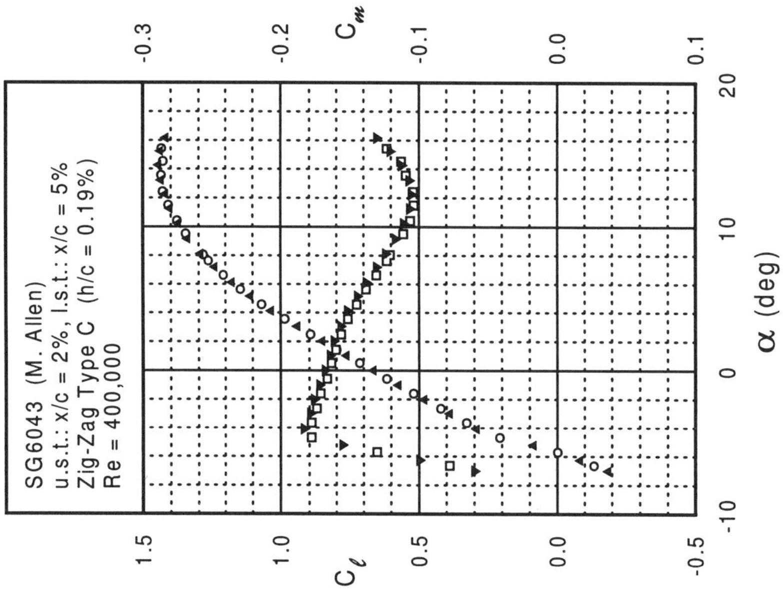
SG6043 (M. Allen)
u.s.t.: x/c = 2%, l.s.t.: x/c = 5%
Zig-Zag Type C (h/c = 0.19%)
Re = 400,000
C_l
C_m
α (deg)

Trainer 60

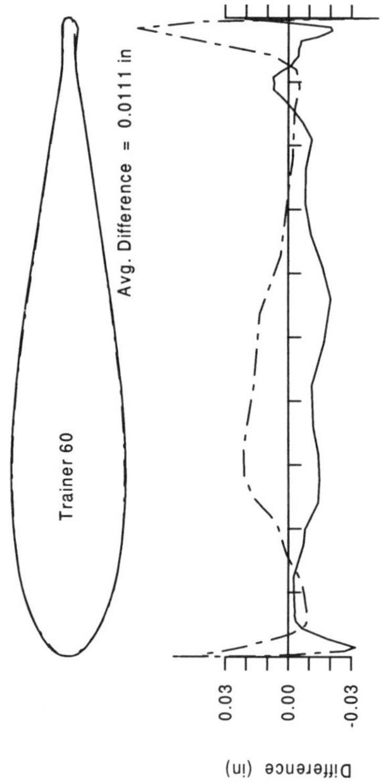
Trainer 60
Avg. Difference = 0.0111 in
0.03
0.00
-0.03
Difference (in)

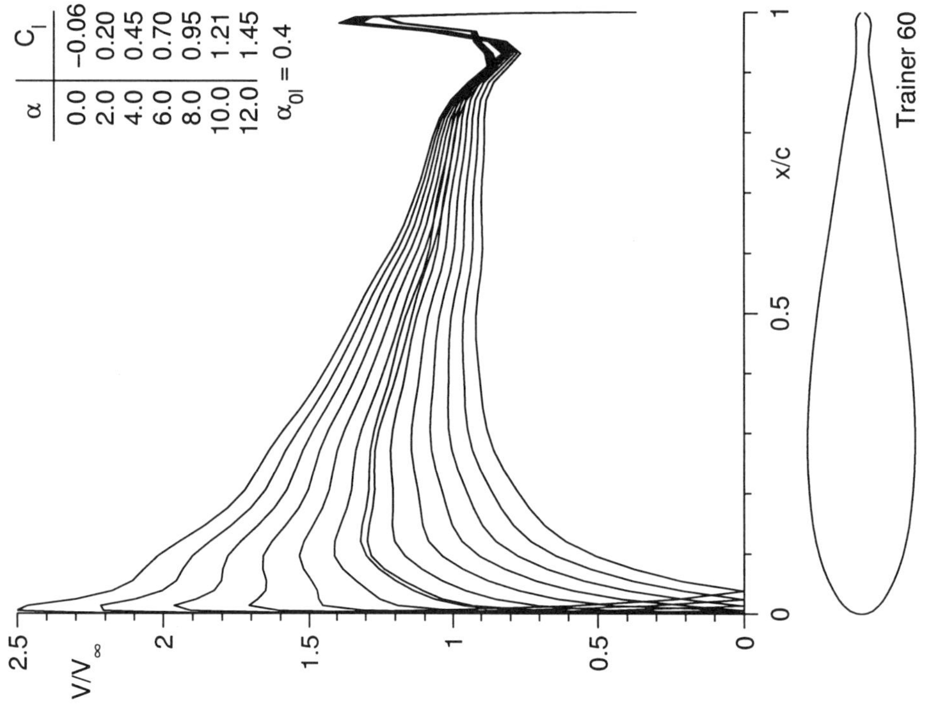
α
0.0
2.0
4.0
6.0
8.0
10.0
12.0
C_l
−0.06
0.20
0.45
0.70
0.95
1.21
1.45
$\alpha_{0l} = 0.4$
V/V_∞
2.5
2
1.5
1
0.5
0
0.5
1
x/c
Trainer 60

Trainer 60

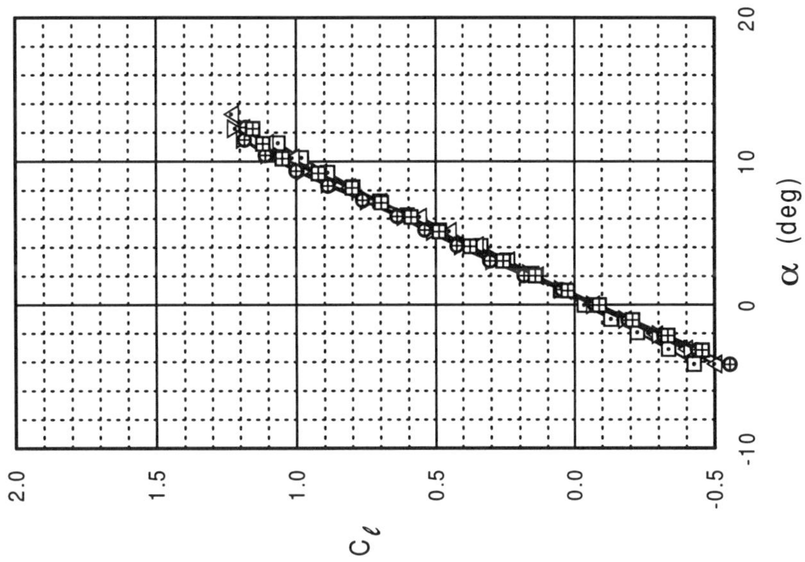
20
10
0
-10
α (deg)
2.0
1.5
1.0
0.5
0.0
-0.5
C_ℓ

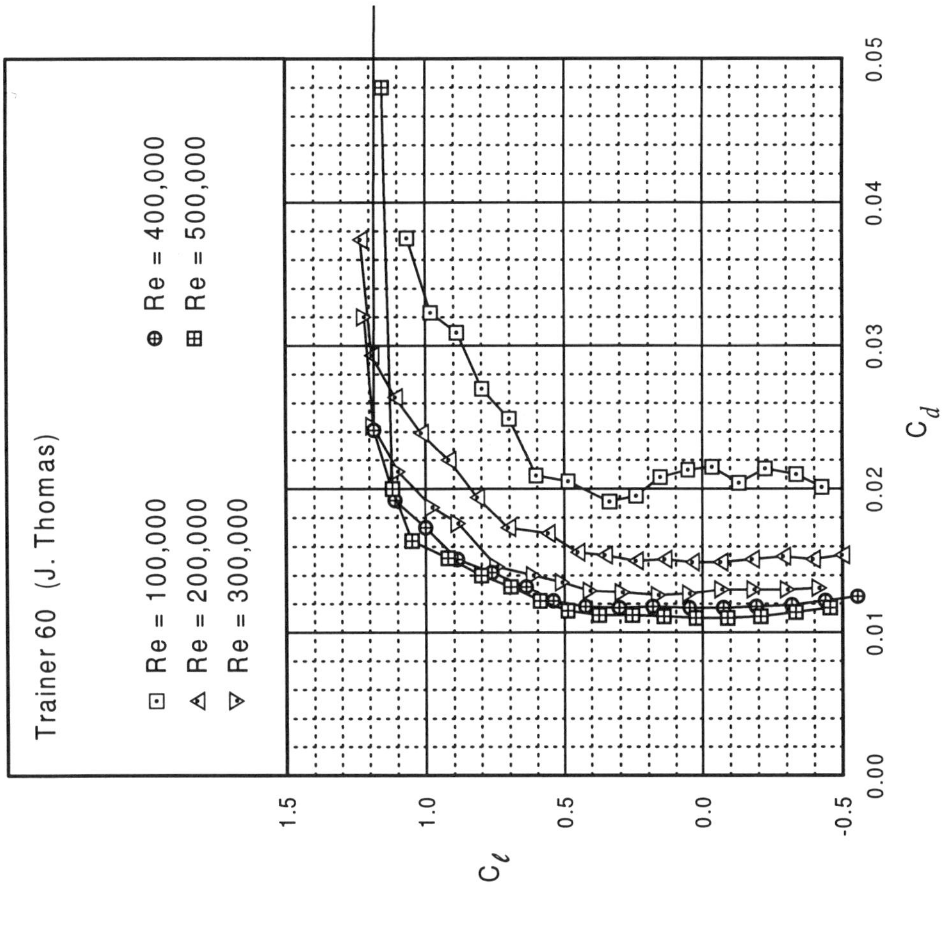
Trainer 60 (J. Thomas)
Re = 100,000
Re = 200,000
Re = 300,000
Re = 400,000
Re = 500,000
1.5
1.0
0.5
0.0
-0.5
C_ℓ
0.00
0.01
0.02
0.03
0.04
0.05
C_d

Trainer 60

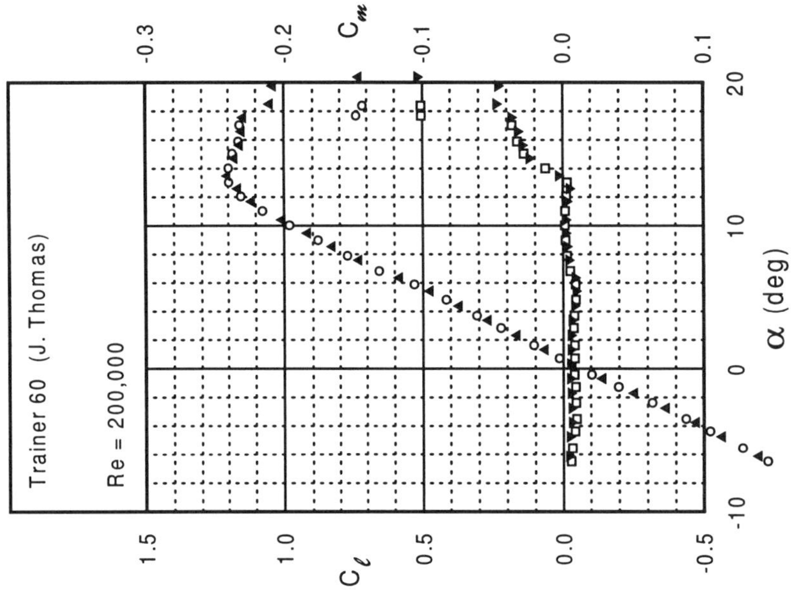
Trainer 60 (J. Thomas)
Re = 200,000
C_ℓ
C_m
α (deg)

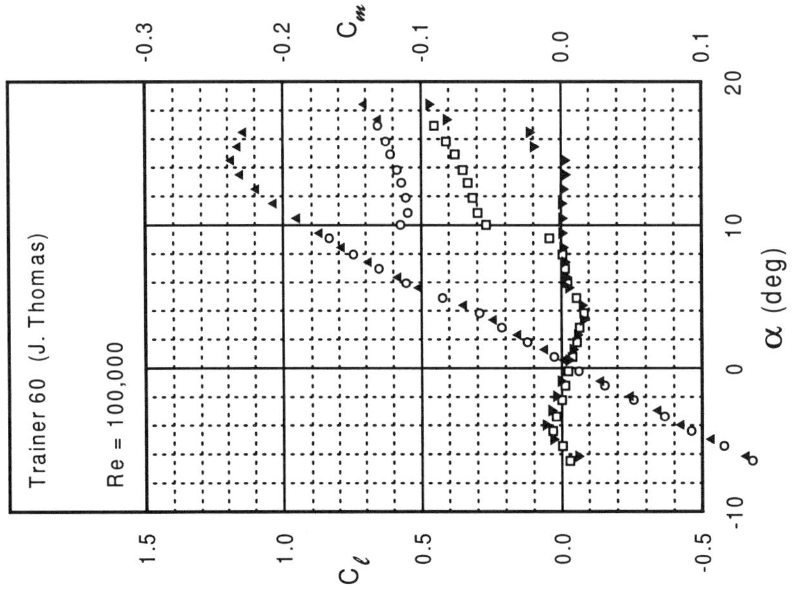
Trainer 60 (J. Thomas)
Re = 100,000
C_ℓ
C_m
α (deg)

Trainer 60

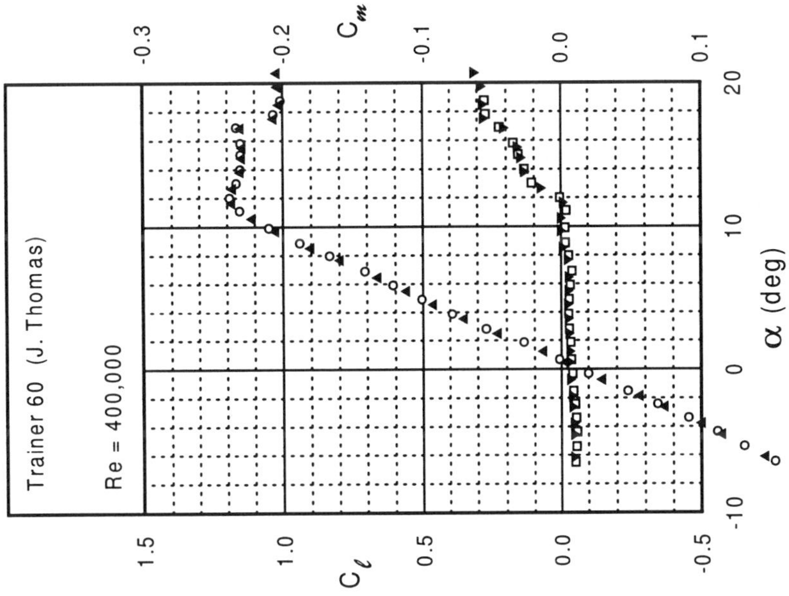
Trainer 60 (J. Thomas)
Re = 400,000
C_ℓ
C_m
α (deg)

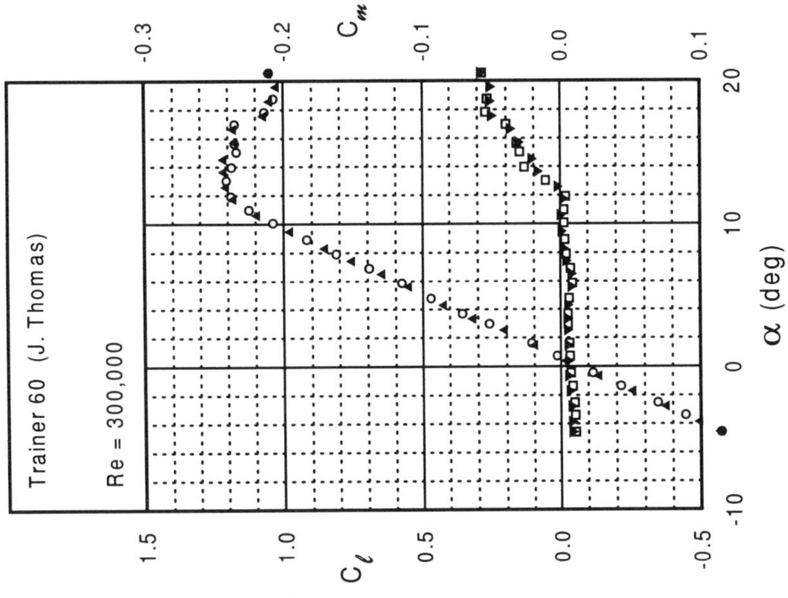
Trainer 60 (J. Thomas)
Re = 300,000
C_ℓ
C_m
α (deg)

Trainer 60

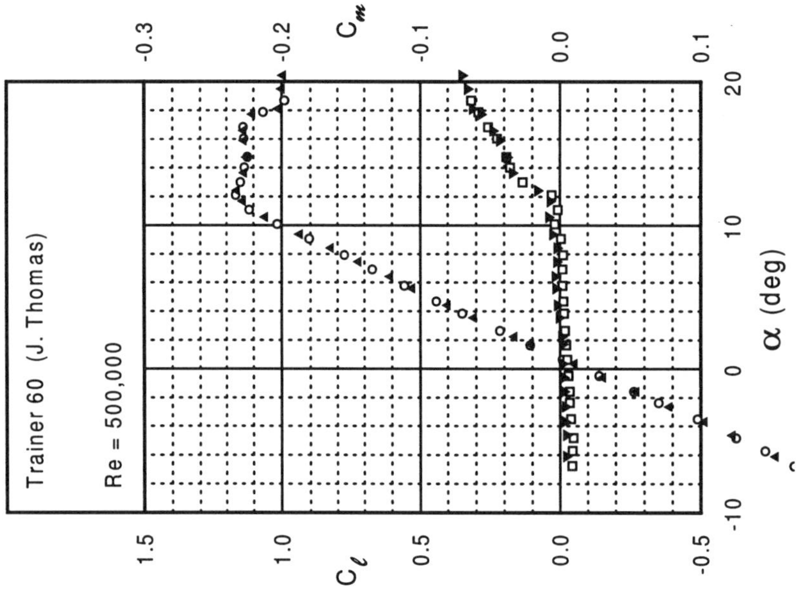

Ultra-Sport 1000

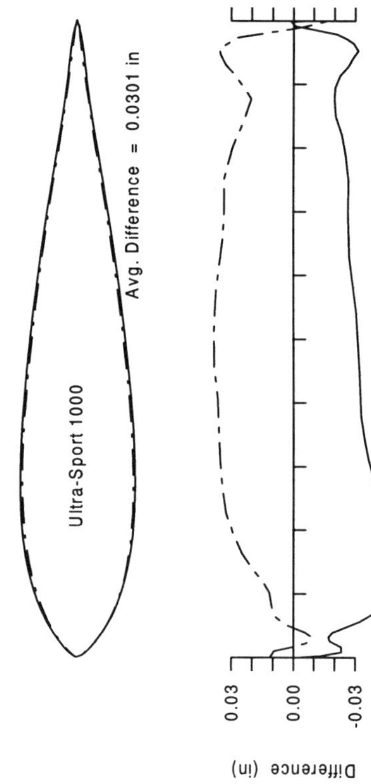

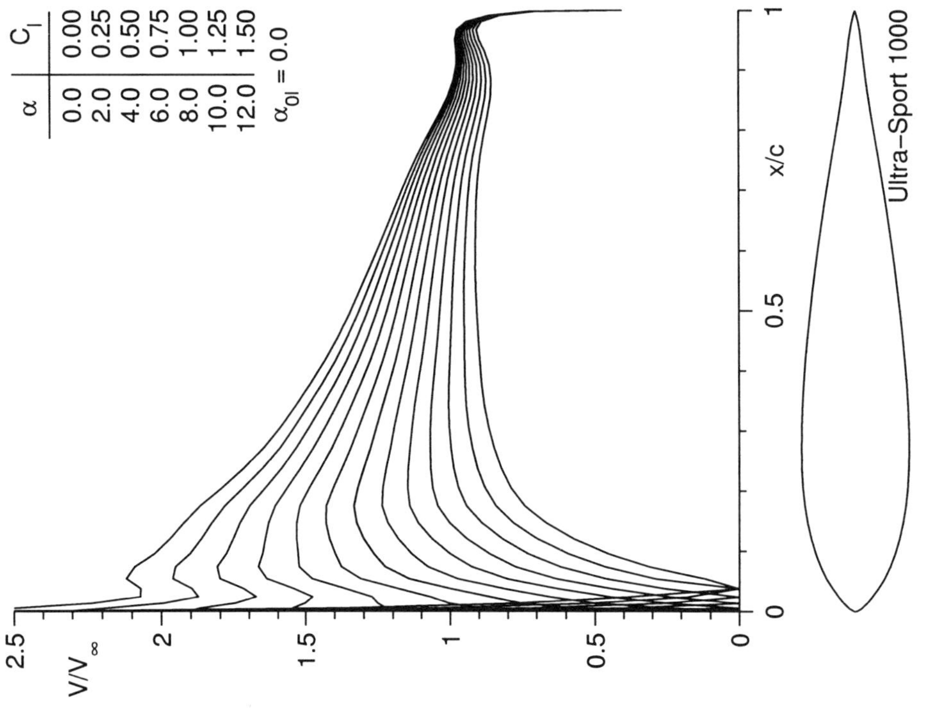

Ultra-Sport 1000

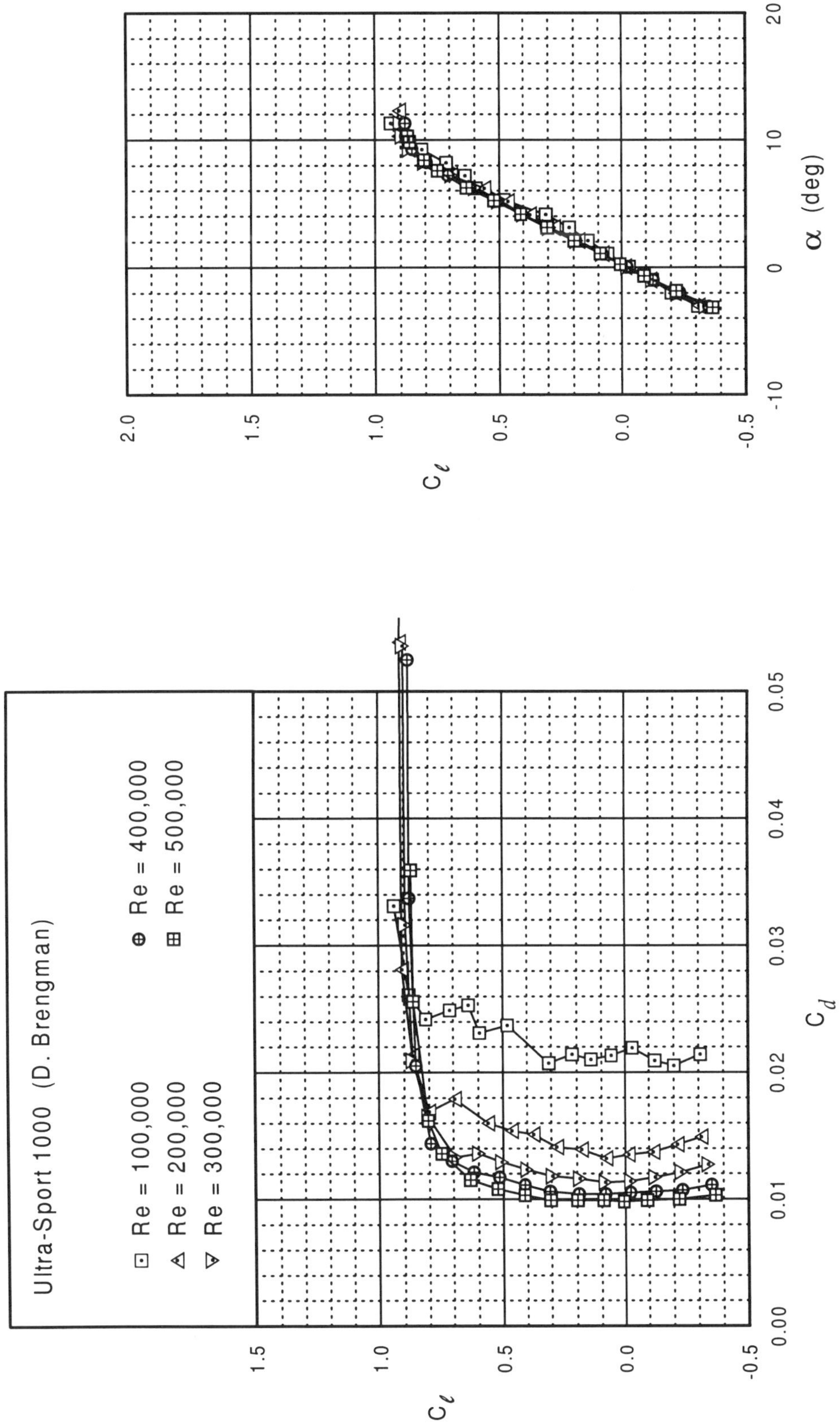

Ultra-Sport 1000

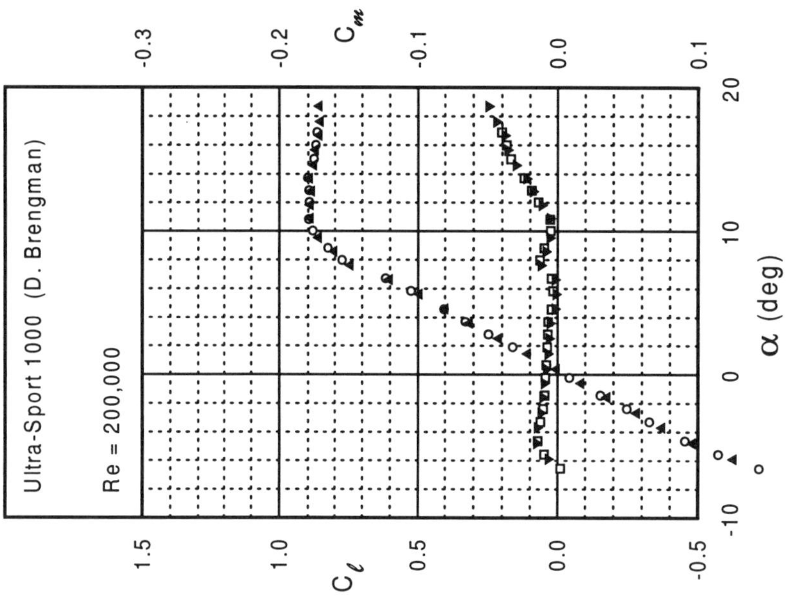

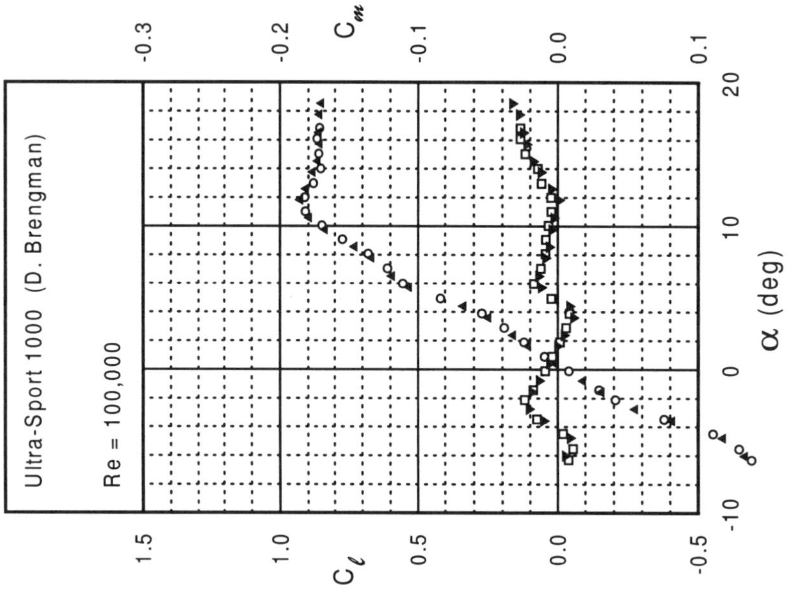

Ultra-Sport 1000

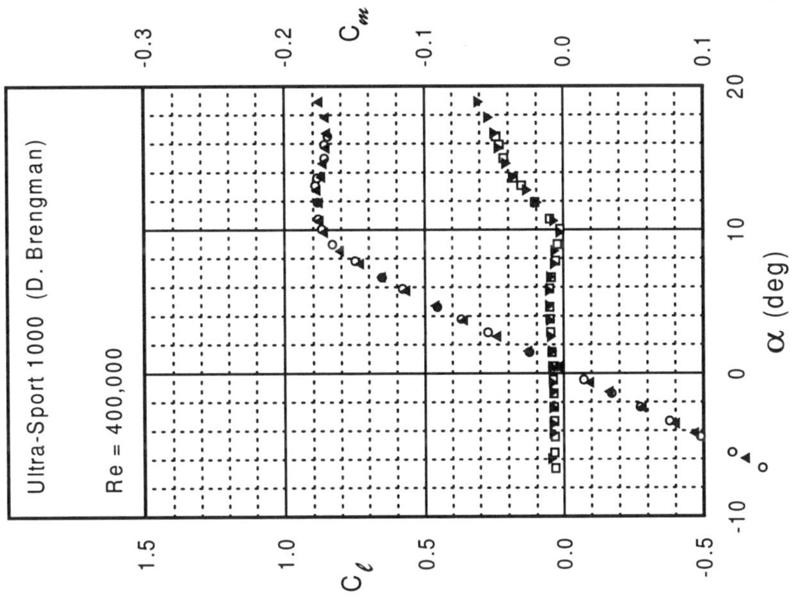

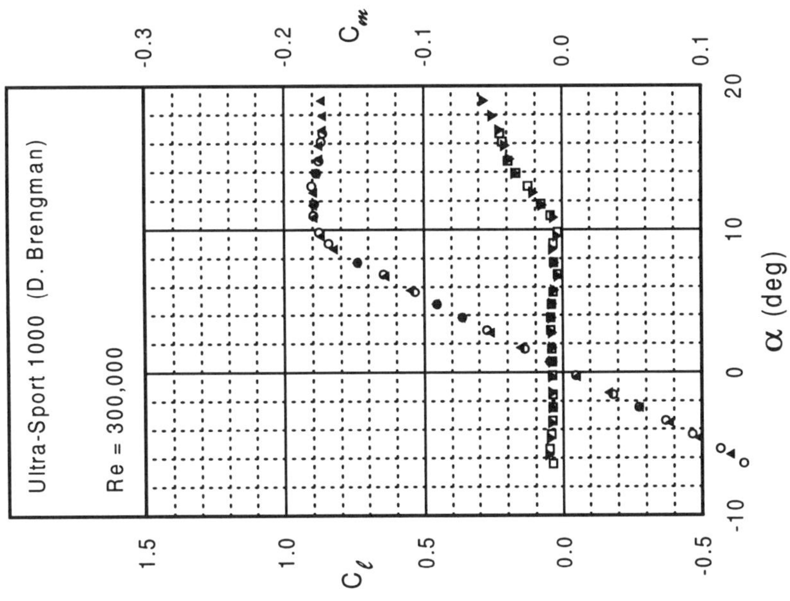

Ultra-Sport 1000

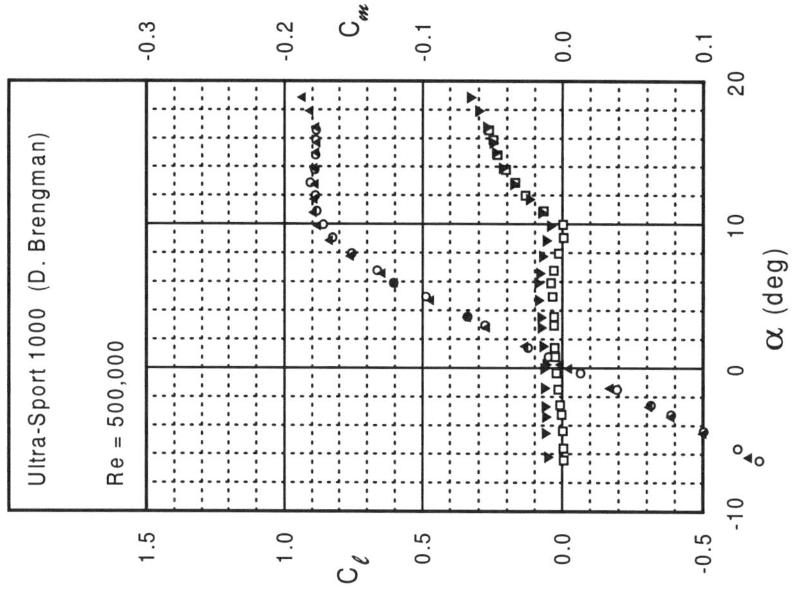

USNPS-4

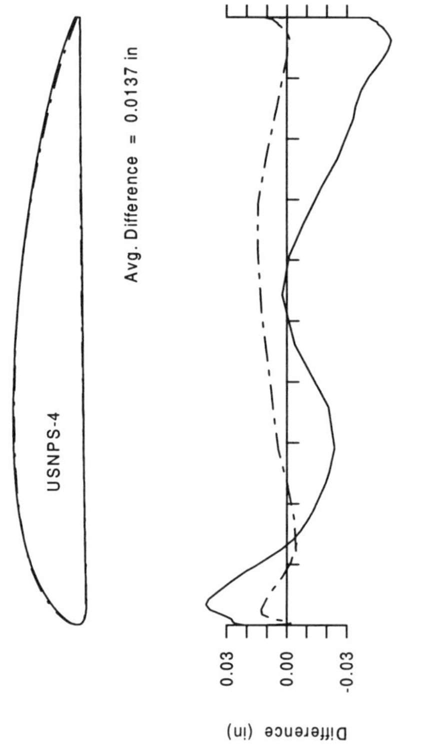
USNPS-4
Avg. Difference = 0.0137 in
Difference (in)
0.03
0.00
-0.03

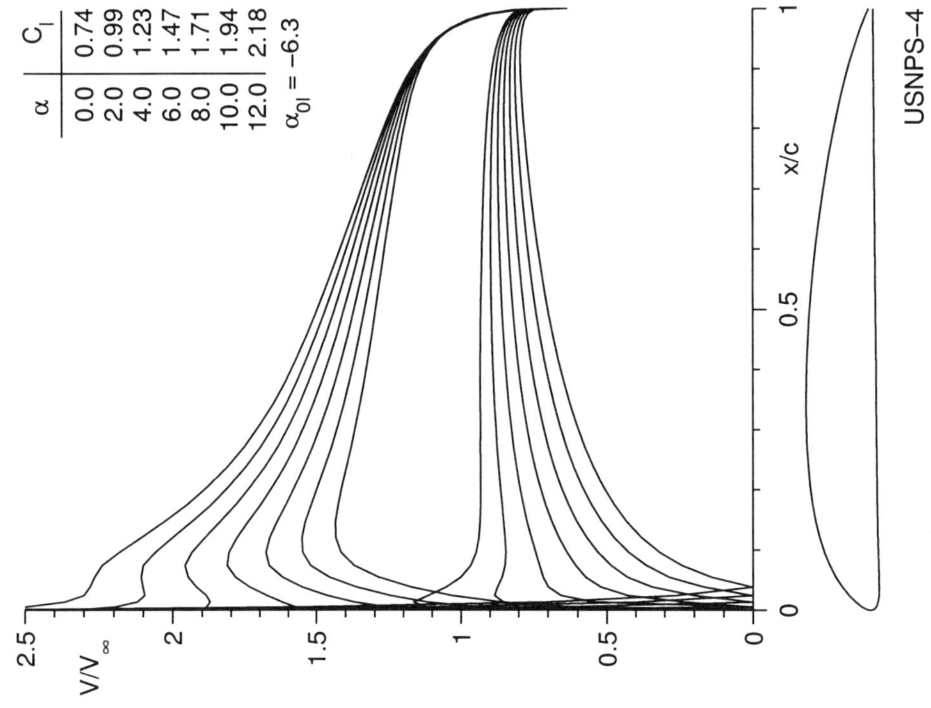
α
0.0
2.0
4.0
6.0
8.0
10.0
12.0
C_l
0.74
0.99
1.23
1.47
1.71
1.94
2.18
α_{0l} = −6.3
V/V_∞
2.5
2
1.5
1
0.5
0
x/c
0
0.5
1
USNPS–4

USNPS-4

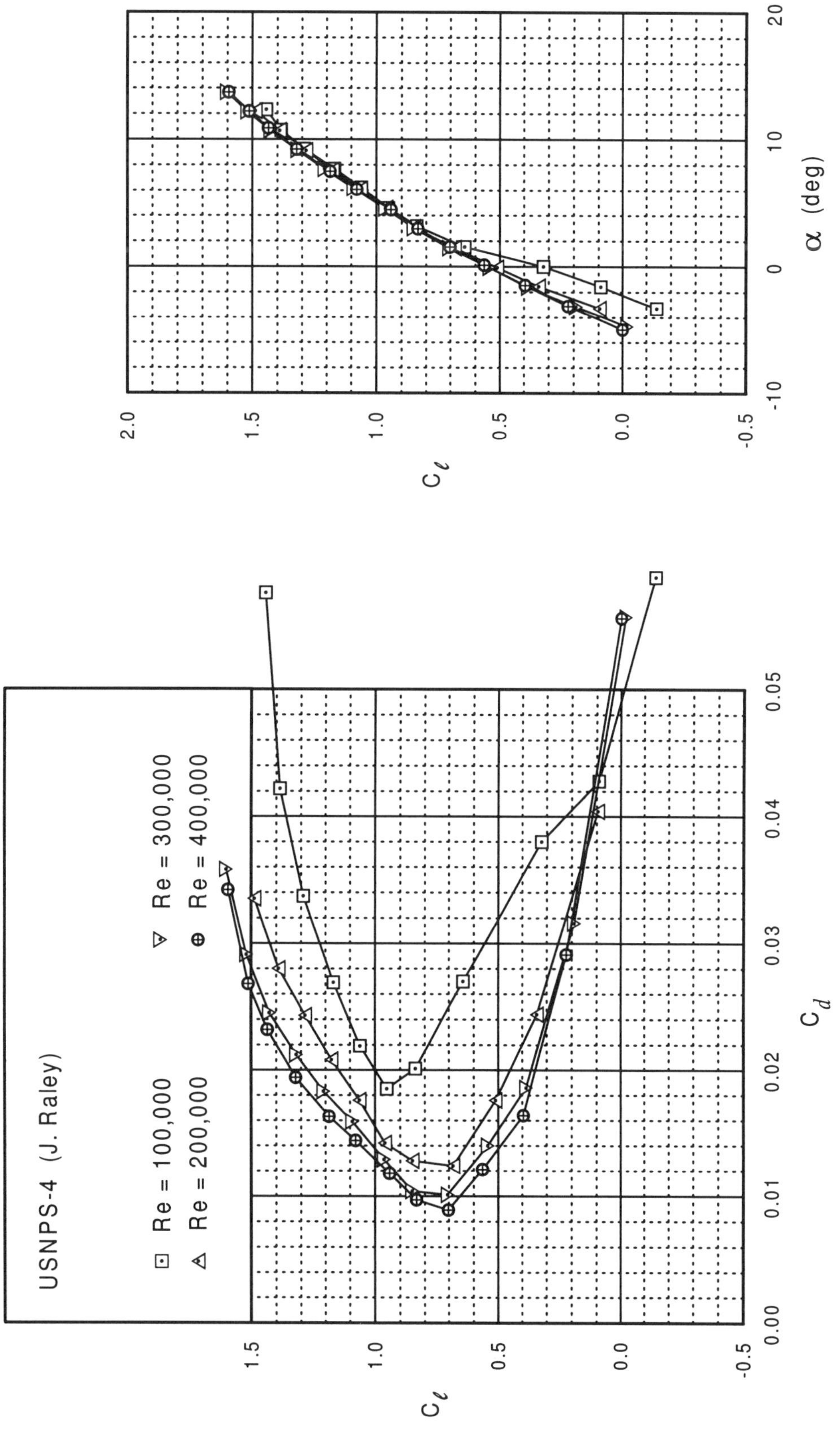
USNPS-4 (J. Raley)
Re = 100,000
Re = 200,000
Re = 300,000
Re = 400,000
C_ℓ
C_d
α (deg)
1.5
1.0
0.5
0.0
-0.5
0.00
0.01
0.02
0.03
0.04
0.05
2.0
-10
0
10
20

USNPS-4

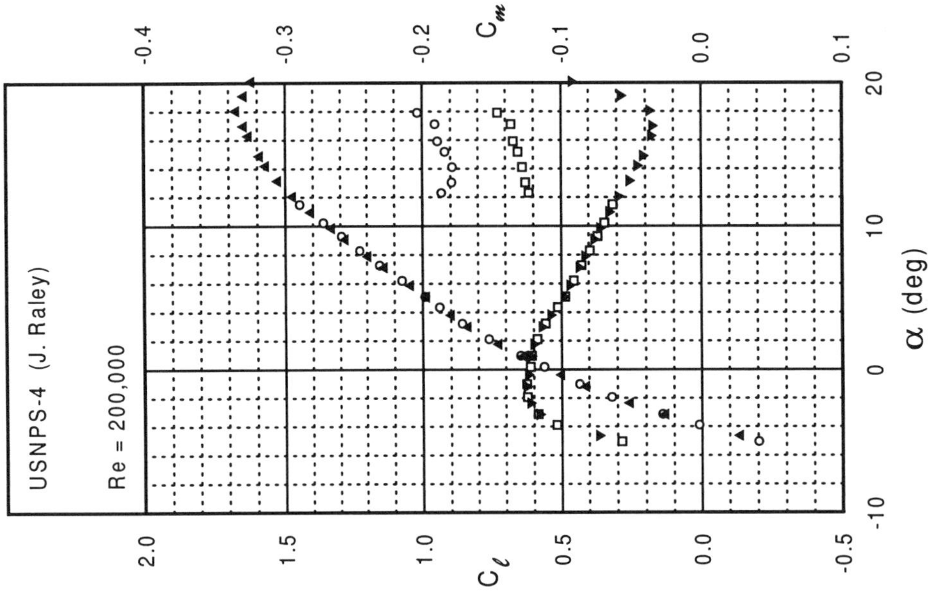
USNPS-4 (J. Raley)
Re = 200,000
C_l
C_m
α (deg)
2.0
1.5
1.0
0.5
0.0
-0.5
-0.4
-0.3
-0.2
-0.1
0.0
0.1
-10
0
10
20

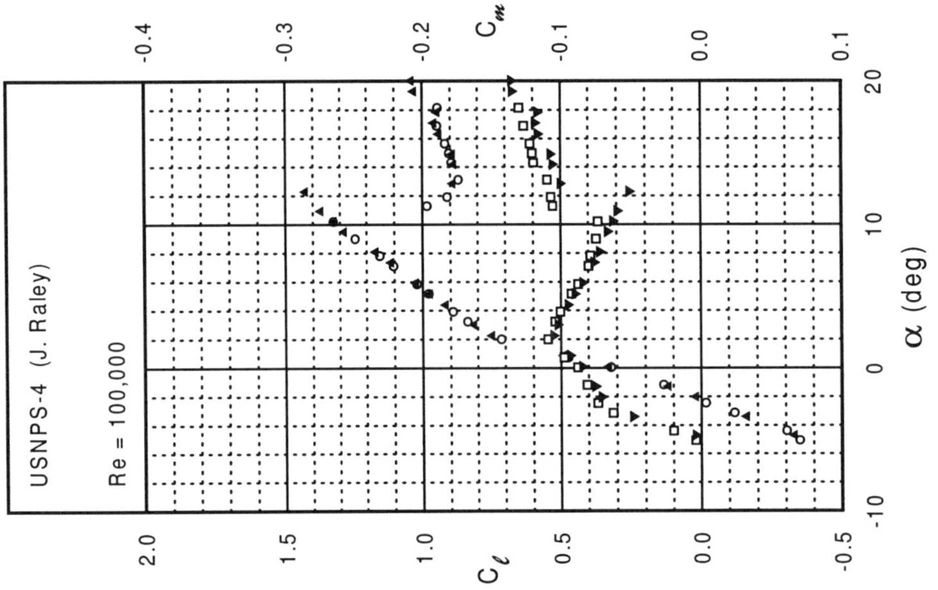
USNPS-4 (J. Raley)
Re = 100,000
C_l
C_m
α (deg)
2.0
1.5
1.0
0.5
0.0
-0.5
-0.4
-0.3
-0.2
-0.1
0.0
0.1
-10
0
10
20

USNPS-4

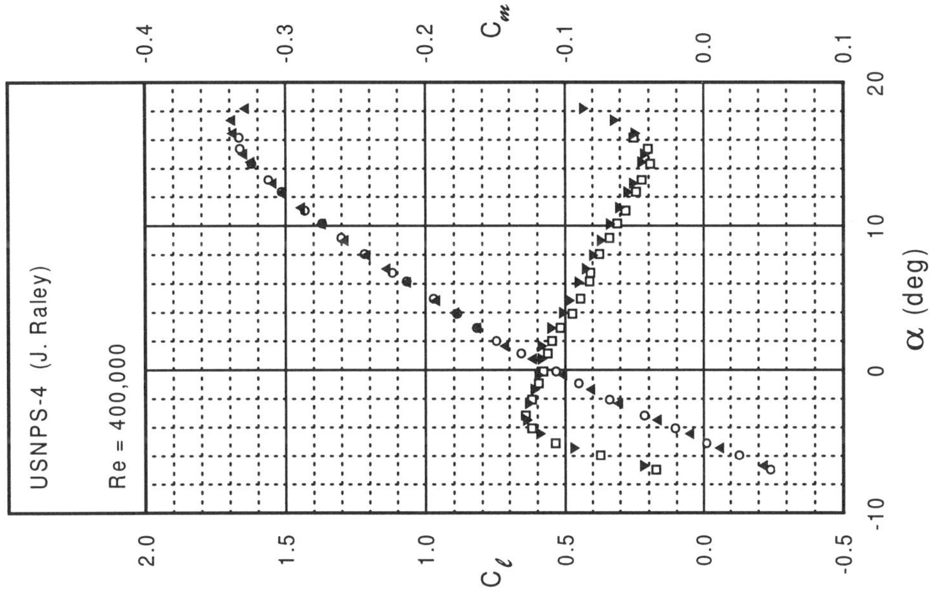
USNPS-4 (J. Raley)
Re = 400,000
C_ℓ
C_m
α (deg)
2.0
1.5
1.0
0.5
0.0
-0.5
-0.4
-0.3
-0.2
-0.1
0.0
0.1
-10
0
10
20

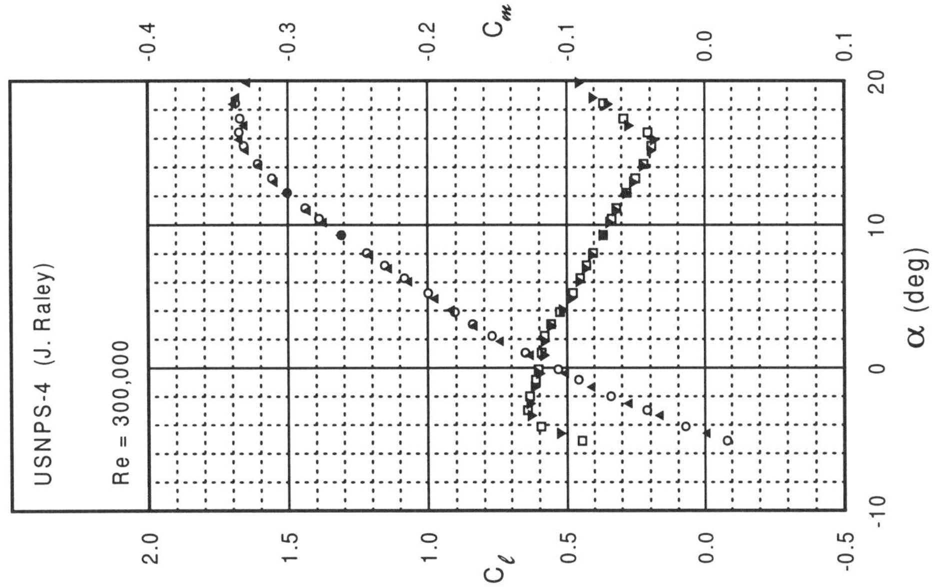
USNPS-4 (J. Raley)
Re = 300,000
C_ℓ
C_m
α (deg)
2.0
1.5
1.0
0.5
0.0
-0.5
-0.4
-0.3
-0.2
-0.1
0.0
0.1
-10
0
10
20

Chapter 6

Boundary-Layer Trips

As a consequence of laminar separation bubbles, low Reynolds number airfoils often develop high drag and extremely nonlinear lift curves. In an effort to prevent these bubbles from forming and subsequently improve low Reynolds number airfoil performance, mechanical turbulators such as trip wires, plain trips, backward facing steps, distributed grit roughness, zigzag tape, triangular patches, vortex generators, etc. are used to promote early transition. Unfortunately, a poor understanding of transition enhancement by means of boundary-layer trips makes trip optimization difficult. Moreover, a particular trip configuration optimized for one airfoil may prove less favorable for another.

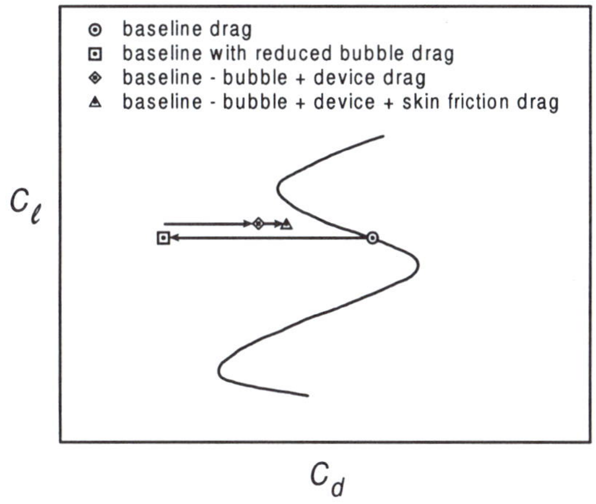

Adding BL Trips:
↓ Bubble drag.
↑ Device Drag.
↑ Skin Fr. Drag.

Fig. 6.1 Conceptual illustration of trip effects.

Essential to unraveling the secrets behind boundary-layer trip performance is an understanding of their three main effects on airfoil performance: added device drag, changes in bubble drag (pressure drag), and changes in skin-friction drag. Figure 6.1 illustrates these effects and shows how their corresponding changes in drag combine to reduce the baseline airfoil drag. Through improper trip design, however, these effects could just as easily produce an increase in baseline drag. Therefore, by determining the relative magnitudes of these effects, a measure of trip effectiveness can be obtained. For example, an extremely effective trip produces large reductions in bubble drag with minimal increases in device or skin-friction drag. Conversely, a poorly performing trip produces large increases in skin-friction and device drag with little reduction in bubble drag. (Figure 6.1 shows only the case where a trip would be beneficial.) As previously mentioned, however, a lack of understanding regarding the influence of trips on transition has hindered accurate predictions of these three effects.

In an effort to provide better insight into the parameters influencing trip performance, an experimental investigation was conducted on a range of trip configurations. In particular, various numbers of trips, trip heights, locations, widths, and geometries were tested on several airfoils with an emphasis placed on obtaining detailed, quantitative data for comparative purposes. Presented here are results obtained for the E374 and SD7037, two airfoils which are fairly typical of the majority of sailplane airfoils currently in use. For the interested reader, additional extensive trip data on the Miley (M06-13-128) airfoil is given in Ref. 12.

Due to the time consuming nature of the experiments, generation of complete drag polars for each trip configuration was impractical. Therefore, the collection of data was limited to a single angle of attack corresponding to the middle of the polar for each airfoil. These were chosen to be 3° and 2.2° for the E374 and SD7037 respectively.

6.1 Trip Locations and Conventions

Figure 6.2 shows each airfoil with its various possible trip locations. The models used in this study were the SD7037 (D) with upper-surface mylar covering and the E374 (B). Both models had 12-in. chords. In order to facilitate comparisons between the two airfoils, all trips were located relative to laminar separation locations as predicted by XFOIL.[4] In this manner, the first trip location (closest to the bubble) was 0.1 in. upstream of the *predicted* bubble with succeeding trips located 1 in. further upstream along the airfoil surface. Unless otherwise noted, the reference line for all trips was the trip trailing edge. For example, the trailing edge of a trip at location 1 would be 0.1 in. upstream of the predicted laminar separation point.

Fig. 6.2 Trip locations on the E374 and SD7037.

6.2 Two-Dimensional Trips

In this section, data will be presented for various 2-D trip configurations tested on both the E374 and SD7037 at various Reynolds numbers. (For the purpose of this study, a trip is considered 2-D when its width, height, and location remain constant along the span of the model.) Trips were constructed by laminating together several layers of 3M Scotch Brand C-4210 Removable Magic Tape. This produced a 0.5-in. wide strip with a thickness that could be varied in 0.002-in. increments. Depending on the desired trip configuration, a specified number of these laminated strips were then located at the various chordwise positions shown in Fig. 6.2. By applying one or more of these strips, single or multiple trips could be created as conceptually illustrated in Fig. 6.3. To avoid confusion when comparing between the numerous configurations, all trips are referred to by their locations as indicated in Fig. 6.2 with a suffix denoting either single or multiple trips: 'S' for single and 'M' for multiple.

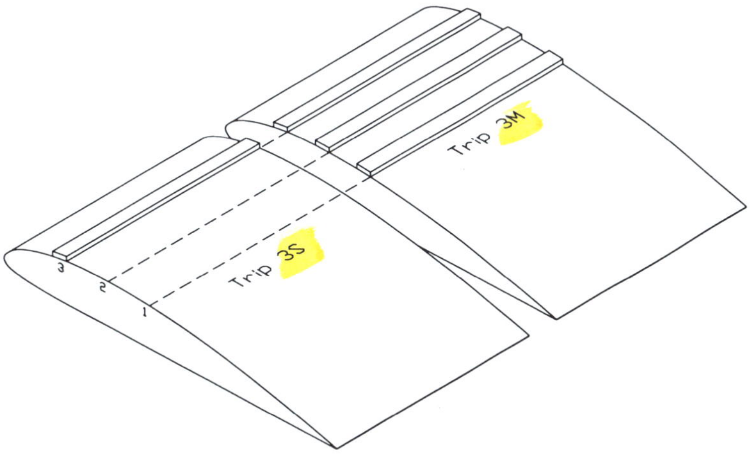

Fig. 6.3 Illustration of single and multiple 2-D trips.

Figure 6.4 shows E374 drag data for several trip heights and locations at $Re = 100{,}000$ for both single and multiple trip configurations. Baseline (untripped) drag values have been plotted as dotted lines for easy reference, and the data flag '1' is used to denote trips 1S or 1M, '2' for trips 2S or 2M, etc. The most striking feature of both the single and multiple trip configurations are the very large reductions in drag for extremely thin trips (excluding trip 1, which was located inside the bubble owing to discrepancies between predicted and actual bubble locations). As an example, trip 3S with $h = 0.002$ in. approximately produces a 25% reduction in drag over the baseline value. For an appreciation of how thin this trip is, a 0.002-in. edge is at the limit of what a person can feel. Knowing this, it seems surprising that a trip this thin is capable of reducing drag by 25%. Such drag reductions suggest that better care should be taken when locating mylar seams during the model covering process.

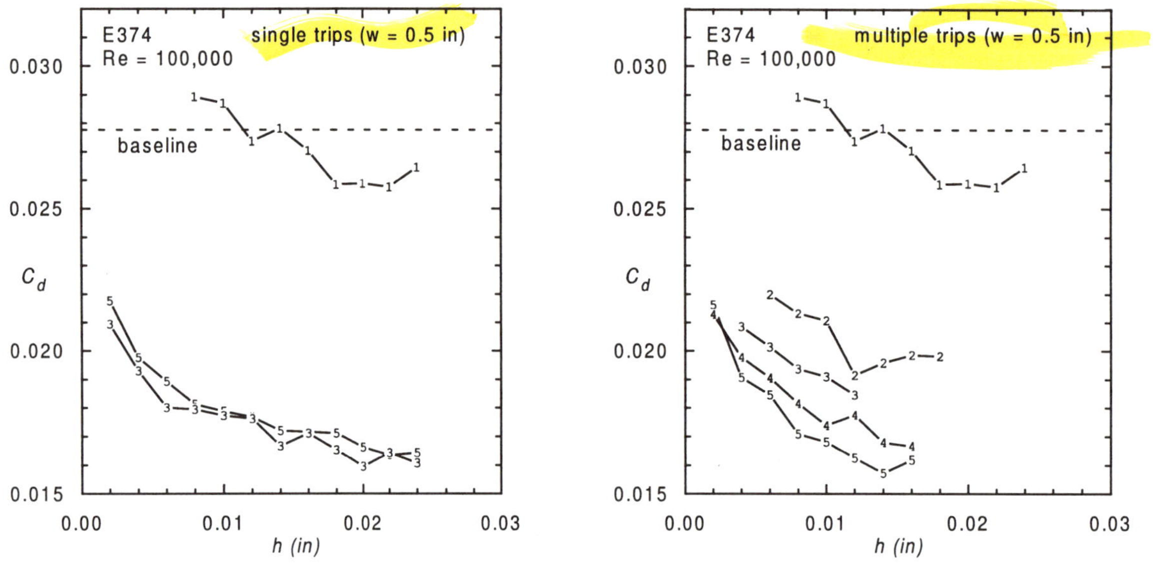

Fig. 6.4 E374 drag data for both single and multiple 2-D trips (Re = 100,000).

While 25% may seem like a large drag reduction, even larger reductions are produced as trip height is increased, with the largest reduction (43%) produced by trip 5M at $h = 0.014$ in. While this trend of continually decreasing drag suggests an increase in trip performance with increasing trip height, it also indicates a failure by the trips to completely eliminate any occurrences of laminar separation. While it is rather common to find trips that do not eliminate laminar separation bubbles, it is surprising that this holds true for some of the thicker trips tested in this study. For example, trip 5S with $h = 0.020$ in. is an extremely thick trip placed at $x/c = 13\%$. One would expect that such a large trip placed so far upstream would be quite effective in eliminating any bubble. The data, however, suggests otherwise since a a slightly thicker trip further reduces drag.

As the Reynolds number is increased to 200,000 (Fig. 6.5), the same trends that are present at Re = 100,000 are seen but on a much smaller scale. This can be attributed to the smaller potential for drag reductions owing to the presence of smaller laminar separation bubbles. Increasing the Reynolds number to 300,000 (Fig. 6.6) results in an even smaller drag reduction, with trip configurations 5S and 5M producing an increase in drag over the baseline for large thicknesses. In instances when drag becomes higher than the baseline value, it is reasonable to speculate that the bubble is either completely eliminated or significantly reduced in size and that any subsequent change in C_d is a result of increasing skin friction and device drag.

Data for the SD7037 was also taken with both single and multiple trips but only at Re = 200,000. These results are shown in Fig. 6.7 and suggest that trips perform poorly on this airfoil at this condition. Realizing that the SD7037 has a smaller bubble than the E374, these results are consistent with the conclusion drawn from

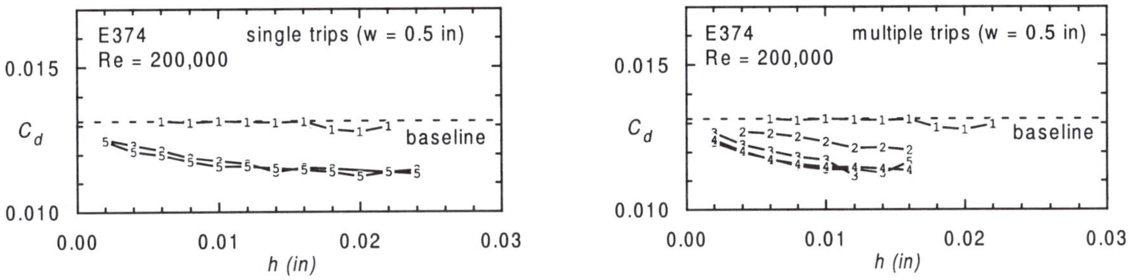

Fig. 6.5 E374 drag data for both single and multiple 2-D trips (Re = 200,000).

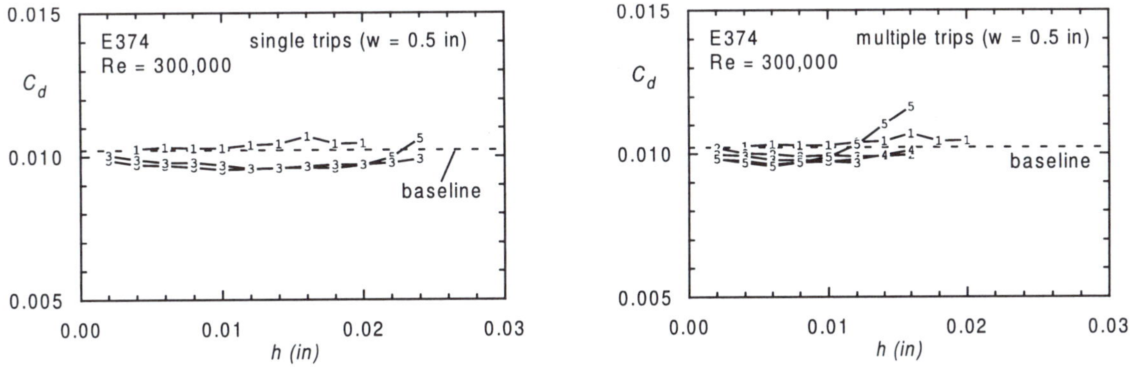

Fig. 6.6 E374 drag data for both single and multiple 2-D trips (Re = 300,000).

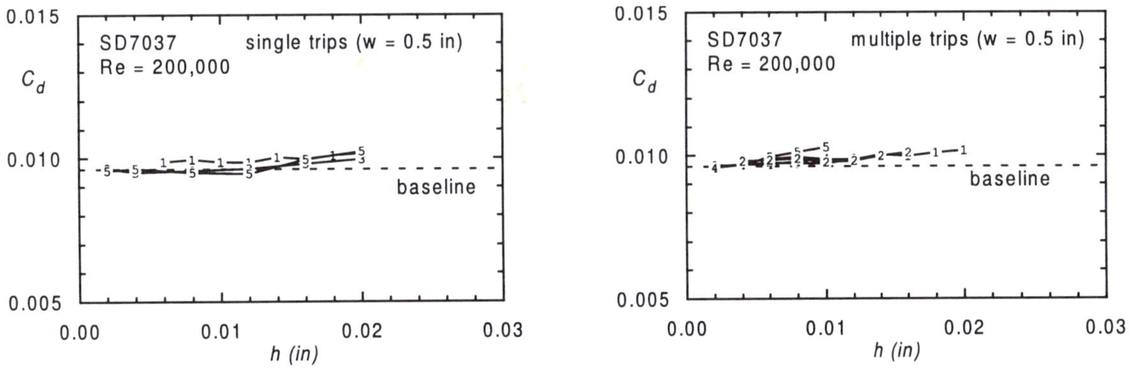

Fig. 6.7 SD7037 drag data for both single and multiple 2-D trips (Re = 200,000).

the E374 data. The presence of smaller bubbles offer less potential for drag reductions. It should be noted that despite the drag reductions trips generate for the E374 at Re = 200,000, the SD7037 still produces lower drag with or without trips.

6.2.1 Effect of Trip Width

The previous data was obtained for a trip width of 0.5 in. From an applications standpoint, this is a much wider trip than is often implemented. Therefore, it was decided to determine the effect of decreasing trip width while maintaining the same trip height and location (i.e., the trailing edge of the trip was held at a fixed chordwise location while the leading edge moved downstream as the trip narrowed). Figure 6.8 shows the effect of various trip widths on the E374 with the trips located at position 3 (see Fig. 6.2). The data supports the conclusion that trip width has little effect on trip performance. In other words, a key parameter in determining trip performance is the location of trip trailing edge.

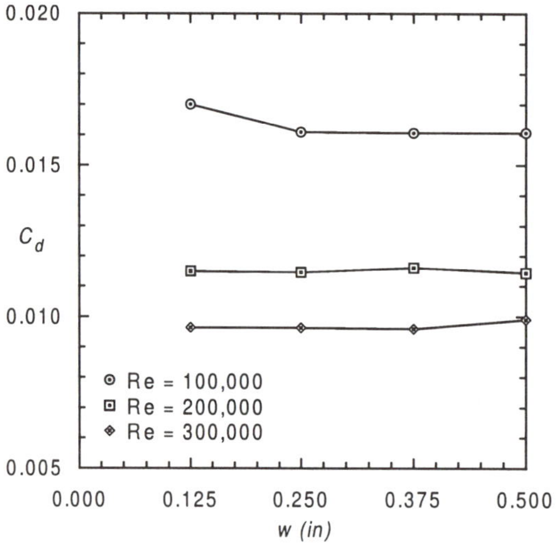

Fig. 6.8 E374 drag data for various widths of single, 2-D trips ($h = 0.013$ in; $Re = 100{,}000$, 200,000, and 300,000).

6.2.2 Effect of Trip Location

For a better understanding of the role that trip location plays in determining trip performance, a study was conducted with a trip of $w = 0.125$ in. and $h = 0.013$ in. placed at various locations along the E374. Figure 6.9 shows the results for $Re = 100{,}000$, 200,000, and 300,000. The x-axis is such that trips positioned furthest downstream are to the right of the graph with s denoting the distance between the trailing edge of the trip and the point 0.1 in. upstream of the XFOIL predicted laminar separation location. Thus, $s = 0$ in. corresponds to 46% chord and $s = 4$ in. corresponds to 13% chord. It should be mentioned that flow visualization showed most of the downstream trips (locations '1' and '2': $s = 0 - 1$ in.) to be submerged within the bubble. As a result, these trips have very little effect on airfoil performance.

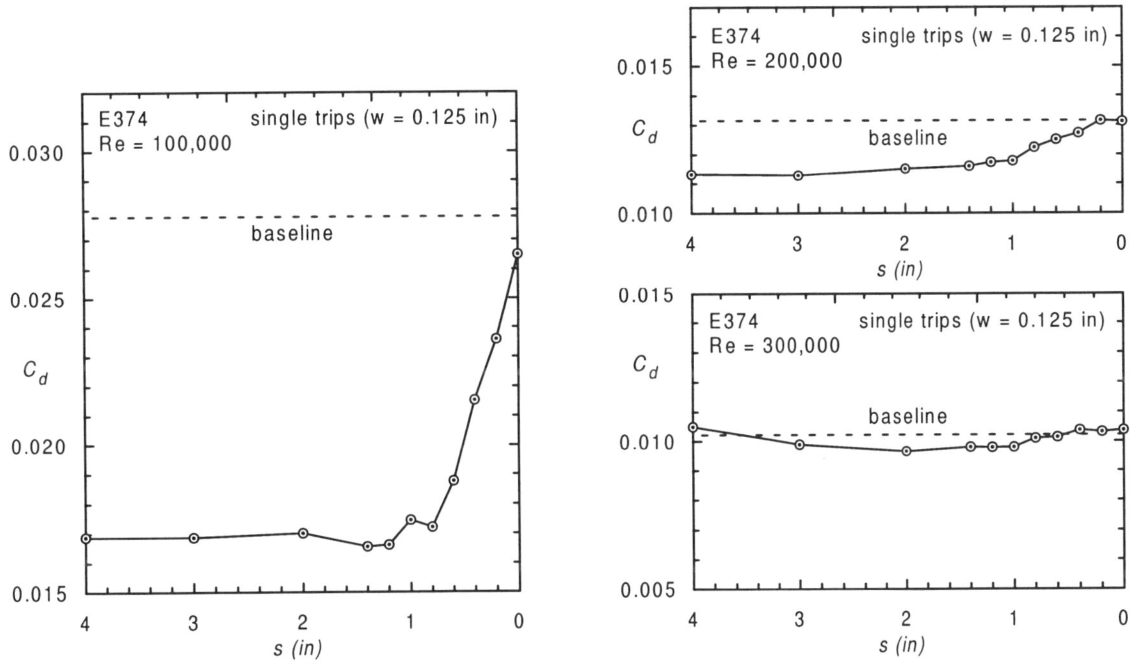

Fig. 6.9 E374 drag data for various locations of single, 2-D trips (h = 0.013 in; w = 0.125 in; Re = 100,000, 200,000, and 300,000).

For Re = 100,000 and 200,000, changes in drag do not occur as the trip moves further upstream past $s = 1$ in. This insensitivity to trip location has been attributed to tradeoffs between a reduction in bubble drag and a corresponding increase in skin-friction and device drag. As the Reynolds number is increased to 300,000 however, upstream trip locations result in drag increases. As expected, the tradeoffs present at the lower Reynolds numbers no longer exist at Re = 300,000. To be more specific, reductions in bubble drag become insignificant relative to increases in skin-friction and device drag. From an applications standpoint, this data suggest two things. First, if the operating condition of the E374 (or similar airfoil) is such that it will remain below Re = 300,000, best performance for $\alpha = 3°$ can be achieved by placing the trip as far upstream as possible. Second, if such an airfoil is going to operate above Re = 300,000, trip placement is critical in achieving the best performance.

Figure 6.10 shows surface oil flow visualization taken at Re = 200,000 for the trip in Fig. 6.9 located at x/c = 22% (s = 3 in.). This location corresponds to the middle of the constant low-drag region shown in Fig. 6.9. Despite being a low drag configuration, a relatively large bubble can be seen from x/c = 45% to near x/c = 61%. (The clean configuration produces a bubble from x/c = 42% to 72%.) Of particular interest is the secondary bubble emanating from the rear face of the trip. It is speculated that the presence of this bubble (separated zone) helps to initiate some of the instabilities that shorten the downstream, primary bubble and thereby reduce overall drag.

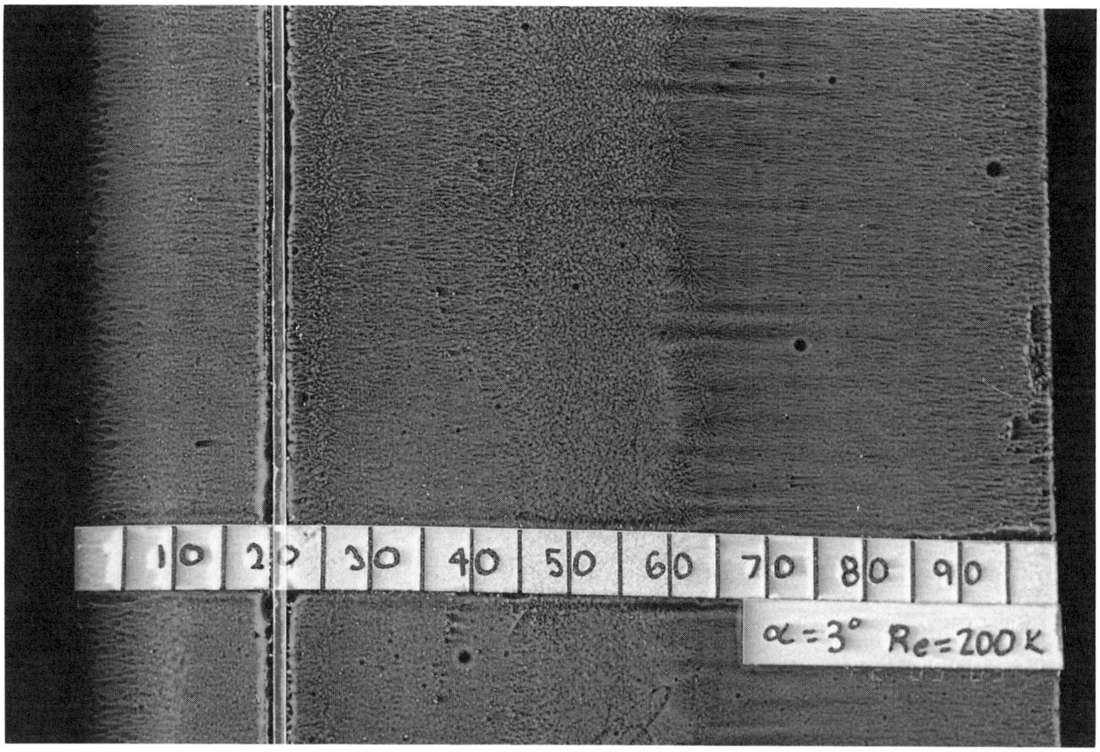

Fig. 6.10 Surface-oil flow visualization of major flow features on a tripped E374 ($\alpha = 3°$; Re = 200,000).

6.3 Three-Dimensional Trips

It has been widely accepted that 3-D trips are better at initiating transition at the higher Reynolds numbers than 2-D trips. For the range of Reynolds numbers used for this study, however, the potential performance advantage of 3-D trips over 2-D trips is unclear. Therefore, in an effort to determine the relative merits of 3-D trips, a study was conducted comparing various geometries of 3-D trips to both the single and multiple trip data previously discussed.

Table 6.1 illustrates the trip geometries tested along with their heights and datum lines. All trips were tested on the E374 at location 3 for a Reynolds number of 200,000, while a few were also tested for Re =100,000 and 300,000. The zigzag trips labeled as "large" and "narrow" correspond to trip types D and E respectively (see Table 4.2). Owing to the difficulty in constructing repeatable 3-D shapes, only a limited range of trip heights were tested.

Figure 6.11 shows the drag each configuration produces as well as baseline drag and drag for trips 3S and 3M. For Re = 100,000 the raised hemispheres produce the largest drag reduction of all trips tested. This can be attributed to the raised

Table 6.1: Three-Dimensional Trips Tested

Trip Name	Trip Geometry†	Trip Heights
small Hama	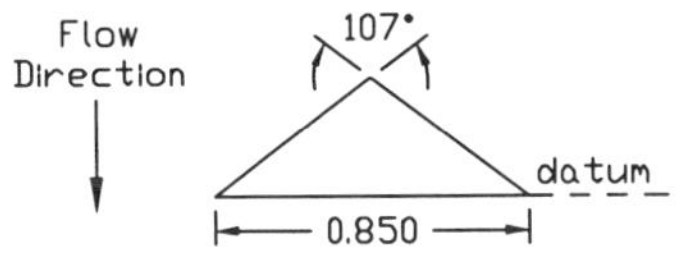	0.013 in. (0.11%c)
large Hama	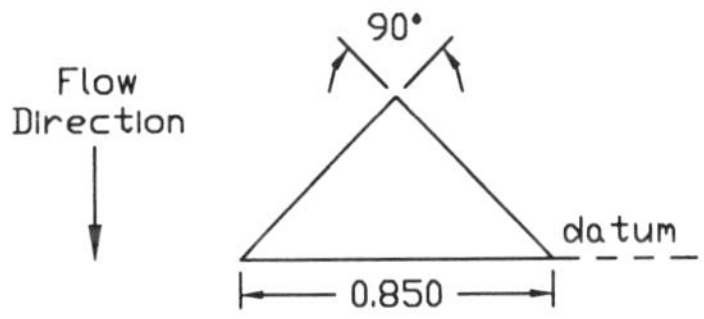	0.013 in. (0.11%c)
narrow zigzag (trip type E)	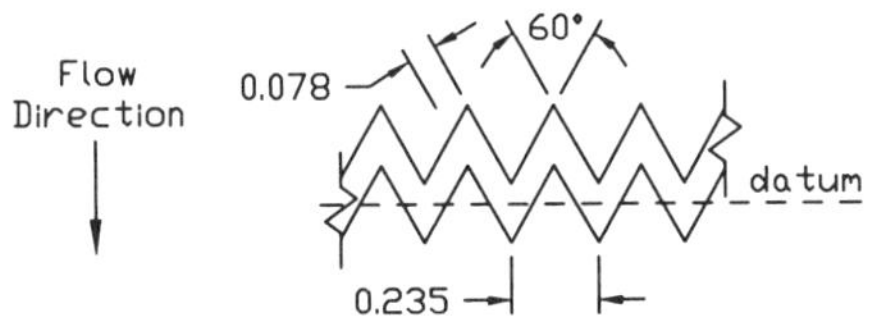	0.018 in. (0.15%c) 0.038 in. (0.32%c)
medium zigzag		0.008 in. (0.07%c) 0.013 in. (0.11%c) 0.026 in. (0.22%c)
wide zigzag (trip type D)	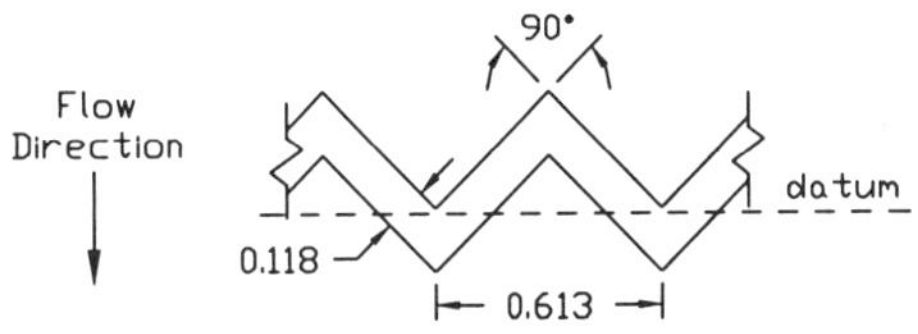	0.015 in. (0.13%c) 0.030 in. (0.25%c)
crescent zigzag	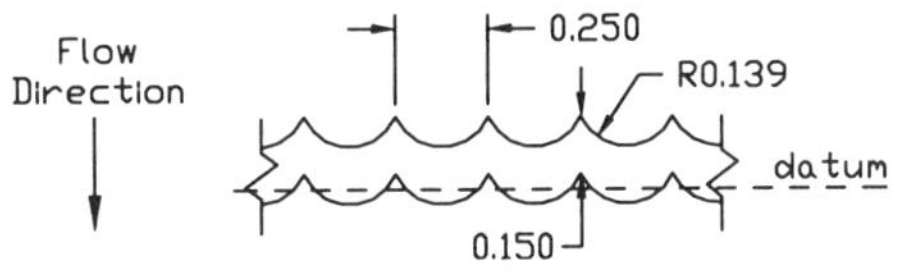	0.008 in. (0.07%c) 0.013 in. (0.11%c) 0.026 in. (0.22%c)
raised hemispheres	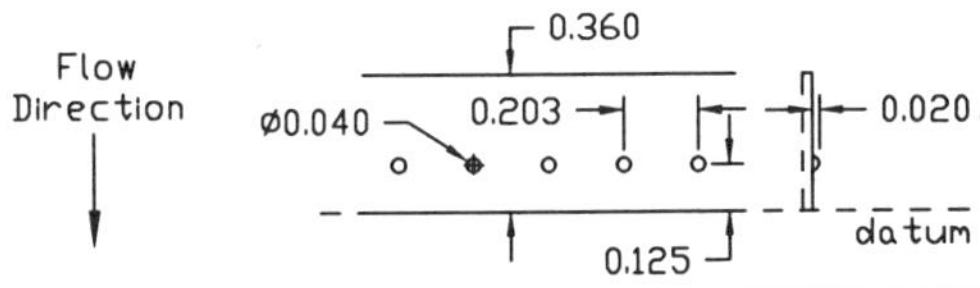	0.012 in. (0.10%c)‡ 0.017 in. (0.14%c)‡ 0.022 in. (0.18%c)‡

† All drawings actual size. All dimensions in inches.
‡ Trip heights do not include heights of hemispheres.

bumps which protrude an additional 0.020 in. above the upper surface of the trip. Furthermore, the wide zigzag appears to produce little advantage over single 2-D trips, while the narrow zigzag shows an advantage for only the most extreme thickness tested. As the Reynolds number is increased to 200,000, the performance of the various trips tends to become geometry independent for $0.008 < h < 0.022$. This lack

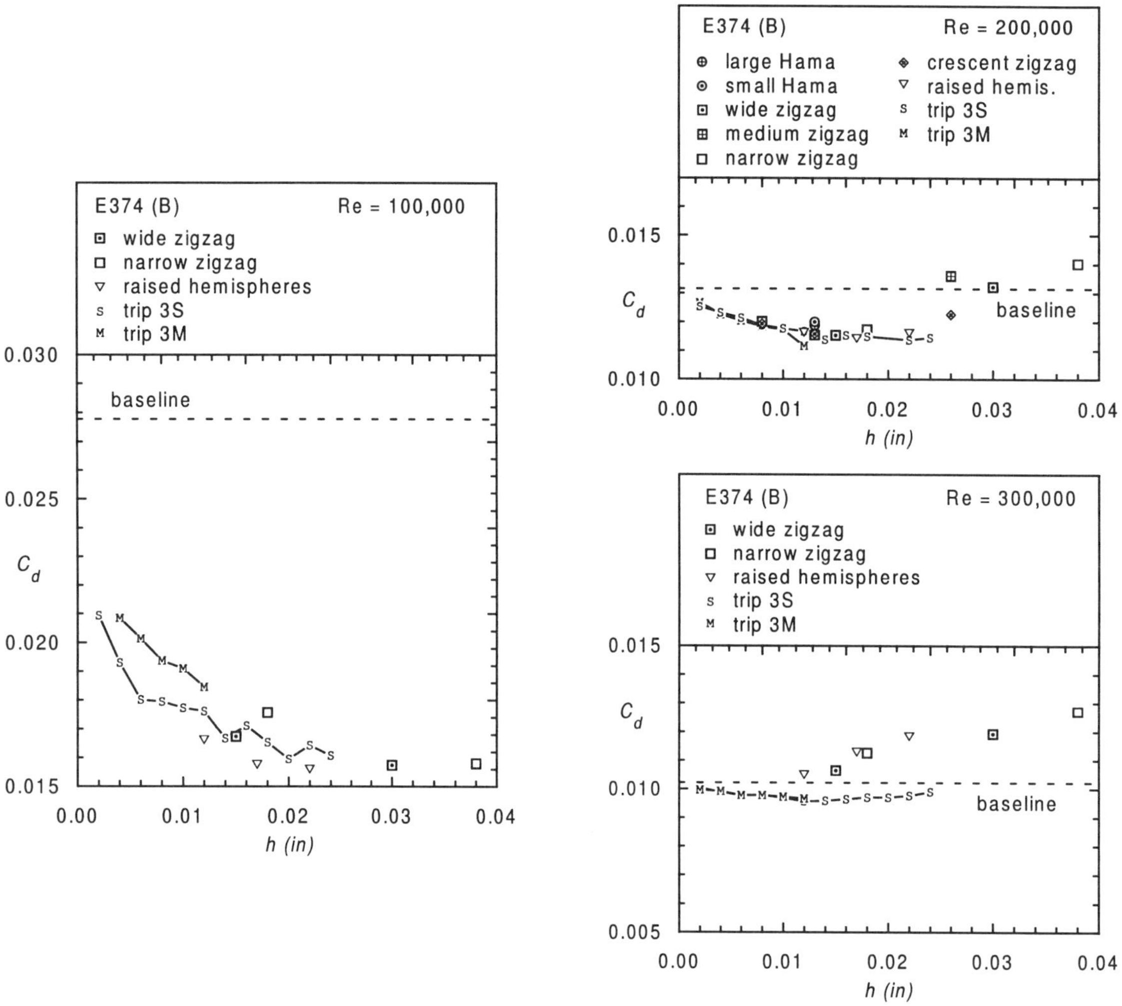

Fig. 6.11 E374 drag data for 3-D trips of varying thickness (Re = 100,000, 200,000, and 300,000).

of geometry dependence on trip performance, however, says nothing regarding the mechanisms by which trips generate drag reductions. The only definitive statement to be made is that any trade-offs made between bubble drag, skin friction drag, and device drag for the various trip geometries tested results in similar performance for Re = 200,000. By increasing the Reynolds number to 300,000, the results show a definite difference in performance between the 2-D and 3-D trips as evident from the increase in drag for the 3-D trips. This drag increase suggests either a complete elimination or a major reduction of the laminar separation bubble with increases in drag caused by either additional skin friction and/or device drag.

While the previous results suggest that 3-D trips are an improvement over 2-D trips at eliminating laminar separation bubbles for Re = 300,000, it is not clear from Fig. 6.11 if this holds true for lower Reynolds numbers. To answer this question, surface oil flow visualization was performed at Re = 200,000 for the wide zigzag trip located at x/c = 30% (position 3), with h = 0.015 in. and h = 0.030 in. A closeup of these results are shown in Figs. 6.12 and 6.13. For the h = 0.015 in. case (Fig. 6.12), it is unclear whether a bubble exists behind the trip. The flow pattern, however, is a distinctively different flow pattern than that produced by a 2-D trip which does not eliminate the bubble (see Fig. 6.10). While the difference in flow patterns indicates a change in performance between 2-D and 3-D trips, it is difficult to quantify this change. Trip effects become easier to identify as the trip height is increased to 0.030 in. (Fig. 6.13). What appears to be turbulent wedges have formed downstream of the trip, possibly as a result of instabilities introduced by the isolated pockets of reversed flow immediately behind the trip's downstream spikes. In between the turbulent wedges, it appears as if regions of attached laminar flow have formed where a laminar separation bubble originally existed.

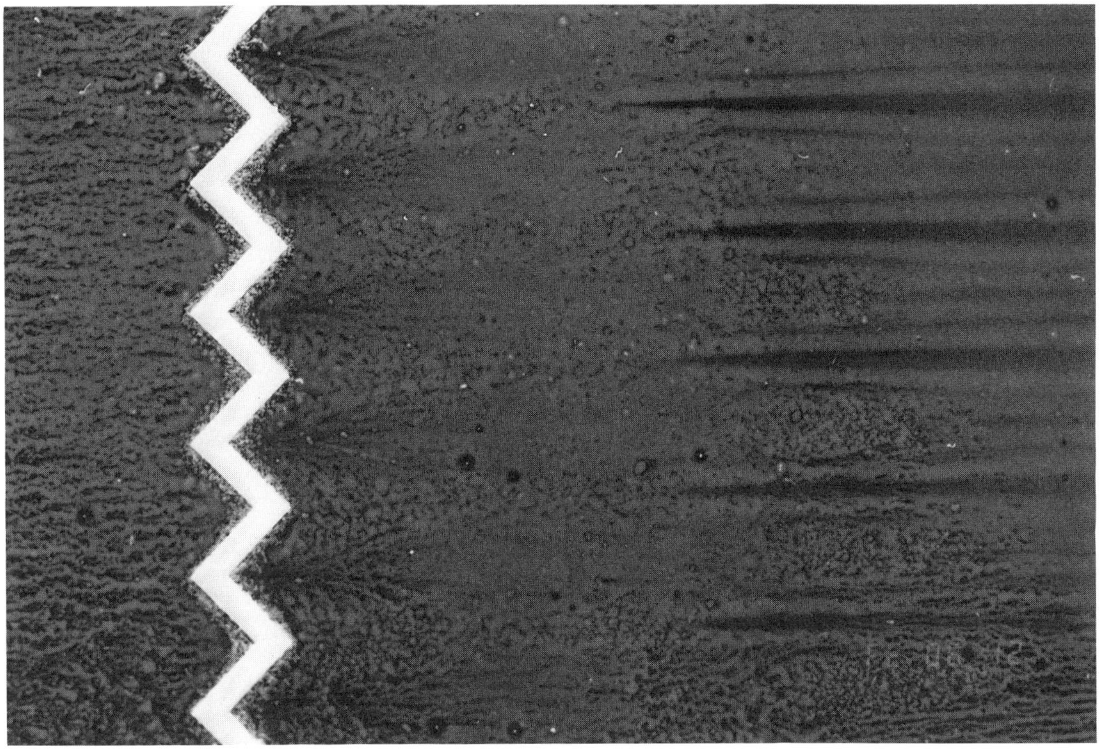

Fig. 6.12 Close up of flow visualization performed on E374 with wide zigzag trip [x/c = 30% (position 3); h = 0.015 in.; Re = 200,000].

In summary, the previous comparisons between 2-D and 3-D trip performance, as supported by surface oil flow visualization, suggests a definite difference in performance between the two types of trips. This data alone, however, is insufficient to make conclusions regarding the specific effects of 3-D trips. In general, however, it

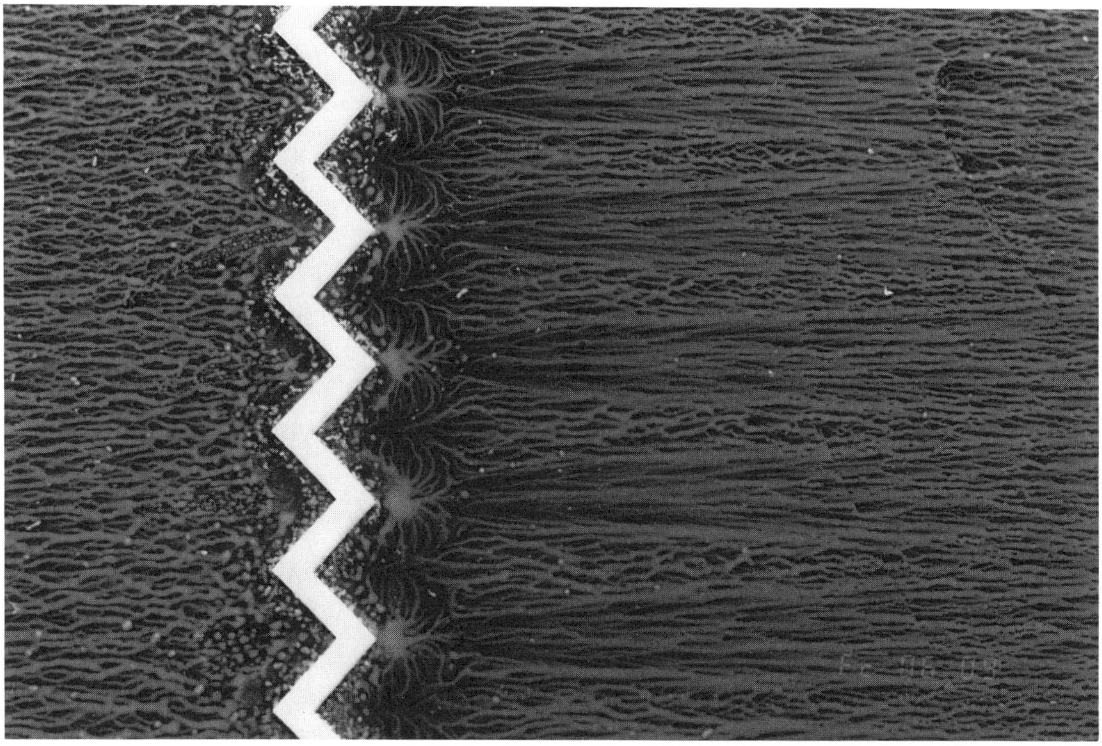

Fig. 6.13 Close up of flow visualization performed on E374 with wide zigzag trip [$x/c = 30\%$ (position 3); $h = 0.030$ in.; $Re = 200{,}000$].

has been shown that 3-D trips have demonstrated a potential for improvement over 2-D trip performance. Those readers interested in a more detailed discussion of this and other trip data are referred to Ref. 12.

References

[1] Selig, M.S., Donovan, J.F. and Fraser, D.B., *Airfoils at Low Speeds,* SoarTech Publications, Virginia Beach, VA, 1989.

[2] Selig, M.S., Guglielmo, J.J., Broeren, A.P., and Giguère, P., *Summary of Low-Speed Airfoil Data - Vol. 1*, SoarTech Publications, Virginia Beach, VA, 1995.

[3] Selig, M.S., Lyon, C.A., Giguère, P., Ninham, C.P., and Guglielmo, J.J., *Summary of Low-Speed Airfoil Data - Vol. 2*, SoarTech Publications, Virginia Beach, VA, 1996.

[4] Drela, M., "XFOIL: An Analysis and Design System for Low Reynolds Number Airfoils," *Lecture Notes in Engineering: Low Reynolds Number Aerodynamics*, T.J. Mueller (ed.), Vol. 54, Springer-Verlag, New York, June 1989.

[5] Drela, M. and Giles, M.B., "ISES: A Two-Dimensional Viscous Aerodynamic Design and Analysis Code," AIAA Paper 87-0424, January 1987.

[6] Eppler, R. and Somers, D.M., "A Computer Program for the Design and Analysis of Low-Speed Airfoils," NASA TM 80210, August 1980.

[7] Giguère, P. and Selig, M.S., "Freestream Velocity Corrections for Two-Dimensional Testing with Splitter Plates," *AIAA Journal*, Vol. 35, No. 7, July 1997, pp. 1195–1200.

[8] Evangelista, R., McGhee, R.J., and Walker, B.S., "Correlation of Theory to Wind-Tunnel Data at Reynolds Numbers below 500,000," *Lecture Notes in Engineering: Low Reynolds Number Aerodynamics*, T.J. Mueller (ed.), Vol. 54, Springer-Verlag, New York, June 1989, pp. 131–145.

[9] Guglielmo, J.J., "Spanwise Variations in Profile Drag for Airfoils at Low Reynolds Numbers," Master's Thesis, University of Illinois at Urbana-Champaign, Dept. of Aeronautical and Astronautical Engineering, May 1996.

[10] Morgan, H.L., Jr., "Computer Program for Smoothing and Scaling Airfoil Coordinates," NASA TM 84666, July 1983.

[11] Selig, M.S. and Maughmer, M.D., "Generalized Multipoint Inverse Airfoil Design," *AIAA Journal*, Vol. 30, No. 11, November 1992, pp. 2618–2625.

[12] Lyon, C.A., Selig, M.S., and Broeren, A.P., "Boundary Layer Trips on Airfoils at Low Reynolds Numbers," AIAA 97-0511, 35th Aerospace Sciences Meeting, Reno, Nevada, January 1997.

[13] Girsberger, R., "New Airfoils for R/C-Gliders," (in german), Proceedings of the International RC-Segelflug Forum, Winterthur, Switzerland, December 1983.

[14] Selig, M.S., Guglielmo, J.J., Broeren, A.P., Giguère, P., "Experiments on Airfoils at Low Reynolds Numbers," AIAA Paper 96-0062, 34th Aerospace Sciences Meeting, Reno, NV, January 1996.

[15] Eppler, R., *Airfoil Design and Data*, Springer-Verlag, New York, 1990.

[16] Giguère, P. and Selig, M.S., "Aerodynamic Performance of Small Wind Turbines Operating at Low Reynolds Numbers," WindPower '96 Conference, Denver, CO, June 23–27, 1996.

[17] Clausen, P.D., Ebert, P., Koh, S.G. and Wood, D.H., "The Aerodynamic Design of a 5 kW Wind Turbine," 11th Australian Fluid Mechanics Conference, December 1992.

[18] Giguère, P. and Selig, M.S., "New Airfoils for Small HAWTS," WindPower '97 Conference, June 16–18, Austin, TX, 1997. (Accepted for publication in the *ASME Journal of Solar Energy Engineering.*)

[19] Pfenninger, W., and Vemuru, C.S., "Design of Low-Reynolds Number Airfoils – I," AIAA Paper 88-2572-CP, AIAA 6th Applied Aerodynamics Conference, Williamsburg, VA, June 6–8, 1988, pp. 639–655.

[20] Pfenninger, W., Vemuru, C.S., Mangalam, S.M., and Evangelista, R., "Design of Low-Reynolds Number Airfoils – II," AIAA Paper 88-3764-CP, AIAA 6th Applied Aerodynamics Conference, Williamsburg, VA, June 6–8, 1988, pp. 1305–1319.

[21] Mangalam, S.M. and Pfenninger, W., "Wind-Tunnel Tests on a High Performance Low-Reynolds Number Airfoil," AIAA Paper 84-0628, March, 1984.

[22] Fisher, S.S. and Abbitt, J.D., "A Smoke-Wire Study of the Low-Reynolds Number Flow Over a NASA LRN(1)-1007 Airfoil Section," Proceedings of the Royal Aeronautical Society Aerodynamics at Low-Reynolds Numbers Conference, Oct. 15–18, 1986, Vol. 1, pp. 5.1–5.28.

[23] Mangalam, S.M., Bar-Sever, A., Zaman, K.B.M.Q., and Harvey, W.D., "Transition and Separation Control on a Low-Reynolds Number Airfoil," Proceedings of the Royal Aeronautical Society Aerodynamics at Low-Reynolds Numbers Conference, Oct. 15–18, 1986, Vol. 1, pp. 10.1–10.19.

[24] Zaman, K.B.M.Q., McKinzie, D.J., and Rumsey, C.L., "A Natural Low-Frequency Oscillation Over Airfoils Near Stalling Conditions," *Journal of Fluid Mechanics*, Vol. 202, 1989, pp. 403–442.

[25] Bragg, M.B., Heinrich, D.C., Balow, F.A., and Zaman, K.B.M.Q., "Flow Oscillation Over an Airfoil Near Stall," *AIAA Journal*, Vol. 34, No. 1, Jan. 1996, pp. 199–201.

[26] Broeren, A.P., and Bragg, M.B., "Phase-Averaged LDV Flowfield Measurements About an Airfoil in Unsteady Stall," AIAA Paper 96-2494-CP, AIAA 14th Applied Aerodynamics Conference, New Orleans, LA, June 17–20, 1996, pp. 921–931.

[27] Bragg, M.B., Heinrich, D.C. and Zaman, K.B.M.Q., "Flow Oscillation Over Airfoils Near Stall," ICAS Paper 94-4.5.2, Proceedings of the 19th Congress of the International Council of the Aeronautical Sciences Conferences, Vol. 2, Sept. 1994, pp. 1639–1648.

Glossary

adverse pressure gradient: A region of constantly increasing pressure (decreasing velocity) that may result in transition to turbulent flow, laminar separation, or turbulent separation. On a velocity distribution plot, an adverse pressure gradient is identified by decreasing values of velocity as x/c increases from leading to trailing edge.

aerodynamic center: The location on an airfoil or aircraft about which the pitching moment does not change with angle of attack. For airfoils, this point is typically near $x/c = 25\%$. The aerodynamic center is often confused with the center of pressure.

boundary layer: The thin layer of flow near the surface of an airfoil that is slowed due to skin friction. At the surface of the airfoil, the local flow velocity is zero. Off the surface of the airfoil, the velocity increases as the distance from the airfoil increases until it eventually reaches the external flow velocity at the edge of the boundary layer. Depending on the nature of the boundary layer (laminar or turbulent), the manner in which the velocity increases, as well as the rate at which the boundary layer thickens, varies. For example, turbulent boundary layers thicken more rapidly than laminar ones.

bubble drag: The portion of airfoil drag that results from a laminar separation bubble. Bubble drag is a large contributor to pressure drag at low Reynolds numbers.

bubble ramp (transition ramp): The region of an airfoil velocity distribution that is designed to ease transition from laminar to turbulent flow, thereby reducing the potential for laminar separation bubbles. As an example, the E387 has a bubble ramp that starts near $x/c = 35\%$ and ends near $x/c = 45\%$.

center of pressure: The location where the airfoil lift force should be located to generate thc proper quarter-chord pitching moment. It is not necessarily located on the airfoil. For example, on a cambered airfoil near zero lift, the center of pressure is behind the trailing edge of the airfoil.

constant-speed wind turbines: Wind turbines that operate at a constant rotational speed. Such wind turbines typically regulate their rated power either by pitching their blades to feather (away from stall) or by having their blades stall without pitch control. Wind turbines of the latter type are known as stall regulated. The blades of a constant-speed wind turbine thus operate over a broad range of angles of attack or lift coefficients.

cut-in velocity: This is the minimum wind speed at which a wind-turbine generates power.

device drag: Drag that results from the presence of an object not typically part of the airfoil geometry (i.e., boundary-layer trips, flap brackets, control horns, etc.). Device drag consists of both pressure drag and skin-friction drag.

displacement thickness: The distance that streamlines are displaced as a result of boundary-layer effects. When an object is analyzed in an inviscid sense, this distance is equivalent to the additional thickness an object requires to generate streamlines similar to the viscous case.

drag bucket: Loosely used in this book to identify the range of lift coefficients over which drag is low and desirable. The width of the bucket, while sometimes obvious (see the DH4009), is often ambiguous and left to individual interpretation.

fixed transition: Used to describe a boundary-layer transition location that would not otherwise occur without an artificial disturbance. Typically, boundary-layer trips are used to produce this disturbance, thereby moving the transition location upstream of the free transition location.

free transition: This is the boundary-layer transition location that naturally occurs for a given set of operating conditions.

inviscid: Having zero viscosity. Equivalent to frictionless flow. Airfoils experiencing steady, uniform flow generate no drag when analyzed in an inviscid manner.

negative stall: Occurs when flow begins to separate from the lower surface of an airfoil. This typically happens at extremely low angles of attack well beyond the operational range of the airfoil and is therefore of little interest for conventional aircraft applications.

pressure drag: Drag that results from relatively low pressure regions on the rearward facing surface of an object. This type of drag is almost entirely caused by regions of separated flow that occur behind blunt-shaped objects, on an airfoil near stall, or on an airfoil with a separation bubble.

profile drag: This is a catch-all term that incorporates all forms of drag produced by an airfoil. This is the drag that is plotted in Chapter 5. When a wing is being considered, this term is used to differentiate between induced drag and that drag resulting from the airfoil alone.

rated power: Maximum power output of a wind turbine.

roughness effects: Detrimental effects on the aerodynamic performance of wind turbine blades caused by blade erosion from environmental factors, insect accumulation, and ice formation at the blade leading edge. Roughness effects typically cause a large increase in drag and can significantly reduce the $C_{l,max}$ of an airfoil. Such a loss in $C_{l,max}$ is particularly detrimental to the energy capture of stall-regulated wind turbines. Consequently, the use of $C_{l,max}$ roughness insensitive airfoils is desired for stall-regulated wind turbines.

skin-friction drag: The component of drag due to friction between the airfoil surface and the fluid flow.

transition: The process by which flow changes from laminar to turbulent. Many problems associated with predicting low Reynolds number airfoil performance stem from inadequate methods of predicting this phenomenon.

variable-speed wind turbines: Wind turbines that operate at a rotational speed proportional to the wind speed. Such wind turbines are the most efficient aerodynamically as their blades operate at more or less a constant angle of attack depending on the atmospheric turbulence. Consequently, blades for variable-speed wind turbines can be designed to operate at the angle of attack or lift condition that provides the optimum lift-to-drag ratio of the airfoil(s) used. The tradeoff, however, is more complex power electronics as compared with constant-speed wind turbines.

viscous: Having viscosity. Flow over a surface is slowed relative to that surface. Viscous effects cause boundary layers to form, airfoils to stall, and are responsible for 100% of the drag produced by an isolated airfoil in steady, uniform flow.

Appendix A

Supplemental Inviscid Velocity Distributions

In an effort to provide more insight into low Reynolds number airfoil performance, the previous volume of this book presented viscous velocity distributions as predicted by XFOIL.[4] While these results are helpful in predicting the size and location of laminar separation bubbles, the results are sometimes difficult to interpret when several distributions are included in a single figure. Additionally, it is often quite helpful to the airfoil designer to see the inviscid results. For these reasons, the decision was made to add this supplemental section that includes inviscid velocity distributions for all of the airfoils presented in *Volume 2* except those that can be found in *Volume 1*. By doing so, a consistent database of inviscid velocity distributions is available for all airfoils tested in *Volumes 1, 2* and *3*.

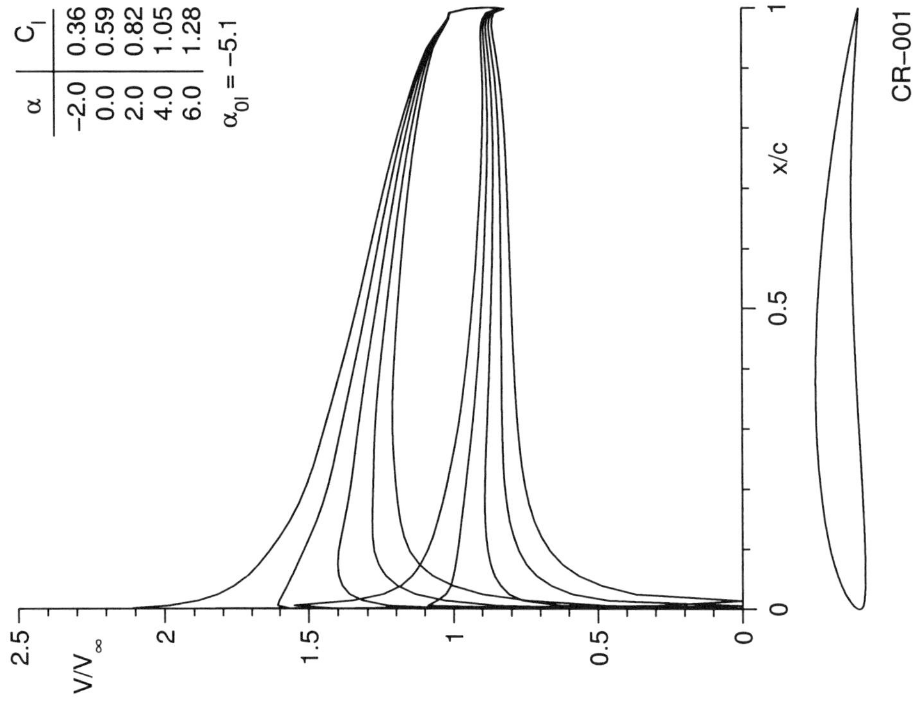
α | C_l
−2.0 | 0.36
0.0 | 0.59
2.0 | 0.82
4.0 | 1.05
6.0 | 1.28
α_{0l} = −5.1
V/V_∞
2.5
2
1.5
1
0.5
0
x/c
0.5
CR–001

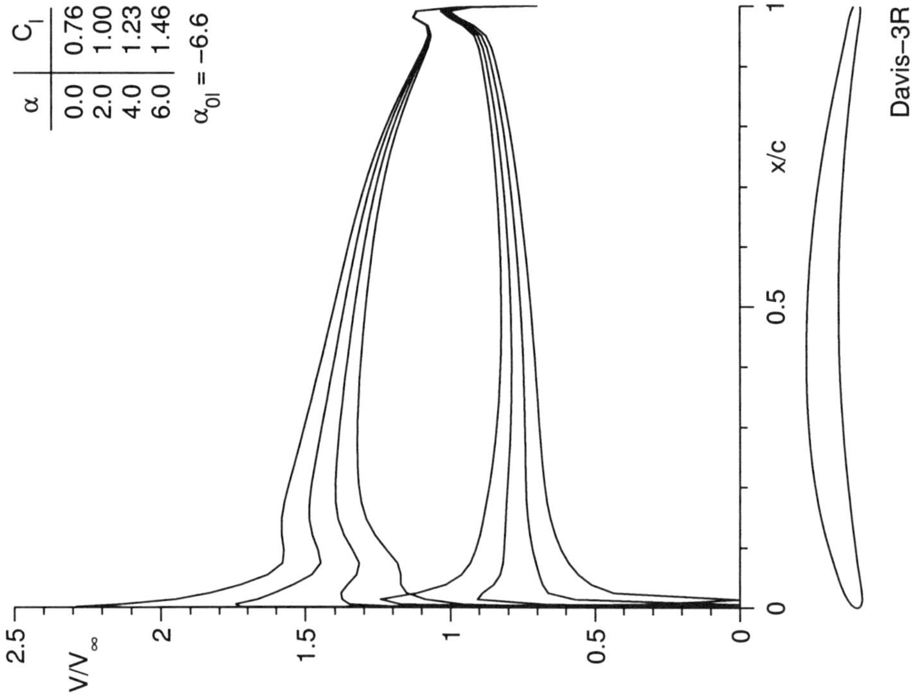
α | C_l
0.0 | 0.76
2.0 | 1.00
4.0 | 1.23
6.0 | 1.46
α_{0l} = −6.6
V/V_∞
2.5
2
1.5
1
0.5
0
x/c
0.5
Davis–3R

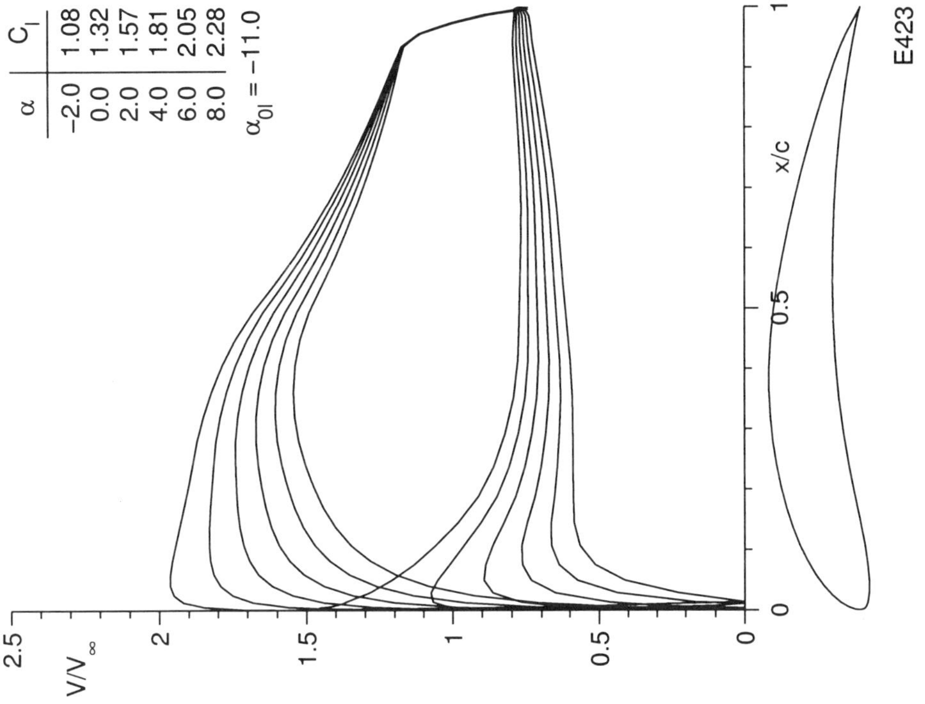
α | C_l
−2.0 | 1.08
0.0 | 1.32
2.0 | 1.57
4.0 | 1.81
6.0 | 2.05
8.0 | 2.28
α_{0l} = −11.0
V/V_∞
0
0.5
1
1.5
2
2.5
x/c
0
0.5
1
E423

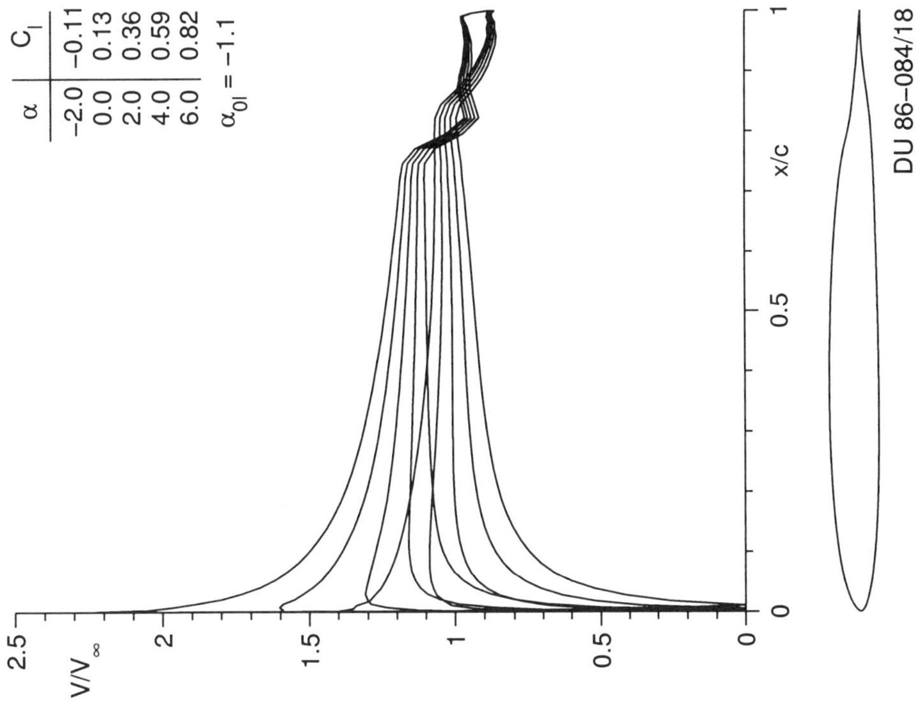
α | C_l
−2.0 | −0.11
0.0 | 0.13
2.0 | 0.36
4.0 | 0.59
6.0 | 0.82
α_{0l} = −1.1
V/V_∞
0
0.5
1
1.5
2
2.5
x/c
0
0.5
1
DU 86-084/18

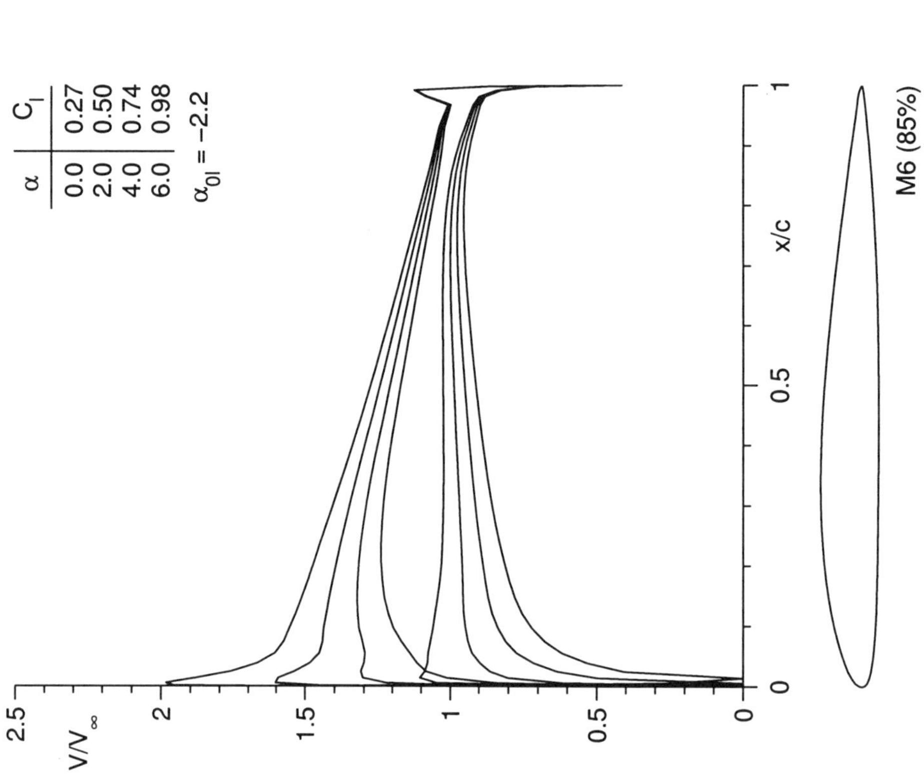

α	C_l
0.0	0.27
2.0	0.50
4.0	0.74
6.0	0.98

$\alpha_{0l} = -2.2$
V/V_∞
2.5
2
1.5
1
0.5
0
x/c
0.5
1
M6 (85%)

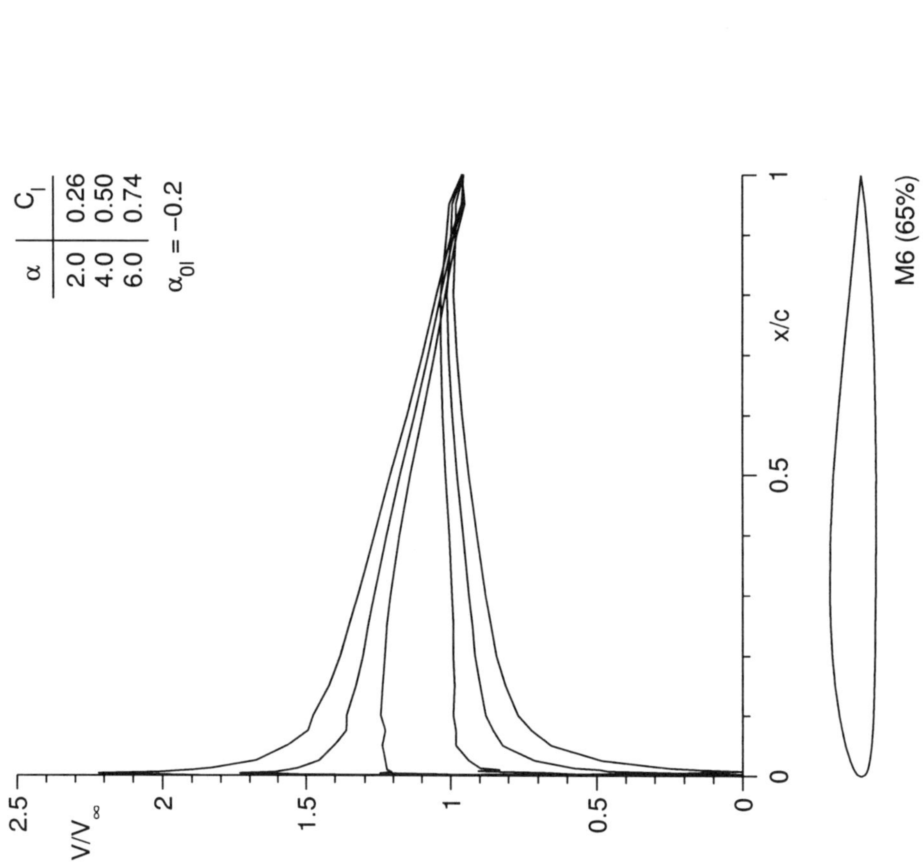

α	C_l
2.0	0.26
4.0	0.50
6.0	0.74

$\alpha_{0l} = -0.2$
V/V_∞
2.5
2
1.5
1
0.5
0
x/c
0.5
1
M6 (65%)

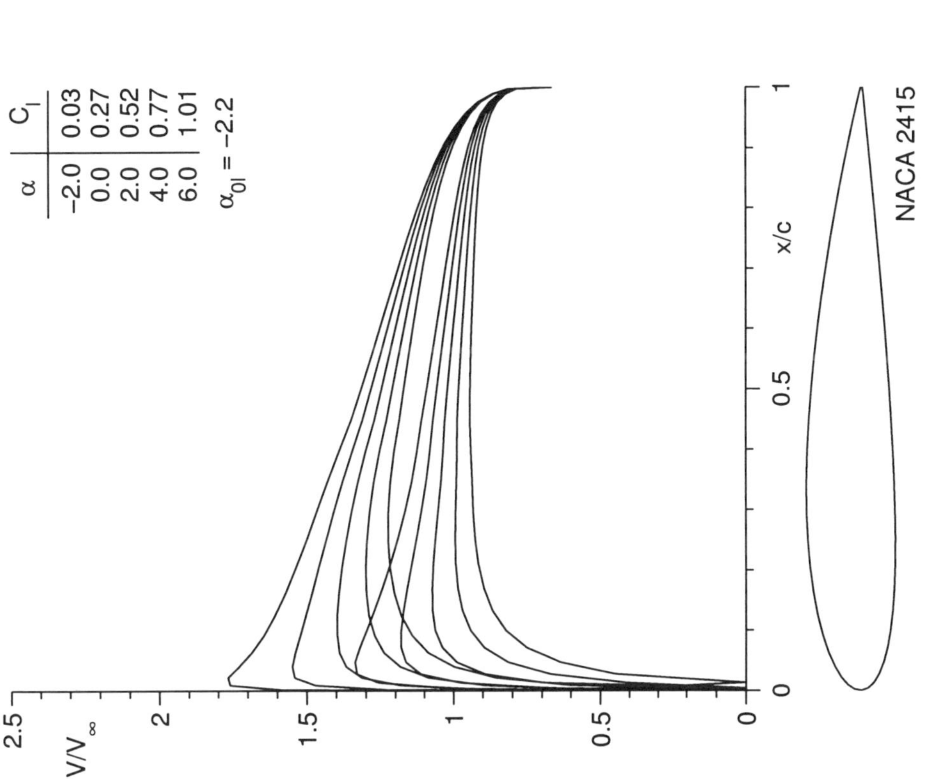
α | C_l
−2.0 | 0.03
0.0 | 0.27
2.0 | 0.52
4.0 | 0.77
6.0 | 1.01
α_{0l} = −2.2
V/V∞
x/c
NACA 2415

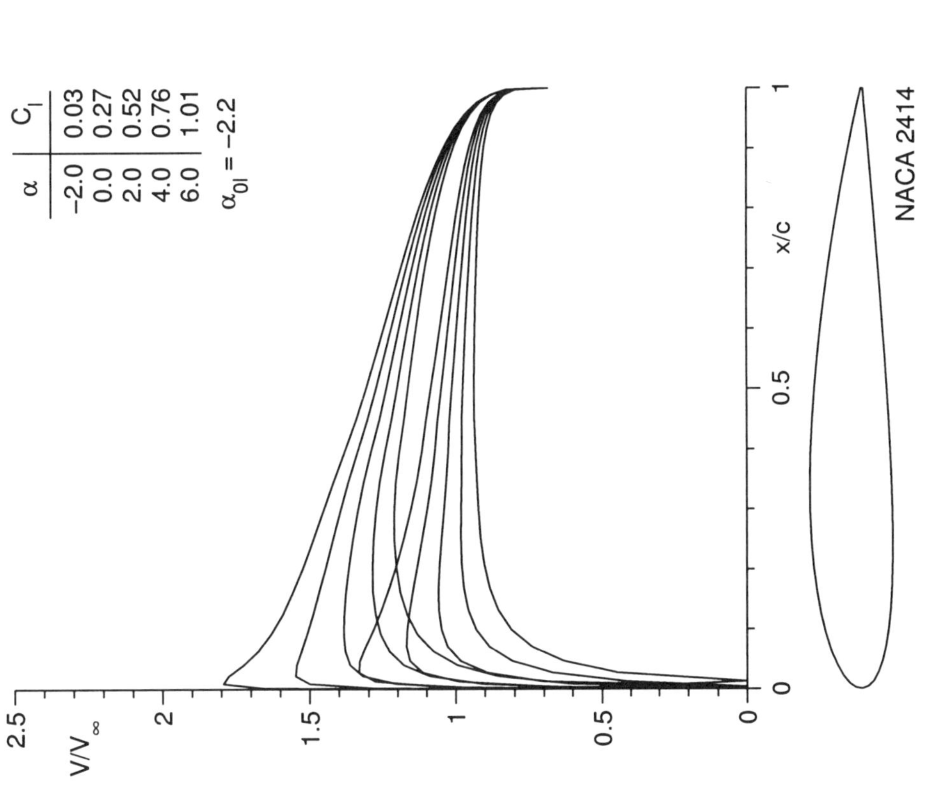
α | C_l
−2.0 | 0.03
0.0 | 0.27
2.0 | 0.52
4.0 | 0.76
6.0 | 1.01
α_{0l} = −2.2
V/V∞
x/c
NACA 2414

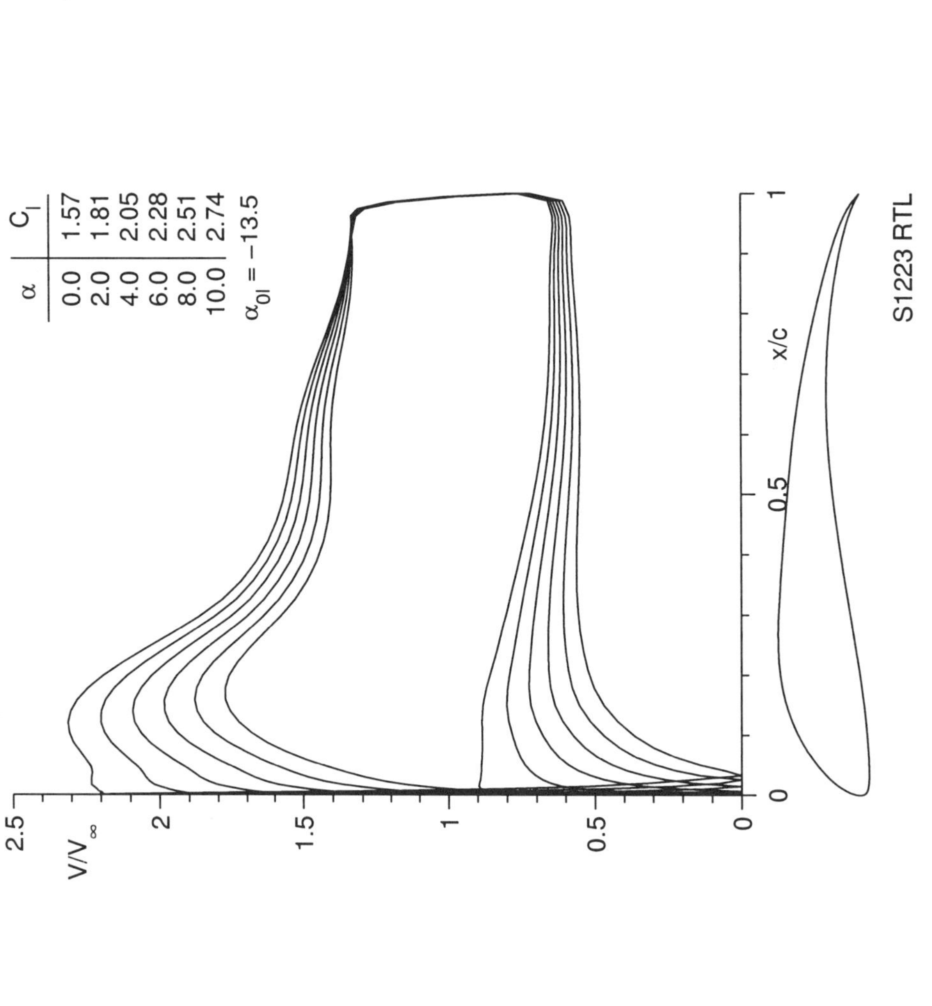
α
C_l
0.0
1.57
2.0
1.81
4.0
2.05
6.0
2.28
8.0
2.51
10.0
2.74
α_0l = −13.5
2.5
V/V_∞
2
1.5
1
0.5
0
0
0.5
x/c
1
S1223 RTL

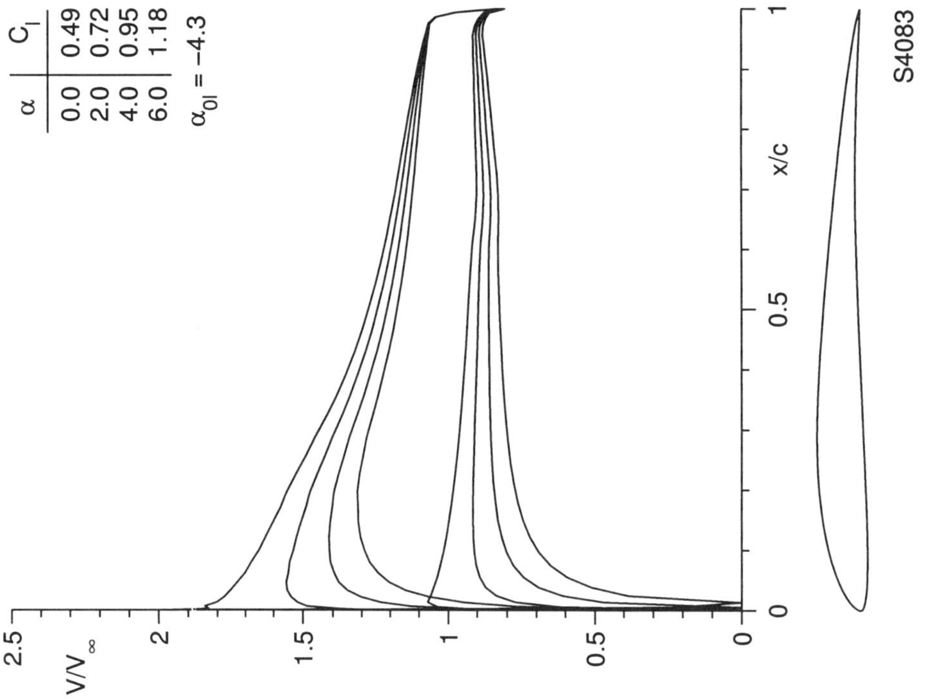
α
C_l
0.0
0.49
2.0
0.72
4.0
0.95
6.0
1.18
α_0l = −4.3
2.5
V/V_∞
2
1.5
1
0.5
0
0
0.5
x/c
1
S4083

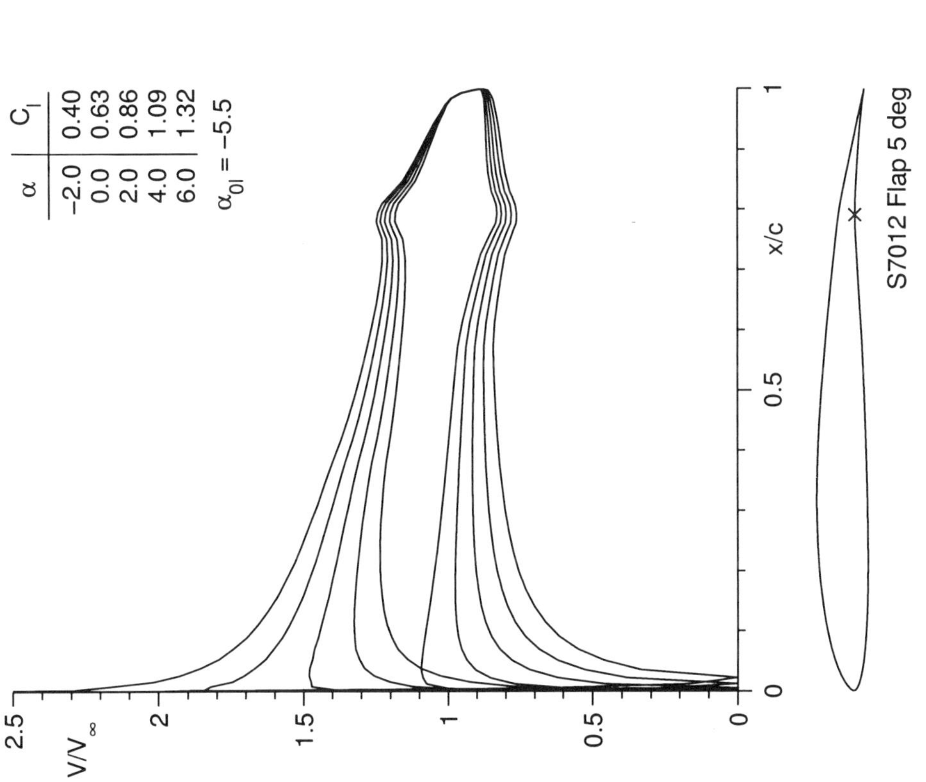
α | C_l
−2.0 | 0.40
0.0 | 0.63
2.0 | 0.86
4.0 | 1.09
6.0 | 1.32
α_{0l} = −5.5
V/V_∞
2.5
2
1.5
1
0.5
0
0
0.5
1
x/c
S7012 Flap 5 deg

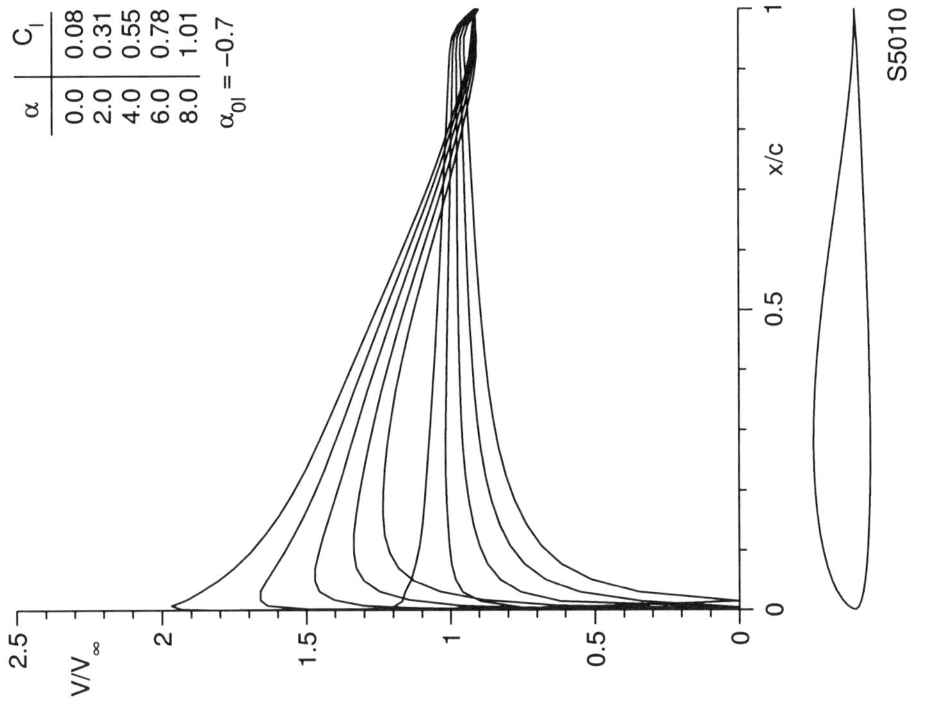
α | C_l
0.0 | 0.08
2.0 | 0.31
4.0 | 0.55
6.0 | 0.78
8.0 | 1.01
α_{0l} = −0.7
V/V_∞
2.5
2
1.5
1
0.5
0
0
0.5
1
x/c
S5010

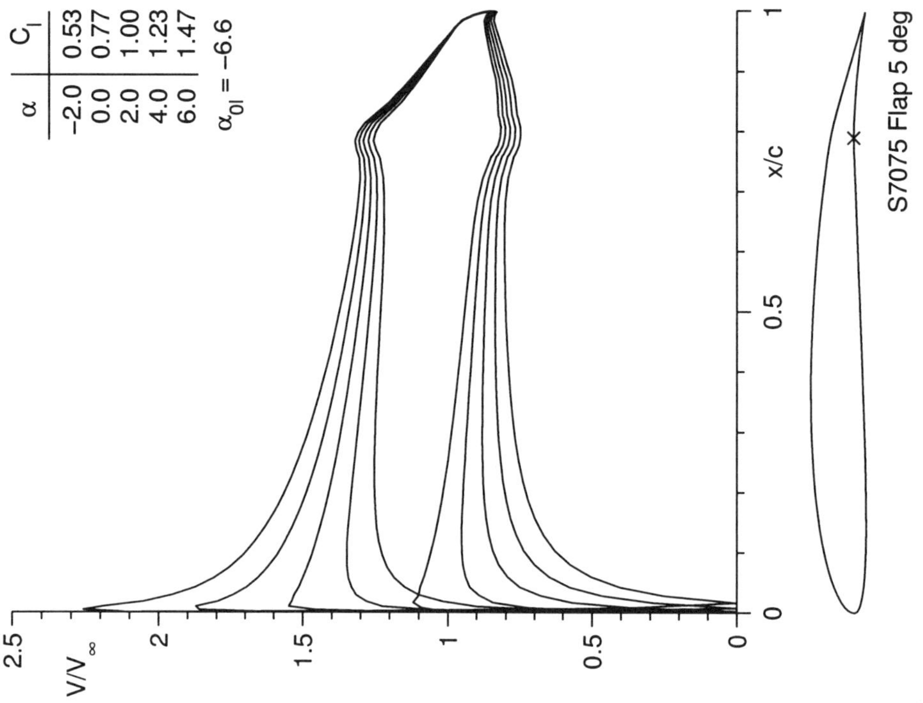
α | C_l
−2.0 | 0.53
0.0 | 0.77
2.0 | 1.00
4.0 | 1.23
6.0 | 1.47
α_{0l} = −6.6
V/V_∞
x/c
S7075 Flap 5 deg

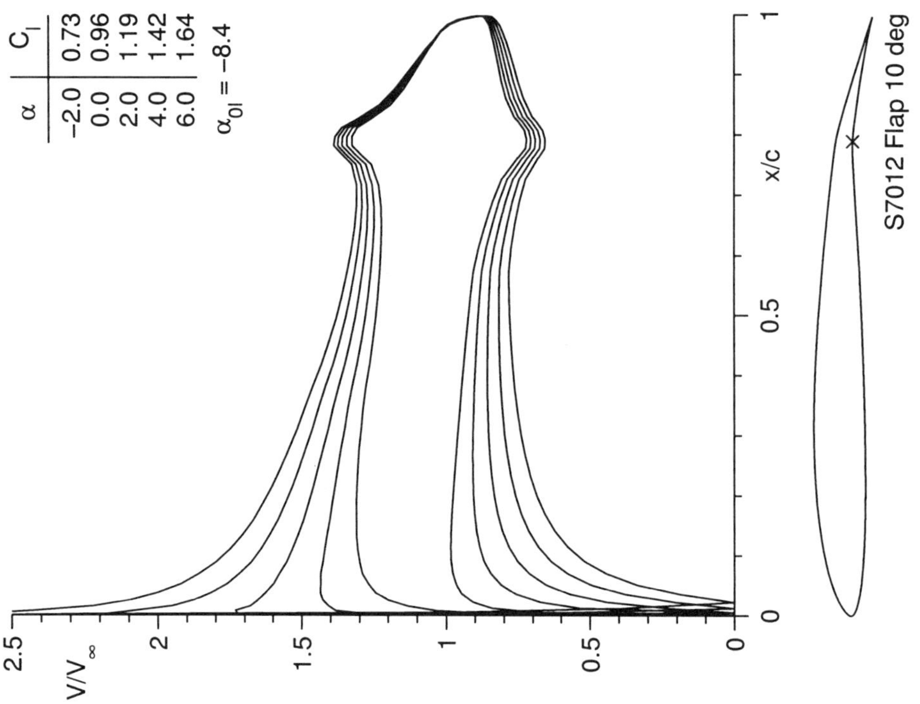
α | C_l
−2.0 | 0.73
0.0 | 0.96
2.0 | 1.19
4.0 | 1.42
6.0 | 1.64
α_{0l} = −8.4
V/V_∞
x/c
S7012 Flap 10 deg

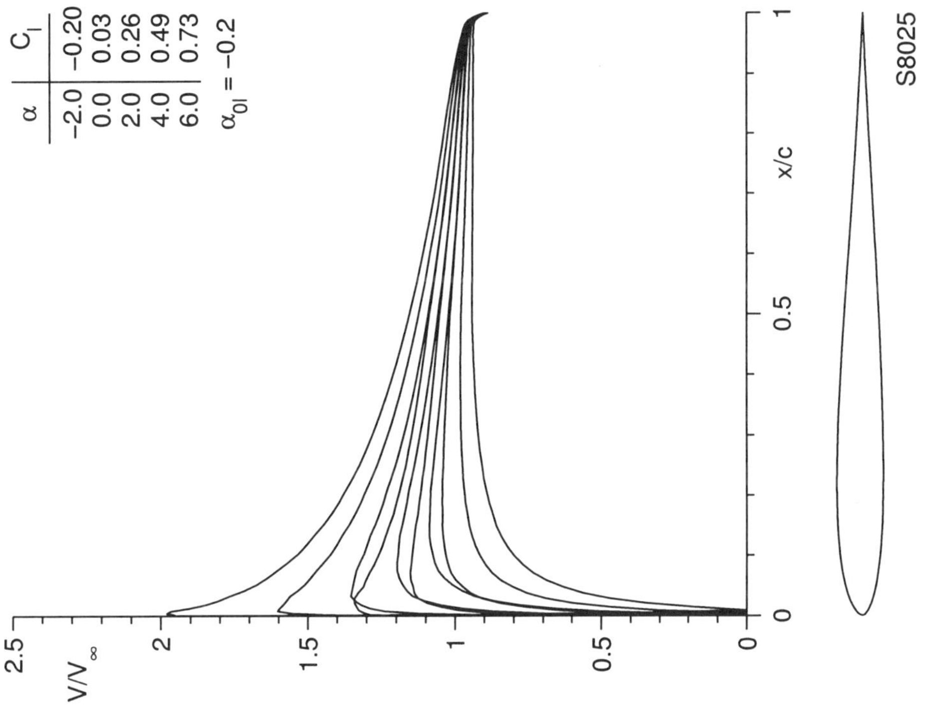
α | C_l
−2.0 | −0.20
0.0 | 0.03
2.0 | 0.26
4.0 | 0.49
6.0 | 0.73
α_{0l} = −0.2
V/V_∞
x/c
S8025

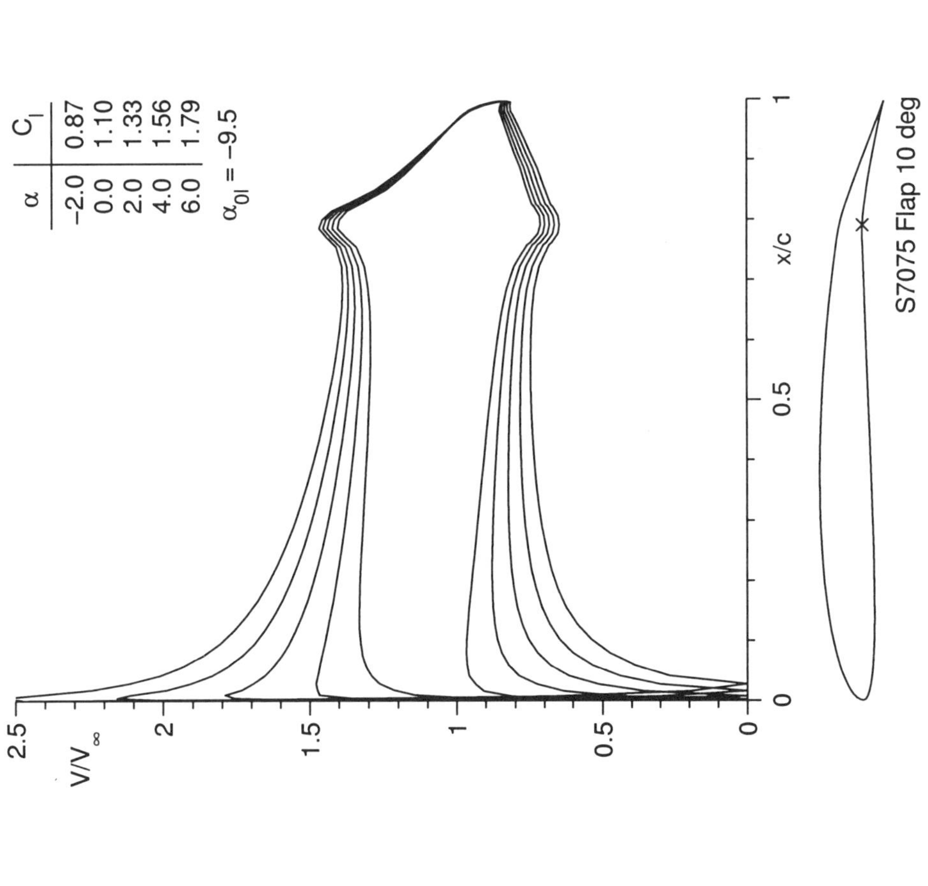
α | C_l
−2.0 | 0.87
0.0 | 1.10
2.0 | 1.33
4.0 | 1.56
6.0 | 1.79
α_{0l} = −9.5
V/V_∞
x/c
S7075 Flap 10 deg

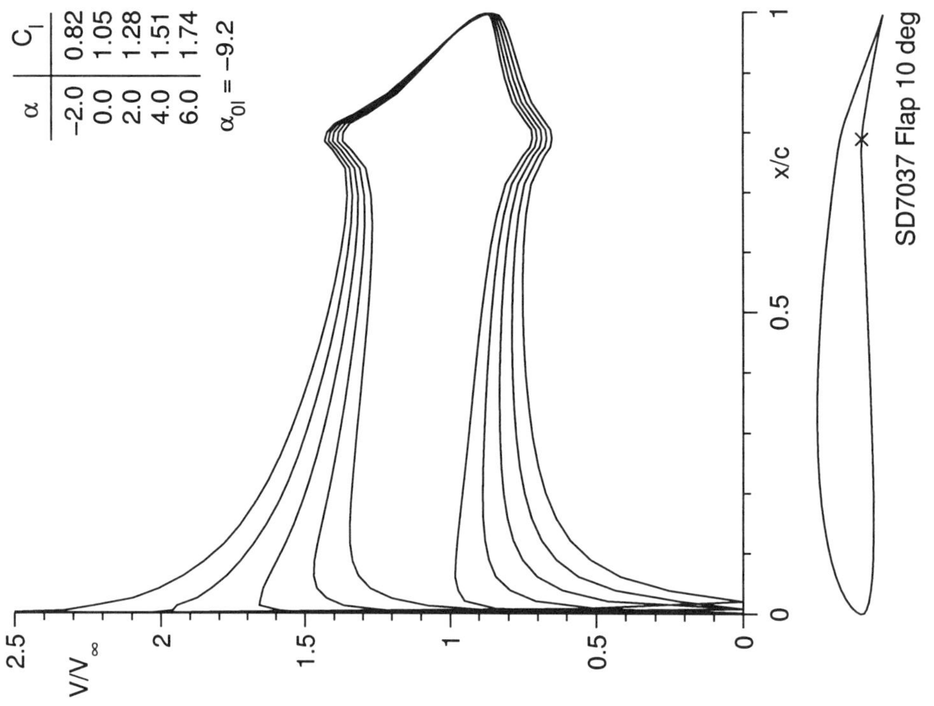
α | C_l
−2.0 | 0.82
0.0 | 1.05
2.0 | 1.28
4.0 | 1.51
6.0 | 1.74
α_{0l} = −9.2
V/V_∞
x/c
SD7037 Flap 10 deg

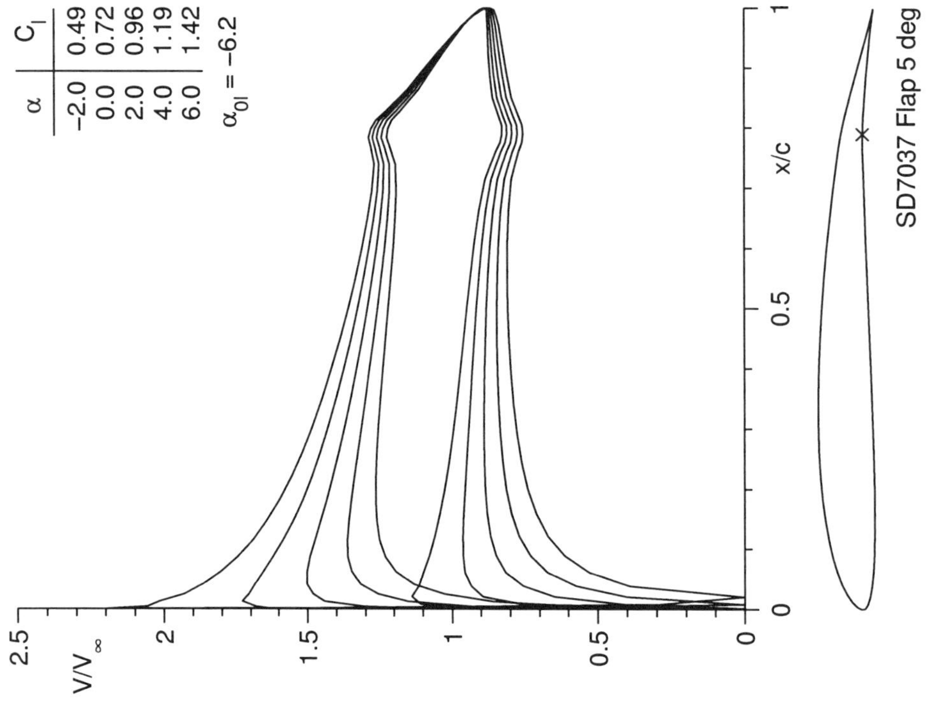
α | C_l
−2.0 | 0.49
0.0 | 0.72
2.0 | 0.96
4.0 | 1.19
6.0 | 1.42
α_{0l} = −6.2
V/V_∞
x/c
SD7037 Flap 5 deg

Appendix B

Tabulated Airfoil Coordinates

This appendix lists both the true (as designed) and actual (as built) airfoil coordinates. For any given airfoil, the true airfoil coordinates are listed first. If the original true coordinates were not 'mathematically' smooth, or if there were too few points defining the airfoil (especially in the vicinity of the leading and trailing edges), the airfoil coordinates were smoothed with the computer program AFSMO[10] developed at NASA Langley.

A18

smoothed

x/c	y/c
1.00000	0.00307
0.99754	0.00365
0.99070	0.00530
0.98037	0.00771
0.96698	0.01070
0.95044	0.01421
0.93064	0.01819
0.90775	0.02253
0.88202	0.02712
0.85370	0.03185
0.82309	0.03661
0.79048	0.04132
0.75616	0.04591
0.72043	0.05029
0.68359	0.05440
0.64594	0.05819
0.60778	0.06160
0.56937	0.06458
0.53099	0.06710
0.49265	0.06911
0.45435	0.07058
0.41638	0.07143
0.37887	0.07157
0.34204	0.07094
0.30609	0.06951
0.27120	0.06733
0.23760	0.06451
0.20549	0.06114
0.17504	0.05734
0.14648	0.05315
0.11999	0.04860
0.09576	0.04368
0.07395	0.03839
0.05468	0.03277
0.03811	0.02694
0.02433	0.02103
0.01338	0.01513
0.00548	0.00932
0.00098	0.00376
0.00000	0.00000
0.00098	-0.00348
0.00548	-0.00770
0.01338	-0.01118
0.02433	-0.01388
0.03811	-0.01581
0.05468	-0.01699
0.07395	-0.01740
0.09576	-0.01708
0.11999	-0.01611
0.14648	-0.01461
0.17504	-0.01269
0.20549	-0.01047
0.23760	-0.00805
0.27120	-0.00550
0.30609	-0.00290
0.34204	-0.00033
0.37887	0.00209
0.41638	0.00429
0.45435	0.00617
0.49265	0.00773
0.53099	0.00895
0.56937	0.00984
0.60778	0.01042
0.64594	0.01067
0.68359	0.01060
0.72043	0.01022
0.75616	0.00954
0.79048	0.00858
0.82309	0.00738
0.85370	0.00600
0.88202	0.00450
0.90775	0.00297
0.93064	0.00149
0.95044	0.00015
0.96698	-0.00098
0.98037	-0.00187
0.99070	-0.00252
0.99754	-0.00292
1.00000	-0.00306

A18

actual

x/c	y/c
1.00000	0.00033
0.99690	0.00099
0.99191	0.00224
0.98482	0.00374
0.97861	0.00511
0.96785	0.00735
0.95536	0.00972
0.94364	0.01198
0.93398	0.01379
0.92071	0.01632
0.90411	0.01925
0.89181	0.02141
0.86400	0.02621
0.83715	0.03060
0.80818	0.03498
0.76407	0.04104
0.71992	0.04669
0.66487	0.05312
0.60663	0.05896
0.54910	0.06333
0.49645	0.06654
0.43814	0.06926
0.38347	0.06938
0.33236	0.06958
0.29351	0.06837
0.24769	0.06500
0.20920	0.06121
0.17697	0.05768
0.14758	0.05409
0.11892	0.04994
0.10641	0.04776
0.09459	0.04542
0.08092	0.04235
0.07075	0.03975
0.06073	0.03692
0.04549	0.03203
0.03643	0.02853
0.02720	0.02440
0.02028	0.02095
0.01302	0.01674
0.00710	0.01159
0.00109	0.00346
0.00000	-0.00001
0.00489	-0.00732
0.01059	-0.01050
0.01487	-0.01187
0.02173	-0.01345
0.02945	-0.01488
0.04403	-0.01646
0.05870	-0.01696
0.07258	-0.01694
0.08798	-0.01669
0.10106	-0.01628
0.11648	-0.01556
0.14251	-0.01381
0.16858	-0.01185
0.19992	-0.00957
0.24110	-0.00652
0.28536	-0.00401
0.34013	-0.00031
0.40095	0.00398
0.45970	0.00579
0.51571	0.00740
0.57735	0.00844
0.63480	0.00884
0.67858	0.00911
0.71898	0.00854
0.76508	0.00691
0.80917	0.00532
0.83733	0.00423
0.87102	0.00289
0.90385	0.00148
0.92078	0.00075
0.93589	0.00012
0.95129	-0.00055
0.96349	-0.00080
0.97274	-0.00121
0.98152	-0.00116
0.98959	-0.00119
0.99507	-0.00100
1.00000	0.00033

Avistar

from plans

x/c	y/c
1.00000	0.00158
0.99754	0.00213
0.99070	0.00365
0.98037	0.00588
0.96698	0.00866
0.95044	0.01197
0.93064	0.01582
0.90775	0.02021
0.88202	0.02516
0.85370	0.03067
0.82309	0.03672
0.79048	0.04321
0.75616	0.04997
0.72043	0.05676
0.68359	0.06330
0.64594	0.06933
0.60778	0.07474
0.56937	0.07949
0.53099	0.08359
0.49265	0.08712
0.45435	0.09013
0.41638	0.09250
0.37887	0.09399
0.34204	0.09427
0.30609	0.09316
0.27120	0.09093
0.23760	0.08790
0.20549	0.08413
0.17504	0.07948
0.14648	0.07377
0.11999	0.06691
0.09576	0.05906
0.07395	0.05059
0.05468	0.04195
0.03811	0.03381
0.02433	0.02668
0.01338	0.02039
0.00548	0.01384
0.00098	0.00558
0.00000	-0.00152
0.00098	-0.00874
0.00548	-0.01753
0.01338	-0.02392
0.02433	-0.02781
0.03811	-0.03036
0.05468	-0.03298
0.07395	-0.03610
0.09576	-0.03932
0.11999	-0.04225
0.14648	-0.04474
0.17504	-0.04686
0.20549	-0.04869
0.23760	-0.05008
0.27120	-0.05092
0.30609	-0.05120
0.34204	-0.05095
0.37887	-0.05021
0.41638	-0.04904
0.45435	-0.04755
0.49265	-0.04584
0.53099	-0.04392
0.56937	-0.04170
0.60778	-0.03906
0.64594	-0.03603
0.68359	-0.03283
0.72043	-0.02967
0.75616	-0.02669
0.79048	-0.02393
0.82309	-0.02130
0.85370	-0.01860
0.88202	-0.01586
0.90775	-0.01337
0.93064	-0.01153
0.95044	-0.01037
0.96698	-0.00941
0.98037	-0.00793
0.99070	-0.00556
0.99754	-0.00289
1.00000	-0.00163

Avistar

actual

x/c	y/c
1.00000	-0.00055
0.99905	0.00378
0.99679	0.00457
0.99454	0.00506
0.99217	0.00547
0.98878	0.00622
0.98565	0.00687
0.98218	0.00778
0.97804	0.00862
0.97314	0.00976
0.95995	0.01243
0.95462	0.01362
0.94308	0.01619
0.92790	0.01946
0.90619	0.02410
0.88014	0.02956
0.85740	0.03425
0.82509	0.04070
0.79616	0.04657
0.77264	0.05118
0.74161	0.05707
0.70440	0.06376
0.64879	0.07286
0.58772	0.08135
0.53261	0.08774
0.47089	0.09329
0.41707	0.09670
0.37800	0.09804
0.32912	0.09761
0.29481	0.09618
0.25242	0.09301
0.21969	0.08925
0.20221	0.08673
0.18966	0.08476
0.16705	0.08038
0.13596	0.07270
0.11181	0.06536
0.09757	0.06029
0.07154	0.04978
0.05035	0.03988
0.03026	0.02949
0.01895	0.02314
0.01712	0.02205
0.00854	0.01628
0.00202	0.00874
0.00031	0.00372
0.00000	-0.00017
0.00020	-0.00301
0.00147	-0.00861
0.00295	-0.01171
0.00352	-0.01245
0.00645	-0.01537
0.00956	-0.01785
0.01523	-0.02054
0.01894	-0.02181
0.02255	-0.02293
0.02869	-0.02451
0.03493	-0.02601
0.04575	-0.02851
0.05424	-0.03030
0.06983	-0.03337
0.08009	-0.03520
0.09569	-0.03764
0.11501	-0.04022
0.13519	-0.04258
0.15880	-0.04458
0.18685	-0.04665
0.22024	-0.04835
0.25721	-0.04926
0.29196	-0.04914
0.32785	-0.04883
0.36661	-0.04828
0.40743	-0.04757
0.47299	-0.04586
0.55162	-0.04220
0.61190	-0.03806
0.67481	-0.03301
0.73518	-0.02803
0.77709	-0.02461
0.81101	-0.02176
0.83992	-0.01921
0.86141	-0.01719
0.89819	-0.01363
0.93829	-0.00977
0.95983	-0.00764
0.97878	-0.00570
0.99623	-0.00378
1.00000	-0.00055

BW-3

from blade

x/c	y/c
1.00000	0.00350
0.99754	0.00538
0.99070	0.00983
0.98037	0.01483
0.96698	0.01922
0.95044	0.02281
0.93064	0.02587
0.90775	0.02892
0.88202	0.03237
0.85370	0.03637
0.82309	0.04084
0.79048	0.04562
0.75616	0.05052
0.72043	0.05536
0.68359	0.05998
0.64594	0.06426
0.60778	0.06809
0.56937	0.07136
0.53099	0.07402
0.49265	0.07600
0.45435	0.07727
0.41638	0.07783
0.37887	0.07771
0.34204	0.07695
0.30609	0.07560
0.27120	0.07364
0.23760	0.07105
0.20549	0.06776
0.17504	0.06374
0.14648	0.05898
0.11999	0.05360
0.09576	0.04781
0.07395	0.04188
0.05468	0.03611
0.03811	0.03075
0.02433	0.02568
0.01338	0.02038
0.00548	0.01438
0.00098	0.00758
0.00000	0.00237
0.00098	-0.00268
0.00548	-0.00851
0.01338	-0.01220
0.02433	-0.01395
0.03811	-0.01422
0.05468	-0.01261
0.07395	-0.00830
0.09576	-0.00176
0.11999	0.00532
0.14648	0.01175
0.17504	0.01742
0.20549	0.02241
0.23760	0.02661
0.27120	0.03005
0.30609	0.03288
0.34204	0.03515
0.37887	0.03688
0.41638	0.03804
0.45435	0.03852
0.49265	0.03818
0.53099	0.03705
0.56937	0.03525
0.60778	0.03298
0.64594	0.03038
0.68359	0.02749
0.72043	0.02432
0.75616	0.02090
0.79048	0.01734
0.82309	0.01376
0.85370	0.01029
0.88202	0.00704
0.90775	0.00403
0.93064	0.00127
0.95044	-0.00141
0.96698	-0.00458
0.98037	-0.00742
0.99070	-0.00752
0.99754	-0.00490
1.00000	-0.00312

BW-3

actual

x/c	y/c
1.00000	-0.00070
0.99888	0.00732
0.99788	0.00826
0.99598	0.00958
0.99373	0.01069
0.99027	0.01181
0.98543	0.01296
0.97586	0.01487
0.96422	0.01713
0.95477	0.01887
0.94428	0.02077
0.93154	0.02297
0.92024	0.02486
0.90383	0.02738
0.88567	0.03017
0.86848	0.03294
0.84678	0.03626
0.81825	0.04046
0.77628	0.04639
0.70636	0.05561
0.61641	0.06597
0.52895	0.07352
0.46801	0.07598
0.41742	0.07638
0.36977	0.07592
0.33731	0.07514
0.30721	0.07399
0.27038	0.07186
0.24358	0.06976
0.22162	0.06767
0.20094	0.06534
0.17800	0.06228
0.16002	0.05943
0.14121	0.05602
0.12749	0.05325
0.11149	0.04971
0.09608	0.04598
0.08564	0.04329
0.07677	0.04087
0.06786	0.03831
0.05345	0.03385
0.03157	0.02621
0.00441	0.01064
0.00120	0.00552
0.00070	0.00413
0.00004	0.00098
0.00024	-0.00246
0.00146	-0.00564
0.00337	-0.00828
0.00809	-0.01182
0.01190	-0.01340
0.00954	-0.01254
0.01643	-0.01450
0.02081	-0.01521
0.02520	-0.01575
0.02932	-0.01605
0.03137	-0.01616
0.05126	-0.01476
0.05592	-0.01395
0.06074	-0.01298
0.06727	-0.01156
0.07642	-0.00926
0.08429	-0.00708
0.09888	-0.00278
0.10924	0.00028
0.12134	0.00368
0.13801	0.00795
0.15410	0.01158
0.17418	0.01555
0.20170	0.02009
0.23746	0.02472
0.28936	0.02979
0.36702	0.03494
0.42553	0.03665
0.48611	0.03686
0.55179	0.03530

0.59662	0.03311
0.63664	0.03065
0.68117	0.02753
0.72092	0.02437
0.74649	0.02211
0.76942	0.01987
0.79449	0.01726
0.81715	0.01482
0.83286	0.01307
0.86542	0.00942
0.89865	0.00560
0.93156	0.00164
0.94816	-0.00027
0.96913	-0.00295
0.97297	-0.00339
0.98225	-0.00432
0.98749	-0.00483
0.99110	-0.00488
0.99400	-0.00453
1.00000	-0.00070

CG Ultimate

from plans

x/c	y/c
1.00000	0.00289
0.99754	0.00286
0.99070	0.00309
0.98037	0.00403
0.96698	0.00581
0.95044	0.00827
0.93064	0.01110
0.90775	0.01398
0.88202	0.01697
0.85370	0.02045
0.82309	0.02458
0.79048	0.02905
0.75616	0.03344
0.72043	0.03760
0.68359	0.04151
0.64594	0.04523
0.60778	0.04879
0.56937	0.05216
0.53099	0.05521
0.49265	0.05787
0.45435	0.06011
0.41638	0.06192
0.37887	0.06317
0.34204	0.06372
0.30609	0.06346
0.27120	0.06239
0.23760	0.06056
0.20549	0.05808
0.17504	0.05503
0.14648	0.05147
0.11999	0.04739
0.09576	0.04277
0.07395	0.03767
0.05468	0.03228
0.03811	0.02686
0.02433	0.02154
0.01338	0.01623
0.00548	0.01072
0.00098	0.00491
0.00000	0.00053
0.00098	-0.00390
0.00548	-0.00994
0.01338	-0.01581
0.02433	-0.02154
0.03811	-0.02726
0.05468	-0.03298
0.07395	-0.03857
0.09576	-0.04382
0.11999	-0.04860
0.14648	-0.05284
0.17504	-0.05651
0.20549	-0.05960
0.23760	-0.06203
0.27120	-0.06373
0.30609	-0.06462
0.34204	-0.06475
0.37887	-0.06425
0.41638	-0.06325
0.45435	-0.06187
0.49265	-0.06012
0.53099	-0.05795
0.56937	-0.05532
0.60778	-0.05221
0.64594	-0.04869
0.68359	-0.04488
0.72043	-0.04093
0.75616	-0.03696
0.79048	-0.03306
0.82309	-0.02926
0.85370	-0.02551
0.88202	-0.02198
0.90775	-0.01895
0.93064	-0.01657
0.95044	-0.01466
0.96698	-0.01268
0.98037	-0.01017
0.99070	-0.00709
0.99754	-0.00418
0.99998	-0.00292

CG Ultimate

actual

x/c	y/c
1.00000	-0.00067
0.99979	0.00422
0.99763	0.00477
0.99452	0.00520
0.99146	0.00573
0.98731	0.00636
0.98286	0.00713
0.97735	0.00785
0.97155	0.00879
0.96656	0.00957
0.95995	0.01058
0.95269	0.01168
0.94530	0.01263
0.93725	0.01383
0.92832	0.01498
0.91680	0.01659
0.90705	0.01798
0.88595	0.02072
0.86357	0.02373
0.83799	0.02703
0.79691	0.03219
0.76136	0.03633
0.70765	0.04198
0.65575	0.04709
0.61239	0.05103
0.55774	0.05543
0.49904	0.05945
0.42592	0.06331
0.37154	0.06497
0.31313	0.06522
0.28340	0.06450
0.26475	0.06374
0.24421	0.06258
0.23050	0.06162
0.21256	0.06018
0.20421	0.05945
0.19807	0.05885
0.19166	0.05822
0.18425	0.05743
0.17556	0.05645
0.16193	0.05480
0.15177	0.05344
0.14464	0.05243
0.13653	0.05121
0.12856	0.04992
0.12265	0.04891
0.11347	0.04723
0.10466	0.04547
0.09782	0.04402
0.08723	0.04160
0.07698	0.03913
0.06785	0.03686
0.06020	0.03486
0.05107	0.03220
0.04335	0.02975
0.03636	0.02730
0.02376	0.02208
0.02027	0.02037
0.01187	0.01574
0.01036	0.01480
0.00556	0.01113
0.00410	0.00966
0.00278	0.00803
0.00156	0.00605
0.00051	0.00349
0.00015	-0.00187
0.00160	-0.00614
0.00525	-0.01077
0.00949	-0.01411
0.01547	-0.01783
0.01747	-0.01897
0.02568	-0.02309
0.03176	-0.02561
0.04262	-0.02961
0.05183	-0.03258
0.06325	-0.03587
0.07456	-0.03892
0.08063	-0.04045
0.09114	-0.04285
0.10026	-0.04482
0.11320	-0.04742
0.12016	-0.04872
0.13373	-0.05107
0.14544	-0.05290
0.16125	-0.05514
0.19044	-0.05851
0.28035	-0.06468
0.34365	-0.06558
0.40042	-0.06432
0.46258	-0.06150
0.52631	-0.05732
0.60849	-0.05069
0.66893	-0.04486
0.72441	-0.03886
0.76846	-0.03373
0.80638	-0.02901
0.83540	-0.02548
0.86930	-0.02109
0.88681	-0.01899
0.90308	-0.01693
0.91379	-0.01556
0.92362	-0.01432
0.93274	-0.01314
0.94073	-0.01209
0.94892	-0.01106
0.95879	-0.00975
0.96243	-0.00919
0.96481	-0.00900
0.96866	-0.00840
0.97555	-0.00758
0.97819	-0.00718
0.98686	-0.00614
0.98907	-0.00584
0.99503	-0.00495
0.99893	-0.00452
0.99931	-0.00422
1.00000	-0.00067

Clark-Y

smoothed

x/c	y/c
1.00000	0.00000
0.99572	0.00115
0.98296	0.00448
0.96194	0.00972
0.93301	0.01656
0.89668	0.02475
0.85355	0.03400
0.80438	0.04394
0.75000	0.05412
0.69134	0.06405
0.62941	0.07319
0.56526	0.08105
0.50000	0.08719
0.43474	0.09128
0.37059	0.09312
0.33928	0.09318
0.30866	0.09266
0.27886	0.09158
0.25000	0.08996
0.22221	0.08774
0.19562	0.08483
0.17033	0.08113
0.14645	0.07660
0.12408	0.07134
0.10332	0.06552
0.08427	0.05939

0.06699	0.05313
0.05156	0.04677
0.03806	0.04027
0.02653	0.03352
0.01704	0.02652
0.00961	0.01943
0.00428	0.01254
0.00107	0.00616
0.00000	0.00047
0.00107	-0.00453
0.00428	-0.00898
0.00961	-0.01296
0.01704	-0.01651
0.02653	-0.01959
0.03806	-0.02214
0.05156	-0.02414
0.06699	-0.02567
0.08427	-0.02680
0.10332	-0.02763
0.12408	-0.02816
0.14645	-0.02839
0.17033	-0.02832
0.19562	-0.02795
0.22221	-0.02734
0.25000	-0.02653
0.27886	-0.02559
0.30866	-0.02458
0.33928	-0.02351
0.37059	-0.02242
0.43474	-0.02018
0.50000	-0.01792
0.56526	-0.01566
0.62941	-0.01345
0.69134	-0.01131
0.75000	-0.00928
0.80438	-0.00741
0.85355	-0.00575
0.89668	-0.00429
0.93301	-0.00302
0.96194	-0.00190
0.98296	-0.00094
0.99572	-0.00025
1.00000	0.00000

Clark-Y (B)

actual

x/c	y/c
1.00000	0.00000
0.99961	-0.00043
0.99774	0.00001
0.99467	0.00081
0.99156	0.00148
0.98763	0.00244
0.98350	0.00334
0.97814	0.00468
0.97252	0.00597
0.96542	0.00757
0.95783	0.00930
0.94909	0.01130
0.93818	0.01377
0.92835	0.01594
0.91681	0.01853
0.90400	0.02135
0.88834	0.02480
0.87125	0.02846
0.85036	0.03283
0.82967	0.03703
0.80484	0.04190
0.77547	0.04749
0.74489	0.05309
0.71664	0.05800
0.67623	0.06452
0.64177	0.06955
0.61554	0.07307
0.57880	0.07754
0.53950	0.08170
0.50036	0.08518
0.45728	0.08808
0.36404	0.09119
0.32935	0.09119
0.27714	0.09001
0.24352	0.08819
0.20253	0.08441
0.19295	0.08320
0.18518	0.08216
0.16792	0.07949
0.14871	0.07596
0.11924	0.06928
0.11055	0.06697
0.10236	0.06467
0.08071	0.05782
0.05895	0.04958
0.03623	0.03891
0.02872	0.03459
0.02102	0.02940
0.00617	0.01527
0.00530	0.01402
0.00309	0.01042
0.00122	0.00613
0.00061	0.00408
0.00008	-0.00144
0.00135	-0.00604
0.00372	-0.00956
0.00726	-0.01283
0.01204	-0.01566
0.01734	-0.01785
0.02505	-0.02028
0.03602	-0.02274
0.04741	-0.02460
0.05961	-0.02611
0.07489	-0.02751
0.08742	-0.02833
0.10435	-0.02908
0.11851	-0.02948
0.13635	-0.02976
0.15529	-0.02980
0.17772	-0.02965
0.20105	-0.02932
0.22456	-0.02886
0.25333	-0.02820
0.29103	-0.02719
0.33195	-0.02596
0.36476	-0.02489
0.41121	-0.02327
0.46886	-0.02120
0.51708	-0.01937
0.55623	-0.01793
0.60864	-0.01591
0.66329	-0.01388
0.72282	-0.01171
0.77867	-0.00952
0.79730	-0.00904
0.82767	-0.00771
0.84193	-0.00703
0.85663	-0.00666
0.86700	-0.00598
0.87514	-0.00579
0.88372	-0.00530
0.89414	-0.00492
0.92106	-0.00366
0.93663	-0.00303
0.94772	-0.00263
0.95148	-0.00229
0.96048	-0.00180
0.96605	-0.00152
0.97737	-0.00094
0.98555	-0.00038
0.99747	0.00043
1.00000	0.00000

DH4009

true

x/c	y/c
1.00000	0.00027
0.99754	0.00167
0.99070	0.00430
0.98037	0.00618
0.96698	0.00756
0.95044	0.00903
0.93064	0.01081
0.90775	0.01289
0.88202	0.01524
0.85370	0.01783
0.82309	0.02066
0.79048	0.02369
0.75616	0.02685
0.72043	0.03006
0.68359	0.03320
0.64594	0.03614
0.60778	0.03874
0.56937	0.04089
0.53099	0.04253
0.49265	0.04364
0.45435	0.04424
0.41638	0.04439
0.37887	0.04415
0.34204	0.04355
0.30609	0.04259
0.27120	0.04125
0.23760	0.03952
0.20549	0.03743
0.17504	0.03501
0.14648	0.03229
0.11999	0.02939
0.09576	0.02627
0.07395	0.02300
0.05468	0.01975
0.03811	0.01663
0.02433	0.01357
0.01338	0.01041
0.00548	0.00704
0.00098	0.00320
0.00000	-0.00001
0.00098	-0.00285
0.00548	-0.00570
0.01338	-0.00805
0.02433	-0.01051
0.03811	-0.01311
0.05468	-0.01594
0.07395	-0.01903
0.09576	-0.02227
0.11999	-0.02551
0.14648	-0.02861
0.17504	-0.03147
0.20549	-0.03416
0.23760	-0.03673
0.27120	-0.03916
0.30609	-0.04133
0.34204	-0.04311
0.37887	-0.04438
0.41638	-0.04506
0.45435	-0.04509
0.49265	-0.04446
0.53099	-0.04318
0.56937	-0.04130
0.60778	-0.03887
0.64594	-0.03598
0.68359	-0.03272
0.72043	-0.02922
0.75616	-0.02559
0.79048	-0.02195
0.82309	-0.01843
0.85370	-0.01513
0.88202	-0.01215
0.90775	-0.00952
0.93064	-0.00726
0.95044	-0.00536
0.96698	-0.00382
0.98037	-0.00255
0.99070	-0.00146
0.99754	-0.00047
1.00000	-0.00001

DH4009

actual

x/c	y/c
1.00000	-0.00102
0.99963	0.00121
0.99565	0.00188
0.99174	0.00237
0.98744	0.00289
0.97928	0.00369
0.96193	0.00536
0.94718	0.00686
0.90523	0.01099
0.87281	0.01411
0.82423	0.01885
0.74299	0.02650
0.66030	0.03399
0.57363	0.04012
0.49265	0.04328
0.40745	0.04385
0.32343	0.04298
0.29106	0.04196

0.24435	0.03976
0.22567	0.03867
0.20593	0.03740
0.18350	0.03568
0.16038	0.03364
0.14535	0.03216
0.13021	0.03059
0.11785	0.02919
0.10335	0.02736
0.08860	0.02532
0.07742	0.02362
0.06516	0.02166
0.05480	0.01987
0.04722	0.01848
0.03577	0.01626
0.02901	0.01480
0.02206	0.01312
0.01584	0.01132
0.00816	0.00847
0.00515	0.00695
0.00273	0.00528
0.00113	0.00354
0.00004	0.00068
0.00013	-0.00118
0.00123	-0.00317
0.00384	-0.00491
0.00912	-0.00672
0.01306	-0.00777
0.01987	-0.00934
0.02525	-0.01042
0.02967	-0.01126
0.04531	-0.01399
0.05428	-0.01546
0.06386	-0.01701
0.07545	-0.01878
0.08523	-0.02026
0.09601	-0.02180
0.11196	-0.02394
0.12463	-0.02554
0.13963	-0.02727
0.15430	-0.02886
0.17546	-0.03091
0.19710	-0.03277
0.23506	-0.03584
0.27911	-0.03903
0.32199	-0.04145
0.40432	-0.04452
0.48932	-0.04424
0.57080	-0.04085
0.65354	-0.03497
0.73935	-0.02702
0.82042	-0.01843
0.85975	-0.01450
0.90212	-0.01025
0.92694	-0.00782
0.94658	-0.00609
0.95929	-0.00497
0.97645	-0.00345
0.98577	-0.00262
0.99369	-0.00187
0.99945	-0.00121
1.00000	-0.00102

E231

true

x/c	y/c
1.00000	0.00000
0.99630	0.00039
0.98554	0.00187
0.96845	0.00477
0.94569	0.00903
0.91774	0.01460
0.88524	0.02148
0.84897	0.02952
0.80968	0.03837
0.76810	0.04759
0.72481	0.05653
0.68007	0.06458
0.63395	0.07143
0.58671	0.07701
0.53865	0.08126
0.49016	0.08419
0.44164	0.08584
0.39357	0.08628
0.34646	0.08556
0.30081	0.08370
0.25710	0.08074
0.21576	0.07671
0.17721	0.07166
0.14182	0.06565
0.10989	0.05877
0.08171	0.05111
0.05750	0.04282
0.03741	0.03407
0.02158	0.02508
0.01005	0.01613
0.00284	0.00764
0.00018	0.00152
0.00000	0.00000
0.00020	-0.00139
0.00093	-0.00259
0.00217	-0.00382
0.00325	-0.00466
0.01291	-0.00966
0.02804	-0.01448
0.04843	-0.01890
0.07381	-0.02288
0.10394	-0.02638
0.13848	-0.02938
0.17704	-0.03189
0.21921	-0.03389
0.26450	-0.03541
0.31240	-0.03644
0.36238	-0.03700
0.41385	-0.03711
0.46624	-0.03677
0.51892	-0.03599
0.57130	-0.03477
0.62279	-0.03312
0.67278	-0.03104
0.72067	-0.02851
0.76591	-0.02537
0.80826	-0.02146
0.84766	-0.01692
0.88392	-0.01224
0.91658	-0.00796
0.94488	-0.00453
0.96803	-0.00209
0.98539	-0.00063
0.99628	-0.00007
1.00000	0.00000

E231

actual

x/c	y/c
1.00000	0.00013
0.99968	0.00056
0.99641	0.00113
0.99327	0.00151
0.98958	0.00175
0.98673	0.00205
0.98385	0.00232
0.98005	0.00243
0.97572	0.00298
0.97046	0.00350
0.96590	0.00381
0.96021	0.00460
0.95553	0.00521
0.94336	0.00687
0.93537	0.00824
0.92510	0.01004
0.90788	0.01344
0.89606	0.01578
0.88044	0.01916
0.86054	0.02343
0.84238	0.02741
0.81498	0.03345
0.79001	0.03901
0.76191	0.04512
0.72908	0.05192
0.69991	0.05757
0.66518	0.06349
0.61601	0.07020
0.50620	0.08009
0.42967	0.08315
0.37433	0.08348
0.34242	0.08293
0.30615	0.08160
0.27828	0.08009
0.22934	0.07626
0.20377	0.07356
0.20021	0.07313
0.17556	0.06988
0.15976	0.06748
0.14445	0.06486
0.13424	0.06298
0.12694	0.06154
0.11226	0.05846
0.09080	0.05334
0.06006	0.04400
0.04670	0.03868
0.03495	0.03311
0.02109	0.02514
0.01506	0.02071
0.01221	0.01825
0.00991	0.01598
0.00844	0.01438
0.00664	0.01227
0.00585	0.01128
0.00406	0.00886
0.00283	0.00707
0.00159	0.00505
0.00046	0.00267
0.00004	-0.00081
0.00101	-0.00394
0.00593	-0.00913
0.00868	-0.01081
0.01029	-0.01167
0.01402	-0.01337
0.01915	-0.01527
0.02599	-0.01731
0.03498	-0.01953
0.04835	-0.02211
0.05974	-0.02382
0.08094	-0.02645
0.10054	-0.02848
0.12679	-0.03082
0.15952	-0.03311
0.19425	-0.03479
0.20897	-0.03537
0.27278	-0.03729
0.35115	-0.03868
0.43503	-0.03906
0.49053	-0.03855
0.53408	-0.03773
0.58309	-0.03634
0.63576	-0.03424
0.69982	-0.03106
0.76319	-0.02659
0.82054	-0.02089
0.85066	-0.01741
0.86765	-0.01525
0.88503	-0.01292
0.89629	-0.01140
0.90547	-0.01010
0.91362	-0.00889
0.92047	-0.00804
0.92791	-0.00704
0.93294	-0.00637
0.93823	-0.00568
0.94233	-0.00520
0.94805	-0.00452
0.95126	-0.00418
0.95513	-0.00365
0.96320	-0.00293
0.96761	-0.00255
0.97353	-0.00209
0.98168	-0.00150
0.98544	-0.00128
0.98751	-0.00120
0.99033	-0.00103
0.99254	-0.00092
0.99545	-0.00081
0.99835	-0.00056
1.00000	0.00013

E374

true

x/c	y/c
1.00000	0.00000
0.99640	0.00045
0.98610	0.00204

0.97000	0.00485
0.94864	0.00846
0.92214	0.01264
0.89078	0.01747
0.85508	0.02297
0.81560	0.02905
0.77293	0.03560
0.72769	0.04246
0.68053	0.04944
0.63210	0.05629
0.58309	0.06269
0.53398	0.06821
0.48511	0.07252
0.43682	0.07544
0.38939	0.07685
0.34312	0.07670
0.29824	0.07507
0.25510	0.07217
0.21415	0.06817
0.17583	0.06319
0.14053	0.05734
0.10860	0.05073
0.08036	0.04351
0.05605	0.03581
0.03589	0.02781
0.02004	0.01973
0.00862	0.01186
0.00178	0.00459
0.00014	-0.00121
0.00437	-0.00622
0.01427	-0.01130
0.02935	-0.01600
0.04949	-0.02015
0.07454	-0.02369
0.10428	-0.02660
0.13845	-0.02890
0.17669	-0.03060
0.21861	-0.03175
0.26374	-0.03238
0.31158	-0.03255
0.36159	-0.03228
0.41320	-0.03163
0.46580	-0.03064
0.51877	-0.02931
0.57150	-0.02767
0.62336	-0.02569
0.67382	-0.02333
0.72243	-0.02059
0.76873	-0.01760
0.81228	-0.01450
0.85254	-0.01153
0.88892	-0.00882
0.92085	-0.00643
0.94783	-0.00432
0.96958	-0.00241
0.98594	-0.00091
0.99637	-0.00016
1.00000	0.00000

E374 (B)

actual

x/c	y/c
1.00000	-0.00002
0.99931	0.00133
0.99547	0.00193
0.99075	0.00259
0.98553	0.00323
0.97942	0.00403
0.97449	0.00478
0.96606	0.00620
0.95897	0.00737
0.95051	0.00873
0.94089	0.01035
0.92731	0.01261
0.91006	0.01542
0.88922	0.01879
0.87081	0.02175
0.84125	0.02638
0.80842	0.03160
0.76422	0.03834
0.72768	0.04383
0.66531	0.05294
0.60199	0.06157
0.54234	0.06862
0.41405	0.07751
0.34811	0.07769
0.28951	0.07547
0.22319	0.07015
0.17954	0.06464
0.14136	0.05815
0.11425	0.05246
0.09448	0.04762
0.08072	0.04382
0.06739	0.03970
0.05486	0.03534
0.04468	0.03137
0.03449	0.02693
0.02480	0.02219
0.01653	0.01759
0.00821	0.01213
0.00459	0.00912
0.00059	0.00354
0.00003	0.00062
0.00177	-0.00446
0.00544	-0.00769
0.00958	-0.01000
0.01319	-0.01157
0.02096	-0.01420
0.03009	-0.01668
0.03948	-0.01882
0.05464	-0.02171
0.06817	-0.02379
0.08950	-0.02626
0.10662	-0.02784
0.13833	-0.02995
0.16421	-0.03111
0.19316	-0.03200
0.23635	-0.03279
0.29760	-0.03315
0.35915	-0.03278
0.42077	-0.03187
0.48279	-0.03053
0.54271	-0.02895
0.60527	-0.02673
0.66857	-0.02379
0.73107	-0.02022
0.77240	-0.01746
0.81508	-0.01434
0.85142	-0.01141
0.87900	-0.00911
0.89676	-0.00765
0.91943	-0.00589
0.93686	-0.00452
0.95206	-0.00329
0.96402	-0.00250
0.97295	-0.00191
0.98200	-0.00151
0.99124	-0.00145
0.99865	-0.00133
1.00000	-0.00002

E387

true

x/c	y/c
1.00000	0.00000
0.99677	0.00043
0.98729	0.00180
0.97198	0.00423
0.95128	0.00763
0.92554	0.01184
0.89510	0.01679
0.86035	0.02242
0.82183	0.02866
0.78007	0.03540
0.73567	0.04249
0.68922	0.04975
0.64136	0.05696
0.59272	0.06390
0.54394	0.07020
0.49549	0.07546
0.44767	0.07936
0.40077	0.08173
0.35505	0.08247
0.31078	0.08156
0.26813	0.07908
0.22742	0.07529
0.18906	0.07037
0.15345	0.06448
0.12094	0.05775
0.09185	0.05033
0.06643	0.04238
0.04493	0.03408
0.02748	0.02562
0.01423	0.01726
0.00519	0.00931
0.00044	0.00234
0.00091	-0.00286
0.00717	-0.00682
0.01890	-0.01017
0.03596	-0.01265
0.05827	-0.01425
0.08569	-0.01500
0.11800	-0.01502
0.15490	-0.01441
0.19599	-0.01329
0.24083	-0.01177
0.28892	-0.00998
0.33968	-0.00804
0.39252	-0.00605
0.44679	-0.00410
0.50182	-0.00228
0.55694	-0.00065
0.61147	0.00074
0.66472	0.00186
0.71602	0.00268
0.76475	0.00320
0.81027	0.00342
0.85202	0.00337
0.88944	0.00307
0.92205	0.00258
0.94942	0.00196
0.97118	0.00132
0.98705	0.00071
0.99674	0.00021
1.00000	0.00000

E387 (C)

actual

x/c	y/c
1.00000	-0.00004
0.99775	0.00030
0.99490	0.00083
0.98916	0.00179
0.98377	0.00265
0.97857	0.00350
0.97304	0.00444
0.95471	0.00724
0.94125	0.00943
0.91072	0.01444
0.86734	0.02143
0.81841	0.02919
0.76405	0.03775
0.72165	0.04437
0.67815	0.05103
0.63306	0.05772
0.60125	0.06226
0.55582	0.06826
0.51639	0.07283
0.47426	0.07680
0.43704	0.07942
0.39061	0.08142
0.35110	0.08189
0.31039	0.08107
0.26971	0.07880
0.24664	0.07690
0.22312	0.07447
0.20811	0.07265
0.18639	0.06966
0.16791	0.06677
0.15241	0.06408
0.13544	0.06081
0.11967	0.05739
0.10549	0.05395
0.09129	0.05006
0.07801	0.04599
0.07092	0.04363
0.06102	0.04003
0.05171	0.03634
0.04266	0.03240
0.03186	0.02724
0.02647	0.02440
0.02193	0.02211
0.01698	0.01912
0.01099	0.01489

0.00856	0.01291
0.00502	0.00955
0.00185	0.00578
0.00204	-0.00620
0.00225	-0.00533
0.00754	-0.00813
0.01225	-0.00955
0.01984	-0.01107
0.02940	-0.01217
0.03719	-0.01280
0.05205	-0.01394
0.06821	-0.01476
0.08630	-0.01529
0.10585	-0.01550
0.14193	-0.01531
0.18213	-0.01445
0.22593	-0.01314
0.27282	-0.01157
0.32040	-0.00988
0.37440	-0.00792
0.42511	-0.00606
0.47362	-0.00433
0.52830	-0.00251
0.56702	-0.00131
0.61242	-0.00010
0.65251	0.00078
0.68264	0.00131
0.72101	0.00185
0.75797	0.00216
0.79610	0.00235
0.81817	0.00237
0.83916	0.00224
0.85910	0.00210
0.87580	0.00186
0.89258	0.00166
0.90913	0.00146
0.92618	0.00117
0.93786	0.00097
0.95278	0.00056
0.96676	0.00032
0.97787	0.00014
0.98634	-0.00001
0.99330	-0.00013
0.99909	-0.00030
1.00000	-0.00004

E387 (D)

actual

x/c	y/c
1.00000	-0.00135
0.99763	0.00099
0.98866	0.00242
0.97472	0.00450
0.96466	0.00598
0.95259	0.00776
0.93092	0.01107
0.91072	0.01416
0.89956	0.01589
0.87458	0.01992
0.85208	0.02353
0.82425	0.02807
0.79637	0.03261
0.77128	0.03671
0.70988	0.04650
0.65708	0.05456
0.60373	0.06223
0.55296	0.06884
0.51725	0.07293
0.48003	0.07659
0.45134	0.07878
0.42129	0.08051
0.38985	0.08168
0.36356	0.08209
0.33031	0.08178
0.31100	0.08117
0.28298	0.07976
0.26067	0.07816
0.24359	0.07660
0.22456	0.07461
0.20559	0.07231
0.18338	0.06917
0.16397	0.06602
0.14827	0.06316
0.13243	0.05995
0.11917	0.05700
0.10621	0.05381
0.09442	0.05069
0.07549	0.04502
0.06448	0.04130
0.03646	0.02983
0.02218	0.02241
0.00604	0.01077
0.00000	0.00020
0.00064	-0.00350
0.00293	-0.00708
0.00796	-0.01015
0.01521	-0.01240
0.02236	-0.01374
0.03356	-0.01510
0.04483	-0.01597
0.06992	-0.01699
0.09309	-0.01728
0.12025	-0.01717
0.14782	-0.01683
0.18064	-0.01612
0.22069	-0.01496
0.26316	-0.01360
0.30728	-0.01194
0.34289	-0.01049
0.38454	-0.00889
0.42809	-0.00724
0.48330	-0.00534
0.52653	-0.00403
0.56084	-0.00299
0.59585	-0.00207
0.63046	-0.00130
0.67065	-0.00049
0.70062	0.00005
0.79680	0.00116
0.87820	0.00108
0.93318	0.00006
0.98647	-0.00172
0.99296	-0.00192
1.00000	-0.00135

E472

true

x/c	y/c
1.00000	0.00000
0.99624	0.00041
0.98567	0.00197
0.96952	0.00464
0.94834	0.00764
0.92198	0.01059
0.89050	0.01364
0.85434	0.01691
0.81396	0.02039
0.76988	0.02407
0.72266	0.02791
0.67288	0.03186
0.62116	0.03587
0.56814	0.03987
0.51446	0.04379
0.46078	0.04754
0.40776	0.05103
0.35605	0.05415
0.30628	0.05680
0.25906	0.05884
0.21499	0.06013
0.17462	0.06049
0.13851	0.05960
0.10683	0.05695
0.07938	0.05236
0.05603	0.04599
0.03670	0.03810
0.02133	0.02903
0.00994	0.01924
0.00262	0.00929
0.00000	0.00000
0.00262	-0.00929
0.00994	-0.01924
0.02133	-0.02903
0.03670	-0.03810
0.05603	-0.04599
0.07938	-0.05236
0.10683	-0.05695
0.13851	-0.05960
0.17462	-0.06049
0.21499	-0.06013
0.25906	-0.05884
0.30628	-0.05680
0.35605	-0.05415
0.40776	-0.05103
0.46078	-0.04754
0.51446	-0.04379
0.56814	-0.03987
0.62116	-0.03587
0.67288	-0.03186
0.72266	-0.02791
0.76988	-0.02407
0.81396	-0.02039
0.85434	-0.01691
0.89050	-0.01364
0.92198	-0.01059
0.94834	-0.00764
0.96952	-0.00464
0.98567	-0.00197
0.99624	-0.00041
1.00000	0.00000

E472

actual

x/c	y/c
1.00000	0.00116
0.99967	0.00085
0.99854	0.00145
0.99502	0.00195
0.99199	0.00242
0.98561	0.00330
0.97832	0.00424
0.96990	0.00537
0.96075	0.00646
0.94552	0.00801
0.93057	0.00974
0.90938	0.01216
0.88580	0.01448
0.84508	0.01841
0.80350	0.02224
0.72162	0.02899
0.63926	0.03561
0.55470	0.04197
0.47255	0.04739
0.38930	0.05237
0.30566	0.05692
0.26567	0.05860
0.25048	0.05912
0.22426	0.05983
0.19880	0.06016
0.18379	0.06022
0.16713	0.06010
0.14314	0.05946
0.12824	0.05859
0.11478	0.05739
0.10124	0.05577
0.08729	0.05356
0.07360	0.05065
0.06121	0.04729
0.05305	0.04472
0.04291	0.04104
0.03400	0.03711
0.02632	0.03298
0.01938	0.02845
0.01352	0.02374
0.00744	0.01753
0.00312	0.01146
0.00099	0.00640
0.00326	-0.01202
0.00919	-0.01942
0.01256	-0.02257
0.01791	-0.02668
0.02231	-0.02948
0.02684	-0.03201
0.03501	-0.03589
0.04335	-0.03932
0.05230	-0.04266
0.06116	-0.04569
0.06774	-0.04775
0.08206	-0.05148
0.09567	-0.05426
0.12342	-0.05799
0.14011	-0.05931

0.15569	-0.06002
0.17569	-0.06054
0.19600	-0.06070
0.21619	-0.06040
0.23393	-0.05984
0.27056	-0.05821
0.31284	-0.05595
0.39997	-0.05079
0.48303	-0.04585
0.56586	-0.04051
0.65006	-0.03398
0.73296	-0.02686
0.81498	-0.01949
0.85590	-0.01589
0.89594	-0.01199
0.92246	-0.00931
0.94223	-0.00734
0.96864	-0.00448
0.98092	-0.00324
0.98967	-0.00223
0.99512	-0.00153
0.99914	-0.00085
1.00000	0.00116

ESA

from plans

x/c	y/c
1.00000	0.00445
0.99754	0.00452
0.99070	0.00460
0.98037	0.00452
0.96698	0.00421
0.95044	0.00375
0.93064	0.00317
0.90775	0.00239
0.88202	0.00158
0.85370	0.00213
0.82309	0.00557
0.79048	0.01020
0.75616	0.01505
0.72043	0.02029
0.68359	0.02564
0.64594	0.03080
0.60778	0.03577
0.56937	0.04063
0.53099	0.04535
0.49265	0.04983
0.45435	0.05396
0.41638	0.05769
0.37887	0.06104
0.34204	0.06400
0.30609	0.06643
0.27120	0.06828
0.23760	0.06952
0.20549	0.07019
0.17504	0.07022
0.14648	0.06944
0.11999	0.06760
0.09576	0.06444
0.07395	0.05976
0.05468	0.05353
0.03811	0.04599
0.02433	0.03760
0.01338	0.02828
0.00548	0.01718
0.00098	0.00598
0.00000	0.00002
0.00098	-0.00398
0.00548	-0.00753
0.01338	-0.01041
0.02433	-0.01355
0.03811	-0.01679
0.05468	-0.01972
0.07395	-0.02239
0.09576	-0.02501
0.11999	-0.02748
0.14648	-0.02970
0.17504	-0.03194
0.20549	-0.03441
0.23760	-0.03681
0.27120	-0.03890
0.30609	-0.04067
0.34204	-0.04224
0.37887	-0.04352
0.41638	-0.04430
0.45435	-0.04447
0.49265	-0.04408
0.53099	-0.04325
0.56937	-0.04196
0.60778	-0.04022
0.64594	-0.03816
0.68359	-0.03600
0.72043	-0.03390
0.75616	-0.03188
0.79048	-0.02981
0.82309	-0.02739
0.85370	-0.02413
0.88202	-0.02025
0.90775	-0.01670
0.93064	-0.01369
0.95044	-0.01108
0.96698	-0.00875
0.98037	-0.00679
0.99070	-0.00538
0.99754	-0.00462
0.99991	-0.00442

ESA

actual

x/c	y/c
1.00000	-0.00010
0.99897	0.00549
0.99439	0.00538
0.99096	0.00533
0.98700	0.00520
0.98304	0.00517
0.97752	0.00501
0.97264	0.00501
0.96857	0.00482
0.96213	0.00481
0.95569	0.00474
0.95034	0.00467
0.94258	0.00462
0.93610	0.00456
0.93106	0.00449
0.92055	0.00435
0.91058	0.00421
0.90128	0.00426
0.89113	0.00420
0.88060	0.00425
0.86936	0.00458
0.86014	0.00509
0.84711	0.00607
0.83379	0.00697
0.81882	0.00847
0.80549	0.01017
0.78308	0.01302
0.75868	0.01637
0.72839	0.02058
0.69270	0.02562
0.65003	0.03162
0.60577	0.03738
0.54712	0.04471
0.41555	0.05874
0.36553	0.06278
0.32560	0.06524
0.28662	0.06707
0.21883	0.06874
0.20011	0.06858
0.17704	0.06821
0.12408	0.06468
0.08955	0.05889
0.06479	0.05218
0.04187	0.04290
0.02900	0.03598
0.01124	0.02080
0.00267	0.00760
0.00309	0.00838
0.00114	0.00464
0.00040	-0.00271
0.00165	-0.00552
0.00733	-0.01026
0.01209	-0.01251
0.02173	-0.01539
0.02580	-0.01635
0.03404	-0.01798
0.04242	-0.01940
0.04603	-0.01996
0.05482	-0.02125
0.06150	-0.02203
0.07006	-0.02295
0.07993	-0.02380
0.09049	-0.02472
0.10186	-0.02577
0.11425	-0.02685
0.12511	-0.02798
0.14094	-0.02954
0.15286	-0.03060
0.17459	-0.03234
0.19159	-0.03356
0.21325	-0.03500
0.23570	-0.03621
0.27044	-0.03782
0.29456	-0.03873
0.34624	-0.04075
0.39783	-0.04186
0.48158	-0.04167
0.56006	-0.03967
0.64219	-0.03602
0.72671	-0.03096
0.81320	-0.02482
0.86339	-0.01918
0.94637	-0.00909
0.97338	-0.00677
0.98804	-0.00549
1.00000	-0.00010

Falcon 56 Mk II

from plans

x/c	y/c
1.00000	0.00134
0.99754	0.00277
0.99070	0.00563
0.98037	0.00801
0.96698	0.00969
0.95044	0.01148
0.93064	0.01407
0.90775	0.01752
0.88202	0.02159
0.85370	0.02607
0.82309	0.03077
0.79048	0.03559
0.75616	0.04048
0.72043	0.04538
0.68359	0.05023
0.64594	0.05496
0.60778	0.05955
0.56937	0.06397
0.53099	0.06816
0.49265	0.07204
0.45435	0.07551
0.41638	0.07844
0.37887	0.08074
0.34204	0.08234
0.30609	0.08316
0.27120	0.08313
0.23760	0.08215
0.20549	0.08017
0.17504	0.07713
0.14648	0.07309
0.11999	0.06811
0.09576	0.06216
0.07395	0.05518
0.05468	0.04708
0.03811	0.03805
0.02433	0.02871
0.01338	0.01979
0.00548	0.01171
0.00098	0.00438
0.00000	-0.00070
0.00098	-0.00567
0.00548	-0.01217
0.01338	-0.01802
0.02433	-0.02327
0.03811	-0.02802
0.05468	-0.03227
0.07395	-0.03591
0.09576	-0.03898
0.11999	-0.04163
0.14648	-0.04405
0.17504	-0.04638
0.20549	-0.04868
0.23760	-0.05085
0.27120	-0.05259

0.30609	-0.05361
0.34204	-0.05392
0.37887	-0.05368
0.41638	-0.05299
0.45435	-0.05190
0.49265	-0.05041
0.53099	-0.04857
0.56937	-0.04641
0.60778	-0.04398
0.64594	-0.04129
0.68359	-0.03829
0.72043	-0.03502
0.75616	-0.03156
0.79048	-0.02805
0.82309	-0.02461
0.85370	-0.02131
0.88202	-0.01819
0.90775	-0.01528
0.93064	-0.01270
0.95044	-0.01063
0.96698	-0.00920
0.98037	-0.00799
0.99070	-0.00586
0.99754	-0.00301
1.00000	-0.00155

Falcon 56 Mk II

actual

x/c	y/c
1.00000	0.00072
0.99981	0.00256
0.99831	0.00376
0.99692	0.00471
0.99591	0.00518
0.99399	0.00587
0.99163	0.00644
0.98873	0.00716
0.98547	0.00780
0.98092	0.00851
0.97388	0.00966
0.96503	0.01099
0.95791	0.01209
0.94723	0.01381
0.93534	0.01566
0.92157	0.01773
0.91018	0.01946
0.89707	0.02151
0.88403	0.02351
0.86371	0.02660
0.84116	0.03002
0.81547	0.03380
0.79392	0.03694
0.76512	0.04109
0.73871	0.04482
0.69344	0.05073
0.65156	0.05598
0.60767	0.06126
0.55577	0.06717
0.48963	0.07405
0.42395	0.07946
0.38681	0.08176
0.36952	0.08263
0.34393	0.08364
0.30216	0.08440
0.27804	0.08434
0.24149	0.08350
0.21127	0.08186
0.18708	0.07967
0.14487	0.07378
0.12413	0.06982
0.11346	0.06746
0.09589	0.06296
0.08531	0.05986
0.07386	0.05599
0.06100	0.05081
0.04783	0.04444
0.03438	0.03670
0.02217	0.02845
0.00710	0.01541
0.00215	0.00828
0.00003	-0.00100
0.00027	-0.00289
0.00176	-0.00808
0.00205	-0.00871
0.00423	-0.01227
0.00724	-0.01528
0.01045	-0.01762
0.01340	-0.01934
0.01835	-0.02165
0.02222	-0.02313
0.02611	-0.02446
0.03186	-0.02617
0.03596	-0.02733
0.04063	-0.02862
0.04759	-0.03035
0.05628	-0.03221
0.06232	-0.03334
0.07128	-0.03485
0.08676	-0.03707
0.09555	-0.03812
0.10630	-0.03926
0.12292	-0.04085
0.14230	-0.04254
0.16882	-0.04471
0.19853	-0.04695
0.22602	-0.04879
0.26486	-0.05051
0.31892	-0.05136
0.37937	-0.05085
0.43291	-0.04948
0.50522	-0.04656
0.56837	-0.04325
0.68111	-0.03514
0.81276	-0.02295
0.86843	-0.01727
0.90779	-0.01313
0.95055	-0.00843
0.96346	-0.00693
0.98420	-0.00435
0.99540	-0.00257
1.00000	0.00072

Goe 417a

true

x/c	y/c
1.00000	0.00000
0.90000	0.02400
0.80000	0.04250
0.70000	0.05450
0.60000	0.06400
0.50000	0.07000
0.40000	0.07350
0.30000	0.07150
0.20000	0.06300
0.15000	0.05700
0.10000	0.04850
0.07500	0.04150
0.05000	0.03250
0.02500	0.02200
0.01250	0.01550
0.00000	0.00000
0.01250	-0.01400
0.02500	-0.01000
0.05000	0.00100
0.07500	0.01050
0.10000	0.01850
0.15000	0.02750
0.20000	0.03400
0.30000	0.04250
0.40000	0.04450
0.50000	0.04100
0.60000	0.03500
0.70000	0.02550
0.80000	0.01350
0.90000	-0.00150
1.00000	0.00000

Goe 417a

actual

x/c	y/c
1.00000	-0.00021
0.99906	0.00069
0.99576	0.00147
0.99137	0.00230
0.98628	0.00340
0.98148	0.00437
0.97608	0.00567
0.97053	0.00679
0.96164	0.00854
0.95444	0.01001
0.94677	0.01162
0.93872	0.01333
0.92934	0.01544
0.92083	0.01708
0.91390	0.01867
0.90286	0.02104
0.89253	0.02316
0.87927	0.02582
0.86475	0.02874
0.85053	0.03140
0.82730	0.03553
0.80023	0.04011
0.76884	0.04490
0.72573	0.05057
0.68543	0.05526
0.65287	0.05867
0.61406	0.06238
0.56415	0.06626
0.51887	0.06905
0.47655	0.07114
0.42474	0.07287
0.38244	0.07308
0.33922	0.07245
0.27353	0.06952
0.22092	0.06520
0.18006	0.06044
0.14635	0.05575
0.13479	0.05393
0.10807	0.04911
0.08381	0.04351
0.06495	0.03824
0.04334	0.03083
0.02246	0.02173
0.00798	0.01250
0.00002	-0.00056
0.00063	-0.00350
0.00122	-0.00527
0.00252	-0.00791
0.00594	-0.01149
0.00875	-0.01312
0.01117	-0.01380
0.01577	-0.01367
0.02046	-0.01233
0.02669	-0.00985
0.03355	-0.00701
0.04039	-0.00434
0.04698	-0.00177
0.05359	0.00082
0.06037	0.00335
0.07488	0.00856
0.08588	0.01219
0.09960	0.01618
0.11425	0.01973
0.13054	0.02291
0.14767	0.02565
0.16376	0.02798
0.17788	0.02993
0.19910	0.03262
0.22028	0.03487
0.24889	0.03727
0.29111	0.04055
0.31096	0.04138
0.35635	0.04269
0.41360	0.04259
0.46197	0.04109
0.52150	0.03821
0.57636	0.03461
0.62646	0.03063
0.66527	0.02729
0.70160	0.02348
0.72401	0.02073
0.75093	0.01744
0.78073	0.01384
0.85409	0.00428
0.89354	0.00061
0.91439	-0.00005
0.92792	-0.00036
0.94803	-0.00038
0.96566	-0.00046
0.97495	-0.00055
0.98745	-0.00066
0.99566	-0.00070
1.00000	-0.00021

LRN1007

true

x/c	y/c
1.00000	0.00000
0.96050	0.00780
0.92050	0.01620
0.88050	0.02509
0.84050	0.03467
0.80050	0.04498
0.76050	0.05585
0.72050	0.06679
0.68050	0.07608
0.63686	0.08280
0.58975	0.08837
0.54195	0.09203
0.49384	0.09433
0.44577	0.09527
0.39818	0.09491
0.35152	0.09334
0.30628	0.09056
0.26288	0.08666
0.22176	0.08167
0.18330	0.07570
0.14786	0.06881
0.11573	0.06113
0.08722	0.05281
0.06252	0.04400
0.04183	0.03492
0.02626	0.02635
0.01288	0.01690
0.00471	0.00864
0.00050	0.00240
0.00050	0.00005
0.00471	0.00100
0.01288	0.00322
0.02626	0.00646
0.04183	0.00922
0.06252	0.01225
0.08722	0.01499
0.11573	0.01703
0.14786	0.01863
0.18330	0.01948
0.22176	0.02009
0.26288	0.02052
0.30628	0.02081
0.35152	0.02151
0.39818	0.02217
0.44577	0.02277
0.49384	0.02310
0.54195	0.02354
0.58975	0.02378
0.63696	0.02353
0.68050	0.02242
0.72050	0.02067
0.76050	0.01839
0.80050	0.01595
0.84050	0.01286
0.88050	0.00972
0.92050	0.00649
0.96050	0.00320
1.00000	0.00000

LRN1007 (B)

actual

x/c	y/c
1.00000	-0.00005
0.99676	0.00206
0.99217	0.00298
0.98818	0.00377
0.98352	0.00474
0.97992	0.00547
0.97203	0.00705
0.96428	0.00869
0.95552	0.01054
0.94499	0.01292
0.93613	0.01488
0.92249	0.01795
0.90941	0.02091
0.89648	0.02391
0.87562	0.02890
0.85283	0.03454
0.82725	0.04110
0.79976	0.04852
0.75591	0.06053
0.70080	0.07432
0.64414	0.08413
0.55453	0.09332
0.47822	0.09689
0.39869	0.09709
0.36344	0.09606
0.32424	0.09388
0.28325	0.09038
0.23795	0.08524
0.21674	0.08236
0.20433	0.08050
0.19116	0.07838
0.17707	0.07592
0.16103	0.07283
0.14501	0.06942
0.13441	0.06695
0.11777	0.06268
0.10745	0.05975
0.09590	0.05620
0.08765	0.05344
0.06308	0.04407
0.04991	0.03818
0.02843	0.02669
0.01450	0.00312
0.01834	0.00395
0.02475	0.00524
0.02976	0.00621
0.03501	0.00722
0.03904	0.00797
0.04478	0.00894
0.05222	0.01008
0.05844	0.01095
0.06367	0.01163
0.06894	0.01226
0.07574	0.01301
0.08406	0.01380
0.09459	0.01469
0.11009	0.01579
0.12870	0.01677
0.15319	0.01763
0.17086	0.01806
0.19082	0.01844
0.21042	0.01875
0.22582	0.01896
0.24709	0.01923
0.26916	0.01950
0.30436	0.01988
0.46088	0.02173
0.57764	0.02247
0.66801	0.02168
0.70924	0.02013
0.78989	0.01533
0.87446	0.00858
0.92190	0.00470
0.95511	0.00195
0.97424	0.00035
0.99702	-0.00141
1.00000	-0.00005

PT-40

from plans

x/c	y/c
1.00000	0.00136
0.99754	0.00196
0.99070	0.00369
0.98037	0.00619
0.96698	0.00928
0.95044	0.01291
0.93064	0.01700
0.90775	0.02145
0.88202	0.02618
0.85370	0.03115
0.82309	0.03632
0.79048	0.04163
0.75616	0.04703
0.72043	0.05243
0.68359	0.05775
0.64594	0.06286
0.60778	0.06761
0.56937	0.07188
0.53099	0.07560
0.49265	0.07879
0.45435	0.08143
0.41638	0.08344
0.37887	0.08474
0.34204	0.08525
0.30609	0.08499
0.27120	0.08403
0.23760	0.08247
0.20549	0.08039
0.17504	0.07778
0.14648	0.07458
0.11999	0.07066
0.09576	0.06589
0.07395	0.06013
0.05468	0.05339
0.03811	0.04606
0.02433	0.03870
0.01338	0.03072
0.00548	0.02062
0.00098	0.00850
0.00000	-0.00074
0.00098	-0.00982
0.00548	-0.02134
0.01338	-0.03082
0.02433	-0.03745
0.03811	-0.04067
0.05468	-0.04114
0.07395	-0.04030
0.09576	-0.03911
0.11999	-0.03798
0.14648	-0.03690
0.17504	-0.03579
0.20549	-0.03459
0.23760	-0.03330
0.27120	-0.03192
0.30609	-0.03047
0.34204	-0.02896
0.37887	-0.02741
0.41638	-0.02584
0.45435	-0.02425
0.49265	-0.02265
0.53099	-0.02105
0.56937	-0.01943
0.60778	-0.01780
0.64594	-0.01617
0.68359	-0.01456
0.72043	-0.01299
0.75616	-0.01147
0.79048	-0.01002
0.82309	-0.00865
0.85370	-0.00738
0.88202	-0.00622
0.90775	-0.00518
0.93064	-0.00425
0.95044	-0.00344
0.96698	-0.00275
0.98037	-0.00219
0.99070	-0.00175
0.99754	-0.00145
1.00000	-0.00135

PT-40 (A)

actual

x/c	y/c
1.00000	0.00026
0.99410	0.00433
0.98774	0.00574
0.97648	0.00818
0.96281	0.01097
0.95203	0.01315
0.93932	0.01588
0.92436	0.01889
0.90804	0.02201
0.89054	0.02527
0.87111	0.02865
0.85123	0.03226
0.83090	0.03581
0.81232	0.03893
0.78171	0.04407
0.74995	0.04911
0.71294	0.05458
0.66888	0.06056
0.62465	0.06598
0.57534	0.07153
0.52641	0.07621
0.47407	0.08014

0.40779	0.08345
0.37067	0.08423
0.33109	0.08440
0.30633	0.08442
0.28160	0.08418
0.25631	0.08355
0.24290	0.08299
0.22637	0.08217
0.20934	0.08115
0.19601	0.08020
0.18228	0.07909
0.16787	0.07771
0.15716	0.07643
0.14567	0.07477
0.13939	0.07380
0.13122	0.07244
0.12047	0.07049
0.09517	0.06511
0.06797	0.05773
0.04442	0.04921
0.02628	0.04043
0.00765	0.02406
0.00538	0.02027
0.00280	0.01482
0.00130	0.01011
0.00048	0.00640
0.00028	-0.00453
0.00130	-0.00998
0.00364	-0.01656
0.00668	-0.02204
0.00846	-0.02448
0.01046	-0.02683
0.01242	-0.02885
0.01751	-0.03319
0.02252	-0.03627
0.02715	-0.03799
0.03333	-0.03883
0.03765	-0.03896
0.04418	-0.03904
0.05434	-0.03899
0.06589	-0.03900
0.08060	-0.03881
0.10547	-0.03826
0.13870	-0.03733
0.18027	-0.03601
0.25373	-0.03328
0.33254	-0.02999
0.40946	-0.02687
0.49229	-0.02373
0.55476	-0.02142
0.61981	-0.01889
0.67195	-0.01681
0.72496	-0.01476
0.77299	-0.01284
0.81986	-0.01090
0.85270	-0.00949
0.87816	-0.00841
0.89822	-0.00749
0.91953	-0.00659
0.96876	-0.00429
1.00000	0.00026

PT-40 (B)

actual

x/c	y/c
1.00000	-0.00053
0.99942	0.00113
0.99641	0.00180
0.99353	0.00227
0.99064	0.00296
0.98778	0.00356
0.98550	0.00402
0.98151	0.00485
0.97805	0.00549
0.97175	0.00677
0.96450	0.00838
0.95639	0.01001
0.94604	0.01212
0.93394	0.01468
0.92163	0.01733
0.90981	0.01989
0.89499	0.02309
0.88464	0.02513
0.87171	0.02800
0.85664	0.03108
0.82359	0.03691
0.77593	0.04476
0.74351	0.04992
0.69895	0.05666
0.65895	0.06229
0.61452	0.06788
0.55655	0.07425
0.48689	0.08049
0.42497	0.08432
0.37654	0.08580
0.33697	0.08644
0.29786	0.08559
0.27466	0.08533
0.26151	0.08496
0.24912	0.08450
0.23003	0.08368
0.21262	0.08261
0.19880	0.08172
0.16697	0.07866
0.14216	0.07613
0.11771	0.07303
0.09783	0.06746
0.06952	0.05895
0.06358	0.05710
0.05777	0.05530
0.05187	0.05341
0.03887	0.04811
0.03269	0.04502
0.02148	0.03804
0.01040	0.02803
0.00702	0.02368
0.00279	0.01463
0.00000	0.00014
0.00001	-0.00074
0.00015	-0.00424
0.00068	-0.00773
0.00112	-0.00956
0.00214	-0.01274
0.00357	-0.01587
0.00613	-0.02018
0.00889	-0.02365
0.01111	-0.02581
0.01518	-0.02910
0.01993	-0.03236
0.02449	-0.03448
0.03204	-0.03740
0.04013	-0.03887
0.04528	-0.03936
0.04789	-0.03942
0.05806	-0.03932
0.05795	-0.03932
0.08655	-0.03810
0.11214	-0.03692
0.13219	-0.03602
0.16778	-0.03466
0.20103	-0.03336
0.24527	-0.03172
0.27746	-0.03012
0.29565	-0.02972
0.31646	-0.02896
0.33522	-0.02820
0.36902	-0.02677
0.41424	-0.02495
0.46249	-0.02308
0.50915	-0.02117
0.56019	-0.01905
0.61393	-0.01685
0.71984	-0.01241
0.84531	-0.00732
0.89903	-0.00505
0.93448	-0.00380
0.96878	-0.00253
0.98703	-0.00175
0.99905	-0.00113
1.00000	-0.00053

RG14

true

x/c	y/c
1.00000	0.00000
0.99667	0.00045
0.98707	0.00195
0.97194	0.00446
0.95169	0.00760
0.92645	0.01112
0.89647	0.01506
0.86218	0.01937
0.82405	0.02394
0.78255	0.02867
0.73817	0.03344
0.69145	0.03811
0.64292	0.04256
0.59310	0.04666
0.54254	0.05029
0.49178	0.05334
0.44133	0.05570
0.39172	0.05727
0.34343	0.05800
0.29692	0.05780
0.25262	0.05666
0.21095	0.05456
0.17226	0.05151
0.13689	0.04755
0.10513	0.04275
0.07725	0.03721
0.05344	0.03104
0.03388	0.02439
0.01867	0.01744
0.00786	0.01048
0.00150	0.00390
0.00000	0.00000
0.00022	-0.00139
0.00465	-0.00595
0.01447	-0.01055
0.02938	-0.01468
0.04931	-0.01824
0.07412	-0.02116
0.10363	-0.02346
0.13756	-0.02515
0.17559	-0.02629
0.21730	-0.02690
0.26227	-0.02705
0.30998	-0.02680
0.35989	-0.02619
0.41144	-0.02527
0.46401	-0.02409
0.51701	-0.02267
0.56979	-0.02105
0.62175	-0.01921
0.67227	-0.01718
0.72073	-0.01480
0.76689	-0.01194
0.81057	-0.00890
0.85121	-0.00612
0.88811	-0.00382
0.92063	-0.00209
0.94817	-0.00091
0.97026	-0.00019
0.98654	0.00012
0.99659	0.00008
1.00000	0.00000

RG14

actual

x/c	y/c
1.00000	-0.00037
0.99971	0.00008
0.99595	0.00054
0.99171	0.00112
0.98457	0.00197
0.97631	0.00321
0.96893	0.00425
0.96007	0.00560
0.94838	0.00745
0.93720	0.00929
0.92454	0.01122
0.91242	0.01311
0.89847	0.01514
0.87959	0.01772
0.85576	0.02086
0.82437	0.02458
0.78593	0.02897
0.73786	0.03395
0.67689	0.03975
0.60536	0.04526
0.52881	0.05057

0.44510	0.05455
0.39215	0.05641
0.35660	0.05727
0.32859	0.05758
0.29732	0.05763
0.26664	0.05717
0.23909	0.05629
0.22037	0.05544
0.19633	0.05403
0.17800	0.05268
0.15492	0.05059
0.14112	0.04911
0.12668	0.04733
0.11117	0.04508
0.09562	0.04248
0.08190	0.03984
0.07038	0.03732
0.05918	0.03448
0.04027	0.02845
0.02247	0.02081
0.00848	0.01238
0.00193	0.00536
0.00032	0.00214
0.00001	-0.00039
0.00059	-0.00293
0.00422	-0.00691
0.01012	-0.00971
0.01815	-0.01233
0.02501	-0.01411
0.03093	-0.01548
0.03907	-0.01708
0.04591	-0.01822
0.05488	-0.01949
0.06193	-0.02032
0.06972	-0.02115
0.07565	-0.02170
0.08711	-0.02262
0.09557	-0.02319
0.10406	-0.02375
0.11256	-0.02430
0.11956	-0.02472
0.12703	-0.02515
0.13820	-0.02564
0.14592	-0.02589
0.15898	-0.02609
0.17414	-0.02616
0.19435	-0.02631
0.20608	-0.02646
0.22402	-0.02675
0.24654	-0.02703
0.27323	-0.02703
0.29338	-0.02685
0.31462	-0.02658
0.33159	-0.02640
0.39295	-0.02571
0.50544	-0.02298
0.55591	-0.02077
0.59593	-0.01866
0.72028	-0.01250
0.77963	-0.00884
0.83494	-0.00582
0.86148	-0.00454
0.91064	-0.00234
0.94488	-0.00135
0.97090	-0.00052
0.98521	-0.00017
0.99518	-0.00008
1.00000	-0.00037

RG15

true

x/c	y/c
1.00000	0.00000
0.99671	0.00054
0.98726	0.00229
0.97237	0.00514
0.95248	0.00865
0.92764	0.01254
0.89810	0.01685
0.86427	0.02152
0.82660	0.02644
0.78557	0.03149
0.74165	0.03654
0.69537	0.04146
0.64723	0.04612
0.59778	0.05039
0.54753	0.05414
0.49702	0.05727
0.44676	0.05966
0.39727	0.06123
0.34902	0.06190
0.30248	0.06162
0.25809	0.06036
0.21624	0.05810
0.17730	0.05486
0.14161	0.05068
0.10945	0.04564
0.08108	0.03985
0.05673	0.03343
0.03658	0.02654
0.02076	0.01935
0.00932	0.01214
0.00235	0.00526
0.00000	0.00000
0.00002	-0.00048
0.00336	-0.00534
0.01247	-0.01006
0.02670	-0.01436
0.04596	-0.01811
0.07010	-0.02123
0.09896	-0.02372
0.13224	-0.02559
0.16963	-0.02688
0.21073	-0.02762
0.25509	-0.02785
0.30221	-0.02762
0.35156	-0.02696
0.40257	-0.02590
0.45463	-0.02446
0.50713	-0.02262
0.55944	-0.02025
0.61128	-0.01717
0.66244	-0.01366
0.71237	-0.01015
0.76037	-0.00691
0.80575	-0.00413
0.84779	-0.00192
0.88583	-0.00034
0.91925	0.00062
0.94748	0.00101
0.97003	0.00097
0.98652	0.00064
0.99660	0.00021
1.00000	0.00000

RG15 (C)

actual

x/c	y/c
1.00000	-0.00024
0.99647	0.00094
0.99433	0.00204
0.98701	0.00312
0.98064	0.00415
0.97295	0.00525
0.96264	0.00683
0.95233	0.00844
0.94060	0.01030
0.92945	0.01192
0.91047	0.01499
0.89599	0.01717
0.87631	0.02024
0.86324	0.02220
0.83897	0.02612
0.82833	0.02765
0.81670	0.02921
0.80412	0.03037
0.76364	0.03545
0.70294	0.04250
0.64030	0.04899
0.57900	0.05383
0.51810	0.05775
0.46034	0.06075
0.39616	0.06279
0.33166	0.06310
0.27013	0.06166
0.20295	0.05776
0.16678	0.05440
0.13881	0.05103
0.10659	0.04599
0.09000	0.04263
0.07935	0.04011
0.06859	0.03733
0.05791	0.03421
0.04626	0.03036
0.03259	0.02465
0.02689	0.02169
0.02093	0.01841
0.01508	0.01471
0.00921	0.01079
0.00007	-0.00094
0.00382	-0.00693
0.01256	-0.01065
0.02571	-0.01441
0.03395	-0.01618
0.04233	-0.01763
0.05243	-0.01904
0.06081	-0.02010
0.07531	-0.02161
0.09147	-0.02299
0.10610	-0.02398
0.12471	-0.02492
0.14900	-0.02572
0.17770	-0.02629
0.21123	-0.02657
0.27086	-0.02624
0.33679	-0.02518
0.39699	-0.02360
0.46204	-0.02161
0.52766	-0.01922
0.58622	-0.01632
0.65052	-0.01211
0.71310	-0.00773
0.77826	-0.00357
0.82242	-0.00114
0.83722	-0.00077
0.86207	-0.00038
0.88093	-0.00037
0.90115	-0.00046
0.91952	-0.00064
0.93417	-0.00087
0.94813	-0.00094
0.95746	-0.00095
0.96431	-0.00084
0.97202	-0.00097
0.97874	-0.00109
0.98505	-0.00119
0.99343	-0.00094
1.00000	-0.00024

S6063

true

x/c	y/c
1.00000	0.00000
0.99687	0.00017
0.98758	0.00078
0.97238	0.00197
0.95160	0.00383
0.92565	0.00638
0.89500	0.00956
0.86012	0.01328
0.82151	0.01739
0.77969	0.02175
0.73513	0.02617
0.68835	0.03050
0.63984	0.03457
0.59007	0.03827
0.53954	0.04152
0.48881	0.04424
0.43838	0.04633
0.38877	0.04774
0.34047	0.04838
0.29396	0.04823
0.24968	0.04725
0.20803	0.04544
0.16941	0.04280
0.13413	0.03938
0.10252	0.03523
0.07481	0.03044
0.05123	0.02513
0.03194	0.01943
0.01708	0.01355
0.00671	0.00772
0.00088	0.00242
0.00058	-0.00190

0.00605	-0.00588
0.01660	-0.00982
0.03215	-0.01333
0.05261	-0.01630
0.07786	-0.01867
0.10771	-0.02042
0.14190	-0.02156
0.18013	-0.02215
0.22201	-0.02222
0.26710	-0.02184
0.31491	-0.02109
0.36490	-0.02001
0.41652	-0.01865
0.46919	-0.01706
0.52232	-0.01530
0.57532	-0.01342
0.62759	-0.01147
0.67854	-0.00952
0.72758	-0.00761
0.77415	-0.00581
0.81769	-0.00418
0.85768	-0.00278
0.89359	-0.00164
0.92495	-0.00081
0.95132	-0.00026
0.97231	0.00002
0.98758	0.00009
0.99688	0.00004
1.00000	0.00000

S6063

actual

x/c	y/c
1.00000	0.00075
0.99983	0.00032
0.99796	0.00061
0.99529	0.00073
0.99281	0.00081
0.99059	0.00103
0.98841	0.00115
0.98422	0.00133
0.98192	0.00148
0.97870	0.00163
0.97157	0.00210
0.96729	0.00231
0.96187	0.00274
0.95431	0.00329
0.94692	0.00381
0.93793	0.00450
0.92891	0.00521
0.91664	0.00616
0.90491	0.00712
0.89025	0.00841
0.87467	0.00976
0.85415	0.01150
0.82537	0.01404
0.79531	0.01673
0.76178	0.01973
0.72610	0.02289
0.69336	0.02575
0.66259	0.02841
0.56169	0.03619
0.51709	0.03897
0.44542	0.04256
0.41060	0.04377
0.36795	0.04490
0.30150	0.04540
0.28217	0.04520
0.23795	0.04423
0.19174	0.04251
0.15639	0.04045
0.12190	0.03719
0.10666	0.03528
0.08798	0.03247
0.08173	0.03137
0.06845	0.02882
0.05188	0.02494
0.03040	0.01822
0.02321	0.01547
0.01485	0.01188
0.00217	0.00491
0.00039	0.00243
0.00004	0.00075
0.00040	-0.00246
0.00119	-0.00432
0.00338	-0.00736
0.00730	-0.01034
0.01213	-0.01255
0.01636	-0.01398
0.01861	-0.01460
0.02578	-0.01632
0.03320	-0.01781
0.04165	-0.01911
0.05100	-0.02031
0.06720	-0.02183
0.08408	-0.02295
0.10242	-0.02385
0.13096	-0.02476
0.15854	-0.02532
0.18591	-0.02571
0.21745	-0.02592
0.25372	-0.02600
0.30536	-0.02582
0.36075	-0.02508
0.43181	-0.02369
0.49801	-0.02223
0.57195	-0.02030
0.61542	-0.01898
0.67469	-0.01709
0.72840	-0.01524
0.78573	-0.01310
0.80400	-0.01234
0.82565	-0.01143
0.83949	-0.01087
0.85035	-0.01038
0.85833	-0.01002
0.86731	-0.00961
0.88061	-0.00891
0.88741	-0.00862
0.89511	-0.00824
0.90303	-0.00785
0.90795	-0.00759
0.91303	-0.00734
0.91697	-0.00712
0.91850	-0.00703
0.92541	-0.00668
0.93309	-0.00622
0.93559	-0.00608
0.94168	-0.00571
0.94817	-0.00531
0.95430	-0.00484
0.95723	-0.00463
0.96337	-0.00416
0.97315	-0.00327
0.98403	-0.00210
0.99402	-0.00083
0.99726	-0.00032
1.00000	0.00075

S7012

true

x/c	y/c
1.00000	0.00000
0.99819	0.00024
0.99294	0.00104
0.98451	0.00231
0.97296	0.00392
0.95833	0.00586
0.94077	0.00817
0.92044	0.01080
0.89749	0.01372
0.87210	0.01689
0.84445	0.02029
0.81474	0.02386
0.78318	0.02757
0.74999	0.03138
0.71540	0.03524
0.67965	0.03909
0.64298	0.04287
0.60563	0.04654
0.56785	0.05002
0.52988	0.05324
0.49193	0.05615
0.45426	0.05870
0.41710	0.06081
0.38070	0.06241
0.34520	0.06335
0.31071	0.06360
0.27741	0.06319
0.24543	0.06209
0.21486	0.06033
0.18590	0.05795
0.15865	0.05496
0.13326	0.05143
0.10983	0.04738
0.08846	0.04287
0.06926	0.03798
0.05229	0.03274
0.03759	0.02728
0.02525	0.02167
0.01528	0.01605
0.00773	0.01058
0.00266	0.00542
0.00007	0.00087
0.00069	-0.00278
0.00486	-0.00600
0.01208	-0.00918
0.02223	-0.01220
0.03523	-0.01495
0.05102	-0.01739
0.06952	-0.01950
0.09065	-0.02126
0.11428	-0.02265
0.14032	-0.02368
0.16861	-0.02436
0.19900	-0.02471
0.23131	-0.02474
0.26535	-0.02448
0.30091	-0.02396
0.33778	-0.02318
0.37572	-0.02220
0.41449	-0.02101
0.45383	-0.01965
0.49348	-0.01813
0.53316	-0.01644
0.57261	-0.01453
0.61171	-0.01227
0.65044	-0.00978
0.68859	-0.00732
0.72582	-0.00504
0.76182	-0.00300
0.79628	-0.00128
0.82889	0.00010
0.85936	0.00112
0.88743	0.00178
0.91283	0.00211
0.93533	0.00214
0.95473	0.00192
0.97084	0.00153
0.98352	0.00105
0.99266	0.00055
0.99816	0.00016
1.00000	0.00000

S7012 (B)

actual

x/c	y/c
1.00000	0.00045
0.99960	0.00001
0.99669	0.00037
0.99294	0.00079
0.98842	0.00130
0.98375	0.00163
0.98021	0.00179
0.97454	0.00222
0.96582	0.00324
0.95379	0.00447
0.94289	0.00586
0.92317	0.00810
0.89969	0.01080
0.85933	0.01547
0.81885	0.02027
0.72430	0.03165
0.65062	0.03941
0.56792	0.04762
0.48218	0.05482
0.40016	0.05987
0.35677	0.06158
0.31806	0.06223
0.27607	0.06193
0.23573	0.06066
0.21576	0.05950
0.19547	0.05795
0.17212	0.05577

0.15611	0.05399
0.14086	0.05204
0.12842	0.05018
0.11345	0.04764
0.09970	0.04497
0.08488	0.04171
0.07186	0.03851
0.05937	0.03499
0.05303	0.03298
0.04160	0.02897
0.03516	0.02640
0.02900	0.02369
0.02277	0.02065
0.01803	0.01807
0.01372	0.01544
0.01036	0.01313
0.00691	0.01039
0.00396	0.00746
0.00111	0.00315
0.00065	0.00185
0.00031	0.00127
0.00204	-0.00328
0.00499	-0.00575
0.00772	-0.00748
0.01262	-0.00981
0.01756	-0.01160
0.02682	-0.01423
0.03026	-0.01506
0.03536	-0.01618
0.04633	-0.01816
0.06809	-0.02082
0.08995	-0.02263
0.12249	-0.02450
0.15033	-0.02555
0.18373	-0.02627
0.22489	-0.02629
0.26479	-0.02578
0.33064	-0.02459
0.38760	-0.02311
0.45196	-0.02111
0.54060	-0.01747
0.62609	-0.01264
0.70349	-0.00776
0.75671	-0.00464
0.82717	-0.00154
0.87128	-0.00003
0.91142	0.00083
0.95396	0.00123
0.96538	0.00084
0.97772	0.00051
0.98739	0.00016
0.99328	-0.00010
0.99939	-0.00001
1.00000	0.00045

S7075

true

x/c	y/c
1.00000	0.00000
0.99811	0.00030
0.99262	0.00130
0.98384	0.00301
0.97189	0.00532
0.95689	0.00825
0.93906	0.01183
0.91865	0.01599
0.89590	0.02064
0.87105	0.02565
0.84435	0.03084
0.81590	0.03603
0.78585	0.04114
0.75438	0.04596
0.72152	0.05039
0.68736	0.05452
0.65213	0.05833
0.61605	0.06181
0.57936	0.06492
0.54228	0.06762
0.50502	0.06988
0.46782	0.07165
0.43086	0.07290
0.39436	0.07362
0.35853	0.07377
0.32355	0.07334
0.28961	0.07232
0.25690	0.07072
0.22556	0.06854
0.19578	0.06579
0.16769	0.06249
0.14143	0.05868
0.11714	0.05438
0.09490	0.04963
0.07485	0.04451
0.05704	0.03906
0.04155	0.03336
0.02846	0.02751
0.01783	0.02158
0.00974	0.01572
0.00426	0.00983
0.00112	0.00396
0.00012	-0.00127
0.00205	-0.00532
0.00740	-0.00861
0.01586	-0.01154
0.02736	-0.01403
0.04182	-0.01616
0.05915	-0.01785
0.07927	-0.01912
0.10208	-0.01997
0.12746	-0.02041
0.15527	-0.02049
0.18536	-0.02022
0.21756	-0.01965
0.25167	-0.01880
0.28749	-0.01774
0.32478	-0.01649
0.36331	-0.01510
0.40283	-0.01360
0.44306	-0.01205
0.48375	-0.01047
0.52460	-0.00889
0.56534	-0.00734
0.60568	-0.00584
0.64532	-0.00438
0.68397	-0.00291
0.72149	-0.00140
0.75769	0.00009
0.79235	0.00160
0.82543	0.00301
0.85679	0.00399
0.88582	0.00420
0.91190	0.00378
0.93478	0.00312
0.95435	0.00237
0.97053	0.00163
0.98322	0.00099
0.99243	0.00051
0.99808	0.00016
1.00000	0.00000

S7075 (A)

actual

x/c	y/c
1.00000	-0.00094
0.99880	-0.00018
0.99573	0.00071
0.99003	0.00183
0.98361	0.00306
0.97892	0.00407
0.97211	0.00546
0.96259	0.00758
0.94762	0.01058
0.93380	0.01333
0.91794	0.01650
0.90252	0.01953
0.88967	0.02200
0.85223	0.02910
0.81262	0.03612
0.72316	0.04930
0.64651	0.05807
0.56353	0.06542
0.48020	0.07034
0.39574	0.07258
0.35700	0.07272
0.31528	0.07211
0.27103	0.07051
0.23201	0.06801
0.21012	0.06617
0.19376	0.06457
0.17123	0.06201
0.14505	0.05840
0.13058	0.05602
0.11998	0.05410
0.10489	0.05105
0.09076	0.04781
0.07660	0.04411
0.06308	0.04008
0.05197	0.03634
0.04408	0.03338
0.03647	0.03025
0.03025	0.02744
0.01964	0.02185
0.01273	0.01733
0.00542	0.01099
0.00146	0.00538
0.00021	0.00202
0.00074	-0.00388
0.00443	-0.00786
0.00722	-0.00938
0.01242	-0.01132
0.02049	-0.01338
0.02870	-0.01496
0.03540	-0.01596
0.04184	-0.01684
0.05288	-0.01802
0.06188	-0.01880
0.07265	-0.01956
0.08391	-0.02018
0.11360	-0.02115
0.14319	-0.02145
0.17676	-0.02128
0.21366	-0.02072
0.26263	-0.01957
0.31707	-0.01795
0.40010	-0.01522
0.48486	-0.01211
0.56908	-0.00895
0.65267	-0.00577
0.73469	-0.00254
0.82017	0.00097
0.86467	0.00249
0.90402	0.00282
0.93582	0.00241
0.96612	0.00152
0.97825	0.00100
0.99095	0.00046
0.99950	0.00018
1.00000	-0.00094

S7075 (B)

actual

x/c	y/c
1.00000	0.00018
0.99824	0.00040
0.99737	0.00091
0.99129	0.00151
0.98450	0.00248
0.97371	0.00392
0.96111	0.00598
0.94657	0.00821
0.93151	0.01051
0.91847	0.01250
0.90383	0.01480
0.89029	0.01698
0.85440	0.02265
0.80416	0.02991
0.71546	0.04355
0.64805	0.05212
0.56423	0.06027
0.48340	0.06597
0.39636	0.06926
0.31389	0.06961
0.27336	0.06844
0.23433	0.06634
0.21456	0.06480
0.19055	0.06253
0.17320	0.06052
0.15512	0.05810
0.13958	0.05574
0.12759	0.05373
0.11461	0.05139
0.09668	0.04783
0.08151	0.04425
0.06785	0.04057

0.05766	0.03749
0.04515	0.03318
0.03759	0.03022
0.02896	0.02640
0.02160	0.02253
0.01709	0.01968
0.01206	0.01584
0.00874	0.01287
0.00365	0.00752
0.00174	0.00518
0.00473	-0.00869
0.00703	-0.01106
0.01091	-0.01338
0.01618	-0.01491
0.02114	-0.01613
0.02532	-0.01704
0.03219	-0.01829
0.04092	-0.01948
0.05730	-0.02100
0.07905	-0.02239
0.10532	-0.02354
0.13410	-0.02421
0.16315	-0.02427
0.20550	-0.02371
0.24315	-0.02266
0.32792	-0.01992
0.40933	-0.01721
0.49307	-0.01435
0.57704	-0.01165
0.65876	-0.00892
0.73955	-0.00629
0.82431	-0.00238
0.86455	-0.00079
0.90509	0.00013
0.93527	0.00052
0.94952	0.00061
0.96477	0.00015
0.97852	0.00016
0.99012	-0.00021
0.99842	-0.00040
1.00000	0.00018

S8036

true

x/c	y/c
1.00000	0.00000
0.99759	0.00046
0.99095	0.00217
0.98105	0.00501
0.96827	0.00843
0.95247	0.01204
0.93354	0.01594
0.91163	0.02027
0.88697	0.02505
0.85984	0.03027
0.83047	0.03590
0.79916	0.04189
0.76618	0.04816
0.73183	0.05463
0.69641	0.06119
0.66022	0.06771
0.62357	0.07402
0.58668	0.07986
0.54969	0.08509
0.51279	0.08963
0.47611	0.09328
0.43972	0.09598
0.40381	0.09776
0.36848	0.09853
0.33380	0.09838
0.29999	0.09735
0.26722	0.09547
0.23564	0.09280
0.20545	0.08939
0.17681	0.08528
0.14988	0.08053
0.12484	0.07520
0.10184	0.06934
0.08104	0.06296
0.06251	0.05610
0.04638	0.04885
0.03271	0.04115
0.02141	0.03305
0.01256	0.02483
0.00612	0.01649
0.00193	0.00823
0.00001	0.00054
0.00098	-0.00588
0.00548	-0.01153
0.01338	-0.01720
0.02433	-0.02273
0.03811	-0.02804
0.05468	-0.03304
0.07395	-0.03773
0.09576	-0.04208
0.11999	-0.04605
0.14648	-0.04964
0.17504	-0.05283
0.20549	-0.05558
0.23760	-0.05786
0.27120	-0.05962
0.30609	-0.06084
0.34204	-0.06146
0.37887	-0.06147
0.41638	-0.06087
0.45435	-0.05959
0.49265	-0.05765
0.53099	-0.05502
0.56937	-0.05165
0.60778	-0.04772
0.64594	-0.04348
0.68359	-0.03907
0.72043	-0.03464
0.75616	-0.03028
0.79048	-0.02609
0.82309	-0.02215
0.85370	-0.01852
0.88202	-0.01523
0.90775	-0.01230
0.93064	-0.00970
0.95044	-0.00739
0.96698	-0.00519
0.98037	-0.00302
0.99070	-0.00120
0.99754	-0.00020
1.00000	0.00000

S8036

actual

x/c	y/c
1.00000	-0.00026
0.99897	0.00304
0.99291	0.00429
0.98565	0.00593
0.97868	0.00726
0.97266	0.00851
0.96611	0.00991
0.95916	0.01136
0.94957	0.01324
0.93602	0.01594
0.92491	0.01810
0.90688	0.02160
0.88140	0.02668
0.84304	0.03403
0.80502	0.04131
0.71565	0.05789
0.63511	0.07175
0.55378	0.08427
0.46505	0.09423
0.38389	0.09858
0.33819	0.09843
0.29908	0.09737
0.25859	0.09487
0.21732	0.09057
0.19628	0.08782
0.17830	0.08520
0.15908	0.08222
0.13937	0.07872
0.12485	0.07558
0.11301	0.07273
0.10113	0.06952
0.08795	0.06550
0.07625	0.06150
0.06396	0.05680
0.05320	0.05200
0.04720	0.04915
0.04181	0.04648
0.03364	0.04228
0.02895	0.03940
0.02527	0.03691
0.01976	0.03238
0.01428	0.02701
0.01048	0.02251
0.00614	0.01636
0.00302	0.00983
0.00064	0.00216
0.00039	0.00180
0.00138	-0.00341
0.00508	-0.00809
0.00936	-0.01213
0.01627	-0.01690
0.02418	-0.02107
0.02966	-0.02354
0.03915	-0.02687
0.04644	-0.02910
0.06348	-0.03370
0.08179	-0.03790
0.11119	-0.04344
0.13457	-0.04669
0.16462	-0.05001
0.20293	-0.05351
0.24214	-0.05647
0.28959	-0.05894
0.36578	-0.06063
0.44394	-0.05973
0.53300	-0.05493
0.61840	-0.04703
0.70395	-0.03757
0.78547	-0.02805
0.82445	-0.02353
0.86379	-0.01897
0.90768	-0.01382
0.92909	-0.01137
0.94771	-0.00914
0.95999	-0.00757
0.96884	-0.00664
0.97899	-0.00542
0.98637	-0.00455
0.99890	-0.00304
1.00000	-0.00026

S8037

true

x/c	y/c
1.00000	0.00000
0.99758	0.00043
0.99089	0.00205
0.98084	0.00478
0.96782	0.00814
0.95171	0.01178
0.93244	0.01580
0.91019	0.02034
0.88520	0.02543
0.85776	0.03105
0.82813	0.03716
0.79662	0.04370
0.76352	0.05059
0.72913	0.05772
0.69378	0.06496
0.65776	0.07216
0.62140	0.07911
0.58488	0.08550
0.54828	0.09116
0.51175	0.09607
0.47544	0.10007
0.43941	0.10309
0.40389	0.10515
0.36897	0.10618
0.33473	0.10623
0.30138	0.10533
0.26907	0.10349
0.23795	0.10079
0.20820	0.09726
0.17997	0.09295
0.15344	0.08790
0.12873	0.08214
0.10594	0.07574
0.08522	0.06875
0.06662	0.06121
0.05020	0.05327
0.03605	0.04497
0.02418	0.03644
0.01470	0.02787

0.00761	0.01929
0.00277	0.01090
0.00023	0.00310
0.00026	-0.00341
0.00381	-0.00893
0.01113	-0.01425
0.02158	-0.01941
0.03494	-0.02434
0.05116	-0.02893
0.07018	-0.03319
0.09182	-0.03711
0.11596	-0.04064
0.14245	-0.04381
0.17109	-0.04659
0.20169	-0.04897
0.23403	-0.05091
0.26794	-0.05237
0.30320	-0.05334
0.33959	-0.05378
0.37693	-0.05366
0.41501	-0.05301
0.45362	-0.05177
0.49262	-0.04997
0.53173	-0.04765
0.57084	-0.04480
0.60985	-0.04155
0.64846	-0.03809
0.68640	-0.03450
0.72339	-0.03089
0.75914	-0.02733
0.79336	-0.02390
0.82577	-0.02064
0.85609	-0.01759
0.88405	-0.01478
0.90939	-0.01222
0.93188	-0.00987
0.95129	-0.00769
0.96747	-0.00552
0.98059	-0.00327
0.99076	-0.00134
0.99755	-0.00023
1.00000	0.00000

S8037

actual

x/c	y/c
1.00000	0.00037
0.99946	0.00184
0.99654	0.00277
0.99433	0.00322
0.99235	0.00371
0.98963	0.00423
0.98600	0.00497
0.98229	0.00590
0.97816	0.00663
0.97458	0.00750
0.97065	0.00831
0.96543	0.00932
0.95654	0.01113
0.94830	0.01297
0.93693	0.01524
0.92383	0.01821
0.90909	0.02140
0.89427	0.02434
0.87498	0.02850
0.85992	0.03177
0.84159	0.03559
0.82130	0.03980
0.80134	0.04386
0.78050	0.04803
0.76241	0.05175
0.74218	0.05566
0.71633	0.06088
0.69393	0.06508
0.66736	0.07026
0.62017	0.07903
0.54722	0.09111
0.46114	0.10083
0.39578	0.10448
0.34717	0.10543
0.30425	0.10464
0.24700	0.10073
0.21865	0.09769
0.18191	0.09264
0.14676	0.08600
0.12286	0.08005
0.09925	0.07288
0.08257	0.06683
0.07743	0.06470
0.06691	0.06009
0.04228	0.04726
0.03678	0.04383
0.03266	0.04108
0.02684	0.03693
0.01853	0.02990
0.01251	0.02366
0.00338	0.01082
0.00077	0.00491
0.00000	-0.00020
0.00031	-0.00309
0.00152	-0.00668
0.00375	-0.00983
0.00518	-0.01125
0.00744	-0.01301
0.01093	-0.01514
0.01443	-0.01691
0.01804	-0.01847
0.02217	-0.02021
0.02619	-0.02176
0.03082	-0.02337
0.03562	-0.02491
0.04106	-0.02648
0.04724	-0.02816
0.05879	-0.03085
0.07304	-0.03364
0.08201	-0.03516
0.09468	-0.03722
0.11028	-0.03942
0.12913	-0.04175
0.16087	-0.04528
0.18119	-0.04733
0.20053	-0.04909
0.22365	-0.05115
0.24886	-0.05306
0.28120	-0.05500
0.32395	-0.05609
0.35803	-0.05619
0.39216	-0.05594
0.43811	-0.05473
0.48035	-0.05290
0.53271	-0.04971
0.62200	-0.04198
0.74686	-0.02959
0.79808	-0.02427
0.83419	-0.02052
0.86790	-0.01699
0.89590	-0.01390
0.91819	-0.01161
0.93587	-0.00958
0.96622	-0.00630
0.98772	-0.00373
0.99410	-0.00279
0.99968	-0.00184
1.00000	0.00037

S8052

true

x/c	y/c
1.00000	0.00000
0.99803	0.00018
0.99231	0.00079
0.98301	0.00181
0.97022	0.00330
0.95412	0.00533
0.93491	0.00790
0.91280	0.01104
0.88804	0.01472
0.86087	0.01891
0.83155	0.02357
0.80036	0.02863
0.76759	0.03400
0.73351	0.03957
0.69841	0.04524
0.66259	0.05083
0.62635	0.05617
0.58984	0.06088
0.55299	0.06482
0.51588	0.06814
0.47873	0.07081
0.44169	0.07283
0.40501	0.07423
0.36887	0.07498
0.33344	0.07511
0.29896	0.07461
0.26560	0.07351
0.23357	0.07182
0.20304	0.06957
0.17422	0.06676
0.14727	0.06341
0.12234	0.05951
0.09956	0.05510
0.07907	0.05012
0.06085	0.04458
0.04493	0.03864
0.03140	0.03234
0.02025	0.02579
0.01165	0.01916
0.00552	0.01240
0.00163	0.00569
0.00001	-0.00037
0.00147	-0.00540
0.00651	-0.00985
0.01471	-0.01427
0.02583	-0.01860
0.03977	-0.02271
0.05642	-0.02654
0.07569	-0.03003
0.09749	-0.03316
0.12169	-0.03590
0.14818	-0.03824
0.17678	-0.04018
0.20735	-0.04171
0.23970	-0.04283
0.27364	-0.04354
0.30896	-0.04385
0.34545	-0.04377
0.38288	-0.04331
0.42100	-0.04248
0.45957	-0.04127
0.49834	-0.03967
0.53702	-0.03758
0.57556	-0.03494
0.61389	-0.03185
0.65181	-0.02843
0.68916	-0.02481
0.72568	-0.02116
0.76113	-0.01762
0.79518	-0.01433
0.82751	-0.01136
0.85782	-0.00876
0.88581	-0.00656
0.91121	-0.00475
0.93377	-0.00330
0.95325	-0.00218
0.96950	-0.00128
0.98246	-0.00056
0.99204	-0.00011
0.99798	0.00002
1.00000	0.00000

S8052

actual

x/c	y/c
1.00000	-0.00026
0.99961	0.00083
0.99610	0.00112
0.99056	0.00142
0.98404	0.00204
0.97717	0.00245
0.97056	0.00317
0.96281	0.00391
0.95139	0.00519
0.94293	0.00630
0.93416	0.00738
0.92308	0.00859
0.90954	0.01040
0.89705	0.01207
0.88482	0.01392
0.86897	0.01620
0.85454	0.01832
0.83627	0.02107
0.81911	0.02377
0.79081	0.02833
0.75912	0.03344

0.70192	0.04257
0.61600	0.05514
0.52831	0.06478
0.46741	0.06934
0.40690	0.07238
0.37239	0.07325
0.33189	0.07347
0.28160	0.07268
0.23989	0.07111
0.21531	0.06963
0.19898	0.06837
0.18083	0.06666
0.16079	0.06437
0.13091	0.06007
0.10127	0.05428
0.08581	0.05056
0.05523	0.04116
0.03790	0.03424
0.02417	0.02716
0.01113	0.01729
0.00169	0.00576
0.00022	0.00207
0.00029	-0.00243
0.00159	-0.00536
0.00449	-0.00858
0.00945	-0.01221
0.01427	-0.01477
0.02017	-0.01724
0.02595	-0.01929
0.03209	-0.02131
0.04118	-0.02393
0.05306	-0.02687
0.06141	-0.02869
0.06936	-0.03025
0.08295	-0.03255
0.09829	-0.03468
0.11202	-0.03626
0.12842	-0.03786
0.14015	-0.03885
0.16107	-0.04044
0.17842	-0.04158
0.20021	-0.04274
0.21523	-0.04344
0.23677	-0.04428
0.27360	-0.04521
0.31565	-0.04561
0.35348	-0.04550
0.40073	-0.04481
0.48630	-0.04157
0.56388	-0.03686
0.65240	-0.02927
0.73246	-0.02134
0.82373	-0.01237
0.86644	-0.00856
0.91006	-0.00530
0.92687	-0.00412
0.94114	-0.00302
0.97009	-0.00161
0.99354	-0.00083
1.00000	-0.00026

SA7035

true

x/c	y/c
1.00000	0.00000
0.99815	0.00017
0.99265	0.00079
0.98367	0.00197
0.97138	0.00375
0.95600	0.00615
0.93777	0.00913
0.91689	0.01264
0.89359	0.01657
0.86806	0.02082
0.84049	0.02528
0.81106	0.02984
0.77993	0.03442
0.74727	0.03894
0.71326	0.04334
0.67810	0.04760
0.64203	0.05168
0.60526	0.05549
0.56801	0.05899
0.53053	0.06214
0.49301	0.06485
0.45566	0.06712
0.41872	0.06887
0.38237	0.07007
0.34681	0.07071
0.31224	0.07073
0.27881	0.07012
0.24673	0.06888
0.21613	0.06699
0.18713	0.06448
0.15991	0.06134
0.13456	0.05759
0.11118	0.05331
0.08989	0.04850
0.07072	0.04323
0.05378	0.03760
0.03911	0.03165
0.02670	0.02552
0.01662	0.01937
0.00895	0.01331
0.00365	0.00752
0.00075	0.00224
0.00000	0.00000
0.00371	-0.00497
0.01052	-0.00821
0.02026	-0.01131
0.03286	-0.01407
0.04832	-0.01642
0.06658	-0.01840
0.08751	-0.02002
0.11103	-0.02126
0.13700	-0.02212
0.16529	-0.02262
0.19574	-0.02278
0.22816	-0.02262
0.26238	-0.02216
0.29819	-0.02144
0.33536	-0.02048
0.37368	-0.01932
0.41289	-0.01800
0.45274	-0.01655
0.49296	-0.01498
0.53331	-0.01329
0.57358	-0.01151
0.61352	-0.00973
0.65286	-0.00799
0.69133	-0.00635
0.72863	-0.00485
0.76451	-0.00351
0.79869	-0.00236
0.83091	-0.00140
0.86091	-0.00064
0.88847	-0.00005
0.91340	0.00039
0.93553	0.00069
0.95469	0.00082
0.97069	0.00079
0.98337	0.00062
0.99256	0.00037
0.99813	0.00012
1.00000	0.00000

SA7035

actual

x/c	y/c
1.00000	0.00063
0.99661	0.00161
0.99316	0.00208
0.98715	0.00295
0.98310	0.00360
0.97710	0.00450
0.97029	0.00551
0.96089	0.00688
0.95793	0.00735
0.95118	0.00827
0.94253	0.00970
0.93223	0.01135
0.91998	0.01321
0.90421	0.01575
0.89252	0.01759
0.87334	0.02059
0.85704	0.02315
0.83933	0.02591
0.81996	0.02892
0.80279	0.03154
0.78175	0.03460
0.76922	0.03645
0.75037	0.03906
0.72975	0.04177
0.71418	0.04375
0.69727	0.04581
0.67736	0.04818
0.65784	0.05040
0.63857	0.05251
0.62208	0.05424
0.60252	0.05620
0.58372	0.05801
0.56717	0.05953
0.53745	0.06205
0.47463	0.06646
0.45732	0.06747
0.43689	0.06851
0.40943	0.06968
0.38710	0.07044
0.34930	0.07125
0.31312	0.07139
0.27634	0.07083
0.23837	0.06937
0.16897	0.06362
0.14891	0.06103
0.12157	0.05663
0.10132	0.05256
0.09294	0.05066
0.07591	0.04621
0.05639	0.04005
0.05037	0.03786
0.03996	0.03362
0.02306	0.02510
0.01876	0.02246
0.00876	0.01492
0.00541	0.01158
0.00046	0.00372
0.00013	0.00196
0.00035	0.00323
0.00014	0.00136
0.00018	0.00074
0.00035	-0.00018
0.00108	-0.00166
0.00181	-0.00256
0.00363	-0.00406
0.00522	-0.00501
0.00636	-0.00560
0.00907	-0.00678
0.01145	-0.00767
0.01434	-0.00864
0.01767	-0.00965
0.02160	-0.01073
0.02706	-0.01204
0.03246	-0.01317
0.03948	-0.01445
0.04908	-0.01586
0.05880	-0.01702
0.06933	-0.01808
0.07965	-0.01893
0.08826	-0.01952
0.09705	-0.02004
0.10460	-0.02041
0.11320	-0.02078
0.12336	-0.02114
0.13311	-0.02142
0.14232	-0.02164
0.14874	-0.02178
0.18152	-0.02219
0.21255	-0.02228
0.24862	-0.02204
0.30788	-0.02097
0.36184	-0.01955
0.43021	-0.01734
0.50306	-0.01451
0.58125	-0.01113
0.65500	-0.00794
0.69117	-0.00647
0.70868	-0.00579
0.75373	-0.00415
0.79294	-0.00287
0.82984	-0.00182
0.87156	-0.00078
0.92642	0.00011

0.94661	0.00025
0.95548	0.00025
0.96769	0.00014
0.98136	-0.00026
0.99110	-0.00060
0.99912	-0.00068
1.00000	0.00063

SA7036

true

x/c	y/c
1.00000	0.00000
0.99815	0.00019
0.99266	0.00085
0.98368	0.00210
0.97142	0.00398
0.95608	0.00650
0.93789	0.00963
0.91709	0.01330
0.89388	0.01740
0.86847	0.02182
0.84103	0.02644
0.81174	0.03115
0.78075	0.03588
0.74825	0.04054
0.71441	0.04506
0.67941	0.04942
0.64348	0.05358
0.60686	0.05747
0.56974	0.06103
0.53237	0.06422
0.49495	0.06698
0.45769	0.06926
0.42082	0.07103
0.38453	0.07222
0.34900	0.07285
0.31444	0.07285
0.28101	0.07221
0.24891	0.07094
0.21826	0.06902
0.18921	0.06645
0.16193	0.06326
0.13650	0.05945
0.11301	0.05510
0.09161	0.05022
0.07232	0.04485
0.05524	0.03913
0.04041	0.03309
0.02784	0.02684
0.01758	0.02054
0.00969	0.01434
0.00415	0.00839
0.00101	0.00291
0.00000	0.00000
0.00029	-0.00154
0.00297	-0.00497
0.00936	-0.00810
0.01873	-0.01110
0.03099	-0.01375
0.04613	-0.01596
0.06411	-0.01779
0.08481	-0.01924
0.10813	-0.02030
0.13395	-0.02097
0.16213	-0.02129
0.19250	-0.02127
0.22490	-0.02094
0.25913	-0.02032
0.29498	-0.01945
0.33225	-0.01836
0.37069	-0.01709
0.41006	-0.01569
0.45010	-0.01418
0.49055	-0.01259
0.53113	-0.01093
0.57163	-0.00922
0.61180	-0.00754
0.65136	-0.00592
0.69004	-0.00442
0.72754	-0.00308
0.76360	-0.00190
0.79795	-0.00093
0.83033	-0.00015
0.86048	0.00043
0.88816	0.00084
0.91320	0.00111
0.93542	0.00123
0.95463	0.00120
0.97067	0.00103
0.98337	0.00075
0.99256	0.00042
0.99813	0.00013
1.00000	0.00000

SA7036 (A)

actual

x/c	y/c
1.00000	-0.00071
0.99886	0.00017
0.99696	0.00115
0.98989	0.00254
0.97744	0.00443
0.96546	0.00642
0.95045	0.00897
0.93429	0.01173
0.91427	0.01524
0.89499	0.01857
0.87503	0.02184
0.83482	0.02843
0.78453	0.03605
0.74588	0.04158
0.70887	0.04650
0.66559	0.05172
0.63112	0.05564
0.58372	0.06018
0.54169	0.06393
0.50162	0.06692
0.46295	0.06924
0.42006	0.07111
0.37820	0.07220
0.33676	0.07274
0.31523	0.07271
0.29405	0.07232
0.27247	0.07176
0.25482	0.07102
0.23485	0.06995
0.21535	0.06864
0.19449	0.06690
0.17337	0.06474
0.14775	0.06132
0.12789	0.05796
0.11242	0.05487
0.09293	0.05033
0.07619	0.04573
0.06082	0.04079
0.05110	0.03722
0.04429	0.03438
0.03808	0.03151
0.02981	0.02712
0.02512	0.02428
0.02036	0.02114
0.01466	0.01698
0.01005	0.01330
0.00514	0.00881
0.00232	0.00540
0.00008	0.00102
0.00071	-0.00301
0.00324	-0.00528
0.01071	-0.00871
0.01414	-0.01002
0.01948	-0.01150
0.02603	-0.01310
0.03111	-0.01415
0.03651	-0.01513
0.03969	-0.01563
0.04711	-0.01651
0.05872	-0.01785
0.06825	-0.01876
0.07786	-0.01951
0.09892	-0.02082
0.11933	-0.02173
0.14066	-0.02222
0.16371	-0.02240
0.20656	-0.02241
0.24714	-0.02186
0.29016	-0.02096
0.33079	-0.01976
0.37493	-0.01833
0.41485	-0.01677
0.45318	-0.01518
0.49604	-0.01335
0.53737	-0.01160
0.57817	-0.00992
0.62265	-0.00801
0.66108	-0.00637
0.70959	-0.00472
0.75046	-0.00290
0.78460	-0.00189
0.80713	-0.00123
0.82729	-0.00070
0.84876	-0.00016
0.87204	0.00032
0.88748	0.00045
0.90193	0.00067
0.92348	0.00090
0.94165	0.00091
0.95871	0.00084
0.97238	0.00054
0.98278	0.00018
0.98930	-0.00019
0.99911	-0.00016
1.00000	-0.00071

SA7036 (B)

actual

x/c	y/c
1.00000	0.00004
0.99828	0.00092
0.99209	0.00171
0.98755	0.00236
0.98515	0.00271
0.98213	0.00326
0.97848	0.00400
0.97474	0.00474
0.97135	0.00530
0.96778	0.00592
0.96385	0.00658
0.95965	0.00727
0.95674	0.00775
0.95193	0.00855
0.94712	0.00947
0.94122	0.01061
0.93529	0.01172
0.92768	0.01322
0.91873	0.01488
0.90831	0.01688
0.89914	0.01866
0.88588	0.02114
0.87566	0.02305
0.86491	0.02507
0.85415	0.02700
0.84920	0.02794
0.83616	0.03047
0.82669	0.03200
0.81664	0.03381
0.80239	0.03621
0.78858	0.03840
0.76954	0.04146
0.74990	0.04427
0.73186	0.04691
0.70931	0.05007
0.69237	0.05194
0.67331	0.05410
0.65469	0.05620
0.63589	0.05822
0.61871	0.06008
0.59967	0.06188
0.58045	0.06363
0.56403	0.06503
0.54402	0.06652
0.52982	0.06754
0.51147	0.06884
0.49278	0.07012
0.47128	0.07138
0.44237	0.07299
0.42346	0.07390
0.40220	0.07474
0.38044	0.07544
0.36427	0.07578
0.33849	0.07595
0.32704	0.07593
0.31217	0.07579
0.29491	0.07546
0.27461	0.07489

0.25776	0.07429
0.23871	0.07332
0.22110	0.07205
0.20464	0.07065
0.18515	0.06873
0.16635	0.06655
0.15058	0.06454
0.12858	0.06108
0.12254	0.06000
0.11280	0.05811
0.10337	0.05612
0.09455	0.05406
0.08625	0.05193
0.07851	0.04984
0.07085	0.04762
0.06318	0.04527
0.05234	0.04170
0.04590	0.03926
0.03464	0.03429
0.02366	0.02834
0.01941	0.02558
0.01418	0.02166
0.01171	0.01954
0.00867	0.01654
0.00611	0.01333
0.00381	0.00988
0.00142	0.00553
0.00055	0.00320
0.00010	0.00123
0.00006	0.00064
0.00000	-0.00002
0.00187	-0.00369
0.00152	-0.00319
0.00291	-0.00424
0.00478	-0.00526
0.00795	-0.00644
0.01089	-0.00749
0.01310	-0.00827
0.01677	-0.00947
0.02052	-0.01061
0.02577	-0.01204
0.03058	-0.01312
0.03779	-0.01445
0.04700	-0.01579
0.05549	-0.01661
0.06499	-0.01726
0.07590	-0.01784
0.08398	-0.01815
0.09364	-0.01840
0.10278	-0.01865
0.11190	-0.01885
0.12461	-0.01915
0.13043	-0.01927
0.13680	-0.01936
0.15397	-0.01951
0.16869	-0.01942
0.18092	-0.01933
0.18962	-0.01912
0.20239	-0.01893
0.22460	-0.01846
0.24297	-0.01805
0.25573	-0.01780

SA7038

true

x/c	y/c
1.00000	0.00000
0.99815	0.00022
0.99266	0.00097
0.98370	0.00235
0.97147	0.00443
0.95619	0.00720
0.93810	0.01061
0.91744	0.01458
0.89440	0.01901
0.86920	0.02375
0.84201	0.02869
0.81298	0.03371
0.78227	0.03873
0.75007	0.04367
0.71655	0.04844
0.68186	0.05300
0.64622	0.05732
0.60986	0.06135
0.57300	0.06502
0.53584	0.06830
0.49861	0.07112
0.46151	0.07343
0.42477	0.07522
0.38857	0.07641
0.35310	0.07700
0.31858	0.07696
0.28515	0.07625
0.25298	0.07490
0.22226	0.07288
0.19310	0.07020
0.16566	0.06690
0.14006	0.06298
0.11638	0.05847
0.09475	0.05344
0.07523	0.04791
0.05786	0.04199
0.04274	0.03574
0.02987	0.02924
0.01925	0.02268
0.01095	0.01619
0.00499	0.00990
0.00138	0.00405
0.00008	-0.00099
0.00190	-0.00474
0.00735	-0.00774
0.01597	-0.01059
0.02753	-0.01306
0.04202	-0.01505
0.05943	-0.01661
0.07964	-0.01777
0.10255	-0.01851
0.12803	-0.01885
0.15596	-0.01882
0.18616	-0.01846
0.21847	-0.01779
0.25269	-0.01685
0.28862	-0.01568
0.32603	-0.01433
0.36470	-0.01283
0.40436	-0.01125
0.44476	-0.00961
0.48563	-0.00796
0.52669	-0.00632
0.56767	-0.00473
0.60831	-0.00322
0.64832	-0.00184
0.68742	-0.00061
0.72533	0.00044
0.76177	0.00129
0.79647	0.00193
0.82917	0.00236
0.85961	0.00259
0.88756	0.00264
0.91282	0.00254
0.93520	0.00231
0.95454	0.00196
0.97065	0.00152
0.98337	0.00102
0.99257	0.00054
0.99814	0.00016
1.00000	0.00000

SA7038

actual

x/c	y/c
1.00000	-0.00060
0.99888	0.00078
0.99425	0.00148
0.98822	0.00219
0.96840	0.00540
0.95274	0.00813
0.93567	0.01142
0.91740	0.01495
0.90154	0.01791
0.88092	0.02173
0.85968	0.02568
0.83850	0.02950
0.81397	0.03382
0.79364	0.03718
0.75582	0.04310
0.71522	0.04889
0.67313	0.05431
0.63532	0.05868
0.59118	0.06320
0.55032	0.06709
0.51034	0.07027
0.46260	0.07349
0.42036	0.07554
0.38346	0.07685
0.34234	0.07765
0.32089	0.07766
0.29999	0.07749
0.27379	0.07689
0.25530	0.07603
0.23747	0.07501
0.21915	0.07389
0.19605	0.07196
0.17817	0.07004
0.15981	0.06773
0.13663	0.06397
0.12055	0.06088
0.09941	0.05599
0.08378	0.05172
0.06956	0.04719
0.06237	0.04465
0.05394	0.04140
0.04665	0.03826
0.04053	0.03543
0.03341	0.03174
0.02691	0.02778
0.01972	0.02277
0.01539	0.01938
0.01209	0.01649
0.00706	0.01132
0.00554	0.00945
0.00315	0.00496
0.00233	0.00427
0.00857	-0.00829
0.01187	-0.00917
0.01706	-0.01073
0.02356	-0.01223
0.03005	-0.01344
0.03971	-0.01484
0.06218	-0.01663
0.08208	-0.01746
0.11510	-0.01789
0.14440	-0.01774
0.18030	-0.01726
0.22345	-0.01649
0.25222	-0.01579
0.31579	-0.01395
0.38039	-0.01155
0.43503	-0.00951
0.49957	-0.00719
0.56241	-0.00473
0.62257	-0.00242
0.67718	-0.00054
0.72907	0.00088
0.77377	0.00180
0.82235	0.00248
0.85928	0.00271
0.88077	0.00268
0.90314	0.00247
0.92383	0.00227
0.94187	0.00179
0.96299	0.00108
0.99103	-0.00038
0.99969	-0.00078
1.00000	-0.00060

SD7032

true

x/c	y/c
1.00000	0.00000
0.99674	0.00048
0.98712	0.00204
0.97155	0.00485
0.95054	0.00894
0.92464	0.01420
0.89436	0.02041
0.86021	0.02731
0.82264	0.03460
0.78208	0.04199
0.73892	0.04925
0.69356	0.05620
0.64646	0.06270

0.59812	0.06861
0.54902	0.07381
0.49967	0.07816
0.45058	0.08154
0.40222	0.08385
0.35506	0.08500
0.30953	0.08493
0.26604	0.08359
0.22499	0.08096
0.18671	0.07703
0.15146	0.07182
0.11948	0.06548
0.09105	0.05809
0.06627	0.04976
0.04524	0.04078
0.02812	0.03145
0.01502	0.02206
0.00606	0.01293
0.00115	0.00448
0.00038	-0.00223
0.00532	-0.00701
0.01649	-0.01088
0.03308	-0.01403
0.05491	-0.01635
0.08180	-0.01787
0.11351	-0.01862
0.14974	-0.01867
0.19010	-0.01810
0.23420	-0.01699
0.28153	-0.01547
0.33154	-0.01363
0.38364	-0.01152
0.43724	-0.00922
0.49176	-0.00678
0.54659	-0.00430
0.60112	-0.00190
0.65469	0.00030
0.70664	0.00224
0.75634	0.00379
0.80313	0.00485
0.84635	0.00535
0.88534	0.00526
0.91942	0.00458
0.94797	0.00350
0.97054	0.00226
0.98684	0.00113
0.99670	0.00030
1.00000	0.00000

SD7032 (D)

actual

x/c	y/c
1.00000	-0.00035
0.99864	0.00040
0.99665	0.00110
0.99353	0.00185
0.98888	0.00276
0.98259	0.00379
0.97541	0.00481
0.96809	0.00616
0.96252	0.00712
0.94859	0.00980
0.93704	0.01181
0.92652	0.01367
0.91821	0.01508
0.90582	0.01711
0.89075	0.01972
0.85677	0.02607
0.82499	0.03159
0.79355	0.03686
0.75266	0.04431
0.71060	0.05136
0.66862	0.05770
0.62429	0.06346
0.56871	0.06980
0.51073	0.07533
0.45314	0.07952
0.39660	0.08205
0.34046	0.08289
0.28613	0.08208
0.24396	0.07994
0.19704	0.07609
0.17147	0.07271
0.14238	0.06773
0.11459	0.06170
0.09965	0.05761
0.08841	0.05421
0.07620	0.05025
0.06239	0.04531
0.04837	0.03948
0.03818	0.03453
0.02912	0.02965
0.01986	0.02401
0.01227	0.01814
0.00787	0.01368
0.00240	0.00633
0.00023	-0.00192
0.00280	-0.00679
0.00928	-0.01148
0.01264	-0.01250
0.01846	-0.01399
0.02881	-0.01573
0.04723	-0.01785
0.06062	-0.01903
0.07656	-0.02001
0.09187	-0.02064
0.10349	-0.02099
0.12003	-0.02144
0.14868	-0.02176
0.17685	-0.02145
0.20472	-0.02073
0.24431	-0.01943
0.28925	-0.01764
0.33962	-0.01591
0.40103	-0.01330
0.45698	-0.01036
0.51631	-0.00740
0.57476	-0.00438
0.62875	-0.00132
0.67810	0.00055
0.71802	0.00157
0.75935	0.00282
0.80501	0.00420
0.83600	0.00481
0.86820	0.00511
0.89605	0.00491
0.90948	0.00457
0.92313	0.00380
0.94075	0.00277
0.95357	0.00173
0.96655	0.00116
0.97922	0.00045
0.99036	-0.00013
0.99484	-0.00029
0.99844	-0.00040
1.00000	-0.00035

SD7037

true

x/c	y/c
1.00000	0.00000
0.99672	0.00042
0.98707	0.00180
0.97146	0.00436
0.95041	0.00811
0.92450	0.01295
0.89425	0.01865
0.86015	0.02490
0.82261	0.03141
0.78201	0.03788
0.73865	0.04413
0.69294	0.05011
0.64539	0.05572
0.59655	0.06085
0.54693	0.06538
0.49706	0.06917
0.44745	0.07211
0.39862	0.07410
0.35101	0.07504
0.30508	0.07488
0.26125	0.07358
0.21989	0.07113
0.18137	0.06754
0.14601	0.06286
0.11410	0.05715
0.08586	0.05049
0.06146	0.04300
0.04102	0.03486
0.02462	0.02632
0.01232	0.01770
0.00418	0.00936
0.00021	0.00185
0.00127	-0.00393
0.00806	-0.00839
0.02038	-0.01227
0.03800	-0.01541
0.06074	-0.01777
0.08844	-0.01934
0.12084	-0.02017
0.15765	-0.02032
0.19850	-0.01987
0.24296	-0.01891
0.29055	-0.01754
0.34071	-0.01586
0.39288	-0.01396
0.44643	-0.01190
0.50074	-0.00976
0.55519	-0.00760
0.60914	-0.00549
0.66197	-0.00349
0.71305	-0.00168
0.76178	-0.00014
0.80752	0.00104
0.84964	0.00182
0.88756	0.00220
0.92071	0.00218
0.94859	0.00185
0.97077	0.00132
0.98690	0.00071
0.99671	0.00021
1.00000	0.00000

SD7037 (D)

actual

x/c	y/c
1.00000	0.00053
0.99979	-0.00046
0.99724	0.00038
0.99384	0.00099
0.98800	0.00184
0.98381	0.00256
0.97449	0.00398
0.96222	0.00599
0.95144	0.00801
0.93069	0.01186
0.90848	0.01618
0.86705	0.02391
0.82631	0.03077
0.74166	0.04337
0.65505	0.05377
0.57209	0.06196
0.48673	0.06829
0.40301	0.07213
0.35879	0.07295
0.31837	0.07297
0.27730	0.07241
0.23531	0.07045
0.21166	0.06875
0.19381	0.06714
0.17297	0.06487
0.15370	0.06235
0.13932	0.06014
0.12442	0.05759
0.11202	0.05510
0.09845	0.05208
0.08377	0.04834
0.06664	0.04318
0.05855	0.04044
0.04925	0.03694
0.03949	0.03281
0.03136	0.02893
0.02425	0.02488
0.01904	0.02155
0.01272	0.01708
0.00891	0.01386
0.00588	0.01088
0.00300	0.00748
0.00042	0.00277
0.00080	-0.00391
0.00057	-0.00296
0.00174	-0.00516
0.00463	-0.00819
0.00864	-0.01026
0.01271	-0.01177

0.01874	-0.01351
0.02366	-0.01473
0.03041	-0.01609
0.04605	-0.01848
0.06753	-0.02055
0.09228	-0.02183
0.12032	-0.02263
0.15059	-0.02288
0.18433	-0.02288
0.22547	-0.02240
0.29085	-0.02055
0.37381	-0.01751
0.45059	-0.01429
0.54248	-0.01013
0.63290	-0.00625
0.71381	-0.00271
0.79662	0.00018
0.84123	0.00134
0.88338	0.00214
0.90668	0.00226
0.92670	0.00219
0.94914	0.00208
0.96759	0.00195
0.98050	0.00160
0.98754	0.00100
0.99322	0.00084
0.99975	0.00046
1.00000	0.00053

SD7037 (E)

actual

x/c	y/c
1.00000	0.00004
1.00000	-0.00069
0.99607	0.00010
0.99201	0.00095
0.98859	0.00154
0.98366	0.00267
0.97775	0.00378
0.96938	0.00540
0.96110	0.00699
0.95216	0.00868
0.94288	0.01038
0.93119	0.01245
0.92096	0.01431
0.90939	0.01620
0.90118	0.01778
0.89025	0.01969
0.88121	0.02127
0.86980	0.02327
0.85481	0.02571
0.83959	0.02826
0.82720	0.03031
0.81388	0.03240
0.80240	0.03416
0.79007	0.03612
0.77875	0.03779
0.77031	0.03910
0.71895	0.04649
0.70612	0.04820
0.68038	0.05149
0.65019	0.05509
0.61343	0.05920
0.56882	0.06368
0.52210	0.06770
0.48901	0.07015
0.44989	0.07255
0.39515	0.07491
0.34472	0.07594
0.31517	0.07591
0.30042	0.07572
0.28373	0.07535
0.25769	0.07444
0.21906	0.07224
0.18246	0.06894
0.14377	0.06399
0.13519	0.06257
0.11263	0.05824
0.09024	0.05291
0.06835	0.04640
0.04405	0.03697
0.04005	0.03513
0.02326	0.02640
0.00633	0.01274
0.00274	0.00746
0.00093	0.00344
0.00030	-0.00193
0.00094	-0.00342
0.00191	-0.00486
0.00297	-0.00600
0.00580	-0.00810
0.00742	-0.00895
0.01327	-0.01105
0.01722	-0.01208
0.01954	-0.01266
0.02259	-0.01332
0.02438	-0.01368
0.02817	-0.01439
0.03379	-0.01531
0.03877	-0.01599
0.04765	-0.01702
0.05241	-0.01749
0.06095	-0.01821
0.06798	-0.01871
0.07864	-0.01933
0.08449	-0.01964
0.09902	-0.02022
0.10489	-0.02041
0.17150	-0.02107
0.17954	-0.02102
0.19239	-0.02092
0.19982	-0.02084
0.21413	-0.02066
0.29543	-0.01885
0.36976	-0.01642
0.42421	-0.01436
0.44981	-0.01336
0.54334	-0.00964
0.58692	-0.00795
0.69661	-0.00392
0.72684	-0.00288
0.77843	-0.00105
0.80118	-0.00058
0.81928	-0.00031
0.83659	0.00000
0.85241	0.00025
0.87209	0.00040
0.88810	0.00048
0.90803	0.00084
0.92122	0.00059
0.93998	0.00074
0.94945	0.00060
0.95609	0.00056
0.96252	0.00050
0.96954	0.00043
0.97761	0.00038
0.98085	0.00039
0.98710	0.00034
0.98926	0.00024
0.99306	0.00037
0.99585	0.00035
0.99812	0.00041
0.99977	0.00069
1.00000	0.00004

SD7062

true

x/c	y/c
1.00000	0.00000
0.99652	0.00057
0.98634	0.00242
0.97004	0.00571
0.94818	0.01036
0.92127	0.01615
0.88967	0.02289
0.85381	0.03049
0.81427	0.03885
0.77166	0.04778
0.72662	0.05702
0.67973	0.06626
0.63154	0.07517
0.58254	0.08347
0.53322	0.09089
0.48402	0.09723
0.43540	0.10229
0.38779	0.10592
0.34159	0.10801
0.29720	0.10846
0.25496	0.10722
0.21521	0.10428
0.17825	0.09965
0.14432	0.09340
0.11363	0.08562
0.08637	0.07647
0.06269	0.06616
0.04272	0.05491
0.02650	0.04300
0.01410	0.03081
0.00562	0.01871
0.00103	0.00711
0.00027	-0.00327
0.00425	-0.01115
0.01414	-0.01701
0.02991	-0.02205
0.05099	-0.02623
0.07710	-0.02948
0.10798	-0.03179
0.14328	-0.03319
0.18265	-0.03367
0.22570	-0.03327
0.27199	-0.03206
0.32105	-0.03010
0.37238	-0.02747
0.42547	-0.02426
0.47979	-0.02064
0.53477	-0.01680
0.58975	-0.01295
0.64403	-0.00927
0.69692	-0.00593
0.74768	-0.00307
0.79557	-0.00079
0.83990	0.00087
0.87998	0.00189
0.91518	0.00230
0.94491	0.00216
0.96864	0.00163
0.98593	0.00092
0.99646	0.00027
1.00000	0.00000

SD7062 (B)

actual

x/c	y/c
1.00000	-0.00051
0.99707	0.00103
0.99490	0.00163
0.99093	0.00219
0.98845	0.00270
0.98575	0.00336
0.98230	0.00421
0.97715	0.00541
0.97194	0.00664
0.96552	0.00812
0.96147	0.00904
0.95393	0.01080
0.94841	0.01206
0.94183	0.01352
0.93532	0.01497
0.92889	0.01638
0.92203	0.01788
0.91274	0.01985
0.90346	0.02186
0.89292	0.02415
0.88219	0.02643
0.87232	0.02855
0.86251	0.03064
0.85094	0.03314
0.84034	0.03539
0.82901	0.03783
0.81934	0.03989
0.80969	0.04196
0.79647	0.04475
0.78422	0.04730
0.77139	0.04997
0.75938	0.05245
0.73907	0.05661
0.72198	0.06008
0.70518	0.06347
0.68629	0.06722
0.66737	0.07086
0.65061	0.07401
0.63009	0.07770
0.61389	0.08049
0.59651	0.08337

0.57561	0.08662
0.55934	0.08905
0.54296	0.09140
0.52496	0.09379
0.50626	0.09611
0.48898	0.09812
0.46845	0.10030
0.45271	0.10183
0.43445	0.10341
0.40688	0.10540
0.39154	0.10630
0.37340	0.10720
0.35076	0.10803
0.33580	0.10843
0.32457	0.10861
0.30545	0.10871
0.29103	0.10856
0.27997	0.10830
0.27070	0.10799
0.25946	0.10747
0.25134	0.10702
0.24194	0.10637
0.23506	0.10582
0.22428	0.10487
0.21511	0.10392
0.19699	0.10172
0.17669	0.09875
0.16818	0.09731
0.15024	0.09385
0.14185	0.09201
0.12565	0.08807
0.10639	0.08257
0.09512	0.07886
0.08289	0.07434
0.07645	0.07173
0.07147	0.06956
0.06481	0.06646
0.05969	0.06393
0.04976	0.05849
0.04138	0.05325
0.03231	0.04672
0.02750	0.04286
0.02441	0.04018
0.01708	0.03297
0.01155	0.02629
0.00773	0.02086
0.00499	0.01624
0.00283	0.01171
0.00125	0.00726
0.00023	0.00308
0.00073	-0.00565
0.00394	-0.01073
0.00519	-0.01199
0.00773	-0.01402
0.01135	-0.01620
0.01441	-0.01762
0.01828	-0.01910
0.02307	-0.02063
0.02556	-0.02132
0.03040	-0.02259
0.03452	-0.02354
0.03892	-0.02446
0.04366	-0.02533
0.04794	-0.02610
0.05293	-0.02691
0.05791	-0.02762
0.06179	-0.02814
0.06707	-0.02879
0.07293	-0.02944
0.07918	-0.03007
0.08484	-0.03056
0.09368	-0.03126
0.10153	-0.03181
0.10970	-0.03231
0.12119	-0.03288
0.13121	-0.03331
0.14214	-0.03367
0.15597	-0.03397
0.16317	-0.03408

SD7080

true

x/c	y/c
1.00000	0.00000
0.99671	0.00037
0.98701	0.00162
0.97133	0.00395
0.95018	0.00743
0.92413	0.01195
0.89372	0.01728
0.85943	0.02313
0.82169	0.02919
0.78084	0.03519
0.73720	0.04097
0.69117	0.04647
0.64326	0.05163
0.59402	0.05635
0.54399	0.06052
0.49371	0.06404
0.44370	0.06680
0.39448	0.06871
0.34656	0.06969
0.30040	0.06966
0.25644	0.06858
0.21509	0.06641
0.17670	0.06312
0.14157	0.05875
0.10996	0.05334
0.08210	0.04699
0.05814	0.03983
0.03821	0.03203
0.02237	0.02383
0.01068	0.01558
0.00322	0.00763
0.00003	0.00061
0.00216	-0.00492
0.01016	-0.00948
0.02347	-0.01354
0.04198	-0.01694
0.06549	-0.01962
0.09383	-0.02156
0.12676	-0.02279
0.16397	-0.02337
0.20508	-0.02338
0.24962	-0.02289
0.29710	-0.02194
0.34700	-0.02059
0.39877	-0.01891
0.45183	-0.01696
0.50559	-0.01480
0.55945	-0.01250
0.61281	-0.01015
0.66505	-0.00783
0.71557	-0.00564
0.76376	-0.00366
0.80900	-0.00201
0.85068	-0.00072
0.88820	0.00017
0.92104	0.00069
0.94871	0.00087
0.97079	0.00076
0.98689	0.00047
0.99671	0.00015
1.00000	0.00000

SD7080

actual

x/c	y/c
1.00000	0.00044
0.99968	0.00011
0.99744	0.00055
0.99374	0.00110
0.98848	0.00179
0.98389	0.00224
0.97774	0.00308
0.97486	0.00347
0.97051	0.00406
0.96734	0.00446
0.96154	0.00520
0.95740	0.00589
0.95173	0.00681
0.94615	0.00765
0.93952	0.00871
0.93404	0.00953
0.92717	0.01055
0.92027	0.01164
0.91142	0.01297
0.90480	0.01409
0.89558	0.01504
0.88599	0.01691
0.87962	0.01803
0.86599	0.02028
0.85458	0.02203
0.84321	0.02383
0.83072	0.02607
0.81794	0.02809
0.80634	0.02992
0.79285	0.03194
0.77478	0.03452
0.75864	0.03699
0.74082	0.03942
0.72295	0.04191
0.70695	0.04418
0.69687	0.04599
0.68707	0.04754
0.67788	0.04848
0.67028	0.04919
0.65488	0.05011
0.63663	0.05210
0.60329	0.05595
0.56728	0.05887
0.53238	0.06114
0.49522	0.06339
0.46018	0.06562
0.42544	0.06758
0.39087	0.06871
0.35078	0.06962
0.31375	0.06946
0.29674	0.06935
0.27803	0.06912
0.26068	0.06874
0.22195	0.06730
0.20425	0.06635
0.16644	0.06268
0.15742	0.06156
0.14886	0.06033
0.13913	0.05894
0.11943	0.05565
0.09871	0.05149
0.07973	0.04678
0.06693	0.04246
0.05280	0.03728
0.04227	0.03278
0.03247	0.02815
0.02344	0.02342
0.01494	0.01836
0.01059	0.01540
0.00348	0.00932
0.00008	0.00142
0.00078	-0.00213
0.00370	-0.00629
0.00956	-0.00949
0.01167	-0.01017
0.01839	-0.01206
0.02379	-0.01325
0.02883	-0.01417
0.03365	-0.01492
0.04016	-0.01585
0.04339	-0.01626
0.04908	-0.01705
0.05423	-0.01770
0.06033	-0.01850
0.06877	-0.01955
0.07790	-0.02057
0.08370	-0.02117
0.09387	-0.02207
0.10486	-0.02294
0.11311	-0.02350
0.12321	-0.02407
0.13152	-0.02450
0.14100	-0.02492
0.14854	-0.02517
0.15792	-0.02548
0.16699	-0.02572
0.17651	-0.02599
0.18766	-0.02625
0.19551	-0.02642
0.20705	-0.02659
0.21423	-0.02675
0.23604	-0.02700
0.24888	-0.02699
0.28903	-0.02626
0.32232	-0.02537
0.39345	-0.02325
0.41959	-0.02236
0.48164	-0.02003

0.54191	-0.01746
0.56166	-0.01666
0.60451	-0.01487
0.65908	-0.01236
0.70854	-0.01042
0.78256	-0.00758
0.80681	-0.00672
0.84773	-0.00538
0.88280	-0.00415
0.91362	-0.00335
0.94975	-0.00213
0.97070	-0.00153

SG6040

true

x/c	y/c
1.00000	0.00000
0.99788	0.00048
0.99183	0.00207
0.98233	0.00477
0.96977	0.00845
0.95442	0.01283
0.93637	0.01766
0.91559	0.02285
0.89221	0.02845
0.86647	0.03446
0.83861	0.04080
0.80887	0.04739
0.77752	0.05415
0.74482	0.06094
0.71104	0.06759
0.67632	0.07394
0.64081	0.07992
0.60477	0.08541
0.56837	0.09032
0.53180	0.09461
0.49526	0.09815
0.45891	0.10089
0.42297	0.10279
0.38758	0.10378
0.35286	0.10386
0.31903	0.10299
0.28619	0.10116
0.25449	0.09843
0.22405	0.09479
0.19496	0.09028
0.16734	0.08506
0.14137	0.07921
0.11719	0.07283
0.09496	0.06598
0.07481	0.05875
0.05686	0.05123
0.04123	0.04347
0.02798	0.03559
0.01725	0.02771
0.00913	0.01987
0.00362	0.01212
0.00072	0.00449
0.00027	-0.00268
0.00289	-0.00884
0.00906	-0.01450
0.01829	-0.02023
0.03027	-0.02580
0.04483	-0.03103
0.06202	-0.03566
0.08193	-0.03971
0.10446	-0.04336
0.12939	-0.04657
0.15655	-0.04937
0.18574	-0.05172
0.21677	-0.05361
0.24943	-0.05503
0.28349	-0.05594
0.31873	-0.05632
0.35490	-0.05613
0.39177	-0.05528
0.42916	-0.05371
0.46691	-0.05132
0.50479	-0.04801
0.54298	-0.04368
0.58164	-0.03863
0.62049	-0.03323
0.65925	-0.02772
0.69762	-0.02231
0.73526	-0.01718
0.77181	-0.01250
0.80690	-0.00840
0.84016	-0.00497
0.87120	-0.00228
0.89963	-0.00033
0.92511	0.00090
0.94728	0.00148
0.96587	0.00154
0.98063	0.00122
0.99132	0.00070
0.99782	0.00022
1.00000	0.00000

SG6040

actual

x/c	y/c
1.00000	0.00001
0.99026	0.00343
0.97825	0.00666
0.96770	0.00942
0.95302	0.01312
0.93704	0.01716
0.91693	0.02213
0.89850	0.02662
0.87755	0.03160
0.85809	0.03613
0.83721	0.04090
0.81315	0.04622
0.78966	0.05128
0.75435	0.05861
0.71621	0.06618
0.67265	0.07411
0.63628	0.08021
0.59026	0.08711
0.54781	0.09257
0.50735	0.09692
0.46615	0.10044
0.42477	0.10295
0.38318	0.10421
0.34263	0.10418
0.32023	0.10362
0.30228	0.10287
0.27671	0.10127
0.25630	0.09951
0.23757	0.09751
0.21762	0.09492
0.19505	0.09144
0.17574	0.08798
0.15308	0.08333
0.13315	0.07865
0.11657	0.07426
0.09529	0.06783
0.08014	0.06257
0.07130	0.05920
0.06031	0.05466
0.04958	0.04973
0.04205	0.04592
0.03565	0.04235
0.02724	0.03704
0.01751	0.02942
0.01330	0.02533
0.00910	0.02046
0.00648	0.01683
0.00330	0.01126
0.00088	0.00510
0.00041	-0.00339
0.00274	-0.00886
0.00466	-0.01136
0.00933	-0.01574
0.01509	-0.01967
0.02337	-0.02388
0.02826	-0.02597
0.03746	-0.02943
0.05109	-0.03361
0.06913	-0.03795
0.08931	-0.04177
0.10567	-0.04434
0.12106	-0.04638
0.14706	-0.04927
0.16682	-0.05109
0.19151	-0.05296
0.22196	-0.05478
0.25750	-0.05625
0.29978	-0.05715
0.33138	-0.05732
0.36920	-0.05702
0.40781	-0.05594
0.44622	-0.05399
0.49340	-0.05037
0.53287	-0.04611
0.57214	-0.04097
0.61800	-0.03461
0.65462	-0.02950
0.69723	-0.02358
0.73566	-0.01844
0.76570	-0.01462
0.78780	-0.01200
0.82366	-0.00821
0.84294	-0.00632
0.86463	-0.00441
0.88336	-0.00302
0.90244	-0.00187
0.91910	-0.00106
0.93819	-0.00036
0.95190	-0.00002
0.96780	0.00011
0.97925	-0.00006
0.99169	-0.00052
1.00000	0.00001

SG6041

true

x/c	y/c
1.00000	0.00000
0.99801	0.00029
0.99227	0.00132
0.98311	0.00310
0.97084	0.00560
0.95568	0.00867
0.93778	0.01217
0.91722	0.01600
0.89412	0.02018
0.86872	0.02469
0.84127	0.02946
0.81200	0.03438
0.78119	0.03933
0.74901	0.04408
0.71552	0.04849
0.68087	0.05261
0.64526	0.05635
0.60884	0.05967
0.57186	0.06258
0.53448	0.06502
0.49692	0.06699
0.45939	0.06847
0.42211	0.06944
0.38529	0.06991
0.34915	0.06987
0.31389	0.06931
0.27971	0.06824
0.24683	0.06666
0.21540	0.06458
0.18563	0.06201
0.15767	0.05896
0.13169	0.05546
0.10783	0.05149
0.08618	0.04710
0.06690	0.04231
0.05002	0.03711
0.03555	0.03159
0.02361	0.02578
0.01411	0.01971
0.00702	0.01366
0.00241	0.00767
0.00016	0.00194
0.00029	-0.00270
0.00378	-0.00645
0.01080	-0.01010
0.02073	-0.01346
0.03361	-0.01637
0.04948	-0.01888
0.06829	-0.02104
0.08990	-0.02294
0.11413	-0.02461
0.14081	-0.02605
0.16976	-0.02728
0.20077	-0.02828
0.23363	-0.02907
0.26812	-0.02964

0.30402	-0.02998
0.34108	-0.03010
0.37907	-0.02997
0.41772	-0.02959
0.45679	-0.02894
0.49604	-0.02797
0.53524	-0.02663
0.57423	-0.02493
0.61284	-0.02285
0.65090	-0.02037
0.68832	-0.01763
0.72492	-0.01482
0.76041	-0.01210
0.79448	-0.00956
0.82683	-0.00728
0.85718	-0.00532
0.88524	-0.00370
0.91073	-0.00243
0.93339	-0.00149
0.95300	-0.00083
0.96938	-0.00036
0.98244	-0.00005
0.99204	0.00008
0.99798	0.00006
1.00000	0.00000

SG6041

actual

x/c	y/c
1.00000	-0.00005
0.99903	0.00138
0.99281	0.00267
0.98058	0.00514
0.96764	0.00770
0.95437	0.01025
0.93773	0.01344
0.91996	0.01678
0.89869	0.02067
0.87722	0.02453
0.85737	0.02803
0.83957	0.03107
0.81550	0.03507
0.79511	0.03832
0.75833	0.04367
0.71566	0.04925
0.67182	0.05438
0.63316	0.05842
0.59023	0.06224
0.54655	0.06533
0.50446	0.06770
0.46048	0.06952
0.41900	0.07053
0.38485	0.07084
0.33824	0.07066
0.31870	0.07033
0.29917	0.06982
0.27246	0.06883
0.25512	0.06801
0.23745	0.06700
0.21942	0.06579
0.19607	0.06390
0.17595	0.06195
0.15855	0.05996
0.12966	0.05588
0.11112	0.05269
0.09281	0.04902
0.07705	0.04532
0.06838	0.04303
0.05642	0.03946
0.04827	0.03673
0.03787	0.03273
0.03143	0.02987
0.02544	0.02686
0.01851	0.02280
0.01320	0.01906
0.00962	0.01611
0.00694	0.01343
0.00325	0.00868
0.00117	0.00487
0.00009	-0.00133
0.00125	-0.00500
0.00472	-0.00856
0.00846	-0.01057
0.01509	-0.01297
0.02604	-0.01588
0.03420	-0.01755
0.04163	-0.01879
0.04959	-0.01991
0.06933	-0.02212
0.08064	-0.02315
0.09912	-0.02455
0.12494	-0.02596
0.14361	-0.02676
0.15983	-0.02736
0.18025	-0.02802
0.20450	-0.02862
0.23094	-0.02908
0.25074	-0.02936
0.27001	-0.02958
0.29408	-0.02978
0.30730	-0.02988
0.32459	-0.02996
0.37102	-0.02991
0.41429	-0.02951
0.45669	-0.02883
0.49854	-0.02779
0.54158	-0.02623
0.58134	-0.02438
0.62675	-0.02175
0.66718	-0.01901
0.70714	-0.01605
0.74404	-0.01328
0.77558	-0.01095
0.82686	-0.00742
0.84875	-0.00607
0.89087	-0.00368
0.92944	-0.00199
0.98754	-0.00129
0.99944	-0.00138
1.00000	-0.00005

SG6042

true

x/c	y/c
1.00000	0.00000
0.99806	0.00048
0.99249	0.00203
0.98372	0.00463
0.97210	0.00811
0.95784	0.01223
0.94102	0.01672
0.92161	0.02148
0.89968	0.02657
0.87546	0.03195
0.84915	0.03756
0.82098	0.04333
0.79117	0.04914
0.75995	0.05489
0.72757	0.06038
0.69410	0.06544
0.65962	0.07006
0.62435	0.07417
0.58844	0.07772
0.55208	0.08071
0.51546	0.08310
0.47877	0.08486
0.44221	0.08600
0.40597	0.08648
0.37024	0.08634
0.33522	0.08556
0.30110	0.08415
0.26808	0.08215
0.23633	0.07956
0.20603	0.07639
0.17735	0.07269
0.15044	0.06847
0.12544	0.06379
0.10249	0.05866
0.08169	0.05313
0.06316	0.04726
0.04698	0.04105
0.03315	0.03458
0.02179	0.02795
0.01284	0.02119
0.00619	0.01446
0.00197	0.00797
0.00011	0.00182
0.00035	-0.00341
0.00371	-0.00730
0.01068	-0.01042
0.02068	-0.01311
0.03369	-0.01516
0.04982	-0.01659
0.06901	-0.01751
0.09117	-0.01805
0.11611	-0.01826
0.14365	-0.01821
0.17358	-0.01792
0.20570	-0.01744
0.23979	-0.01679
0.27559	-0.01599
0.31286	-0.01506
0.35135	-0.01401
0.39078	-0.01286
0.43089	-0.01159
0.47140	-0.01016
0.51208	-0.00853
0.55277	-0.00671
0.59330	-0.00475
0.63345	-0.00273
0.67298	-0.00077
0.71161	0.00102
0.74904	0.00255
0.78494	0.00377
0.81900	0.00462
0.85093	0.00510
0.88041	0.00521
0.90718	0.00496
0.93097	0.00442
0.95153	0.00364
0.96868	0.00273
0.98224	0.00178
0.99204	0.00090
0.99800	0.00026
1.00000	0.00000

SG6042

actual

x/c	y/c
1.00000	-0.00065
0.99820	0.00117
0.99562	0.00192
0.99282	0.00269
0.98748	0.00408
0.98398	0.00502
0.98019	0.00601
0.97557	0.00721
0.97112	0.00837
0.96673	0.00951
0.96175	0.01078
0.95713	0.01195
0.95062	0.01358
0.94081	0.01599
0.93585	0.01721
0.92897	0.01893
0.92329	0.02028
0.91560	0.02213
0.90857	0.02379
0.90049	0.02569
0.89336	0.02733
0.88614	0.02899
0.87604	0.03122
0.86787	0.03303
0.86009	0.03472
0.85066	0.03673
0.83979	0.03901
0.82794	0.04143
0.81423	0.04417
0.79742	0.04741
0.77945	0.05077
0.76279	0.05374
0.74191	0.05740
0.72474	0.06023
0.70708	0.06296
0.68959	0.06555
0.66833	0.06833
0.65026	0.07059
0.63207	0.07268
0.61519	0.07450
0.59681	0.07634
0.57946	0.07798
0.56150	0.07943
0.54348	0.08080
0.52649	0.08195

0.50796	0.08305
0.48829	0.08402
0.47137	0.08475
0.45186	0.08547
0.43298	0.08594
0.41741	0.08619
0.40169	0.08638
0.38085	0.08638
0.35813	0.08616
0.34442	0.08589
0.32186	0.08518
0.30805	0.08461
0.29836	0.08415
0.29028	0.08374
0.27927	0.08313
0.27211	0.08268
0.26171	0.08200
0.25453	0.08148
0.24367	0.08062
0.22520	0.07896
0.21725	0.07817
0.19849	0.07605
0.18031	0.07368
0.16061	0.07074
0.14398	0.06794
0.12392	0.06410
0.11625	0.06248
0.09760	0.05810
0.09247	0.05677
0.07485	0.05173
0.07064	0.05040
0.06117	0.04716
0.05438	0.04464
0.03896	0.03801
0.03025	0.03352
0.02657	0.03140
0.01884	0.02632
0.01258	0.02133
0.00760	0.01627
0.00268	0.00906
0.00074	0.00439
0.00040	0.00305
0.00004	0.00095
0.00029	-0.00261
0.00255	-0.00619
0.00417	-0.00738
0.00658	-0.00863
0.00951	-0.00977
0.01279	-0.01085
0.01664	-0.01191
0.02074	-0.01290
0.02587	-0.01394
0.03015	-0.01465
0.03583	-0.01542
0.03899	-0.01579
0.04608	-0.01645
0.05489	-0.01708
0.05790	-0.01725
0.06734	-0.01768
0.07755	-0.01802
0.08602	-0.01821
0.09435	-0.01835
0.10406	-0.01843
0.11521	-0.01848
0.12556	-0.01848
0.13355	-0.01846
0.14340	-0.01841
0.16156	-0.01831
0.17788	-0.01815
0.19547	-0.01795
0.23315	-0.01733
0.25052	-0.01700
0.26803	-0.01662
0.28799	-0.01619
0.30513	-0.01581

SG6043

true

x/c	y/c
1.00000	0.00000
0.99811	0.00066
0.99274	0.00271
0.98439	0.00607
0.97343	0.01047
0.96007	0.01552
0.94429	0.02092
0.92597	0.02655
0.90516	0.03248
0.88207	0.03868
0.85688	0.04510
0.82979	0.05165
0.80101	0.05824
0.77074	0.06478
0.73922	0.07114
0.70666	0.07717
0.67320	0.08268
0.63889	0.08761
0.60396	0.09190
0.56854	0.09551
0.53276	0.09842
0.49685	0.10059
0.46096	0.10201
0.42528	0.10269
0.39000	0.10260
0.35527	0.10175
0.32131	0.10017
0.28827	0.09788
0.25633	0.09491
0.22567	0.09127
0.19645	0.08702
0.16882	0.08220
0.14293	0.07684
0.11890	0.07100
0.09687	0.06473
0.07694	0.05809
0.05920	0.05117
0.04376	0.04398
0.03062	0.03659
0.01983	0.02916
0.01142	0.02171
0.00528	0.01436
0.00150	0.00741
0.00002	0.00094
0.00059	-0.00468
0.00393	-0.00860
0.01077	-0.01120
0.02087	-0.01316
0.03409	-0.01430
0.05050	-0.01460
0.07015	-0.01422
0.09292	-0.01329
0.11864	-0.01197
0.14712	-0.01034
0.17815	-0.00848
0.21150	-0.00645
0.24693	-0.00430
0.28417	-0.00210
0.32296	0.00011
0.36301	0.00232
0.40401	0.00450
0.44571	0.00669
0.48785	0.00893
0.53029	0.01117
0.57286	0.01327
0.61525	0.01504
0.65704	0.01635
0.69785	0.01721
0.73735	0.01760
0.77520	0.01751
0.81109	0.01695
0.84468	0.01592
0.87567	0.01447
0.90377	0.01265
0.92871	0.01056
0.95024	0.00828
0.96814	0.00591
0.98215	0.00361
0.99209	0.00170
0.99803	0.00045
1.00000	0.00000

SG6043

actual

x/c	y/c
1.00000	-0.00060
0.99461	0.00293
0.99069	0.00446
0.97948	0.00863
0.96794	0.01276
0.95533	0.01712
0.93942	0.02241
0.92235	0.02774
0.90345	0.03321
0.88485	0.03831
0.86186	0.04435
0.84111	0.04956
0.81957	0.05476
0.79355	0.06078
0.75733	0.06854
0.71816	0.07608
0.67908	0.08264
0.63708	0.08862
0.59262	0.09384
0.54989	0.09781
0.50986	0.10052
0.46512	0.10250
0.42717	0.10340
0.38588	0.10348
0.34483	0.10238
0.32838	0.10159
0.30169	0.09995
0.28060	0.09833
0.26205	0.09663
0.24234	0.09451
0.22209	0.09195
0.20014	0.08871
0.17931	0.08515
0.16075	0.08162
0.13618	0.07631
0.11822	0.07182
0.09761	0.06583
0.08409	0.06143
0.07008	0.05637
0.06123	0.05285
0.05576	0.05051
0.04896	0.04738
0.03945	0.04253
0.03205	0.03825
0.02617	0.03441
0.01973	0.02965
0.01475	0.02540
0.01047	0.02116
0.00415	0.01271
0.00064	0.00492
0.00065	-0.00517
0.00267	-0.00735
0.00633	-0.00950
0.01576	-0.01226
0.02321	-0.01340
0.03224	-0.01422
0.03944	-0.01456
0.04744	-0.01472
0.05529	-0.01472
0.06834	-0.01442
0.08901	-0.01360
0.11369	-0.01239
0.13685	-0.01113
0.15612	-0.01003
0.17620	-0.00888
0.19877	-0.00757
0.21944	-0.00636
0.24561	-0.00481
0.25981	-0.00399
0.29006	-0.00222
0.32265	-0.00037
0.36729	0.00213
0.40911	0.00443
0.45071	0.00666
0.49537	0.00903
0.53587	0.01108
0.57123	0.01282
0.62660	0.01519
0.66208	0.01631
0.70379	0.01717
0.71968	0.01732
0.74512	0.01737
0.78090	0.01713
0.80726	0.01679
0.83145	0.01617
0.85258	0.01529
0.87543	0.01411
0.89460	0.01289
0.91201	0.01174
0.93116	0.01021
0.95012	0.00819
0.96119	0.00666

0.97688	0.00407
0.98817	0.00165
0.99987	-0.00071
1.00000	-0.00060

Trainer 60

from plans

x/c	y/c
1.00000	0.00319
0.99754	0.00571
0.99070	0.01001
0.98037	0.01148
0.96698	0.01039
0.95044	0.01025
0.93064	0.01068
0.90775	0.01247
0.88202	0.01661
0.85370	0.02179
0.82309	0.02711
0.79048	0.03233
0.75616	0.03756
0.72043	0.04296
0.68359	0.04861
0.64594	0.05446
0.60778	0.06051
0.56937	0.06657
0.53099	0.07224
0.49265	0.07732
0.45435	0.08183
0.41638	0.08573
0.37887	0.08895
0.34204	0.09138
0.30609	0.09282
0.27120	0.09303
0.23760	0.09197
0.20549	0.08987
0.17504	0.08698
0.14648	0.08324
0.11999	0.07844
0.09576	0.07237
0.07395	0.06505
0.05468	0.05688
0.03811	0.04840
0.02433	0.03986
0.01338	0.03102
0.00548	0.02118
0.00098	0.00979
0.00000	0.00117
0.00098	-0.00708
0.00548	-0.01740
0.01338	-0.02674
0.02433	-0.03560
0.03811	-0.04426
0.05468	-0.05277
0.07395	-0.06095
0.09576	-0.06841
0.11999	-0.07476
0.14648	-0.07979
0.17504	-0.08367
0.20549	-0.08667
0.23760	-0.08886
0.27120	-0.09005
0.30609	-0.09004
0.34204	-0.08885
0.37887	-0.08666
0.41638	-0.08371
0.45435	-0.08002
0.49265	-0.07553
0.53099	-0.07030
0.56937	-0.06462
0.60778	-0.05894
0.64594	-0.05345
0.68359	-0.04801
0.72043	-0.04250
0.75616	-0.03704
0.79048	-0.03186
0.82309	-0.02699
0.85370	-0.02222
0.88202	-0.01747
0.90775	-0.01318
0.93064	-0.01064
0.95044	-0.01112
0.96698	-0.01359
0.98037	-0.01481
0.99070	-0.01216
0.99754	-0.00659
1.00000	-0.00350

Trainer 60

actual

x/c	y/c
1.00000	-0.00225
0.99918	0.00494
0.99847	0.00778
0.99729	0.00857
0.99636	0.00907
0.99302	0.01031
0.98985	0.01075
0.98598	0.01066
0.98022	0.01092
0.97630	0.01094
0.97054	0.01075
0.96444	0.01090
0.95796	0.01080
0.95245	0.01085
0.94659	0.01089
0.94102	0.01093
0.93333	0.01128
0.92568	0.01175
0.91727	0.01266
0.90758	0.01392
0.89865	0.01510
0.88831	0.01680
0.87318	0.01903
0.85707	0.02156
0.83568	0.02485
0.80926	0.02876
0.76401	0.03578
0.70902	0.04386
0.65956	0.05102
0.61040	0.05808
0.55971	0.06544
0.49191	0.07474
0.42401	0.08248
0.35842	0.08749
0.31659	0.08920
0.27683	0.08953
0.24098	0.08845
0.19907	0.08596
0.17480	0.08352
0.13903	0.07870
0.12407	0.07598
0.10649	0.07190
0.08904	0.06682
0.07700	0.06263
0.06921	0.05957
0.06440	0.05758
0.05431	0.05297
0.04344	0.04726
0.02780	0.03709
0.01466	0.02596
0.01494	0.02622
0.00890	0.02019
0.00370	0.01381
0.00177	0.00973
0.00108	0.00783
0.00002	0.00095
0.00088	-0.00712
0.00177	-0.00994
0.00365	-0.01429
0.00438	-0.01578
0.00609	-0.01888
0.00901	-0.02324
0.01395	-0.02899
0.02549	-0.03917
0.03019	-0.04275
0.03624	-0.04687
0.04818	-0.05376
0.05421	-0.05674
0.06171	-0.06009
0.06998	-0.06348
0.08105	-0.06756
0.08957	-0.07039
0.09898	-0.07320
0.10458	-0.07471
0.11326	-0.07686
0.13150	-0.08053
0.14641	-0.08285
0.15837	-0.08442
0.17635	-0.08644
0.18858	-0.08760
0.20586	-0.08893
0.22188	-0.08977
0.23233	-0.09012
0.24938	-0.09048
0.28216	-0.09072
0.32261	-0.08995
0.37047	-0.08749
0.42196	-0.08343
0.53676	-0.06936
0.62855	-0.05621
0.77987	-0.03345
0.85493	-0.02171
0.89223	-0.01538
0.91972	-0.01099
0.94238	-0.00896
0.98344	-0.00723
0.99451	-0.00493
1.00000	-0.00225

Ultra-Sport

from plans

x/c	y/c
1.00000	0.00005
0.99754	0.00093
0.99070	0.00309
0.98037	0.00568
0.96698	0.00820
0.95044	0.01056
0.93064	0.01292
0.90775	0.01562
0.88202	0.01898
0.85370	0.02318
0.82309	0.02823
0.79048	0.03392
0.75616	0.03995
0.72043	0.04605
0.68359	0.05205
0.64594	0.05787
0.60778	0.06346
0.56937	0.06874
0.53099	0.07363
0.49265	0.07810
0.45435	0.08216
0.41638	0.08573
0.37887	0.08873
0.34204	0.09100
0.30609	0.09240
0.27120	0.09283
0.23760	0.09221
0.20549	0.09038
0.17504	0.08712
0.14648	0.08224
0.11999	0.07592
0.09576	0.06843
0.07395	0.05990
0.05468	0.05008
0.03811	0.03923
0.02433	0.02873
0.01338	0.01948
0.00548	0.01089
0.00098	0.00338
0.00000	0.00131
0.00098	-0.00338
0.00548	-0.01089
0.01338	-0.01948
0.02433	-0.02873
0.03811	-0.03923
0.05468	-0.05008
0.07395	-0.05990
0.09576	-0.06843
0.11999	-0.07592
0.14648	-0.08224
0.17504	-0.08712
0.20549	-0.09038
0.23760	-0.09221
0.27120	-0.09283
0.30609	-0.09240
0.34204	-0.09100
0.37887	-0.08873
0.41638	-0.08573
0.45435	-0.08216
0.49265	-0.07810

0.53099	-0.07363
0.56937	-0.06874
0.60778	-0.06346
0.64594	-0.05787
0.68359	-0.05205
0.72043	-0.04605
0.75616	-0.03995
0.79048	-0.03392
0.82309	-0.02823
0.85370	-0.02318
0.88202	-0.01898
0.90775	-0.01562
0.93064	-0.01292
0.95044	-0.01056
0.96698	-0.00820
0.98037	-0.00568
0.99070	-0.00309
0.99754	-0.00093
1.00000	-0.00005

Ultra-Sport

actual

x/c	y/c
1.00000	-0.00027
0.99822	0.00112
0.99476	0.00221
0.99127	0.00318
0.98613	0.00384
0.97873	0.00462
0.97187	0.00556
0.96396	0.00658
0.95229	0.00805
0.94062	0.00970
0.92821	0.01132
0.91345	0.01339
0.89495	0.01589
0.87669	0.01840
0.84938	0.02247
0.82073	0.02699
0.78976	0.03225
0.74725	0.03955
0.70039	0.04744
0.62768	0.05860
0.54354	0.06980
0.45179	0.07999
0.38000	0.08614
0.34132	0.08837
0.31177	0.08942
0.29012	0.08975
0.27075	0.08974
0.25354	0.08946
0.23960	0.08901
0.22398	0.08823
0.21080	0.08730
0.19540	0.08589
0.18407	0.08460
0.17324	0.08315
0.15928	0.08098
0.14574	0.07848
0.13314	0.07574
0.12371	0.07342
0.11832	0.07198
0.11147	0.07000
0.10149	0.06687
0.09703	0.06536
0.09191	0.06355
0.08660	0.06160
0.08117	0.05950
0.07633	0.05756
0.05625	0.04844
0.03894	0.03841
0.03156	0.03315
0.02555	0.02841
0.01621	0.02027
0.00802	0.01223
0.00268	0.00561
0.00038	0.00211
0.00249	-0.00545
0.00429	-0.00828
0.01005	-0.01526
0.01389	-0.01944
0.01759	-0.02314
0.02235	-0.02751
0.02702	-0.03143
0.03211	-0.03541
0.03895	-0.04016
0.05243	-0.04821
0.06435	-0.05441
0.07332	-0.05860
0.08736	-0.06432
0.10232	-0.06950
0.12145	-0.07487
0.13888	-0.07879
0.16362	-0.08314
0.18979	-0.08635
0.22649	-0.08883
0.26817	-0.08974
0.30871	-0.08916
0.35726	-0.08692
0.40604	-0.08333
0.45816	-0.07833
0.51283	-0.07233
0.57426	-0.06469
0.61979	-0.05848
0.66071	-0.05250
0.69844	-0.04659
0.75094	-0.03776
0.80113	-0.02928
0.83765	-0.02343
0.87665	-0.01772
0.89472	-0.01496
0.90667	-0.01321
0.92053	-0.01128
0.93031	-0.00990
0.94000	-0.00864
0.94778	-0.00760
0.95590	-0.00649
0.97395	-0.00441
0.98971	-0.00335
0.99988	-0.00112
1.00000	-0.00027

USNPS-4

smoothed

x/c	y/c
1.00000	0.00391
0.99754	0.00501
0.99070	0.00807
0.98037	0.01241
0.96698	0.01764
0.95044	0.02357
0.93064	0.03005
0.90775	0.03688
0.88202	0.04388
0.85370	0.05091
0.82309	0.05784
0.79048	0.06459
0.75616	0.07110
0.72043	0.07728
0.68359	0.08307
0.64594	0.08841
0.60778	0.09324
0.56937	0.09751
0.53099	0.10119
0.49265	0.10426
0.45435	0.10670
0.41638	0.10848
0.37887	0.10956
0.34204	0.10990
0.30609	0.10945
0.27120	0.10817
0.23760	0.10599
0.20549	0.10283
0.17504	0.09860
0.14648	0.09324
0.11999	0.08664
0.09576	0.07874
0.07395	0.06953
0.05468	0.05911
0.03811	0.04794
0.02433	0.03654
0.01338	0.02542
0.00548	0.01503
0.00098	0.00584
0.00000	0.00017
0.00098	-0.00449
0.00548	-0.00931
0.01338	-0.01237
0.02433	-0.01394
0.03811	-0.01440
0.05468	-0.01414
0.07395	-0.01353
0.09576	-0.01285
0.11999	-0.01223
0.14648	-0.01169
0.17504	-0.01122
0.20549	-0.01082
0.23760	-0.01046
0.27120	-0.01013
0.30609	-0.00982
0.34204	-0.00952
0.37887	-0.00921
0.41638	-0.00890
0.45435	-0.00857
0.49265	-0.00824
0.53099	-0.00790
0.56937	-0.00755
0.60778	-0.00721
0.64594	-0.00686
0.68359	-0.00652
0.72043	-0.00619
0.75616	-0.00587
0.79048	-0.00557
0.82309	-0.00529
0.85370	-0.00503
0.88202	-0.00479
0.90775	-0.00458
0.93064	-0.00440
0.95044	-0.00424
0.96698	-0.00411
0.98037	-0.00401
0.99070	-0.00392
0.99754	-0.00387
1.00000	-0.00385

USNPS-4

actual

x/c	y/c
1.00000	0.00054
0.99984	0.00174
0.99629	0.00323
0.99280	0.00457
0.98957	0.00596
0.98500	0.00772
0.98013	0.00959
0.97485	0.01158
0.97101	0.01300
0.96294	0.01589
0.95575	0.01851
0.94591	0.02195
0.93247	0.02648
0.91782	0.03120
0.90298	0.03572
0.88070	0.04197
0.85394	0.04876
0.83381	0.05341
0.80317	0.06019
0.76712	0.06749
0.72247	0.07585
0.65753	0.08635
0.60657	0.09331
0.54260	0.10016
0.46132	0.10556
0.35822	0.10738
0.29120	0.10610
0.25307	0.10430
0.23326	0.10291
0.20907	0.10073
0.18661	0.09805
0.17059	0.09574
0.15513	0.09312
0.14079	0.09032
0.12474	0.08675
0.09662	0.07889
0.08843	0.07607
0.07004	0.06837
0.05279	0.05924
0.04256	0.05276
0.03341	0.04598
0.02890	0.04216
0.02500	0.03862
0.02106	0.03469
0.01481	0.02784
0.01349	0.02628

0.01029	0.02231
0.00533	0.01526
0.00275	0.01060
0.00011	0.00205
0.00012	-0.00229
0.00036	-0.00374
0.00140	-0.00685
0.00269	-0.00891
0.00502	-0.01093
0.00928	-0.01253
0.01263	-0.01324
0.01755	-0.01393
0.02574	-0.01471
0.04283	-0.01522
0.06143	-0.01504
0.07832	-0.01476
0.10352	-0.01431
0.13745	-0.01369
0.12181	-0.01401
0.17347	-0.01285
0.23080	-0.01168
0.28742	-0.01058
0.34642	-0.00969
0.42195	-0.00862
0.51388	-0.00724
0.61260	-0.00591
0.69463	-0.00494
0.77594	-0.00433
0.85682	-0.00383
0.90865	-0.00350
0.93621	-0.00335
0.94497	-0.00326
0.95077	-0.00321
0.95554	-0.00316
0.96423	-0.00314
0.96960	-0.00304
0.97463	-0.00291
0.97722	-0.00283
0.97858	-0.00280
0.98009	-0.00271
0.98298	-0.00262
0.98541	-0.00254
0.98690	-0.00246
0.99133	-0.00231
0.99419	-0.00218
0.99681	-0.00207
0.99925	-0.00174
1.00000	0.00054

Appendix C

Tabulated Polar Data

All of the polar data shown in Chapter 5 is listed in this appendix and identified by airfoil name, figure number, and run number. This same data, in addition to the spanwise C_d values used to obtain the average C_d's, are included with the data distribution disk. Also included on the disk but not listed here is the lift and moment data presented in the C_l and C_m-α curves of Chapter 5.

A18

Fig. 5.3

Run: PG02559
Re = 103018

α	C_l	C_d
-4.19	-0.219	0.0443
-3.65	-0.157	0.0321
-2.28	-0.028	0.0209
-1.26	0.081	0.0201
-0.31	0.228	0.0181
0.84	0.374	0.0147
1.39	0.432	0.0149
2.80	0.531	0.0166
3.89	0.629	0.0189
4.81	0.712	0.0208
5.91	0.809	0.0235
6.90	0.892	0.0261
7.86	0.969	0.0289
9.01	1.050	0.0331
9.81	1.104	0.0362
10.83	1.156	0.0421

Run: PG02561
Re = 199845

α	C_l	C_d
-4.42	-0.198	0.0328
-2.77	0.016	0.0150
-1.26	0.209	0.0135
0.22	0.331	0.0153
1.97	0.503	0.0170
3.42	0.638	0.0188
4.84	0.764	0.0211
6.10	0.873	0.0234
8.03	1.034	0.0271
9.42	1.140	0.0304
10.94	1.222	0.0366
12.50	1.253	0.0524

Run: PG02563
Re = 300494

α	C_l	C_d
-4.66	-0.199	0.0166
-2.88	0.032	0.0125
-1.34	0.181	0.0136
0.15	0.323	0.0146
1.68	0.474	0.0156
3.27	0.632	0.0174
4.89	0.785	0.0200
6.60	0.934	0.0233
7.42	1.001	0.0250
10.01	1.185	0.0334
10.63	1.207	0.0369

Run: PG02564
Re = 298924

α	C_l	C_d
11.71	1.224	0.0500

Avistar

Fig. 5.7

Run: CL01118
Re = 100000

α	C_l	C_d
-3.32	-0.067	0.0207
-2.33	0.037	0.0203
-1.30	0.183	0.0210
-0.24	0.387	0.0225
0.77	0.453	0.0217
1.78	0.527	0.0218
2.90	0.612	0.0223
3.81	0.683	0.0223
4.93	0.777	0.0227
5.88	0.850	0.0216
6.89	0.925	0.0210
7.91	0.970	0.0237
8.94	0.987	0.0315
9.90	1.008	0.0364

Run: CL01120
Re = 200600

α	C_l	C_d
-5.44	-0.210	0.0206
-4.32	-0.140	0.0174
-3.33	-0.047	0.0146
-2.29	0.032	0.0122
-1.33	0.130	0.0119
-0.25	0.245	0.0115
0.77	0.389	0.0124
1.78	0.523	0.0125
2.78	0.617	0.0130
3.82	0.726	0.0137
4.87	0.821	0.0145
5.90	0.888	0.0159
6.89	0.923	0.0203
7.89	0.950	0.0262

Run: CL01122
Re = 300900

α	C_l	C_d
-5.37	-0.215	0.0173
-4.38	-0.154	0.0153
-3.32	-0.054	0.0134
-2.31	0.042	0.0105
-1.23	0.144	0.0095
-0.22	0.237	0.0098
0.78	0.416	0.0108
1.83	0.575	0.0112
2.85	0.679	0.0111
3.84	0.771	0.0116
4.86	0.853	0.0127
5.90	0.907	0.0169
6.93	0.944	0.0213
7.89	0.980	0.0257
8.91	1.011	0.0320

Run: CL01124
Re = 402100

α	C_l	C_d
-5.42	-0.236	0.0158
-4.37	-0.144	0.0138
-3.35	-0.045	0.0122
-2.34	0.054	0.0100
-1.34	0.149	0.0084
-0.26	0.284	0.0087
0.77	0.408	0.0096
1.80	0.546	0.0100
2.85	0.686	0.0090
3.87	0.784	0.0109
4.95	0.850	0.0130
5.88	0.886	0.0163
6.87	0.923	0.0206
7.91	0.964	0.0255

BW-3

Fig. 5.11

Run: PG01723
Re = 59400

α	C_l	C_d
-2.11	-0.085	0.0747
-0.45	0.156	0.0463
1.06	0.386	0.0337
2.69	0.609	0.0324
4.24	0.872	0.0279
5.78	1.033	0.0230
7.32	1.141	0.0218
8.77	1.207	0.0325
10.30	1.289	0.0410
11.83	1.321	0.0548

Run: PG01515
Re = 99400

α	C_l	C_d
-2.16	-0.166	0.0681
-0.48	0.123	0.0414
1.18	0.456	0.0282
2.71	0.699	0.0251
4.26	0.866	0.0203
5.77	0.993	0.0181
7.30	1.100	0.0214
8.70	1.183	0.0296
10.35	1.284	0.0423
11.73	1.341	0.0589

Run: PG01517
Re = 198900

α	C_l	C_d
-2.58	-0.084	0.0575
-0.85	0.294	0.0317
0.48	0.530	0.0192
2.24	0.728	0.0159
3.75	0.882	0.0151
5.39	1.034	0.0149
6.82	1.129	0.0221
8.41	1.240	0.0291
9.73	1.317	0.0386
11.34	1.393	0.0689

Run: PG01519
Re = 297600

α	C_l	C_d
-2.58	0.014	0.0515
-0.98	0.350	0.0275
0.55	0.555	0.0182
2.10	0.734	0.0141
3.66	0.904	0.0129
5.21	1.048	0.0149
6.73	1.155	0.0203
8.29	1.269	0.0264
9.92	1.363	0.0391
11.22	1.432	0.0703

Run: PG01521
Re = 397300

α	C_l	C_d
-3.16	-0.060	0.0581
-1.41	0.312	0.0305
0.30	0.546	0.0180
1.60	0.701	0.0136
3.16	0.872	0.0121
4.68	1.012	0.0134
6.20	1.134	0.0185
7.77	1.255	0.0233
9.29	1.348	0.0309
10.90	1.433	0.0586

BW-3

Fig. 5.13

Run: PG01773
Re = 99700

α	C_l	C_d
-2.14	-0.153	0.0732
-0.46	0.182	0.0400
1.14	0.520	0.0265
2.59	0.666	0.0255
4.09	0.811	0.0250
5.68	0.949	0.0270
7.19	1.071	0.0311
8.76	1.191	0.0366
10.29	1.293	0.0479

Run: PG01775
Re = 199200

α	C_l	C_d
-3.64	-0.286	0.0820
-2.11	0.024	0.0477
-0.63	0.261	0.0306
1.26	0.565	0.0219
2.76	0.733	0.0216
4.33	0.882	0.0236
5.80	1.007	0.0275
7.34	1.130	0.0324
8.87	1.243	0.0398
10.25	1.344	0.0550

Run: PG01777
Re = 298300

α	C_l	C_d
-3.59	-0.180	0.0704
-2.07	0.060	0.0484
-0.49	0.365	0.0297
1.08	0.586	0.0196
2.63	0.747	0.0202
4.13	0.890	0.0221
5.66	1.026	0.0256
7.28	1.156	0.0314
8.93	1.268	0.0396
10.42	1.361	0.0505

Run: PG01779
Re = 402400

α	C_l	C_d
-3.60	-0.182	0.0687
-2.06	0.143	0.0441

Run: PG01780
Re = 397200

α	C_l	C_d
-0.52	0.409	0.0251
0.92	0.597	0.0186
2.50	0.765	0.0195
4.33	0.941	0.0221
6.00	1.087	0.0259
7.64	1.209	0.0314
9.42	1.318	0.0392

CG Ultimate

Fig. 5.17

Run: AB01671
Re = 100500

α	C_l	C_d
-2.95	-0.414	0.0168
-1.86	-0.341	0.0174
-0.92	-0.256	0.0159
0.18	-0.053	0.0170
1.23	0.131	0.0178
2.26	0.294	0.0172
3.29	0.371	0.0159
4.27	0.440	0.0165
5.30	0.521	0.0168
6.31	0.604	0.0172
7.32	0.671	0.0218
8.33	0.722	0.0284
9.36	0.792	0.0346
10.36	0.845	0.0437

Run: AB01672
Re = 102000

α	C_l	C_d
-4.99	-0.578	0.0171
-3.93	-0.492	0.0170

Run: AB01674
Re = 201200

α	C_l	C_d
-5.05	-0.605	0.0126
-3.96	-0.513	0.0121
-2.90	-0.408	0.0122
-1.86	-0.261	0.0130
-0.84	-0.108	0.0099
0.21	0.001	0.0095
1.16	0.078	0.0106
2.23	0.189	0.0105
3.25	0.326	0.0113
4.27	0.462	0.0121
5.29	0.548	0.0125
6.28	0.625	0.0139
7.34	0.692	0.0199
8.34	0.764	0.0236
9.36	0.829	0.0290
10.36	0.879	0.0390

Run: CL01676
Re = 301400

α	C_l	C_d
-4.98	-0.611	0.0115
-3.94	-0.515	0.0103
-2.97	-0.389	0.0108
-1.92	-0.234	0.0102
-0.82	-0.099	0.0090
0.09	-0.014	0.0088
1.19	0.081	0.0085
2.21	0.176	0.0090
3.26	0.322	0.0105
4.28	0.458	0.0106
5.30	0.576	0.0115
6.30	0.640	0.0140
7.33	0.707	0.0176
8.42	0.785	0.0213
9.37	0.846	0.0261
10.44	0.902	0.0370

Run: CL01678
Re = 402300

α	C_l	C_d
-5.16	-0.625	0.0119
-3.85	-0.480	0.0108
-2.80	-0.337	0.0097
-1.65	-0.197	0.0085
-0.81	-0.111	0.0080
0.23	-0.002	0.0078
1.17	0.089	0.0078
2.18	0.209	0.0081
3.25	0.324	0.0089
4.25	0.441	0.0082
5.28	0.564	0.0113
6.30	0.647	0.0136
7.44	0.729	0.0168
8.37	0.799	0.0195
9.42	0.870	0.0239
10.66	0.944	0.0340

Run: CL01760
Re = 502600

α	C_l	C_d
-4.00	-0.466	0.0085
-2.82	-0.317	0.0073
-1.74	-0.196	0.0078
-0.61	-0.079	0.0075
0.43	0.030	0.0075
1.33	0.127	0.0076
2.36	0.235	0.0079
3.33	0.334	0.0082
4.23	0.425	0.0090
5.35	0.558	0.0110
6.27	0.639	0.0129
7.31	0.718	0.0152
8.39	0.803	0.0179
9.77	0.892	0.0229
10.78	0.939	0.0297

Clark-Y (B)

Fig. 5.21

Run: PG01709
Re = 60700

α	C_l	C_d
-2.07	-0.084	0.0265
-0.45	0.076	0.0232
1.02	0.222	0.0295
2.54	0.389	0.0333
4.16	0.604	0.0339
5.71	0.744	0.0357
7.19	0.884	0.0389
8.74	1.016	0.0390
10.26	1.127	0.0349

Run: PG01710
Re = 59500

α	C_l	C_d
11.31	1.169	0.0355
12.80	1.221	0.0374

Run: PG01712
Re = 100000

α	C_l	C_d
-3.07	-0.125	0.0249
-1.42	0.070	0.0220
0.01	0.235	0.0229
1.64	0.436	0.0240
3.17	0.605	0.0225
4.64	0.759	0.0206
6.20	0.907	0.0193
7.72	1.033	0.0207
9.29	1.152	0.0255

Run: PG01713
Re = 99000

α	C_l	C_d
12.30	1.299	0.0326

Run: PG01715
Re = 200000

α	C_l	C_d
-4.64	-0.177	0.0213
-2.92	0.038	0.0156
-1.57	0.166	0.0132
0.10	0.338	0.0105
1.74	0.526	0.0111
3.11	0.654	0.0115
4.76	0.810	0.0125
6.24	0.944	0.0147
7.78	1.073	0.0170
9.30	1.197	0.0192
10.78	1.290	0.0225
12.35	1.309	0.0349

Run: PG01717
Re = 299800

α	C_l	C_d
-4.56	-0.111	0.0156
-2.97	0.032	0.0124
-1.44	0.177	0.0102
0.14	0.348	0.0088
1.71	0.541	0.0090
3.28	0.694	0.0097
4.95	0.846	0.0108
6.28	0.963	0.0125
7.87	1.104	0.0148
9.26	1.219	0.0168
10.77	1.304	0.0212
12.32	1.334	0.0316

Run: PG01719
Re = 399900

α	C_l	C_d
-4.70	-0.132	0.0142
-3.00	0.027	0.0111
-1.48	0.181	0.0095
0.10	0.337	0.0074
1.69	0.550	0.0079
3.21	0.696	0.0086
4.78	0.836	0.0098
6.51	0.990	0.0116
8.06	1.129	0.0138
9.56	1.245	0.0165
10.96	1.303	0.0223
12.39	1.332	0.0307

Clark-Y (B)

Fig. 5.23

Run: PG01752
Re = 99800

α	C_l	C_d
-3.10	-0.115	0.0253
-1.42	0.080	0.0241
0.09	0.303	0.0253
1.67	0.530	0.0164
3.18	0.643	0.0157
4.77	0.739	0.0181
6.29	0.842	0.0228
7.78	0.948	0.0279
9.22	1.040	0.0321

Run: PG01753
Re = 99600

α	C_l	C_d
10.74	1.103	0.0383

Run: PG01755
Re = 199900

α	C_l	C_d
-4.70	-0.179	0.0199
-3.08	0.027	0.0166
-1.29	0.196	0.0138
0.22	0.334	0.0132
1.76	0.467	0.0148
3.22	0.586	0.0175
4.78	0.711	0.0198
6.22	0.817	0.0216
7.82	0.922	0.0258
9.25	1.004	0.0319
10.86	1.065	0.0437

Run: PG01757
Re = 300100

α	C_l	C_d
-4.53	-0.109	0.0153
-3.05	0.027	0.0138
-1.43	0.164	0.0132

0.20	0.329	0.0136
1.81	0.470	0.0154
3.36	0.608	0.0172
4.81	0.729	0.0190
6.14	0.832	0.0210
7.85	0.953	0.0250
9.17	1.025	0.0294
10.77	1.084	0.0394

Run: PG01759
Re = 400700

α	C_l	C_d
-4.60	-0.124	0.0139
-3.03	0.013	0.0130
-1.49	0.151	0.0130
0.10	0.304	0.0136
1.65	0.450	0.0150
3.25	0.598	0.0165
4.90	0.739	0.0184
6.47	0.862	0.0209
7.92	0.958	0.0244
9.21	1.022	0.0295
10.82	1.081	0.0406

DH4009

Fig. 5.27

Run: CL01082
Re = 100000

α	C_l	C_d
-3.66	-0.339	0.0162
-2.62	-0.295	0.0129
-1.59	-0.085	0.0155
-0.53	0.046	0.0122
0.43	0.051	0.0139
1.47	0.174	0.0138
2.88	0.356	0.0121
3.51	0.389	0.0159
4.54	0.462	0.0217
5.58	0.549	0.0293
6.60	0.636	0.0424
7.61	0.717	0.0637
8.68	0.784	0.1076
10.16	0.821	0.1478

Run: CL01084
Re = 200800

α	C_l	C_d
-3.65	-0.352	0.0149
-2.63	-0.278	0.0119
-1.56	-0.193	0.0112
-0.60	-0.063	0.0107
0.43	0.063	0.0095
1.46	0.187	0.0089
2.49	0.317	0.0090
3.52	0.386	0.0138
4.55	0.465	0.0176
5.56	0.548	0.0246
6.57	0.646	0.0415
7.65	0.737	0.0715
8.61	0.799	0.1097

Run: CL01086
Re = 301500

α	C_l	C_d
-3.65	-0.362	0.0139
-2.61	-0.271	0.0120
-1.60	-0.149	0.0087
-0.58	-0.035	0.0083
1.20	0.125	0.0075
1.58	0.163	0.0078
2.53	0.309	0.0086
3.57	0.412	0.0135
4.55	0.490	0.0169
5.63	0.584	0.0272
6.60	0.668	0.0439

Run: CL01088
Re = 401700

α	C_l	C_d
-3.71	-0.379	0.0138
-2.66	-0.266	0.0104
-1.62	-0.149	0.0067
-0.58	-0.039	0.0067
0.42	0.049	0.0068
1.56	0.186	0.0065
2.52	0.289	0.0078
3.53	0.400	0.0115
4.54	0.497	0.0155
5.61	0.592	0.0266

Run: CL02072
Re = 501500

α	C_l	C_d
-3.81	-0.387	0.0148
-2.71	-0.250	0.0106
-1.69	-0.146	0.0064
-0.58	-0.034	0.0060
0.45	0.068	0.0060
1.50	0.175	0.0063
2.54	0.279	0.0070
3.53	0.380	0.0109
4.54	0.486	0.0132
5.59	0.578	0.0210
6.62	0.659	0.0386

E231

Fig. 5.31

Run: CL01252
Re = 60900

α	C_l	C_d
-2.51	-0.223	0.0197
-1.44	-0.124	0.0204
-0.44	-0.021	0.0230
0.59	0.071	0.0275
1.61	0.141	0.0310
2.62	0.214	0.0391
3.67	0.282	0.0437
4.67	0.365	0.0499
5.75	0.504	0.0515
6.75	0.734	0.0362
7.81	0.840	0.0318
8.79	0.901	0.0333
9.82	0.953	0.0394
10.81	0.992	0.0442
11.86	1.027	0.0539

Run: CL01256
Re = 60800

α	C_l	C_d
-4.57	-0.452	0.0276
-3.53	-0.363	0.0219

Run: CL01254
Re = 99900

α	C_l	C_d
-2.50	-0.210	0.0188
-1.44	-0.024	0.0219
-0.36	0.067	0.0225
0.65	0.120	0.0253
1.66	0.185	0.0293
2.62	0.260	0.0331
3.70	0.380	0.0367
4.77	0.608	0.0302
5.73	0.734	0.0231
6.78	0.839	0.0189
7.82	0.929	0.0198
8.82	0.994	0.0242
9.87	1.040	0.0303
10.83	1.080	0.0359

Run: CL01255
Re = 101900

α	C_l	C_d
-4.54	-0.371	0.0264
-3.53	-0.283	0.0219

Run: CL01258
Re = 200600

α	C_l	C_d
-4.56	-0.238	0.0195
-3.48	-0.159	0.0175
-2.35	-0.095	0.0148
-1.38	-0.021	0.0163
-0.29	0.153	0.0185
0.61	0.306	0.0149
1.78	0.425	0.0160
2.73	0.506	0.0155
3.69	0.590	0.0147
4.72	0.703	0.0131
5.75	0.795	0.0124
6.82	0.882	0.0135
7.84	0.943	0.0163
8.85	0.999	0.0204
9.83	1.052	0.0244
10.84	1.099	0.0294

Run: CL01260
Re = 301000

α	C_l	C_d
-4.57	-0.284	0.0168
-3.39	-0.177	0.0150
-2.26	-0.090	0.0120
-1.51	-0.039	0.0113
-0.44	0.050	0.0123
0.57	0.160	0.0112
1.72	0.382	0.0117
2.78	0.518	0.0108
3.75	0.627	0.0102
4.72	0.723	0.0100
5.74	0.820	0.0106
6.83	0.902	0.0126
7.85	0.954	0.0151
8.86	1.019	0.0180
9.83	1.080	0.0216
10.85	1.130	0.0262

Run: CL01262
Re = 400400

α	C_l	C_d
-4.53	-0.302	0.0146
-3.48	-0.203	0.0133
-2.47	-0.110	0.0113
-1.44	-0.034	0.0089
-0.34	0.064	0.0094
0.64	0.182	0.0097
1.69	0.314	0.0088
2.66	0.439	0.0084
3.75	0.589	0.0093
4.78	0.724	0.0085
5.83	0.835	0.0094
6.80	0.905	0.0120
7.80	0.945	0.0143
8.83	1.013	0.0169
9.86	1.085	0.0200
10.86	1.128	0.0243

E374 (B)

Fig. 5.35

Run: CL01013
Re = 60200

α	C_l	C_d
-5.46	-0.382	0.0225
-4.38	-0.334	0.0168
-3.19	-0.258	0.0216
-2.17	-0.151	0.0226
-1.11	0.001	0.0228
-0.26	0.122	0.0282
0.94	0.213	0.0337
1.96	0.284	0.0363
2.98	0.344	0.0423
4.01	0.432	0.0466
5.06	0.619	0.0472
6.12	0.771	0.0392
7.14	0.877	0.0278
8.13	0.928	0.0346
9.14	0.973	0.0459
9.97	1.015	0.0536

Run: CL01015
Re = 100100

α	C_l	C_d
-5.62	-0.416	0.0236
-4.04	-0.269	0.0182
-3.00	-0.114	0.0156
-2.38	-0.020	0.0153
-1.34	0.123	0.0181
-0.28	0.236	0.0198
0.71	0.307	0.0234
1.71	0.374	0.0263
2.74	0.440	0.0289
3.78	0.529	0.0296
4.81	0.634	0.0239
5.83	0.740	0.0196

6.88	0.846	0.0184
7.86	0.902	0.0247
8.91	0.947	0.0344
9.93	0.986	0.0456

Run: CL01017
Re = 200500

α	C_l	C_d
-5.18	-0.301	0.0174
-3.80	-0.152	0.0119
-3.50	-0.127	0.0115
-2.47	-0.050	0.0103
-1.51	0.036	0.0105
0.48	0.293	0.0112
0.60	0.304	0.0112
2.44	0.471	0.0121
2.71	0.508	0.0124
3.74	0.604	0.0128
4.75	0.736	0.0127
5.76	0.841	0.0127
6.82	0.930	0.0154
8.38	0.996	0.0263
8.85	1.016	0.0292
9.87	1.050	0.0371

Run: CL00983
Re = 301200

α	C_l	C_d
-6.40	-0.470	0.0275
-5.42	-0.347	0.0183
-4.40	-0.242	0.0138
-3.42	-0.163	0.0113
-2.36	-0.071	0.0094
-1.35	0.012	0.0085
-0.40	0.094	0.0089
0.71	0.232	0.0091
1.74	0.396	0.0095
2.79	0.502	0.0100
3.81	0.608	0.0103
4.84	0.715	0.0105
5.87	0.818	0.0110
6.91	0.900	0.0144
7.92	0.947	0.0202
8.94	0.996	0.0252
9.95	1.035	0.0323

E387 (C)

Fig. 5.39

Run: CL02130
Re = 59885

α	C_l	C_d
-4.13	-0.179	0.0331
-3.05	-0.084	0.0261
-2.08	0.053	0.0219
-1.00	0.176	0.0221
-0.02	0.270	0.0256
1.00	0.398	0.0302
2.05	0.462	0.0346
3.07	0.521	0.0400
4.09	0.571	0.0501
5.07	0.637	0.0563
6.20	0.708	0.0717
7.11	0.760	0.0844
8.29	1.064	0.0356
9.17	1.159	0.0299
10.29	1.185	0.0374
11.27	1.206	0.0486

Run: CL01575
Re = 58800

α	C_l	C_d
11.29	1.256	0.0394

Run: CL02132
Re = 99744

α	C_l	C_d
-4.08	-0.119	0.0306
-3.05	0.025	0.0211
-2.04	0.160	0.0176
-0.95	0.267	0.0164
0.05	0.356	0.0190
1.04	0.450	0.0214
2.19	0.547	0.0241
3.06	0.625	0.0275
4.12	0.728	0.0282
5.16	0.837	0.0264
6.16	0.937	0.0227
7.15	1.038	0.0211
8.22	1.133	0.0202
9.24	1.179	0.0258
10.27	1.193	0.0359

Run: CL02134
Re = 199604

α	C_l	C_d
-3.99	-0.046	0.0202
-2.81	0.066	0.0145
-1.93	0.148	0.0126
-0.92	0.245	0.0101
0.06	0.344	0.0106
1.17	0.462	0.0116
2.10	0.562	0.0124
3.10	0.671	0.0134
4.15	0.779	0.0141
5.30	0.899	0.0146
6.17	0.984	0.0147
7.32	1.099	0.0150
8.32	1.172	0.0179
9.27	1.202	0.0262
10.26	1.216	0.0341

Run: CL02136
Re = 299856

α	C_l	C_d
-4.11	-0.075	0.0208
-3.08	0.028	0.0140
-2.04	0.138	0.0114
-0.99	0.249	0.0095
0.11	0.360	0.0086
1.02	0.465	0.0091
2.10	0.580	0.0097
3.11	0.689	0.0105
4.19	0.803	0.0111
5.13	0.907	0.0115
6.23	1.023	0.0119
7.33	1.131	0.0130
8.34	1.198	0.0176
9.32	1.233	0.0248
10.32	1.251	0.0319

Run: CL02138
Re = 458992

α	C_l	C_d
-4.10	-0.078	0.0189
-3.08	0.026	0.0124
-2.03	0.139	0.0102
-1.04	0.248	0.0089
0.02	0.356	0.0072
1.16	0.485	0.0075
2.03	0.578	0.0079
3.36	0.728	0.0087
4.10	0.808	0.0092
5.09	0.917	0.0097
6.42	1.055	0.0105
7.27	1.129	0.0128
8.30	1.196	0.0176
9.36	1.250	0.0229
10.47	1.277	0.0287

E387 (D)

Fig. 5.43

Run: CL02707
Re = 59507

α	C_l	C_d
-4.25	-0.282	0.0556
-3.19	-0.178	0.0379
-2.09	-0.046	0.0278
-1.31	0.070	0.0247
-0.12	0.226	0.0248
0.95	0.356	0.0283
1.97	0.450	0.0331
3.11	0.532	0.0383
3.97	0.598	0.0423
5.02	0.663	0.0496
5.93	0.750	0.0504
6.85	0.937	0.0419
7.70	1.038	0.0353
9.09	1.161	0.0281
10.02	1.189	0.0374
10.94	1.213	0.0489

E472

Fig. 5.47

Run: CL01074
Re = 100300

α	C_l	C_d
-2.95	-0.301	0.0177
-1.94	-0.209	0.0168
-0.90	-0.119	0.0141
0.16	0.047	0.0122
1.16	0.184	0.0146
2.21	0.278	0.0166
3.22	0.374	0.0187
4.24	0.465	0.0207
5.29	0.558	0.0240
6.28	0.639	0.0313
7.33	0.733	0.0380
8.34	0.730	0.1006
9.30	0.687	0.1452
10.32	0.667	0.1640

Run: CL01076
Re = 200100

α	C_l	C_d
-2.94	-0.335	0.0138
-1.90	-0.198	0.0130
-0.90	-0.064	0.0109
0.14	0.032	0.0102
1.15	0.118	0.0117
2.20	0.257	0.0132
3.22	0.367	0.0143
4.26	0.466	0.0156
5.27	0.561	0.0176
6.27	0.656	0.0196
7.31	0.748	0.0221
8.35	0.842	0.0245
9.35	0.929	0.0273
10.40	1.013	0.0307
11.42	1.084	0.0345

Run: CL01857
Re = 200100

α	C_l	C_d
12.26	0.671	0.1673

Run: CL01078
Re = 300900

α	C_l	C_d
-2.89	-0.333	0.0120
-1.88	-0.179	0.0112
-0.85	-0.058	0.0103
0.21	0.036	0.0106
1.15	0.121	0.0102
2.19	0.247	0.0109
3.24	0.384	0.0121
4.33	0.486	0.0131
5.26	0.573	0.0143
6.29	0.669	0.0155
7.34	0.769	0.0169
8.37	0.867	0.0181
9.47	0.969	0.0198
10.58	1.067	0.0224
11.44	1.135	0.0254
12.47	1.202	0.0294

Run: CL01858
Re = 298100

α	C_l	C_d
13.71	1.245	0.0339
14.42	1.257	0.0409

Run: CL01080
Re = 401000

α	C_l	C_d
-2.90	-0.304	0.0105
-2.02	-0.185	0.0098
-0.81	-0.059	0.0101
0.18	0.036	0.0104
1.15	0.133	0.0102
2.19	0.234	0.0098
3.25	0.372	0.0105
4.25	0.484	0.0115
5.28	0.582	0.0125
6.38	0.690	0.0137
7.35	0.787	0.0146
8.45	0.894	0.0159
9.84	1.020	0.0179

10.45	1.063	0.0190
11.64	1.156	0.0225
12.94	1.237	0.0277

Run: CL01859
$Re = 397000$

α	C_l	C_d
13.90	1.271	0.0313
14.71	1.257	0.0732

Run: CL01856
$Re = 501400$

α	C_l	C_d
-3.14	-0.369	0.0099
-1.79	-0.192	0.0093
-0.48	-0.064	0.0098
0.45	0.028	0.0101
1.28	0.109	0.0102
2.28	0.210	0.0099
3.27	0.312	0.0094
4.20	0.433	0.0102
5.31	0.552	0.0112
6.24	0.635	0.0121
7.40	0.750	0.0131
8.40	0.850	0.0140
9.73	0.985	0.0156
10.74	1.079	0.0172
12.02	1.186	0.0207
12.40	1.209	0.0217
13.62	1.285	0.0276

ESA

Fig. 5.51

Run: AB01091
$Re = 100500$

α	C_l	C_d
-4.67	-0.477	0.0295
-3.64	-0.373	0.0226
-2.60	-0.256	0.0182
-1.58	-0.185	0.0167
-0.56	-0.024	0.0166
0.42	0.060	0.0172
1.49	0.140	0.0168
2.52	0.220	0.0173
3.53	0.300	0.0180
4.53	0.396	0.0194
5.60	0.514	0.0217
6.60	0.621	0.0247
7.64	0.718	0.0290
8.64	0.806	0.0325
9.68	0.903	0.0397

Run: AB01093
$Re = 200600$

α	C_l	C_d
-4.67	-0.505	0.0274
-3.67	-0.418	0.0206
-2.65	-0.318	0.0174
-1.61	-0.212	0.0148
-0.56	-0.133	0.0115
0.45	0.019	0.0119
1.49	0.105	0.0115
2.56	0.205	0.0123
3.54	0.305	0.0132
4.59	0.416	0.0144
5.60	0.522	0.0160
6.65	0.648	0.0181
7.66	0.755	0.0197
8.68	0.860	0.0218
9.70	0.963	0.0240
10.82	1.070	0.0269

Run: AB01095
$Re = 301600$

α	C_l	C_d
-4.61	-0.533	0.0281
-3.65	-0.446	0.0192
-2.61	-0.340	0.0155
-1.59	-0.233	0.0135
-0.56	-0.145	0.0095
0.48	-0.027	0.0119
1.59	0.102	0.0107
2.61	0.211	0.0105
3.64	0.338	0.0116
4.71	0.452	0.0128
5.69	0.555	0.0138
6.69	0.663	0.0151
7.88	0.792	0.0166
9.04	0.915	0.0186
10.42	1.062	0.0209
10.94	1.115	0.0219

Run: AB01097
$Re = 403800$

α	C_l	C_d
-3.66	-0.460	0.0177
-1.75	-0.261	0.0126
-1.36	-0.220	0.0121
-0.48	-0.137	0.0086
0.65	-0.029	0.0097
1.55	0.062	0.0106
2.49	0.209	0.0100
3.96	0.368	0.0109
4.55	0.434	0.0114
5.93	0.588	0.0126

Run: CL01107
$Re = 398100$

α	C_l	C_d
5.58	0.542	0.0125
6.73	0.664	0.0137
7.69	0.776	0.0145
8.67	0.884	0.0158
9.77	1.001	0.0174
10.82	1.110	0.0195
11.90	1.203	0.0215

Falcon 56 Mk II

Fig. 5.55

Run: CL01424
$Re = 100200$

α	C_l	C_d
-5.23	-0.436	0.0231
-4.20	-0.385	0.0187
-3.16	-0.276	0.0171
-2.14	-0.118	0.0156
-1.08	0.065	0.0187
-0.03	0.256	0.0183
1.01	0.340	0.0166
2.01	0.405	0.0172
3.03	0.482	0.0182
4.03	0.552	0.0195
5.07	0.634	0.0207
6.07	0.722	0.0213
7.10	0.816	0.0225
8.11	0.897	0.0252
9.26	0.991	0.0277
10.15	1.061	0.0305

Run: CL01451
$Re = 100200$

α	C_l	C_d
11.16	1.059	0.0306
12.22	1.132	0.0322

Run: CL01426
$Re = 200500$

α	C_l	C_d
-5.22	-0.466	0.0186
-4.16	-0.323	0.0129
-3.10	-0.190	0.0107
-2.11	-0.099	0.0100
-1.08	-0.018	0.0092
-0.06	0.081	0.0103
0.96	0.236	0.0116
2.02	0.384	0.0134
3.01	0.465	0.0129
4.03	0.549	0.0135
5.04	0.639	0.0146
6.08	0.734	0.0154
7.15	0.831	0.0167
8.14	0.919	0.0182
9.16	1.005	0.0200
10.16	1.087	0.0216

Run: CL01427
$Re = 198700$

α	C_l	C_d
11.19	1.165	0.0239
12.21	1.233	0.0263

Run: CL01447
$Re = 301300$

α	C_l	C_d
-5.27	-0.526	0.0181
-4.10	-0.377	0.0167
-3.06	-0.237	0.0119
-2.05	-0.145	0.0094
-1.04	-0.062	0.0087
-0.05	0.022	0.0094
0.90	0.121	0.0097
1.97	0.265	0.0103
3.08	0.507	0.0123
4.02	0.593	0.0116
5.12	0.693	0.0125
6.11	0.785	0.0134
7.11	0.877	0.0144
8.15	0.970	0.0155
9.16	1.058	0.0170
10.18	1.139	0.0188
11.23	1.215	0.0215
12.24	1.262	0.0263

Run: CL01450
$Re = 401700$

α	C_l	C_d
-4.17	-0.376	0.0125
-3.06	-0.251	0.0106
-1.99	-0.148	0.0097
-1.18	-0.079	0.0085
-0.05	0.028	0.0081
0.98	0.121	0.0086
2.01	0.344	0.0107
3.09	0.508	0.0132
4.11	0.622	0.0111
5.11	0.716	0.0116
6.10	0.811	0.0123
7.13	0.907	0.0133
8.17	1.004	0.0144
9.20	1.091	0.0158
10.29	1.175	0.0180
11.45	1.246	0.0217

Goe 417a

Fig. 5.59

Run: PG01725
$Re = 59300$

α	C_l	C_d
-2.25	-0.230	0.0773
-0.67	-0.066	0.0607
0.87	0.178	0.0457
2.49	0.479	0.0360
3.95	0.710	0.0364
5.62	0.984	0.0285
7.10	1.131	0.0213
8.65	1.238	0.0226
10.17	1.337	0.0283
11.76	1.398	0.0412

Run: PG01126
$Re = 99500$

α	C_l	C_d
-3.77	-0.332	0.0720
-2.20	-0.220	0.0608
-0.63	0.066	0.0450
0.87	0.341	0.0362
2.43	0.598	0.0296
4.05	0.819	0.0241
5.54	0.943	0.0191
7.06	1.076	0.0166
8.62	1.211	0.0202
10.13	1.304	0.0311
11.65	1.371	0.0441

Run: CL01128
$Re = 198500$

α	C_l	C_d
-3.73	-0.289	0.0672
-2.24	-0.027	0.0454
-0.69	0.283	0.0314
0.94	0.568	0.0209
2.57	0.759	0.0162
4.09	0.923	0.0137
5.64	1.067	0.0140
7.14	1.185	0.0163
8.66	1.272	0.0237

10.21	1.345	0.0320
11.64	1.387	0.0448

Run: CL01130
Re = 297500

α	C_l	C_d
-3.73	-0.127	0.0556
-2.15	0.101	0.0394
-0.39	0.409	0.0252
0.94	0.590	0.0185
2.50	0.778	0.0143
4.02	0.939	0.0123
5.54	1.075	0.0129
7.15	1.199	0.0159
8.71	1.291	0.0225
10.30	1.370	0.0303
11.69	1.408	0.0436

Run: CL01132
Re = 398800

α	C_l	C_d
-2.40	0.172	0.0364
-0.34	0.498	0.0217
1.01	0.672	0.0165
2.49	0.842	0.0133

LRN1007 (B)

Fig. 5.63

Run: AB01697
Re = 59500

α	C_l	C_d
-3.56	-0.293	0.0873
-2.50	-0.128	0.0603
-1.45	0.052	0.0498
-0.40	0.218	0.0459
0.60	0.325	0.0456
1.63	0.417	0.0432
2.64	0.506	0.0457
3.67	0.575	0.0471
4.67	0.630	0.0508
5.70	0.693	0.0572
6.70	0.763	0.0687
7.76	0.862	0.0819
8.89	1.255	0.0323
9.87	1.257	0.0373
10.84	1.250	0.0459
11.86	1.247	0.0546

Run: AB01365
Re = 100200

α	C_l	C_d
-3.52	-0.210	0.0762
-2.47	0.051	0.0543
-1.38	0.223	0.0468
-0.40	0.312	0.0445
0.63	0.397	0.0406
1.65	0.488	0.0412
2.67	0.577	0.0469
3.71	0.666	0.0511
4.72	0.751	0.0596
5.73	0.897	0.0571

Run: AB01698
Re = 98800

α	C_l	C_d
3.67	0.586	0.0459
4.73	0.668	0.0527
5.76	0.750	0.0625
6.79	0.903	0.0587
7.86	1.237	0.0233
8.87	1.276	0.0266
9.86	1.264	0.0381
10.85	1.262	0.0479
11.88	1.268	0.0565

Run: AB01700
Re = 199100

α	C_l	C_d
-4.54	-0.093	0.0673
-3.33	0.003	0.0575
-2.32	0.152	0.0449
-1.34	0.293	0.0375
-0.32	0.421	0.0333
0.71	0.615	0.0309
1.79	0.742	0.0292
2.81	0.859	0.0258
3.93	0.991	0.0231
4.86	1.101	0.0203
5.78	1.205	0.0166
6.79	1.255	0.0190
7.83	1.257	0.0288
8.84	1.263	0.0382
9.87	1.275	0.0477
10.86	1.290	0.0584

Run: AB01363
Re = 298700

α	C_l	C_d
-5.50	-0.139	0.0726
-4.47	-0.016	0.0571
-3.42	0.150	0.0424
-2.27	0.338	0.0306
-1.36	0.456	0.0257
-0.25	0.587	0.0223
0.71	0.695	0.0199
1.73	0.806	0.0182
2.87	0.922	0.0154
3.78	1.017	0.0145
4.78	1.123	0.0133
5.82	1.228	0.0115
6.84	1.239	0.0212
7.86	1.250	0.0285
8.94	1.263	0.0380

PT-40 (A)

Fig. 5.67

Run: CL01021
Re = 100600

α	C_l	C_d
-5.91	-0.353	0.0957
-3.70	-0.261	0.0608
-3.07	-0.170	0.0435
-1.68	0.038	0.0314
-1.04	0.144	0.0300
0.34	0.349	0.0237
1.33	0.445	0.0213
2.27	0.535	0.0179
3.28	0.631	0.0167
4.30	0.714	0.0176
5.31	0.799	0.0208
6.30	0.883	0.0229
7.27	0.964	0.0245
8.57	1.061	0.0284
9.16	1.100	0.0303
10.39	1.173	0.0388

Run: CL01023
Re = 200200

α	C_l	C_d
-4.28	-0.200	0.0602
-3.53	-0.084	0.0425
-2.50	0.049	0.0267
-1.50	0.149	0.0198
0.32	0.333	0.0147
0.63	0.361	0.0149
1.64	0.463	0.0139
2.71	0.567	0.0134
3.70	0.660	0.0119
4.72	0.751	0.0136
5.73	0.843	0.0158
7.24	0.976	0.0183
7.79	1.019	0.0192
8.88	1.096	0.0226
9.87	1.157	0.0259
10.88	1.208	0.0313

Run: CL01025
Re = 299800

α	C_l	C_d
-3.28	-0.025	0.0348
-2.38	0.065	0.0240
-1.35	0.166	0.0173
-0.32	0.271	0.0141
0.71	0.379	0.0130
1.73	0.483	0.0127
3.25	0.629	0.0121
3.78	0.680	0.0113
4.81	0.777	0.0125
5.85	0.873	0.0138
6.78	0.952	0.0150
7.75	1.031	0.0167
8.90	1.119	0.0194
10.35	1.214	0.0246
10.86	1.239	0.0268
11.89	1.284	0.0326

Run: AB01027
Re = 399500

α	C_l	C_d
-3.37	-0.037	0.0353
-2.42	0.056	0.0246
-1.35	0.169	0.0166
-0.31	0.276	0.0134
0.57	0.368	0.0122
1.73	0.489	0.0119
2.76	0.591	0.0120
3.49	0.656	0.0122
4.81	0.788	0.0117
5.58	0.857	0.0126
6.64	0.953	0.0136
7.90	1.062	0.0159
8.91	1.140	0.0182
9.93	1.206	0.0215
10.95	1.258	0.0257
11.93	1.295	0.0312

PT-40 (B)

Fig. 5.71

Run: CL01066
Re = 99700

α	C_l	C_d
-2.84	-0.109	0.0396
-1.71	0.062	0.0275
-0.85	0.174	0.0235
0.35	0.297	0.0197
1.34	0.384	0.0179
2.35	0.475	0.0171
3.36	0.567	0.0158
4.43	0.659	0.0168
5.45	0.746	0.0195
6.48	0.831	0.0215
7.48	0.920	0.0245
8.51	0.995	0.0268
9.50	1.066	0.0326
10.46	1.126	0.0378
11.51	1.183	0.0458

Run: CL01068
Re = 200100

α	C_l	C_d
-2.67	0.006	0.0253
-1.46	0.115	0.0188
-0.85	0.172	0.0168
0.47	0.310	0.0146
1.49	0.414	0.0140
2.48	0.512	0.0140
3.50	0.618	0.0134
4.54	0.711	0.0145
5.54	0.799	0.0162
6.53	0.886	0.0181
7.52	0.973	0.0190
8.45	1.047	0.0209
9.53	1.125	0.0250
10.48	1.189	0.0285
11.54	1.245	0.0332

Run: CL01070
Re = 299800

α	C_l	C_d
-2.81	0.004	0.0226
-1.76	0.090	0.0169
-0.69	0.201	0.0142
0.32	0.322	0.0134
1.36	0.428	0.0132
2.35	0.521	0.0140
3.48	0.626	0.0147
4.59	0.727	0.0141
5.65	0.826	0.0154
6.64	0.908	0.0164
7.65	0.994	0.0184
8.65	1.073	0.0206
9.59	1.145	0.0229
10.59	1.206	0.0267
11.49	1.255	0.0305

12.50	1.293	0.0364

Run: CL01072
Re = 399100

α	C_l	C_d
-2.90	0.019	0.0199
-1.62	0.145	0.0142
-0.72	0.233	0.0130
0.29	0.336	0.0126
1.34	0.444	0.0131
2.28	0.532	0.0141
3.34	0.637	0.0143
4.43	0.743	0.0132
5.47	0.845	0.0139
6.60	0.942	0.0153
7.36	1.007	0.0169
8.75	1.120	0.0197
9.46	1.166	0.0214
10.48	1.233	0.0242
11.55	1.296	0.0286
12.59	1.336	0.0332

RG14

Fig. 5.75

Run: AB01058
Re = 60400

α	C_l	C_d
-5.20	-0.410	0.0321
-4.18	-0.329	0.0250
-3.23	-0.263	0.0180
-2.17	-0.166	0.0159
-1.13	-0.084	0.0126
-0.14	0.021	0.0125
0.93	0.136	0.0159
1.93	0.271	0.0169
3.01	0.485	0.0163
4.65	0.654	0.0162
5.09	0.713	0.0175
6.12	0.804	0.0243
7.13	0.885	0.0261
8.13	0.949	0.0297
9.12	1.006	0.0371
10.13	1.052	0.0456

Run: AB01060
Re = 100600

α	C_l	C_d
-5.20	-0.375	0.0276
-4.14	-0.290	0.0204
-3.17	-0.208	0.0156
-2.12	-0.128	0.0119
-1.13	-0.053	0.0101
-0.09	0.050	0.0123
0.92	0.165	0.0138
1.97	0.336	0.0142
2.99	0.450	0.0125
4.01	0.551	0.0133
5.07	0.646	0.0149
6.11	0.750	0.0178
7.10	0.841	0.0187
8.12	0.918	0.0254
9.13	0.985	0.0327
10.12	1.034	0.0414

Run: AB01062
Re = 199900

α	C_l	C_d
-5.25	-0.395	0.0226
-4.16	-0.308	0.0166
-3.14	-0.218	0.0131
-2.21	-0.129	0.0102
-1.08	0.010	0.0097
-0.07	0.151	0.0096
0.96	0.287	0.0088
1.97	0.399	0.0084
3.00	0.497	0.0092
4.03	0.593	0.0105
5.06	0.692	0.0120
6.06	0.783	0.0144
7.14	0.875	0.0172
8.13	0.956	0.0207
9.13	1.032	0.0255
10.14	1.083	0.0340

Run: CL01064
Re = 300300

α	C_l	C_d
-5.28	-0.401	0.0218
-4.12	-0.304	0.0150
-3.13	-0.202	0.0118
-2.05	-0.057	0.0097
-1.04	0.065	0.0078
0.02	0.186	0.0068
1.07	0.311	0.0065
2.20	0.426	0.0072
3.12	0.516	0.0084
3.96	0.597	0.0093
5.11	0.711	0.0111
6.12	0.804	0.0129
7.15	0.896	0.0151
8.11	0.975	0.0180
9.17	1.058	0.0220
10.12	1.111	0.0290

RG15 (C)

Fig. 5.79

Run: CL02374
Re = 60056

α	C_l	C_d
-6.93	-0.297	0.0390
-6.06	-0.227	0.0286
-5.10	-0.151	0.0238
-4.06	-0.069	0.0163
-2.90	0.004	0.0128
-2.02	0.101	0.0172
-0.97	0.195	0.0170
0.20	0.299	0.0193
1.07	0.411	0.0231
2.32	0.598	0.0217
3.29	0.696	0.0214
4.17	0.758	0.0237
5.32	0.853	0.0256
6.18	0.904	0.0260
7.21	0.973	0.0301
8.16	1.003	0.0390
9.36	1.051	0.0527

Run: CL02376
Re = 99818

α	C_l	C_d
-7.12	-0.399	0.0401
-5.99	-0.296	0.0261
-4.86	-0.189	0.0197
-3.99	-0.098	0.0152
-2.99	-0.023	0.0148
-1.87	0.076	0.0121
-0.95	0.166	0.0153
0.09	0.319	0.0164
1.17	0.478	0.0142
2.17	0.566	0.0146
3.01	0.659	0.0149
4.18	0.772	0.0160
5.17	0.862	0.0180
6.17	0.941	0.0209
7.44	0.995	0.0294
8.44	1.033	0.0377
9.22	1.060	0.0454

Run: CL02378
Re = 199819

α	C_l	C_d
-6.96	-0.325	0.0296
-5.94	-0.236	0.0205
-4.83	-0.126	0.0156
-3.92	-0.034	0.0140
-2.92	0.078	0.0117
-1.78	0.217	0.0104
-0.89	0.322	0.0090
0.20	0.436	0.0096
1.24	0.548	0.0094
2.28	0.652	0.0106
3.20	0.745	0.0118
4.28	0.843	0.0137
5.42	0.935	0.0167
6.28	0.992	0.0202
7.28	1.047	0.0256
8.30	1.094	0.0322
9.25	1.127	0.0412

Run: PG02380
Re = 299791

α	C_l	C_d
-7.14	-0.310	0.0285
-6.14	-0.202	0.0176
-5.09	-0.094	0.0133
-3.83	0.034	0.0110
-2.96	0.130	0.0087
-1.88	0.245	0.0086
-0.96	0.337	0.0082
0.51	0.482	0.0076
1.40	0.578	0.0082
1.75	0.612	0.0085
3.37	0.777	0.0110
4.40	0.873	0.0131
5.32	0.955	0.0156
6.35	1.030	0.0196
7.44	1.098	0.0251
8.43	1.144	0.0311
9.14	1.167	0.0378

S6063

Fig. 5.89

Run: CL01689
Re = 60000

α	C_l	C_d
-4.89	-0.416	0.0328
-3.88	-0.323	0.0210
-2.87	-0.250	0.0181
-1.83	-0.124	0.0122
-0.81	-0.047	0.0108
0.21	0.021	0.0122
1.24	0.119	0.0133
2.26	0.232	0.0160
3.30	0.431	0.0174
4.34	0.519	0.0164
5.33	0.580	0.0176
6.37	0.648	0.0188
7.36	0.722	0.0286
8.38	0.792	0.0374
9.40	0.853	0.0505

Run: AB01691
Re = 100000

α	C_l	C_d
-4.88	-0.375	0.0271
-3.86	-0.294	0.0199
-2.84	-0.213	0.0153
-1.83	-0.117	0.0088
-0.81	-0.034	0.0090
0.22	0.029	0.0101
1.24	0.137	0.0101
2.28	0.335	0.0111
3.32	0.441	0.0119
4.33	0.524	0.0126
5.35	0.607	0.0155
6.37	0.687	0.0207
7.40	0.766	0.0282
8.41	0.836	0.0375
9.43	0.900	0.0559

Run: AB01693
Re = 200100

α	C_l	C_d
-4.91	-0.438	0.0205
-3.90	-0.350	0.0159
-2.87	-0.254	0.0128
-1.83	-0.164	0.0093
-0.82	-0.045	0.0069
0.22	0.084	0.0067
1.24	0.191	0.0077
2.25	0.302	0.0082
3.30	0.392	0.0087
4.34	0.484	0.0103
5.37	0.565	0.0144
6.37	0.646	0.0177
7.37	0.725	0.0227
8.40	0.818	0.0344
9.44	0.896	0.0797

Run: AB01695
Re = 300600

α	C_l	C_d
-4.91	-0.448	0.0179

-3.86	-0.352	0.0135
-2.86	-0.257	0.0115
-1.84	-0.163	0.0094
-0.81	-0.043	0.0064
0.23	0.102	0.0060
1.27	0.188	0.0064
2.26	0.286	0.0068
3.29	0.410	0.0077
4.32	0.507	0.0103
5.37	0.596	0.0139
6.37	0.682	0.0174
7.40	0.764	0.0225
8.43	0.845	0.0342
9.44	0.913	0.0769

S7012 (B)

Fig. 5.93

Run: PG01782
Re = 100500

α	C_l	C_d
-2.79	-0.151	0.0166
-1.28	-0.019	0.0156
0.28	0.184	0.0148
1.82	0.305	0.0113
3.34	0.437	0.0158
4.87	0.580	0.0198
6.42	0.729	0.0223
7.97	0.871	0.0266
9.49	0.998	0.0343

Run: PG01784
Re = 199600

α	C_l	C_d
-2.81	-0.141	0.0135
-1.27	0.042	0.0125
0.31	0.198	0.0131
1.84	0.354	0.0153
3.35	0.511	0.0166
4.92	0.666	0.0178
6.43	0.803	0.0203
7.99	0.940	0.0247
9.50	1.067	0.0296

Run: PG01785
Re = 198500

α	C_l	C_d
11.05	1.157	0.0398

Run: PG01787
Re = 300100

α	C_l	C_d
-2.92	-0.143	0.0126
-1.28	0.022	0.0123
0.37	0.196	0.0134
1.82	0.348	0.0145
3.37	0.512	0.0157
4.90	0.662	0.0174
6.44	0.805	0.0197
7.95	0.937	0.0228
9.51	1.074	0.0266
11.01	1.160	0.0328

S7012 (B)

Fig. 5.96

Run: CL02257
Re = 59938

α	C_l	C_d
-5.68	-0.313	0.0264
-4.57	-0.212	0.0211
-3.59	-0.120	0.0157
-2.54	-0.044	0.0172
-1.56	0.044	0.0145
-0.53	0.159	0.0167
0.51	0.280	0.0163
1.56	0.420	0.0237
2.62	0.536	0.0266
3.60	0.615	0.0245
4.69	0.695	0.0238
5.63	0.757	0.0264
6.64	0.823	0.0273
7.68	0.878	0.0292
8.63	0.919	0.0355
9.66	0.951	0.0470

Run: CL02259
Re = 100182

α	C_l	C_d
-5.68	-0.333	0.0241
-4.64	-0.227	0.0191
-3.59	-0.106	0.0142
-2.53	-0.006	0.0130
-1.53	0.080	0.0102
-0.41	0.218	0.0133
0.55	0.349	0.0134
1.65	0.457	0.0111
2.61	0.549	0.0146
3.65	0.633	0.0160
4.70	0.731	0.0182
5.61	0.825	0.0197
6.74	0.942	0.0223
7.69	0.998	0.0292
8.78	1.055	0.0409
9.77	1.086	0.0532

Run: CL02261
Re = 200048

α	C_l	C_d
-5.52	-0.222	0.0166
-4.56	-0.124	0.0130
-3.40	0.002	0.0103
-2.55	0.089	0.0095
-1.52	0.200	0.0099
-0.47	0.308	0.0101
0.56	0.405	0.0089
1.62	0.511	0.0096
2.59	0.609	0.0110
3.61	0.717	0.0121
4.62	0.823	0.0132
5.62	0.926	0.0146
6.78	1.025	0.0186
7.73	1.086	0.0252
8.73	1.134	0.0323
9.70	1.155	0.0430

Run: CL02263
Re = 299562

α	C_l	C_d
-5.48	-0.217	0.0135
-4.47	-0.110	0.0108
-3.62	-0.022	0.0092
-2.37	0.115	0.0085
-1.39	0.216	0.0084
-0.41	0.311	0.0079
0.74	0.424	0.0074
1.78	0.534	0.0083
2.53	0.615	0.0090
3.89	0.763	0.0107
4.59	0.834	0.0114
5.73	0.949	0.0133
6.68	1.029	0.0163
7.70	1.104	0.0222
8.70	1.160	0.0286
9.77	1.186	0.0395

S7075 (A)

Fig. 5.106

Run: CL01318
Re = 59800

α	C_l	C_d
-4.00	-0.188	0.0277
-2.99	-0.111	0.0226
-1.97	-0.032	0.0165
-0.95	0.076	0.0185
0.08	0.180	0.0208
1.12	0.297	0.0201
2.14	0.445	0.0246
3.17	0.606	0.0286
4.23	0.753	0.0267
5.26	0.847	0.0228
6.27	0.931	0.0248
7.29	1.008	0.0209
8.31	1.073	0.0291
9.31	1.123	0.0341
10.34	1.173	0.0446

Run: CL01320
Re = 99600

α	C_l	C_d
-4.00	-0.177	0.0239
-2.96	-0.081	0.0183
-1.96	0.015	0.0172
-0.95	0.108	0.0150
0.09	0.273	0.0171
1.13	0.422	0.0191
2.17	0.563	0.0212
3.21	0.691	0.0210
4.22	0.790	0.0184
5.26	0.886	0.0165
6.26	0.968	0.0181
7.38	1.060	0.0217
8.32	1.118	0.0261
9.31	1.163	0.0331
10.33	1.206	0.0425

Run: CL01322
Re = 199700

α	C_l	C_d
-4.03	-0.111	0.0188
-2.99	0.006	0.0152
-1.88	0.133	0.0133
-0.86	0.250	0.0118
0.22	0.366	0.0108
1.20	0.472	0.0106
2.23	0.581	0.0101
3.22	0.700	0.0100
4.27	0.807	0.0108
5.28	0.901	0.0127
6.31	0.990	0.0148
7.31	1.068	0.0175
8.32	1.128	0.0220
9.34	1.179	0.0278
10.37	1.219	0.0353

Run: CL01324
Re = 299400

α	C_l	C_d
-3.97	-0.041	0.0144
-2.96	0.059	0.0117
-1.93	0.162	0.0100
-0.91	0.266	0.0085
0.15	0.395	0.0077
1.18	0.510	0.0078
2.24	0.623	0.0083
3.27	0.730	0.0088
4.20	0.818	0.0096
5.26	0.915	0.0113
6.32	1.007	0.0134
7.28	1.080	0.0156
8.32	1.151	0.0200
9.34	1.208	0.0248
10.35	1.245	0.0312

S7075 (A)

Fig. 5.108

Run: CL01806
Re = 60300

α	C_l	C_d
-4.01	-0.255	0.0375
-2.99	-0.162	0.0286
-1.98	-0.069	0.0208
-0.95	0.032	0.0164
0.07	0.146	0.0152
1.13	0.271	0.0182
2.13	0.392	0.0204
3.18	0.531	0.0235
4.21	0.669	0.0271
5.24	0.792	0.0229
6.26	0.883	0.0220
7.29	0.958	0.0205
8.30	1.031	0.0244
9.32	1.090	0.0283
10.33	1.133	0.0330
11.35	1.161	0.0426

Run: CL01808
Re = 99700

α	C_l	C_d
-5.04	-0.322	0.0481
-4.02	-0.228	0.0284
-3.00	-0.120	0.0217
-1.97	0.006	0.0162
-0.94	0.140	0.0087
0.10	0.271	0.0154
1.13	0.385	0.0171
2.22	0.515	0.0206
3.20	0.633	0.0207
4.22	0.739	0.0177
5.25	0.825	0.0163
6.29	0.905	0.0161
7.29	0.984	0.0196
8.32	1.057	0.0219
9.32	1.105	0.0285
10.36	1.146	0.0355
11.35	1.184	0.0466

Run: CL01810
Re = 199400

α	C_l	C_d
-5.08	-0.207	0.0295
-3.87	-0.077	0.0177
-2.92	0.018	0.0139
-1.98	0.107	0.0115
-0.81	0.215	0.0091
0.23	0.334	0.0089
1.13	0.436	0.0098
2.12	0.541	0.0105
3.27	0.657	0.0099
4.27	0.757	0.0104
5.32	0.857	0.0118
6.29	0.941	0.0139
7.33	1.025	0.0163
8.31	1.096	0.0193
9.33	1.151	0.0251
10.34	1.199	0.0311
11.35	1.234	0.0393

Run: CL01812
Re = 299500

α	C_l	C_d
-5.02	-0.207	0.0258
-4.00	-0.109	0.0163
-2.92	-0.001	0.0126
-1.95	0.097	0.0101
-0.94	0.203	0.0081
0.09	0.317	0.0074
1.14	0.438	0.0076
2.15	0.550	0.0078
3.31	0.671	0.0085
4.37	0.780	0.0093
5.35	0.870	0.0108
6.24	0.946	0.0124
7.27	1.031	0.0147
8.29	1.107	0.0177
9.36	1.175	0.0226
10.35	1.222	0.0274
11.36	1.257	0.0355

S7075 (B)

Fig. 5.112

Run: CL01109
Re = 60300

α	C_l	C_d
-5.22	-0.301	0.0269
-4.10	-0.231	0.0236
-3.11	-0.156	0.0180
-2.12	-0.073	0.0164
-1.05	0.030	0.0163
-0.04	0.152	0.0189
0.97	0.314	0.0254
2.00	0.483	0.0287
3.04	0.627	0.0248
4.07	0.734	0.0238
5.08	0.824	0.0221
6.16	0.905	0.0230
7.13	0.988	0.0306
8.18	1.072	0.0354
9.18	1.133	0.0358
10.14	1.181	0.0428

Run: CL01112
Re = 99800

α	C_l	C_d
-5.19	-0.274	0.0350
-4.18	-0.199	0.0250
-3.11	-0.098	0.0208
-1.98	0.051	0.0167
-1.09	0.159	0.0149
-0.06	0.318	0.0158
0.97	0.458	0.0168
2.00	0.580	0.0173
3.05	0.680	0.0171
4.04	0.771	0.0168
5.12	0.869	0.0164
6.10	0.959	0.0174
7.11	1.045	0.0200
8.13	1.101	0.0252
9.15	1.165	0.0318
10.17	1.213	0.0394

Run: CL01114
Re = 199800

α	C_l	C_d
-5.23	-0.250	0.0256
-4.15	-0.125	0.0183
-3.13	0.016	0.0149
-2.10	0.135	0.0117
-1.07	0.231	0.0099
-0.03	0.332	0.0099
1.01	0.443	0.0102
2.01	0.548	0.0099
3.06	0.660	0.0103
4.08	0.762	0.0111
5.10	0.859	0.0124
6.12	0.949	0.0146
7.15	1.029	0.0186
8.24	1.108	0.0227
9.20	1.176	0.0266
10.19	1.224	0.0328

Run: CL01116
Re = 299800

α	C_l	C_d
-5.20	-0.204	0.0216
-4.16	-0.087	0.0146
-3.14	0.024	0.0118
-2.11	0.122	0.0107
-1.09	0.233	0.0086
-0.08	0.334	0.0077
0.98	0.455	0.0077
2.00	0.571	0.0082
3.08	0.681	0.0087
4.06	0.781	0.0096
5.08	0.883	0.0110
6.09	0.975	0.0132
7.14	1.060	0.0164
8.15	1.141	0.0197
9.23	1.219	0.0231
10.21	1.268	0.0292

S8036

Fig. 5.116

Run: CL01029
Re = 100800

α	C_l	C_d
-5.51	-0.374	0.0196
-3.99	-0.198	0.0204
-3.06	-0.130	0.0221
-1.24	0.010	0.0229
-1.03	0.030	0.0238
0.10	0.143	0.0251
1.02	0.257	0.0244
2.01	0.407	0.0254
3.20	0.543	0.0238
4.10	0.598	0.0223
5.12	0.680	0.0213
6.26	0.788	0.0205
7.16	0.867	0.0203
8.20	0.953	0.0214
9.24	1.034	0.0231
10.34	1.099	0.0257

Run: CL01038
Re = 99700

α	C_l	C_d
12.19	1.135	0.0422
12.28	1.140	0.0390

Run: CL01031
Re = 200700

α	C_l	C_d
-3.95	-0.202	0.0132
-2.90	-0.108	0.0137
-1.84	-0.014	0.0138
-0.72	0.085	0.0141
0.02	0.149	0.0140
1.05	0.253	0.0143
2.30	0.384	0.0144
3.24	0.513	0.0143
4.40	0.712	0.0144
5.45	0.800	0.0147
6.18	0.860	0.0149
7.19	0.936	0.0154
8.23	1.007	0.0175
9.38	1.085	0.0205
10.84	1.161	0.0258
11.40	1.181	0.0300

Run: CL01033
Re = 301400

α	C_l	C_d
-3.96	-0.222	0.0110
-2.94	-0.128	0.0109
-1.54	0.007	0.0109
-0.88	0.060	0.0110
0.14	0.171	0.0114
1.18	0.316	0.0117
2.22	0.427	0.0118
3.24	0.539	0.0127
4.27	0.673	0.0120
5.08	0.780	0.0129
6.32	0.887	0.0125
7.33	0.956	0.0140
8.36	1.030	0.0164
9.38	1.104	0.0190
10.28	1.158	0.0215
11.43	1.205	0.0293

Run: CL01034
Re = 304500

α	C_l	C_d
-6.73	-0.458	0.0152
-6.17	-0.414	0.0143

Run: CL01035
Re = 305000

α	C_l	C_d
-4.99	-0.312	0.0127

Run: CL01037
Re = 402400

α	C_l	C_d
-7.02	-0.501	0.0144
-5.99	-0.414	0.0130
-5.13	-0.334	0.0121
-3.83	-0.215	0.0101
-2.91	-0.129	0.0098
-1.92	-0.029	0.0098
-0.90	0.062	0.0098
0.53	0.256	0.0101
0.80	0.285	0.0101
2.20	0.439	0.0106
3.22	0.546	0.0109
4.25	0.665	0.0122
5.09	0.771	0.0115
6.33	0.888	0.0117
7.19	0.944	0.0133
8.21	1.021	0.0154
9.56	1.119	0.0188
10.34	1.162	0.0213
11.37	1.200	0.0275
12.37	1.224	0.0433

Run: CL01771
Re = 502900

α	C_l	C_d
-6.29	-0.459	0.0132
-5.03	-0.345	0.0117
-3.97	-0.247	0.0108

-2.94	-0.159	0.0092
-1.91	-0.056	0.0091
-0.92	0.045	0.0091
0.14	0.171	0.0092
1.28	0.302	0.0094
2.29	0.415	0.0095
3.42	0.537	0.0099
4.13	0.606	0.0102
5.21	0.722	0.0109
6.47	0.859	0.0123
7.44	0.928	0.0127
8.33	0.992	0.0141
9.32	1.066	0.0158
10.34	1.128	0.0191
11.35	1.169	0.0246
12.52	1.197	0.0394

S8037

Fig. 5.120

Run: CL01040
Re = 100500

α	C_l	C_d
-4.14	-0.188	0.0220
-2.90	-0.108	0.0249
-1.59	-0.018	0.0261
-1.07	0.037	0.0269
-0.04	0.183	0.0280
1.00	0.362	0.0282
1.87	0.480	0.0278
3.27	0.567	0.0263
3.84	0.614	0.0260
4.86	0.703	0.0278
5.97	0.811	0.0256
7.30	0.926	0.0236
8.17	0.998	0.0250
9.03	1.069	0.0269
10.07	1.159	0.0294
11.22	1.225	0.0321
12.23	1.258	0.0366

Run: CL01042
Re = 200800

α	C_l	C_d
-4.16	-0.184	0.0147
-3.11	-0.105	0.0138
-2.11	-0.018	0.0140
-1.09	0.069	0.0146
-0.09	0.163	0.0150
0.98	0.281	0.0155
2.00	0.419	0.0150
3.07	0.589	0.0152
4.03	0.686	0.0153
5.09	0.775	0.0152
6.11	0.868	0.0152
7.16	0.960	0.0155
8.14	1.043	0.0171
9.17	1.124	0.0190
10.19	1.194	0.0212
11.20	1.251	0.0243
12.26	1.289	0.0276

Run: CL01044
Re = 300900

α	C_l	C_d
-4.15	-0.178	0.0133
-3.11	-0.101	0.0113
-2.06	-0.010	0.0113
-1.07	0.085	0.0114
-0.03	0.193	0.0112
1.00	0.297	0.0114
2.00	0.409	0.0118
3.00	0.545	0.0121
4.02	0.692	0.0133
5.09	0.808	0.0123
6.09	0.891	0.0126
7.12	0.974	0.0132
8.16	1.053	0.0149
9.18	1.125	0.0167
10.30	1.199	0.0193
11.17	1.248	0.0213
12.25	1.280	0.0257

Run: CL01046
Re = 400800

α	C_l	C_d
-4.17	-0.181	0.0124
-3.03	-0.095	0.0104
-2.13	-0.016	0.0101
-1.09	0.086	0.0099
-0.04	0.199	0.0099
1.00	0.310	0.0097
2.07	0.425	0.0099
3.09	0.545	0.0107
4.11	0.684	0.0107
5.07	0.802	0.0109
6.09	0.893	0.0111
7.11	0.970	0.0121
8.16	1.047	0.0135
9.25	1.122	0.0160
10.37	1.191	0.0184
11.28	1.232	0.0212
12.22	1.253	0.0252

Run: CL01789
Re = 501900

α	C_l	C_d
-4.16	-0.216	0.0120
-3.15	-0.134	0.0107
-2.12	-0.046	0.0090
-1.07	0.055	0.0090
-0.01	0.171	0.0090
1.05	0.288	0.0088
2.11	0.404	0.0088
3.27	0.529	0.0090
4.29	0.655	0.0097
5.28	0.788	0.0110
6.33	0.902	0.0104
7.49	0.987	0.0117
8.27	1.042	0.0128
9.15	1.098	0.0143
10.17	1.165	0.0164
11.18	1.217	0.0189
12.26	1.246	0.0216

S8052

Fig. 5.124

Run: CL01413
Re = 100200

α	C_l	C_d
-3.99	-0.317	0.0148
-2.94	-0.180	0.0165
-1.93	-0.088	0.0160
-0.91	0.002	0.0179
0.14	0.152	0.0190
1.17	0.287	0.0190
2.18	0.355	0.0203
3.21	0.445	0.0188
4.22	0.547	0.0174
5.25	0.643	0.0165
6.29	0.748	0.0179
7.35	0.847	0.0194
8.35	0.939	0.0231
9.41	1.024	0.0271
10.36	1.088	0.0338

Run: CL01414
Re = 101000

α	C_l	C_d
-6.02	-0.465	0.0235
-5.00	-0.414	0.0181

Run: CL01416
Re = 202500

α	C_l	C_d
-6.06	-0.477	0.0177
-4.94	-0.349	0.0141
-4.03	-0.266	0.0126

Run: CL01417
Re = 200400

α	C_l	C_d
-2.99	-0.172	0.0112
-1.87	-0.071	0.0115
-0.86	0.019	0.0114
0.12	0.103	0.0122
1.15	0.190	0.0123
2.16	0.315	0.0108
3.19	0.436	0.0110
4.22	0.543	0.0111
5.27	0.678	0.0120
6.29	0.782	0.0139
7.32	0.880	0.0158
8.34	0.968	0.0189
9.35	1.046	0.0227
10.36	1.116	0.0272

Run: CL01420
Re = 301200

α	C_l	C_d
-6.04	-0.479	0.0168
-5.00	-0.389	0.0142
-3.95	-0.290	0.0118
-2.93	-0.190	0.0100
-1.93	-0.090	0.0096
-0.88	0.014	0.0100
0.13	0.102	0.0098
1.30	0.227	0.0095
2.17	0.340	0.0092
3.20	0.471	0.0088
4.22	0.581	0.0094
5.26	0.688	0.0107
6.29	0.791	0.0123
7.30	0.894	0.0139
8.34	0.992	0.0163
9.41	1.080	0.0198
10.39	1.153	0.0237

Run: CL01422
Re = 401800

α	C_l	C_d
-6.03	-0.498	0.0154
-5.00	-0.401	0.0134
-3.97	-0.296	0.0111
-2.96	-0.193	0.0093
-1.90	-0.083	0.0084
-0.83	0.027	0.0084
0.11	0.115	0.0083
1.23	0.262	0.0083
2.17	0.363	0.0078
3.24	0.482	0.0078
4.26	0.599	0.0089
5.29	0.706	0.0104
6.30	0.811	0.0118
7.33	0.915	0.0134
8.33	1.007	0.0154
9.36	1.092	0.0182
10.37	1.168	0.0219

Run: CL01769
Re = 502300

α	C_l	C_d
-6.03	-0.536	0.0148
-5.01	-0.438	0.0129
-3.99	-0.334	0.0111
-2.91	-0.221	0.0095
-1.88	-0.118	0.0079
-0.81	-0.002	0.0076
0.24	0.113	0.0075
1.37	0.240	0.0075
2.49	0.365	0.0074
3.49	0.470	0.0074
4.38	0.566	0.0080
5.22	0.652	0.0094
6.26	0.761	0.0106
7.31	0.867	0.0120
8.42	0.982	0.0141
9.33	1.058	0.0160
10.35	1.143	0.0189
11.38	1.213	0.0228

SA7035

Fig. 5.128

Run: AG02573
Re = 59893

α	C_l	C_d
-6.53	-0.363	0.0499
-5.33	-0.277	0.0325
-4.27	-0.203	0.0240
-3.35	-0.139	0.0200
-2.52	-0.086	0.0165

-1.42	-0.009	0.0180
-0.46	0.063	0.0155
0.82	0.196	0.0185
1.69	0.335	0.0196
2.77	0.475	0.0266
3.67	0.563	0.0302
4.68	0.658	0.0293
5.69	0.743	0.0272
6.80	0.836	0.0280
7.74	0.913	0.0315
8.85	0.995	0.0331
9.75	1.050	0.0305
10.80	1.103	0.0352
11.84	1.128	0.0801

Run: AG02576
Re = 99882

α	C_l	C_d
-6.50	-0.383	0.0591
-5.63	-0.306	0.0369
-4.38	-0.208	0.0232
-3.34	-0.129	0.0178
-2.32	-0.054	0.0138
-1.37	0.015	0.0129
-0.26	0.151	0.0148
0.70	0.313	0.0170
1.86	0.462	0.0174
2.84	0.547	0.0169
3.82	0.624	0.0170
4.71	0.707	0.0170
5.76	0.799	0.0181
6.77	0.887	0.0204
7.97	0.999	0.0237
8.97	1.074	0.0276
9.90	1.121	0.0325
10.99	1.151	0.0422

Run: AG02578
Re = 150142

α	C_l	C_d
-5.11	-0.280	0.0263
-4.43	-0.230	0.0214
-3.40	-0.136	0.0167
-2.34	-0.022	0.0125
-1.43	0.073	0.0122
-0.22	0.261	0.0120
0.96	0.374	0.0118
1.70	0.443	0.0119
2.88	0.566	0.0120
3.81	0.655	0.0130
4.96	0.761	0.0142
5.77	0.836	0.0159
6.93	0.942	0.0178
7.86	1.026	0.0200
8.88	1.098	0.0232
9.86	1.148	0.0290
10.86	1.172	0.0373

Run: AG02580
Re = 199609

α	C_l	C_d
-5.30	-0.307	0.0255
-4.35	-0.234	0.0180
-3.28	-0.121	0.0138
-2.23	0.027	0.0109
-1.26	0.159	0.0091
-0.16	0.283	0.0093
0.93	0.379	0.0093
1.90	0.473	0.0099
2.64	0.542	0.0104
3.88	0.671	0.0116
4.84	0.766	0.0127
5.97	0.880	0.0144
6.85	0.960	0.0158
7.97	1.056	0.0186
8.98	1.126	0.0222
9.80	1.168	0.0270
10.82	1.192	0.0351

Run: AG02582
Re = 299447

α	C_l	C_d
-5.95	-0.383	0.0435
-4.93	-0.278	0.0196
-3.90	-0.141	0.0142
-2.94	-0.012	0.0107
-1.91	0.104	0.0085
-0.72	0.204	0.0073
-0.12	0.271	0.0072
1.35	0.429	0.0076
2.61	0.563	0.0086
3.38	0.645	0.0095
4.46	0.758	0.0107
5.43	0.859	0.0119
6.48	0.966	0.0135
7.23	1.035	0.0149
8.47	1.141	0.0180
9.62	1.217	0.0233
10.41	1.248	0.0283
11.35	1.251	0.0371

SA7036 (A)

Fig. 5.132

Run: AG02629
Re = 100036

α	C_l	C_d
-5.93	-0.347	0.0455
-4.87	-0.270	0.0288
-4.08	-0.211	0.0219
-2.76	-0.087	0.0157
-1.73	0.013	0.0137
-0.66	0.171	0.0144
0.05	0.281	0.0165
1.24	0.460	0.0171
2.32	0.561	0.0163
3.33	0.641	0.0166
4.19	0.711	0.0162
5.47	0.819	0.0174
6.32	0.898	0.0188
7.47	1.020	0.0214
8.44	1.109	0.0241
9.52	1.163	0.0301
10.42	1.189	0.0384
11.47	1.202	0.0474

Run: AG02635
Re = 149976

α	C_l	C_d
-5.84	-0.339	0.0444
-5.04	-0.264	0.0275
-3.73	-0.131	0.0161
-3.01	-0.046	0.0128
-1.66	0.128	0.0120
-0.64	0.260	0.0106
0.13	0.344	0.0109
1.39	0.470	0.0114
2.40	0.573	0.0116
3.19	0.649	0.0122
4.43	0.774	0.0131
5.38	0.868	0.0146
6.47	0.971	0.0165
7.51	1.063	0.0185
8.51	1.144	0.0219
9.39	1.192	0.0276
10.52	1.218	0.0365
11.51	1.220	0.0487

Run: AG02631
Re = 200059

α	C_l	C_d
-5.77	-0.344	0.0382
-4.69	-0.238	0.0206
-4.04	-0.162	0.0167
-3.13	-0.029	0.0131
-1.46	0.176	0.0100
-0.50	0.273	0.0098
0.45	0.364	0.0095
1.59	0.480	0.0098
2.53	0.577	0.0102
3.10	0.634	0.0105
4.14	0.734	0.0117
5.17	0.834	0.0132
6.71	0.983	0.0156
7.30	1.039	0.0172
8.52	1.126	0.0220
9.43	1.171	0.0285
10.44	1.194	0.0368
11.40	1.194	0.0527

Run: AG02633
Re = 299385

α	C_l	C_d
-6.00	-0.363	0.0465
-4.98	-0.223	0.0195
-3.92	-0.081	0.0118
-2.94	0.037	0.0097
-1.83	0.150	0.0084
-0.83	0.238	0.0077
0.00	0.325	0.0073
1.18	0.453	0.0076
2.35	0.574	0.0085
3.41	0.686	0.0096
4.76	0.823	0.0112
5.49	0.898	0.0121
6.53	0.998	0.0140
7.41	1.077	0.0158
8.53	1.155	0.0208
9.48	1.201	0.0271
10.36	1.222	0.0346
11.34	1.224	0.0505

SA7036 (B)

Fig. 5.136

Run: AG02699
Re = 99750

α	C_l	C_d
-5.92	-0.355	0.0605
-4.72	-0.234	0.0315
-3.76	-0.142	0.0223
-2.52	-0.014	0.0166
-1.67	0.065	0.0160
-0.51	0.221	0.0151
0.40	0.369	0.0158
1.40	0.515	0.0173
2.46	0.613	0.0163
3.39	0.686	0.0161
4.45	0.777	0.0169
5.40	0.864	0.0182
6.52	0.965	0.0203
7.55	1.052	0.0230
8.46	1.113	0.0272
9.58	1.183	0.0333
10.56	1.236	0.0389
11.55	1.272	0.0455

Run: AG02701
Re = 150030

α	C_l	C_d
-4.73	-0.223	0.0269
-3.74	-0.108	0.0201
-2.58	0.041	0.0152
-1.48	0.194	0.0132
-0.53	0.316	0.0112
0.56	0.433	0.0120
1.38	0.517	0.0119
2.33	0.610	0.0113
3.47	0.723	0.0129
4.41	0.815	0.0145
5.70	0.940	0.0165
6.56	1.017	0.0183
7.54	1.092	0.0212
8.65	1.165	0.0252
9.59	1.224	0.0298
10.62	1.274	0.0356
11.56	1.307	0.0436

Run: AG02703
Re = 199817

α	C_l	C_d
-5.10	-0.245	0.0275
-3.38	-0.012	0.0164
-2.59	0.100	0.0143
-1.44	0.234	0.0108
-0.43	0.332	0.0100
0.06	0.384	0.0097
1.57	0.539	0.0097
2.53	0.639	0.0104
3.15	0.701	0.0113
4.61	0.843	0.0130
5.61	0.943	0.0145
6.63	1.037	0.0166
7.55	1.111	0.0191
8.78	1.196	0.0240
9.76	1.258	0.0275

10.55	1.298	0.0315
11.64	1.334	0.0389

Run: AG02705
Re = 299784

α	C_l	C_d
-5.26	-0.191	0.0234
-4.24	-0.056	0.0155
-3.40	0.045	0.0127
-2.32	0.160	0.0114
-1.34	0.254	0.0094
-0.31	0.355	0.0078
0.77	0.466	0.0075
1.78	0.574	0.0082
2.86	0.685	0.0095
3.92	0.796	0.0108
4.66	0.873	0.0118
6.08	1.011	0.0140
7.08	1.102	0.0159
8.30	1.197	0.0197
9.07	1.251	0.0219
10.18	1.322	0.0260
11.09	1.362	0.0304
12.10	1.380	0.0336

SA7038

Fig. 5.140

Run: AG02645
Re = 99713

α	C_l	C_d
-5.71	-0.336	0.0639
-4.66	-0.220	0.0346
-3.57	-0.117	0.0242
-2.49	0.017	0.0180
-1.43	0.127	0.0143
-0.44	0.263	0.0153
0.45	0.397	0.0154
1.57	0.531	0.0157
2.68	0.637	0.0152
3.71	0.730	0.0161
4.52	0.793	0.0166
5.60	0.892	0.0178
6.70	0.990	0.0219
7.64	1.071	0.0236
8.77	1.148	0.0259
9.80	1.216	0.0291
10.74	1.251	0.0323
11.76	1.261	0.0438

Run: AG02647
Re = 149636

α	C_l	C_d
-4.60	-0.174	0.0291
-3.59	-0.044	0.0190
-2.44	0.120	0.0143
-1.36	0.253	0.0120
-0.31	0.345	0.0107
0.53	0.418	0.0117
1.64	0.532	0.0112
2.58	0.635	0.0118
3.66	0.749	0.0133
4.64	0.838	0.0145
5.73	0.945	0.0163
6.70	1.034	0.0185
7.61	1.114	0.0204
8.70	1.204	0.0227
9.68	1.270	0.0250
10.81	1.297	0.0311
11.64	1.298	0.0407

Run: AG02657
Re = 199566

α	C_l	C_d
-4.49	-0.092	0.0201
-3.33	0.057	0.0145
-2.51	0.149	0.0121
-1.89	0.210	0.0108
-0.79	0.297	0.0092
0.73	0.451	0.0094
1.71	0.556	0.0098
2.77	0.669	0.0109
3.63	0.760	0.0120
4.75	0.874	0.0135
5.75	0.973	0.0151
6.74	1.070	0.0169
7.69	1.151	0.0188
8.75	1.229	0.0211
9.79	1.286	0.0243
10.68	1.308	0.0299
11.73	1.300	0.0468

Run: AG02658
Re = 200114

α	C_l	C_d
-5.66	-0.276	0.0482

Run: AG02660
Re = 299315

α	C_l	C_d
-5.86	-0.248	0.0474
-4.76	-0.089	0.0185
-3.73	0.023	0.0128
-2.69	0.124	0.0107
-1.68	0.225	0.0088
-0.61	0.331	0.0076
0.48	0.442	0.0077
1.92	0.594	0.0089
2.57	0.661	0.0097
3.68	0.776	0.0109
4.75	0.888	0.0120
5.80	0.996	0.0133
6.72	1.085	0.0148
7.77	1.180	0.0168
8.84	1.261	0.0191
9.76	1.314	0.0230
10.66	1.342	0.0287
11.61	1.347	0.0419

SD7032 (D)

Fig. 5.144

Run: PG02566
Re = 99943

α	C_l	C_d
-4.48	-0.144	0.0317
-2.87	0.062	0.0196
-1.05	0.316	0.0186
0.26	0.440	0.0154
1.69	0.556	0.0144
3.23	0.690	0.0172
4.86	0.826	0.0211
6.27	0.947	0.0239
7.85	1.062	0.0295
9.51	1.161	0.0357
10.93	1.223	0.0429
12.34	1.257	0.0509

Run: PG02568
Re = 199502

α	C_l	C_d
-5.94	-0.256	0.0508
-4.39	-0.030	0.0163
-2.93	0.139	0.0128
-1.16	0.301	0.0129
0.24	0.442	0.0145
2.06	0.601	0.0174
2.86	0.675	0.0186
5.14	0.890	0.0217
6.43	0.994	0.0241
7.93	1.100	0.0285
9.11	1.167	0.0334
11.08	1.229	0.0460

Run: PG02570
Re = 299568

α	C_l	C_d
-6.22	-0.239	0.0524
-4.57	-0.034	0.0127
-2.96	0.114	0.0127
-1.46	0.273	0.0127
0.11	0.430	0.0142
1.60	0.574	0.0160
3.12	0.727	0.0181
4.96	0.900	0.0210
6.32	1.015	0.0237
8.20	1.151	0.0286
9.65	1.220	0.0344
10.62	1.252	0.0425

SD7037 (B)

Fig. 5.148

Run: PG02638
Re = 99934

α	C_l	C_d
-4.61	-0.332	0.0424
-3.01	-0.191	0.0225
-1.36	-0.017	0.0182
0.09	0.187	0.0175
1.65	0.385	0.0160
3.10	0.521	0.0133
4.63	0.641	0.0153
6.23	0.777	0.0201
7.71	0.886	0.0251
9.21	1.000	0.0306
10.76	1.100	0.0390
12.42	1.193	0.0504

Run: PG02640
Re = 201522

α	C_l	C_d
-4.56	-0.263	0.0231
-2.82	-0.066	0.0136
-1.36	0.113	0.0135
-0.04	0.246	0.0118
1.78	0.411	0.0146
3.29	0.559	0.0168
4.79	0.705	0.0190
6.22	0.873	0.0215
7.84	1.021	0.0247
9.31	1.133	0.0300
10.75	1.210	0.0368
12.24	1.255	0.0539

Run: PG02642
Re = 299298

α	C_l	C_d
-4.49	-0.244	0.0173
-3.02	-0.076	0.0131
-1.50	0.077	0.0125
-0.09	0.231	0.0139
1.78	0.484	0.0153
3.39	0.667	0.0173
4.70	0.801	0.0193
6.35	0.963	0.0224
7.70	1.080	0.0254
9.35	1.195	0.0307
10.93	1.268	0.0421
12.41	1.289	0.0725

SD7037 (D)

Fig. 5.152

Run: AB01100
Re = 59200

α	C_l	C_d
-5.06	-0.258	0.0409
-4.02	-0.182	0.0271
-2.94	-0.099	0.0208
-1.97	-0.023	0.0162
-0.95	0.074	0.0200
0.13	0.181	0.0173
1.15	0.334	0.0262
2.19	0.490	0.0307
3.18	0.610	0.0279
4.26	0.703	0.0279
5.29	0.789	0.0301
6.31	0.883	0.0355
7.26	0.961	0.0374
8.28	1.036	0.0370
9.33	1.112	0.0397
10.31	1.167	0.0464

Run: AB01102
Re = 98500

α	C_l	C_d
-5.07	-0.326	0.0316
-3.99	-0.237	0.0222
-2.94	-0.132	0.0181
-2.00	-0.022	0.0153
-0.97	0.117	0.0147
0.04	0.280	0.0158

1.10	0.419	0.0165
2.29	0.530	0.0166
3.27	0.618	0.0164
4.15	0.693	0.0160
5.33	0.799	0.0173
6.21	0.875	0.0188
7.32	0.974	0.0211
8.19	1.054	0.0237
9.39	1.138	0.0286
10.40	1.188	0.0339

Run: PG01104
Re = 196600

α	C_l	C_d
-4.72	-0.247	0.0209
-4.07	-0.174	0.0180
-2.96	-0.007	0.0132
-1.64	0.130	0.0101
-0.99	0.189	0.0095
0.21	0.299	0.0096
1.11	0.400	0.0094
2.22	0.529	0.0092
3.32	0.661	0.0103
4.14	0.744	0.0114
5.29	0.853	0.0130
6.33	0.949	0.0147
7.24	1.031	0.0167
8.18	1.113	0.0188
9.31	1.194	0.0223
10.32	1.246	0.0275

Run: PG01106
Re = 295100

α	C_l	C_d
-5.09	-0.251	0.0207
-3.95	-0.113	0.0143
-3.00	-0.008	0.0118
-1.92	0.087	0.0096
-0.90	0.198	0.0080
0.17	0.323	0.0071
1.23	0.443	0.0074
2.33	0.564	0.0079
3.30	0.668	0.0090
4.46	0.789	0.0104
5.48	0.895	0.0118
6.63	1.010	0.0134
7.21	1.062	0.0145
8.25	1.148	0.0169
9.45	1.238	0.0210
10.30	1.281	0.0261

SD7037 (D)

Fig. 5.154

Run: CL02018
Re = 196300

α	C_l	C_d
-4.98	-0.308	0.0282
-3.93	-0.183	0.0185
-2.88	-0.037	0.0146
-2.03	0.078	0.0126
-0.79	0.203	0.0097
0.06	0.275	0.0093
1.09	0.390	0.0089
2.21	0.515	0.0094
3.25	0.624	0.0101
4.24	0.721	0.0113
5.28	0.824	0.0126
6.29	0.920	0.0142
7.27	1.014	0.0163
8.32	1.106	0.0186
9.31	1.178	0.0218
10.35	1.236	0.0269
11.35	1.268	0.0351

SD7037 (E)

Fig. 5.158

Run: CL02307
Re = 99678

α	C_l	C_d
-5.19	-0.272	0.0410
-4.18	-0.192	0.0277
-3.11	-0.098	0.0207
-2.11	-0.003	0.0163
-1.06	0.117	0.0142
0.05	0.301	0.0152
0.99	0.429	0.0163
2.10	0.534	0.0162
3.10	0.618	0.0155
4.08	0.714	0.0158
5.11	0.820	0.0170
6.14	0.918	0.0183
7.15	1.003	0.0206
8.25	1.083	0.0225
9.17	1.145	0.0256
10.19	1.190	0.0312
11.22	1.226	0.0395

Run: CL02309
Re = 199620

α	C_l	C_d
-4.48	-0.155	0.0198
-3.65	-0.056	0.0156
-2.34	0.100	0.0122
-1.37	0.208	0.0107
-0.30	0.310	0.0093
0.67	0.410	0.0095
1.78	0.527	0.0100
2.73	0.622	0.0106
3.76	0.722	0.0116
4.89	0.831	0.0128
5.85	0.923	0.0144
6.73	1.002	0.0161
7.73	1.088	0.0181
8.77	1.170	0.0207
9.90	1.232	0.0258
10.81	1.269	0.0317
11.78	1.296	0.0389

Run: CL02311
Re = 299364

α	C_l	C_d
-4.96	-0.175	0.0225
-4.02	-0.070	0.0149
-2.96	0.046	0.0119
-1.95	0.150	0.0097
-0.95	0.250	0.0085
-0.24	0.315	0.0077
0.88	0.437	0.0080
2.20	0.579	0.0087
3.20	0.681	0.0096
4.12	0.775	0.0104
5.14	0.875	0.0118
6.16	0.971	0.0131
7.15	1.063	0.0149
8.20	1.153	0.0170
9.20	1.226	0.0197
10.17	1.277	0.0246
11.21	1.320	0.0305
12.23	1.338	0.0412

SD7037 (E)

Fig. 5.161

Run: CL02323
Re = 105616

α	C_l	C_d
-6.16	-0.415	0.0410
-5.19	-0.346	0.0278
-4.11	-0.266	0.0212
-2.96	-0.176	0.0161
-2.12	-0.099	0.0144
-1.07	0.037	0.0130
0.01	0.228	0.0134
1.02	0.355	0.0137
2.01	0.448	0.0144
3.03	0.539	0.0154
4.13	0.641	0.0158
5.15	0.730	0.0163
6.13	0.820	0.0175
7.16	0.898	0.0207
8.21	0.976	0.0250
9.20	1.037	0.0310
10.22	1.086	0.0385
11.21	1.121	0.0507

Run: CL02315
Re = 200149

α	C_l	C_d
-6.06	-0.429	0.0363
-4.83	-0.331	0.0209
-4.12	-0.272	0.0174
-2.94	-0.115	0.0129
-2.04	-0.002	0.0113
-1.04	0.103	0.0102
0.14	0.217	0.0097
1.00	0.309	0.0099
2.10	0.419	0.0099
3.07	0.524	0.0105
4.08	0.625	0.0113
5.10	0.728	0.0121
6.41	0.846	0.0149
7.22	0.916	0.0175
8.24	0.998	0.0209
9.16	1.062	0.0237
10.33	1.131	0.0297
11.19	1.170	0.0355
12.21	1.193	0.0454

Run: CL02313
Re = 299430

α	C_l	C_d
-6.33	-0.460	0.0498
-5.40	-0.382	0.0248
-3.97	-0.224	0.0158
-3.07	-0.108	0.0110
-2.28	-0.021	0.0096
-1.11	0.096	0.0096
-0.10	0.200	0.0097
0.93	0.310	0.0092
1.87	0.405	0.0090
2.89	0.514	0.0087
3.92	0.623	0.0096
4.92	0.723	0.0106
5.92	0.817	0.0125
7.02	0.917	0.0151
7.97	1.003	0.0173
8.98	1.085	0.0197
9.97	1.151	0.0235
11.06	1.206	0.0295
12.06	1.233	0.0382
13.02	1.233	0.0583

SD7037 (E)

Fig. 5.164

Run: CL02325
Re = 99786

α	C_l	C_d
-5.22	-0.222	0.0536
-4.17	-0.079	0.0306
-3.05	0.045	0.0228
-2.10	0.156	0.0191
-1.03	0.286	0.0165
0.02	0.431	0.0172
1.01	0.572	0.0167
2.03	0.665	0.0156
3.07	0.768	0.0149
4.11	0.856	0.0164
5.16	0.938	0.0179
6.16	1.019	0.0211
7.21	1.111	0.0235
8.20	1.195	0.0265
9.21	1.268	0.0285
10.25	1.326	0.0313
11.21	1.357	0.0352
12.26	1.373	0.0435
13.18	1.382	0.0542
14.18	1.369	0.0711

Run: CL02327
Re = 205042

α	C_l	C_d
-4.96	-0.015	0.0268
-3.97	0.090	0.0189
-2.92	0.196	0.0160
-1.63	0.338	0.0120
-0.86	0.406	0.0109
0.01	0.486	0.0106
1.03	0.603	0.0103
2.05	0.699	0.0105
3.10	0.800	0.0116
4.15	0.897	0.0129

5.11	0.983	0.0150
6.12	1.070	0.0173
7.33	1.178	0.0195
8.29	1.258	0.0210
9.23	1.333	0.0228
10.21	1.382	0.0268
11.33	1.396	0.0369
12.23	1.403	0.0466

Run: CL02329
Re = 298981

α	C_l	C_d
-5.11	-0.032	0.0298
-4.10	0.082	0.0161
-3.08	0.185	0.0126
-1.85	0.315	0.0103
-1.01	0.401	0.0093
0.06	0.508	0.0081
1.27	0.635	0.0087
2.25	0.731	0.0096
3.50	0.855	0.0110
4.45	0.945	0.0123
5.29	1.023	0.0135
6.30	1.113	0.0154
7.26	1.203	0.0169
8.24	1.290	0.0182
9.31	1.369	0.0203
10.23	1.403	0.0258
11.23	1.426	0.0327

SD7037 (E)

Fig. 5.167

Run: AG02368
Re = 100009

α	C_l	C_d
-5.49	-0.013	0.0262
-4.38	0.091	0.0221
-3.44	0.171	0.0200
-2.33	0.234	0.0199
-1.40	0.302	0.0185
-0.44	0.431	0.0218
0.60	0.567	0.0249
1.68	0.743	0.0221
2.90	0.884	0.0185
3.74	0.955	0.0161
4.68	1.018	0.0179
5.73	1.095	0.0209
7.12	1.196	0.0251
7.80	1.237	0.0281
8.75	1.277	0.0312
9.89	1.312	0.0404
10.77	1.341	0.0484

Run: CL02370
Re = 199795

α	C_l	C_d
-7.09	-0.130	0.0455
-5.72	0.038	0.0189
-4.90	0.124	0.0162
-3.77	0.245	0.0156
-2.83	0.336	0.0152
-1.77	0.426	0.0162
-0.92	0.527	0.0158
0.13	0.637	0.0137
1.10	0.736	0.0124
2.11	0.837	0.0114
3.13	0.930	0.0122
4.08	1.014	0.0139
5.10	1.102	0.0159
6.28	1.195	0.0188
7.17	1.260	0.0214
8.38	1.324	0.0260
9.23	1.350	0.0321

Run: CL02372
Re = 299247

α	C_l	C_d
-7.22	-0.148	0.0525
-6.03	0.008	0.0192
-5.09	0.107	0.0147
-4.00	0.226	0.0125
-3.02	0.337	0.0115
-1.95	0.454	0.0111
-0.95	0.556	0.0105
0.07	0.657	0.0096
1.16	0.773	0.0092
2.16	0.864	0.0099
3.24	0.964	0.0115
4.26	1.056	0.0129
5.44	1.159	0.0151
6.14	1.213	0.0166
7.45	1.308	0.0199
8.38	1.356	0.0239
9.55	1.396	0.0304
10.41	1.422	0.0362
11.40	1.438	0.0462

SD7062 (B)

Fig. 5.171

Run: PG01721
Re = 59900

α	C_l	C_d
-3.82	-0.113	0.0312
-2.21	0.040	0.0295
-0.79	0.184	0.0306
0.87	0.337	0.0333
2.43	0.463	0.0465
3.90	0.528	0.0578
5.49	0.585	0.0728
6.88	0.627	0.0958
8.51	0.668	0.1240
9.89	0.702	0.1430
11.47	0.753	0.1614

Run: CF01155
Re = 99400

α	C_l	C_d
-3.23	0.117	0.0231
-2.20	0.227	0.0231
-1.21	0.321	0.0235
-0.16	0.412	0.0220
0.86	0.527	0.0244
1.90	0.642	0.0242
2.96	0.751	0.0246
3.94	0.848	0.0242
4.96	0.935	0.0249
5.98	1.022	0.0257
7.00	1.109	0.0260
8.12	1.205	0.0279
9.04	1.283	0.0309
10.07	1.365	0.0360
11.14	1.442	0.0416
12.11	1.489	0.0451

Run: PG01163
Re = 101500

α	C_l	C_d
-5.32	-0.141	0.0300
-4.24	0.003	0.0260

Run: PG01157
Re = 199800

α	C_l	C_d
-5.33	-0.079	0.0187
-3.74	0.078	0.0150
-2.11	0.261	0.0128
-0.67	0.392	0.0129
0.86	0.551	0.0125
2.41	0.723	0.0132
4.00	0.863	0.0147
5.51	1.018	0.0163
7.02	1.164	0.0183
8.55	1.307	0.0206
10.12	1.430	0.0237
11.60	1.524	0.0285

Run: PG01158
Re = 198400

α	C_l	C_d
13.15	1.600	0.0363
14.60	1.633	0.0500

Run: PG01160
Re = 299500

α	C_l	C_d
-5.35	-0.107	0.0160
-3.73	0.058	0.0129
-2.11	0.236	0.0105
-0.66	0.380	0.0103
1.02	0.582	0.0105
2.57	0.750	0.0111
4.17	0.927	0.0123
5.72	1.085	0.0140
7.21	1.230	0.0156
8.67	1.362	0.0175
10.21	1.480	0.0208
11.69	1.569	0.0258
13.23	1.628	0.0351
14.67	1.652	0.0540

Run: PG01162
Re = 399900

α	C_l	C_d
-5.53	-0.128	0.0151
-3.74	0.053	0.0118
-2.24	0.197	0.0099
-0.70	0.397	0.0091
0.86	0.575	0.0098
2.40	0.740	0.0103
3.96	0.911	0.0109
5.54	1.074	0.0122
7.05	1.222	0.0138
8.72	1.372	0.0162
10.23	1.487	0.0196
11.84	1.577	0.0251
13.09	1.618	0.0332
14.53	1.639	0.0390

SD7062 (B)

Fig. 5.173

Run: PG01744
Re = 100300

α	C_l	C_d
-5.41	-0.213	0.0361
-3.65	0.030	0.0261
-2.15	0.196	0.0250
-0.59	0.343	0.0252
0.90	0.476	0.0223
2.46	0.614	0.0209
3.95	0.776	0.0204
5.58	0.928	0.0204
7.06	1.049	0.0227
8.59	1.146	0.0279
10.04	1.214	0.0344

Run: PG01746
Re = 200400

α	C_l	C_d
-5.36	-0.100	0.0187
-3.79	0.039	0.0157
-2.15	0.204	0.0163
-0.63	0.367	0.0162
0.92	0.531	0.0152
2.47	0.685	0.0160
3.92	0.817	0.0181
5.48	0.941	0.0207
6.97	1.044	0.0237
8.49	1.140	0.0292
10.02	1.191	0.0478
11.55	1.202	0.0877

Run: PG01748
Re = 300300

α	C_l	C_d
-5.27	-0.116	0.0151
-3.77	0.030	0.0139
-2.21	0.194	0.0137
-0.70	0.351	0.0135
0.89	0.530	0.0141
2.34	0.678	0.0158
4.05	0.834	0.0184
5.48	0.956	0.0208
7.10	1.078	0.0247
8.53	1.149	0.0298
10.00	1.185	0.0516

Run: PG01750
Re = 401200

α	C_l	C_d
-5.40	-0.137	0.0137
-3.75	0.028	0.0125
-2.13	0.199	0.0126
-0.69	0.364	0.0130
0.85	0.526	0.0142
2.39	0.678	0.0160

3.95	0.826	0.0179
5.48	0.957	0.0202
7.05	1.071	0.0239
8.49	1.140	0.0295
10.16	1.175	0.0384

SD7080

Fig. 5.177

Run: AG02588
Re = 99943

α	C_l	C_d
-6.02	-0.413	0.0437
-4.93	-0.346	0.0292
-3.91	-0.269	0.0246
-2.97	-0.176	0.0191
-1.95	-0.053	0.0143
-0.76	0.124	0.0138
0.18	0.245	0.0149
1.22	0.355	0.0150
2.28	0.446	0.0156
3.25	0.526	0.0151
4.21	0.633	0.0151
5.26	0.734	0.0170
6.32	0.827	0.0193
7.29	0.905	0.0224
8.32	0.986	0.0242
9.34	1.024	0.0292
10.32	1.052	0.0372
11.24	1.076	0.0441

Run: AG02590
Re = 149842

α	C_l	C_d
-6.01	-0.410	0.0433
-4.81	-0.284	0.0255
-3.90	-0.189	0.0201
-2.86	-0.061	0.0160
-1.81	0.041	0.0125
-0.71	0.172	0.0114
0.40	0.273	0.0113
1.21	0.346	0.0118
2.24	0.457	0.0125
3.25	0.567	0.0116
4.21	0.660	0.0121
5.31	0.755	0.0151
6.22	0.836	0.0170
7.26	0.928	0.0194
8.34	1.005	0.0228
9.29	1.036	0.0295
10.42	1.069	0.0376
11.22	1.083	0.0463

Run: AG02592
Re = 199879

α	C_l	C_d
-5.94	-0.379	0.0376
-5.34	-0.306	0.0277
-3.84	-0.139	0.0175
-2.75	-0.028	0.0150
-1.71	0.056	0.0113
-0.75	0.152	0.0099
0.24	0.259	0.0094
1.41	0.379	0.0098
2.27	0.474	0.0096
3.24	0.573	0.0102
4.40	0.688	0.0120
5.09	0.753	0.0132
6.31	0.866	0.0153
7.25	0.953	0.0173
8.28	1.013	0.0224
9.34	1.056	0.0285
10.23	1.085	0.0352
11.26	1.101	0.0456

Run: AG02594
Re = 299767

α	C_l	C_d
-6.19	-0.390	0.0412
-5.06	-0.265	0.0226
-4.10	-0.169	0.0168
-3.04	-0.071	0.0132
-1.96	0.025	0.0101
-0.77	0.135	0.0083
0.21	0.266	0.0075
1.26	0.376	0.0076
2.16	0.470	0.0082
2.86	0.538	0.0087
4.35	0.689	0.0106
5.19	0.774	0.0119
6.17	0.870	0.0135
7.29	0.970	0.0156
8.34	1.030	0.0217
9.35	1.078	0.0265
10.23	1.110	0.0325
11.26	1.128	0.0432

SG6040

Fig. 5.181

Run: PG02399
Re = 99872

α	C_l	C_d
-3.97	-0.202	0.0264
-3.01	-0.169	0.0302
-1.87	-0.082	0.0331
-0.85	0.025	0.0336
0.25	0.152	0.0331
1.08	0.183	0.0379
2.31	0.292	0.0427
3.30	0.369	0.0471
4.23	0.456	0.0549
5.20	0.657	0.0451
6.37	0.883	0.0298
7.25	0.987	0.0261
8.30	1.085	0.0247
9.38	1.163	0.0253
10.36	1.235	0.0274
11.43	1.248	0.0323
12.40	1.266	0.0408

Run: PG02400
Re = 100031

α	C_l	C_d
-6.06	-0.478	0.0257
-4.87	-0.343	0.0219

Run: CF02402
Re = 149969

α	C_l	C_d
-4.83	-0.227	0.0190
-3.98	-0.115	0.0203
-2.38	0.066	0.0207
-1.77	0.130	0.0209
-0.95	0.210	0.0206
0.37	0.325	0.0202
1.27	0.398	0.0201
2.39	0.506	0.0201
3.43	0.609	0.0200
4.43	0.711	0.0202
5.57	0.829	0.0199
6.87	0.984	0.0199
7.25	1.024	0.0201
8.55	1.142	0.0203
9.39	1.214	0.0213
10.59	1.270	0.0254
11.24	1.266	0.0297
12.51	1.285	0.0399

Run: CF02403
Re = 149553

α	C_l	C_d
-6.93	-0.503	0.0270
-6.10	-0.392	0.0227

Run: CF02405
Re = 199979

α	C_l	C_d
-6.81	-0.395	0.0207
-5.70	-0.254	0.0170
-5.14	-0.192	0.0149
-3.54	-0.037	0.0150
-2.76	0.040	0.0152
-1.61	0.154	0.0157
-0.69	0.246	0.0157
0.53	0.374	0.0156
1.18	0.447	0.0155
2.30	0.563	0.0158
3.34	0.664	0.0159
4.34	0.758	0.0165
5.32	0.851	0.0165
6.48	0.961	0.0164
7.40	1.070	0.0168
8.49	1.173	0.0177
9.51	1.246	0.0194
10.37	1.278	0.0235
11.42	1.284	0.0325

Run: CF02406
Re = 199113

α	C_l	C_d
12.35	1.294	0.0407
13.41	1.311	0.0514

Run: CF02408
Re = 299663

α	C_l	C_d
-8.07	-0.482	0.0201
-7.06	-0.362	0.0164
-6.04	-0.274	0.0144
-4.93	-0.175	0.0119
-3.99	-0.085	0.0105
-2.96	0.022	0.0111
-1.93	0.133	0.0115
-0.54	0.283	0.0117
0.27	0.369	0.0119
1.31	0.489	0.0120
2.26	0.587	0.0125
3.45	0.716	0.0131
4.05	0.779	0.0134
5.02	0.878	0.0137
6.65	1.037	0.0141
7.50	1.114	0.0142
8.63	1.210	0.0160
9.60	1.262	0.0189
10.53	1.287	0.0240
11.39	1.298	0.0303
12.34	1.311	0.0396
13.47	1.330	0.0519
14.49	1.347	0.0621

Run: CF02411
Re = 399758

α	C_l	C_d
-8.07	-0.458	0.0168
-7.09	-0.371	0.0145
-5.95	-0.277	0.0127
-5.00	-0.184	0.0112
-3.99	-0.088	0.0093
-2.97	0.019	0.0093
-1.90	0.137	0.0097
-0.89	0.249	0.0099
0.23	0.373	0.0102
1.14	0.476	0.0104
2.29	0.599	0.0108
3.35	0.713	0.0115
4.26	0.813	0.0119
5.26	0.918	0.0123
6.43	1.034	0.0130
7.51	1.127	0.0135
8.54	1.199	0.0152
9.64	1.250	0.0192
10.62	1.280	0.0243
11.55	1.301	0.0309
12.46	1.320	0.0396
13.43	1.341	0.0487

Run: CF02413
Re = 498293

α	C_l	C_d
-5.92	-0.282	0.0119
-4.96	-0.184	0.0106
-4.04	-0.089	0.0090
-2.97	0.029	0.0083
-1.94	0.151	0.0087
-0.90	0.271	0.0089
0.11	0.385	0.0091
1.13	0.504	0.0094
2.19	0.624	0.0099
3.21	0.737	0.0104
4.25	0.850	0.0110
5.25	0.956	0.0116
6.34	1.066	0.0124
7.12	1.134	0.0131
8.53	1.224	0.0154
9.23	1.257	0.0181
10.30	1.292	0.0232
11.33	1.313	0.0299

12.73	1.339	0.0442

SG6040

Fig. 5.183

Run: PG02505
Re = 137122

α	C_l	C_d
-4.27	-0.124	0.0230
-2.55	0.054	0.0214
-1.45	0.174	0.0205
0.30	0.341	0.0161
2.01	0.500	0.0166
3.37	0.598	0.0192
4.99	0.696	0.0210
6.36	0.792	0.0244
7.87	0.883	0.0302
9.29	0.950	0.0382
10.98	0.992	0.0511
5.25	0.956	0.0116

Run: PG02506
Re = 149726

α	C_l	C_d
-5.90	-0.353	0.0247

Run: PG02510
Re = 299854

α	C_l	C_d
-5.96	-0.267	0.0169
-4.40	-0.119	0.0155
-2.90	0.028	0.0137
-1.21	0.184	0.0145
0.35	0.342	0.0151
2.01	0.490	0.0172
3.65	0.636	0.0195
4.62	0.719	0.0211
6.14	0.839	0.0242
7.63	0.927	0.0288
9.64	1.000	0.0432
10.93	1.032	0.0581

Run: CF02513
Re = 500268

α	C_l	C_d
-5.95	-0.293	0.0145
-4.47	-0.143	0.0138
-2.90	0.012	0.0138
-1.30	0.186	0.0144
0.16	0.333	0.0153
1.71	0.494	0.0163
3.30	0.643	0.0182
4.74	0.763	0.0208
6.44	0.885	0.0250
8.14	0.974	0.0315
9.91	1.037	0.0413

SG6041

Fig. 5.187

Run: AG02331
Re = 99837

α	C_l	C_d
-7.41	-0.498	0.0793
-6.51	-0.443	0.0522
-5.13	-0.339	0.0301
-4.47	-0.281	0.0253
-3.45	-0.187	0.0206
-2.37	-0.094	0.0152
-1.39	-0.013	0.0142
-0.29	0.096	0.0168
0.68	0.225	0.0194
1.74	0.394	0.0199
2.68	0.537	0.0196
3.56	0.647	0.0173
4.94	0.782	0.0161
5.84	0.860	0.0167
7.03	0.943	0.0225
7.85	0.991	0.0267
9.15	1.069	0.0340
9.86	1.106	0.0396
10.89	1.141	0.0474

Run: PG02496
Re = 149940

α	C_l	C_d
-6.36	-0.429	0.0488
-5.20	-0.350	0.0290
-4.26	-0.274	0.0219
-3.10	-0.158	0.0177
-2.18	-0.053	0.0134
-0.82	0.071	0.0151
-0.11	0.200	0.0138
1.03	0.380	0.0132
1.96	0.477	0.0122
2.72	0.526	0.0119
4.09	0.673	0.0117
4.95	0.769	0.0135
6.18	0.876	0.0172
6.84	0.922	0.0194
7.82	0.989	0.0237
9.12	1.075	0.0296
10.06	1.125	0.0360
11.09	1.165	0.0456

Run: AG02333
Re = 199832

α	C_l	C_d
-5.31	-0.359	0.0267
-4.36	-0.269	0.0201
-3.27	-0.151	0.0168
-2.28	-0.021	0.0127
-1.28	0.102	0.0130
-0.11	0.251	0.0123
0.82	0.374	0.0114
1.83	0.471	0.0105
2.83	0.581	0.0107
3.95	0.699	0.0109
4.87	0.781	0.0126
5.95	0.871	0.0152
6.83	0.942	0.0181
7.86	1.020	0.0219
8.84	1.091	0.0257
9.96	1.155	0.0311
11.14	1.199	0.0397

Run: AG02335
Re = 299731

α	C_l	C_d
-5.91	-0.393	0.0346
-4.98	-0.291	0.0208
-3.94	-0.158	0.0157
-2.82	0.000	0.0122
-1.91	0.113	0.0088
-0.83	0.215	0.0080
0.25	0.308	0.0078
1.07	0.413	0.0080
2.56	0.574	0.0087
3.28	0.650	0.0090
4.27	0.745	0.0104
5.26	0.834	0.0125
6.33	0.930	0.0147
7.41	1.020	0.0178
8.47	1.102	0.0210
9.37	1.167	0.0245
10.39	1.223	0.0286
11.38	1.264	0.0353

Run: AG02337
Re = 399121

α	C_l	C_d
-6.54	-0.423	0.0499
-5.44	-0.291	0.0216
-4.47	-0.175	0.0140
-3.44	-0.047	0.0114
-2.40	0.069	0.0091
-1.40	0.162	0.0070
-0.38	0.266	0.0068
0.71	0.371	0.0069
1.67	0.477	0.0069
2.72	0.600	0.0075
3.79	0.702	0.0090
4.83	0.799	0.0110
6.15	0.935	0.0137
6.94	0.994	0.0153
7.89	1.073	0.0175
9.05	1.165	0.0208
9.90	1.221	0.0236
10.97	1.276	0.0283
11.95	1.309	0.0352

Run: AG02340
Re = 498408

α	C_l	C_d
-6.48	-0.383	0.0447
-5.41	-0.259	0.0167
-4.45	-0.150	0.0122
-3.40	-0.045	0.0103
-2.40	0.064	0.0086
-1.34	0.171	0.0067
-0.33	0.279	0.0062
0.79	0.393	0.0065
1.77	0.493	0.0066
2.72	0.608	0.0072
3.83	0.715	0.0089
4.85	0.813	0.0109
6.00	0.925	0.0128
7.24	1.044	0.0155
8.35	1.133	0.0179
9.35	1.208	0.0207
10.16	1.259	0.0233
11.09	1.305	0.0275
12.12	1.338	0.0352

SG6041

Fig. 5.189

Run: PG02528
Re = 144195

α	C_l	C_d
-5.68	-0.401	0.0363
-4.36	-0.297	0.0231
-2.86	-0.151	0.0173
-1.24	0.102	0.0170
0.29	0.289	0.0139
1.88	0.413	0.0169
3.28	0.529	0.0177
5.00	0.675	0.0205
6.38	0.802	0.0241
7.87	0.917	0.0287
9.53	1.008	0.0366
10.92	1.071	0.0449
6.00	0.925	0.0128

Run: PG02531
Re = 299200

α	C_l	C_d
-5.85	-0.408	0.0346
-4.30	-0.217	0.0167
-2.84	-0.018	0.0120
-1.21	0.131	0.0135
0.44	0.296	0.0148
1.94	0.456	0.0162
3.39	0.598	0.0180
4.89	0.736	0.0206
6.39	0.867	0.0236
8.08	0.987	0.0282
9.41	1.064	0.0340
10.90	1.121	0.0439
12.35	1.138	0.0674

Run: PG02534
Re = 498804

α	C_l	C_d
-5.87	-0.333	0.0205
-4.31	-0.175	0.0148
-2.72	-0.017	0.0128
-1.46	0.127	0.0132
0.22	0.306	0.0138
1.71	0.465	0.0150
3.34	0.628	0.0169
4.86	0.769	0.0195
6.60	0.923	0.0236
8.03	1.025	0.0277
9.59	1.103	0.0333
10.87	1.143	0.0372
12.54	1.184	0.0383

SG6042

Fig. 5.193

Run: AG02342
Re = 99856

α	C_l	C_d
-5.13	-0.351	0.0615
-3.78	-0.123	0.0309
-3.02	-0.027	0.0258
-1.97	0.108	0.0208
-0.94	0.203	0.0193
0.08	0.311	0.0230
1.06	0.397	0.0263
2.09	0.488	0.0281
3.14	0.643	0.0277
4.32	0.811	0.0234
5.23	0.900	0.0195
6.20	0.995	0.0184
7.28	1.095	0.0197
8.25	1.160	0.0245

Run: AG02343
Re = 99645

α	C_l	C_d
9.29	1.246	0.0286
10.36	1.278	0.0373
11.39	1.325	0.0436
12.42	1.347	0.0529
13.35	1.347	0.0646

Run: PG02498
Re = 150036

α	C_l	C_d
-4.86	-0.214	0.0495
-3.77	-0.072	0.0273
-2.78	0.053	0.0228
-1.73	0.165	0.0189
-0.65	0.269	0.0168
0.57	0.403	0.0166
1.40	0.492	0.0165
2.41	0.596	0.0164
3.39	0.691	0.0148
4.53	0.800	0.0143
5.43	0.890	0.0149
6.47	0.984	0.0166
7.45	1.048	0.0192
8.53	1.106	0.0242
9.53	1.166	0.0296
10.31	1.211	0.0338
11.45	1.262	0.0422
12.57	1.287	0.0526

Run: CL02345
Re = 199860

α	C_l	C_d
-4.82	-0.156	0.0363
-3.98	-0.031	0.0230
-2.89	0.121	0.0182
-1.86	0.244	0.0160
-0.93	0.349	0.0148
0.32	0.487	0.0137
1.25	0.593	0.0131
2.24	0.700	0.0128
3.29	0.809	0.0125
4.31	0.911	0.0124
5.39	1.011	0.0130
6.07	1.066	0.0138
7.30	1.160	0.0173
8.37	1.228	0.0215
9.28	1.278	0.0254
10.45	1.335	0.0312
11.35	1.369	0.0369
12.49	1.393	0.0477

Run: CL02347
Re = 299571

α	C_l	C_d
-5.09	-0.076	0.0291
-4.01	0.065	0.0169
-3.02	0.168	0.0140
-1.94	0.292	0.0115
-0.90	0.398	0.0098
0.09	0.485	0.0089
1.11	0.602	0.0091
2.11	0.704	0.0093
3.16	0.816	0.0095
4.22	0.921	0.0102
5.42	1.034	0.0116
6.37	1.110	0.0135
7.44	1.192	0.0166
8.41	1.257	0.0194
9.85	1.341	0.0244
10.32	1.365	0.0259
11.50	1.415	0.0325
12.32	1.436	0.0390

Run: CL02349
Re = 399200

α	C_l	C_d
-5.32	-0.048	0.0270
-4.15	0.060	0.0144
-2.92	0.189	0.0111
-1.97	0.292	0.0101
-0.95	0.401	0.0084
0.07	0.499	0.0076
1.06	0.606	0.0074
2.12	0.717	0.0079
3.16	0.829	0.0084
4.17	0.929	0.0092
5.21	1.023	0.0109
6.25	1.109	0.0128
7.29	1.192	0.0152
8.42	1.274	0.0183
9.38	1.335	0.0210
10.49	1.396	0.0253
11.93	1.449	0.0335
12.61	1.463	0.0396

Run: CL02351
Re = 498399

α	C_l	C_d
-4.96	-0.006	0.0147
-3.98	0.078	0.0123
-3.24	0.155	0.0107

Run: CL02352
Re = 498548

α	C_l	C_d
-1.84	0.310	0.0086
-0.90	0.414	0.0074
0.10	0.515	0.0068
1.24	0.633	0.0066
2.15	0.733	0.0071
3.17	0.837	0.0079
4.20	0.938	0.0091
5.21	1.030	0.0108
6.25	1.117	0.0128
7.30	1.202	0.0150
8.31	1.279	0.0173
9.36	1.349	0.0202
10.52	1.412	0.0245
11.66	1.456	0.0305
12.85	1.486	0.0421

SG6042

Fig. 5.195

Run: PG02549
Re = 149755

α	C_l	C_d
-6.25	-0.422	0.0914
-4.44	-0.177	0.0356
-2.81	0.053	0.0219
-1.30	0.246	0.0224
0.33	0.476	0.0144
1.91	0.602	0.0177
3.12	0.695	0.0197
4.91	0.833	0.0215
6.40	0.952	0.0254
7.87	1.059	0.0299
9.43	1.147	0.0368
10.85	1.199	0.0470
12.36	1.231	0.0602

Run: PG02552
Re = 299566

α	C_l	C_d
-6.14	-0.279	0.0699
-4.50	-0.005	0.0175
-2.79	0.175	0.0125
-1.32	0.308	0.0135
0.07	0.443	0.0148
1.71	0.602	0.0168
3.41	0.761	0.0191
4.56	0.860	0.0207
6.69	1.036	0.0252
8.11	1.128	0.0293
9.19	1.187	0.0340
10.77	1.240	0.0436
12.38	1.267	0.0624

Run: PG02553
Re = 298975

α	C_l	C_d
-5.37	-0.133	0.0420
-3.76	0.090	0.0140

Run: PG02556
Re = 499026

α	C_l	C_d
-4.46	-0.003	0.0157
-3.00	0.148	0.0136
-1.31	0.319	0.0135
-0.05	0.445	0.0141
1.60	0.614	0.0155
3.31	0.779	0.0176
4.71	0.904	0.0198
6.10	1.020	0.0225
7.79	1.141	0.0272
9.47	1.228	0.0333
10.96	1.278	0.0416

Run: PG02557
Re = 500070

α	C_l	C_d
-5.60	-0.120	0.0402
-5.02	-0.051	0.0173

SG6042

Fig. 5.197

Run: CL02347
Re = 299571

α	C_l	C_d
-5.09	-0.076	0.0291
-4.01	0.065	0.0169
-3.02	0.168	0.0140
-1.94	0.292	0.0115
-0.90	0.398	0.0098
0.09	0.485	0.0089
1.11	0.602	0.0091
2.11	0.704	0.0093
3.16	0.816	0.0095
4.22	0.921	0.0102
5.42	1.034	0.0116
6.37	1.110	0.0135
7.44	1.192	0.0166
8.41	1.257	0.0194
9.85	1.341	0.0244
10.32	1.365	0.0259
11.50	1.415	0.0325
12.32	1.436	0.0390

Run: PG02649
Re = 299308

α	C_l	C_d
-6.14	-0.276	0.0693
-4.50	-0.013	0.0198
-2.96	0.182	0.0136
-1.41	0.351	0.0111
0.07	0.490	0.0092
1.71	0.659	0.0091
3.17	0.807	0.0095
4.78	0.953	0.0115
6.26	1.079	0.0149
7.64	1.185	0.0185
9.70	1.318	0.0252
10.77	1.369	0.0303
12.38	1.416	0.0418

Run: PG02651
Re = 298882

α	C_l	C_d
-6.06	-0.268	0.0674
-4.52	-0.012	0.0192
-2.93	0.187	0.0136
-1.42	0.354	0.0119
0.19	0.518	0.0104

1.68	0.661	0.0086
3.22	0.795	0.0120
4.72	0.904	0.0165
6.36	1.054	0.0189
7.96	1.184	0.0221
9.50	1.293	0.0262
11.02	1.369	0.0326
12.28	1.406	0.0431

Run: PG02653
Re = 299228

α	C_l	C_d
-6.14	-0.279	0.0693
-4.53	-0.015	0.0185
-2.91	0.188	0.0135
-1.39	0.353	0.0114
0.13	0.497	0.0112
1.60	0.617	0.0142
3.28	0.764	0.0171
4.80	0.900	0.0184
6.24	1.026	0.0199
8.16	1.174	0.0243
9.63	1.266	0.0293
10.92	1.320	0.0365
12.43	1.357	0.0533

Run: PG02655
Re = 299932

α	C_l	C_d
-6.13	-0.273	0.0678
-4.78	-0.047	0.0196
-3.03	0.173	0.0142
-1.36	0.340	0.0118
0.25	0.475	0.0133
1.93	0.631	0.0162
3.13	0.742	0.0177
4.84	0.896	0.0201
6.35	1.025	0.0229
8.08	1.127	0.0266
9.55	1.226	0.0326
11.12	1.281	0.0434
12.36	1.300	0.0592

Run: PG02552
Re = 299566

α	C_l	C_d
-6.14	-0.279	0.0699
-4.50	-0.005	0.0175
-2.79	0.175	0.0125
-1.32	0.308	0.0135
0.07	0.443	0.0148
1.71	0.602	0.0168
3.41	0.761	0.0191
4.56	0.860	0.0207
6.69	1.036	0.0252
8.11	1.128	0.0293
9.19	1.187	0.0340
10.77	1.240	0.0436
12.38	1.267	0.0624

Run: PG02553
Re = 298975

α	C_l	C_d
-5.37	-0.133	0.0420
-3.76	0.090	0.0140

SG6042

Fig. 5.199

Run: PG02695
Re = 299822

α	C_l	C_d
-6.15	-0.292	0.0736
-4.56	-0.020	0.0227
-2.96	0.178	0.0139
-1.41	0.346	0.0114
0.13	0.495	0.0092
1.92	0.677	0.0086
3.17	0.805	0.0096
4.66	0.940	0.0118
6.35	1.087	0.0151
7.98	1.209	0.0192
9.53	1.310	0.0241
11.07	1.392	0.0303
12.51	1.432	0.0418

SG6043

Fig. 5.203

Run: PG02385
Re = 99681

α	C_l	C_d
-4.32	-0.278	0.0888
-3.07	0.047	0.0428
-1.89	0.190	0.0332
-0.87	0.300	0.0338
0.10	0.379	0.0330
1.13	0.483	0.0383
2.25	0.616	0.0428
2.95	0.708	0.0427
4.20	0.961	0.0359
5.10	1.095	0.0298
6.08	1.203	0.0235
7.27	1.285	0.0218
8.39	1.366	0.0230
9.35	1.413	0.0257
10.33	1.424	0.0295
11.30	1.432	0.0364

Run: PG02500
Re = 149877

α	C_l	C_d
-4.01	-0.071	0.0623
-3.18	0.141	0.0391
-2.22	0.269	0.0308
-0.69	0.466	0.0276
0.37	0.601	0.0247
1.34	0.745	0.0236
2.33	0.873	0.0216
2.95	0.946	0.0206
4.56	1.110	0.0182
5.43	1.192	0.0173
6.72	1.313	0.0177
7.12	1.348	0.0184
8.51	1.438	0.0212
9.40	1.473	0.0252
10.41	1.490	0.0339
11.58	1.521	0.0408

12.43	1.538	0.0483

Run: PG02387
Re = 199784

α	C_l	C_d
-4.95	-0.122	0.0750
-3.91	0.142	0.0395
-2.93	0.303	0.0257
-1.85	0.454	0.0210
-0.97	0.570	0.0186
0.08	0.682	0.0162
1.07	0.793	0.0154
2.20	0.915	0.0148
3.36	1.038	0.0145
3.97	1.095	0.0143
5.32	1.228	0.0143
6.46	1.325	0.0153
7.55	1.403	0.0172
8.59	1.458	0.0201
9.56	1.493	0.0245
10.20	1.507	0.0285
11.44	1.542	0.0352

Run: PG02388
Re = 199052

α	C_l	C_d
12.28	1.563	0.0404
13.61	1.576	0.0507

Run: PG02390
Re = 299102

α	C_l	C_d
-6.23	-0.124	0.0864
-5.20	0.052	0.0583
-4.13	0.271	0.0240
-3.05	0.401	0.0158
-2.05	0.517	0.0136
-1.02	0.631	0.0121
-0.02	0.733	0.0109
1.05	0.840	0.0106
2.17	0.958	0.0108
3.12	1.053	0.0108
4.17	1.158	0.0110
5.18	1.249	0.0119
6.35	1.349	0.0135
7.19	1.410	0.0154
8.25	1.467	0.0187
9.34	1.511	0.0230
10.31	1.540	0.0282
11.33	1.570	0.0343
12.32	1.592	0.0408
13.30	1.607	0.0519

Run: PG02391
Re = 302510

α	C_l	C_d
-5.73	-0.051	0.0740

Run: PG02393
Re = 399643

α	C_l	C_d
-6.25	-0.051	0.0776
-5.21	0.118	0.0498
-4.11	0.305	0.0171
-3.09	0.418	0.0124
-2.02	0.539	0.0111

-0.96	0.656	0.0102

Run: PG02394
Re = 398416

α	C_l	C_d
-0.02	0.755	0.0095
1.04	0.858	0.0088
2.03	0.961	0.0090
3.06	1.066	0.0093
4.10	1.168	0.0099
5.13	1.261	0.0109
6.14	1.344	0.0126
7.21	1.417	0.0151
8.18	1.468	0.0182
9.27	1.516	0.0220
10.32	1.554	0.0271
11.33	1.587	0.0328
12.38	1.614	0.0419

Run: PG02397
Re = 498423

α	C_l	C_d
-6.32	-0.036	0.0763
-5.22	0.143	0.0459
-4.17	0.316	0.0134
-3.17	0.433	0.0108
-2.12	0.550	0.0098
-1.08	0.667	0.0092
-0.06	0.782	0.0085
1.00	0.883	0.0080
2.11	0.997	0.0084
3.15	1.101	0.0088
4.20	1.201	0.0096
5.20	1.288	0.0109
6.17	1.360	0.0127
7.29	1.433	0.0156
8.18	1.482	0.0183
9.22	1.530	0.0226
10.21	1.566	0.0269
11.22	1.599	0.0328
12.28	1.627	0.0409

SG6043

Fig. 5.205

Run: PG02537
Re = 138371

α	C_l	C_d
-6.20	-0.358	0.1138
-4.64	-0.255	0.0892
-2.95	0.153	0.0371
-1.33	0.369	0.0286
-0.26	0.550	0.0227
1.99	0.813	0.0183
3.23	0.900	0.0215
4.82	1.008	0.0240
6.22	1.108	0.0272
7.74	1.202	0.0335
9.28	1.268	0.0414
10.74	1.306	0.0535
6.17	1.360	0.0127

Run: PG02540
Re = 299460

α	C_l	C_d
-5.80	-0.090	0.0802
-4.64	0.140	0.0444
-2.99	0.405	0.0115
-1.50	0.531	0.0136
0.00	0.666	0.0157
1.55	0.798	0.0185
3.11	0.932	0.0206
4.55	1.049	0.0234
6.19	1.166	0.0278
7.74	1.255	0.0339
9.34	1.320	0.0439
10.37	1.348	0.0532

Run: PG02541
Re = 298916

α	C_l	C_d
-4.15	0.248	0.0187
-3.71	0.319	0.0160

Run: PG02545
Re = 453329

α	C_l	C_d
-4.68	0.206	0.0303
-3.08	0.392	0.0140
-1.41	0.554	0.0145
0.12	0.700	0.0161
1.55	0.840	0.0178
3.14	0.978	0.0202
4.69	1.102	0.0234
6.32	1.213	0.0287
7.66	1.284	0.0345
9.48	1.348	0.0457
9.34	1.320	0.0439

Run: PG02546
Re = 498927

α	C_l	C_d
-5.25	0.057	0.0569
-3.60	0.345	0.0141

Trainer 60

Fig. 5.209

Run: CL01048
Re = 100800

α	C_l	C_d
-4.15	-0.427	0.0201
-3.10	-0.336	0.0210
-1.92	-0.224	0.0214
-1.00	-0.132	0.0204
-0.02	-0.036	0.0215
1.00	0.054	0.0213
2.15	0.156	0.0208
3.09	0.243	0.0195
4.09	0.335	0.0191
5.18	0.486	0.0205
6.28	0.604	0.0209
7.24	0.701	0.0249
8.24	0.798	0.0270
9.24	0.889	0.0309
10.25	0.980	0.0323
11.28	1.064	0.0375

Run: CL01050
Re = 201600

α	C_l	C_d
-4.15	-0.498	0.0154
-3.11	-0.395	0.0151
-2.09	-0.288	0.0153
-1.07	-0.177	0.0151
-0.02	-0.065	0.0149
0.99	0.039	0.0149
2.01	0.142	0.0151
3.05	0.246	0.0150
4.11	0.353	0.0154
5.12	0.450	0.0156
6.12	0.561	0.0169
7.17	0.706	0.0173
8.20	0.816	0.0194
9.21	0.919	0.0220
10.24	1.015	0.0239
11.26	1.107	0.0264

Run: CL01053
Re = 199200

α	C_l	C_d
12.29	1.190	0.0293
13.27	1.231	0.0374

Run: CL01052
Re = 301800

α	C_l	C_d
-3.11	-0.419	0.0131
-2.08	-0.304	0.0130
-1.04	-0.186	0.0130
-0.03	-0.072	0.0130
1.07	0.049	0.0127
2.16	0.163	0.0126
3.06	0.290	0.0128
4.08	0.404	0.0129
5.10	0.508	0.0135
6.15	0.614	0.0140
7.23	0.744	0.0146
8.46	0.885	0.0176
9.30	0.976	0.0187
10.48	1.097	0.0212
11.53	1.186	0.0244
12.27	1.217	0.0320

Run: AB01055
Re = 404100

α	C_l	C_d
-4.17	-0.553	0.0125
-3.14	-0.439	0.0122
-2.17	-0.319	0.0119
-1.07	-0.193	0.0118
-0.05	-0.076	0.0117
1.06	0.049	0.0117
2.02	0.184	0.0118
3.07	0.303	0.0117
4.11	0.422	0.0118
5.20	0.540	0.0122
6.17	0.641	0.0132
7.32	0.763	0.0142
8.32	0.888	0.0151
9.34	0.999	0.0173
10.42	1.109	0.0192

Run: AB01056
Re = 398700

α	C_l	C_d
11.50	1.183	0.0241
12.35	1.179	0.0658

Run: CL02074
Re = 504000

α	C_l	C_d
-3.18	-0.456	0.0117
-2.14	-0.334	0.0114
-1.06	-0.207	0.0111
-0.02	-0.090	0.0110
1.00	0.026	0.0110
2.02	0.144	0.0111
3.04	0.257	0.0112
4.08	0.373	0.0112
5.10	0.488	0.0115
6.12	0.591	0.0122
7.14	0.696	0.0132
8.17	0.800	0.0140
9.19	0.921	0.0152
10.20	1.049	0.0164
11.23	1.118	0.0200
12.27	1.153	0.0480

Ultra-Sport

Fig. 5.213

Run: CL01680
Re = 101100

α	C_l	C_d
-3.03	-0.307	0.0214
-1.97	-0.202	0.0205
-0.96	-0.125	0.0209
0.05	-0.032	0.0219
1.08	0.059	0.0213
2.09	0.142	0.0210
3.12	0.217	0.0214
4.14	0.308	0.0207
5.28	0.481	0.0237
6.18	0.594	0.0231
7.21	0.639	0.0253
8.23	0.713	0.0249
9.26	0.813	0.0242
10.28	0.881	0.0261
11.30	0.937	0.0331

Run: CL01682
Re = 201500

α	C_l	C_d
-3.02	-0.311	0.0149
-2.00	-0.215	0.0143
-1.00	-0.117	0.0137
0.05	-0.015	0.0135
1.07	0.080	0.0132
2.08	0.180	0.0139
3.11	0.278	0.0141
4.13	0.372	0.0151
5.17	0.464	0.0154
6.18	0.561	0.0160
7.22	0.693	0.0179
8.27	0.799	0.0169
9.29	0.868	0.0213
10.28	0.907	0.0281
11.27	0.914	0.0539
12.27	0.905	0.0691

Run: CL01684
Re = 302100

α	C_l	C_d
-3.04	-0.340	0.0127
-1.99	-0.230	0.0121
-0.97	-0.125	0.0117
0.03	-0.023	0.0114
1.09	0.086	0.0113
2.10	0.189	0.0116
3.13	0.296	0.0118
4.16	0.402	0.0123
5.19	0.505	0.0129
6.22	0.599	0.0136
7.23	0.693	0.0133
8.26	0.805	0.0165
9.28	0.866	0.0209
10.28	0.894	0.0316
11.30	0.900	0.0536

Run: CL01686
Re = 403400

α	C_l	C_d
-3.07	-0.353	0.0111
-1.95	-0.235	0.0107
-0.97	-0.128	0.0106
0.04	-0.023	0.0105
1.07	0.084	0.0104
2.09	0.191	0.0104
3.13	0.305	0.0106
4.16	0.410	0.0111
5.20	0.516	0.0117
6.24	0.619	0.0121
7.23	0.706	0.0130
8.27	0.794	0.0144
9.29	0.854	0.0205
10.28	0.878	0.0337
11.29	0.882	0.0525

Run: CL01861
Re = 504100

α	C_l	C_d
-3.15	-0.368	0.0103
-1.83	-0.221	0.0100
-0.67	-0.093	0.0099
0.26	0.007	0.0098
1.09	0.092	0.0099
2.09	0.197	0.0099
3.12	0.303	0.0099
4.16	0.411	0.0103
5.23	0.522	0.0108
6.24	0.633	0.0115
7.61	0.748	0.0136
8.42	0.804	0.0162
9.84	0.864	0.0256
10.28	0.872	0.0359

USNPS-4

Fig. 5.217

Run: PG01166
Re = 99800

α	C_l	C_d
-3.28	-0.143	0.0588
-1.58	0.091	0.0428
0.01	0.320	0.0380
1.54	0.643	0.0270
3.15	0.839	0.0201
4.59	0.953	0.0185
6.22	1.060	0.0219
7.69	1.167	0.0269
9.26	1.291	0.0337
10.71	1.386	0.0422
12.31	1.445	0.0576

Run: PG01507
Re = 199300

α	C_l	C_d
-3.29	0.096	0.0404
-1.56	0.340	0.0244
-0.09	0.511	0.0176
1.45	0.683	0.0124
3.08	0.850	0.0128
4.60	0.961	0.0142
6.08	1.063	0.0176
7.60	1.176	0.0208
9.11	1.283	0.0243
10.67	1.391	0.0280
12.21	1.488	0.0335

Run: PG01511
Re = 299000

α	C_l	C_d
-4.72	-0.020	0.0557
-3.15	0.191	0.0316
-1.60	0.384	0.0186
-0.07	0.543	0.0140
1.46	0.706	0.0101
3.00	0.851	0.0104
4.53	0.964	0.0129
6.15	1.090	0.0159
7.67	1.208	0.0183
9.17	1.320	0.0212
10.68	1.428	0.0245
12.16	1.520	0.0291
13.68	1.602	0.0358

Run: PG01513
Re = 398300

α	C_l	C_d
-4.97	0.000	0.0556
-3.11	0.222	0.0291
-1.46	0.395	0.0164
0.12	0.564	0.0121
1.54	0.702	0.0089
2.97	0.834	0.0097
4.45	0.942	0.0118
6.07	1.077	0.0144
7.50	1.186	0.0163
9.22	1.322	0.0194
10.87	1.436	0.0232
12.19	1.513	0.0268
13.68	1.594	0.0342

Appendix D

Airfoil Data Distribution

All of the airfoil coordinates and performance data presented in this book (see Chapter 5) are available on IBM and Macintosh compatible diskettes through SoarTech Publications. SoarTech will be returning a portion of the proceeds from all disk and book sales to help support the continuation of these airfoil wind-tunnel tests. For more information, write to:

SoarTech Publications
c/o Herk Stokely
1504 N. Horseshoe Circle
Virginia Beach, VA 23451
e-mail: herkstok@aol.com

The data is also available on the Internet from the host `opus.aae.uiuc.edu` using anonymous FTP. See the file `/pub/lsat/airfoil.data` for directions on which files to copy. The data can also be obtained from the world wide web at `http://www.uiuc.edu/ph/www/m-selig`.

Because the airfoil performance data presented in this book is copyrighted, several stipulations have been placed on its use. Simply put, the data may be freely copied and used in any way (e.g., for personal use, in magazine articles, or with a commercial product). However, all products that make use of this data must state that it was produced under the UIUC LSATs program, and no restrictions can be placed on the recipient with respect to the use of the source data. If the data is used in a magazine article or book, this book must be referenced, and it must be stated where the data can be obtained. There must also be no extra charge for providing the data — other than the cost of reproduction and distribution. Furthermore, the recipient must be allowed to copy the data and freely distribute it as well. It is in this sense that all access to this data is *free*. More details can be found in the GNU General Public License that accompanies the data distribution files. Finally, a copy of the license, the copyright notice, and the UIUC Low-Speed Airfoil Tests Manifesto must be included with each distribution of the data disk.

If you find the airfoil performance data useful, *please send a donation* to support our work. If you have already made a donation, we hope that you will consider renewing your commitment. Your tax deductible donations (see Manifesto) can be mailed to:

Prof. Michael Selig
Dept. of Aeronautical and Astronautical Eng.
University of Illinois at Urbana-Champaign
306 Talbot Laboratory, 104 S. Wright St.
Urbana, IL 61801-2935
e-mail: m-selig@uiuc.edu

Appendix E

UIUC Low-Speed Airfoil Tests Manifesto

The UIUC LSATs Manifesto which appears below is a modified version of the initial announcement of the wind-tunnel test program written in December 1993. For recent information on the UIUC LSATs, please see the latest bulletin available from either the coordinator at the address given at the end of this Manifesto or from the world wide web at `http://www.uiuc.edu/ph/www/m-selig`.

The University of Illinois at Urbana-Champaign Low-Speed Airfoil Tests (UIUC LSATs) team is continuing its search for experienced modelers to build a variety of airfoil wind-tunnel models to be tested in the UIUC Department of Aeronautical and Astronautical Engineering Subsonic Aerodynamics Research Laboratory. A low-speed, low-turbulence wind tunnel has been instrumented to take lift, moment, and drag measurements on airfoils at low speeds over the Reynolds number range from 40,000 to 500,000. The scope of the airfoil wind-tunnel tests is limited only by the number of wind-tunnel models provided and the amount of funding received. While the LSATs program has been self-sustaining since its inception, additional support is needed to continue these tests. If you choose to support the program, your help will be acknowledged in subsequent project reports to be published through SoarTech Publications (Herk Stokely).

At the present time, there is a need for new airfoils for R/C sailplanes. For example, R/C hand launch soaring is booming, but few good airfoils (e.g., E387 and SD7037) presently exist for such sailplanes. Sailplanes for the new F3J competition are just beginning to evolve, and new airfoils will probably be required. What will they look like? In the past, only a few airfoils (e.g., HQ 1.5/8.5, RG15 and SD7003) have been favored for F3B competition. In shape, handling, and performance the SD7003 is quite different from the other airfoils mentioned. These significant differences suggest that it may be possible to design new airfoils that have better overall characteristics for F3B competition. In addition to the design and wind tunnel testing of new airfoils, several existing airfoils should be tested. The SD7037 and RG15 are quite popular and often used with flaps. The flap effectiveness of these airfoils should be quantified through wind-tunnel tests, and the results should be used in the design of new airfoils.

There is also a need for new airfoils for R/C sport, aerobatic, and electric planes, as well as R/C helicopters. Often, NACA airfoils are used for these applications, but as compared with airfoils that could be designed today, many of the NACA airfoils (which were designed decades ago mostly by trial and error) are inferior. At the time the NACA airfoils were designed, little was known about the complex aerodynamics of airfoils operating at low Reynolds numbers. (Airfoils with small chords at low speeds, such as those on model aircraft, are said to operate in the low Reynolds number flight regime). In recent years, much has been learned about low Reynolds number aerodynamics, and this knowledge has successfully been applied to the design of new airfoils for R/C sailplanes, ushering in a new era in R/C soaring. Overall, R/C sailplane performance has improved dramatically. Older airfoils are no longer used. R/C power-aircraft performance could likewise be dramatically improved through the use of newly designed, specially tailored airfoils.

Unique airfoil design requirements also exist for other categories of model aircraft. For example, FAI free flight aircraft (which incorporate both a powered launch segment and gliding flight) operate over a wide range of speeds. In the past, many airfoils with good performance characteristics have been designed for FAI free flight. These airfoils should be wind-tunnel tested to quantify their performance. The results gleaned from the tests could then be applied in the design process in an effort to develop new airfoils with improved performance.

Other topics of interest include the effects of contour accuracy and blended airfoils. While previous tests have shed some light on how accurate airfoils must be in order to achieve expected performance, a more systematic effort should be made to test the best airfoils for sensitivity to contour accuracy. We are also interested in designing and testing families of airfoils for use in "blending" from one airfoil at the root to a different airfoil at the tip. It is unlikely that the best performance can be obtained from a single airfoil used along the entire wing span. This is especially true for flying wings. Companion airfoils for blending should be designed for use with the most popular existing airfoils, e.g., SD7037 and RG15. It is expected that the practice of blending airfoils along the span will become much more popular than it is today.

Overall, the LSATs test objectives are to design and wind-tunnel test new airfoils for each category of aircraft listed above and to examine the effects of flaps, turbulators, and contour accuracy. We are especially interested in testing existing airfoils that are known to have superior performance. Wind-tunnel data on such airfoils will be used during the design of new and better airfoils. If you believe that we have overlooked an important area, we would be interested in your input and may consider expanding the scope of the project. The number of airfoil models to be tested has not been predefined; rather, it will be depend on the level of interest and support from the modeling community.

The wind-tunnel models should have a 33 5/8-in. span with a 12-in. chord and can either be built-up or foam core. We will supply 12-in. chord wing templates to ensure the construction of accurate models. The surface finish should preferably be smooth (fiberglass or heat-shrinkable mylar covering); however, we are interested in the effects of surface finish and will consider testing models with non-smooth surfaces. The models will be attached to the wind-tunnel balance by standard model wing rods. Standard model construction techniques should provide the necessary strength (supporting 15–20 lb of lift when pinned at both ends). The brass tubing and collars for the models will be supplied along with full-scale plots and/or coordinates of the airfoil, if requested. (Please contact us before starting any construction on a wind-tunnel model.)

As previously mentioned, the airfoils will be tested in the UIUC open-circuit 3×4 ft subsonic wind tunnel. The turbulence intensity level is minimal and more than sufficient to ensure good flow integrity at low Reynolds numbers. Lift and pitching moment characteristics will be determined through force-balance measurements, while drag will be evaluated by the momentum method through the use of total-head probes traversed through the airfoil wake at several spanwise locations.

If you are interested in building wind tunnel models for the tests or wish to request information, please write, fax or send e-mail to the coordinator:

UIUC LSATs Coordinator
c/o Prof. Michael Selig
Dept. of Aeronautical and Astronautical Eng.
University of Illinois at Urbana-Champaign
306 Talbot Laboratory, 104 S. Wright St.
Urbana, IL 61801-2935
e-mail: uiuclsats@uiuc.edu
fax: (217) 244-0720

The program will be self-sustaining so long as funds are made available for equipment maintenance/upgrades and graduate student stipend support and tuition and fees (approximately $16,000/yr per student). It is envisioned that a small level of support from a large number of modeling enthusiasts could sustain the airfoil-design/wind-tunnel test program indefinitely. The impact on model aviation could be tremendous. Donations can be mailed to:

Prof. Michael Selig
Dept. of Aeronautical and Astronautical Eng.
University of Illinois at Urbana-Champaign
306 Talbot Laboratory, 104 S. Wright St.
Urbana, IL 61801-2935
e-mail: m-selig@uiuc.edu

Please make checks payable to "University of Illinois, AAE Dept." Also, please write on the check "Selig — Wind Tunnel Testing/AAE Unrestricted Funds," and provide a letter stating that your contribution is to be used by Prof. Selig and his group of students (both undergraduate and graduate) in support of the airfoil wind-tunnel tests. Finally, for a suggested donation of $18 in US, Canada, and Mexico (or $22 in other countries) you can receive a UIUC LSATs white short-sleeve shirt. All proceeds will go toward the continuation of the project.